Microbial Nanotechnology for Sustainable Future

This book covers the fundamentals and methods of biologically producing nanoparticles using microorganisms such as bacteria, fungi, and algae, along with optimization strategies for microbe-mediated nanoparticle production. It explores industrial and agricultural applications of microbial nanoparticles and their use in healthcare and pharmaceuticals, including treatments for multidrug-resistant infections and cancer. Focusing on microbial nanotechnology, this book highlights its applications in food production, pharmaceuticals production, water treatment, and environmental remediation. It provides valuable insights for researchers and students into food sciences, biotechnology, microbiology, and pharmaceuticals. Additionally, it discusses the environmental applications of microbial nanotechnology, emphasizing recent advancements and future research directions, serving for both academic and industrial researchers as a guide to transformative applications in this field.

Features

- Gives an overview of microbial nanotechnology and its applications to the environment
- Deals with the challenging effects of microbial nanotechnology on the environment, human health, safety, and sustainability
- Offers guidelines and cutting-edge methods and trends for environmental remediation
- Examines how nanotechnology can facilitate the detection of minute amounts of viruses, bacteria, and other pollutants in food and other industrial applications
- Incorporates case studies and real-world applications to show how microbial nanotechnology affects contemporary sciences and technologies

Dr. Yugal Kishore Mohanta is an Assistant Professor and Group Leader of Nano-biotechnology and Translational Knowledge Laboratory (NanoBioTraK_Lab) at the Department of Applied Biology, School of Biological Sciences, University of Science and Technology, Meghalaya (USTM), Meghalaya, India.

Dr. Bishwambhar Mishra is an Assistant Professor in the Department of Biotechnology at Chaitanya Bharathi Institute of Technology (CBIT) in Hyderabad, India.

Dr. Tamanna Bhuyan is an Assistant Professor at the Department of Applied Biology, School of Biological Sciences, University of Science and Technology, Meghalaya (USTM), Meghalaya, India.

Microbial Nanotechnology for Sustainable Future

Industrial and Environmental Perspectives

Edited by
Yugal Kishore Mohanta, Bishwambhar Mishra,
and Tamanna Bhuyan

CRC Press
Taylor & Francis Group
Boca Raton London New York

CRC Press is an imprint of the
Taylor & Francis Group, an **informa** business

Designed cover image: Getty

First edition published 2026
by CRC Press
2385 NW Executive Center Drive, Suite 320, Boca Raton FL 33431

and by CRC Press
4 Park Square, Milton Park, Abingdon, Oxon, OX14 4RN

CRC Press is an imprint of Taylor & Francis Group, LLC

ISBN: 978-1-032-78242-3 (hbk)
ISBN: 978-1-032-78276-8 (pbk)
ISBN: 978-1-003-48714-2 (ebk)

DOI: 10.1201/9781003487142

Typeset in Times
by codeMantra

Contents

Preface

The rapid advancements in nanotechnology have revolutionized multiple fields, from medicine and energy to environmental sustainability. Among these, microbial nanotechnology has emerged as a transformative interdisciplinary domain, harnessing the unique capabilities of microorganisms to develop and utilize nanomaterials. This synergy between microbiology and nanotechnology offers promising solutions to some of the most pressing global challenges, including environmental pollution, sustainable energy production, and industrial efficiency.

This edited book, *Microbial Nanotechnology for Sustainable Future: Industrial and Environmental Perspectives*, explores the multifaceted role of microbial nanotechnology in advancing sustainability. Through its diverse and comprehensive chapters, this book provides a holistic view of how microbial systems can be leveraged to create innovative, eco-friendly nanotechnological applications. The opening chapter elaborates on the significance of nanotechnology in microbiology, laying the foundation for understanding its transformative potential. Subsequent chapters explore the diverse applications of photosynthetic microbes and bacterial nanoparticle generation, showcasing practical utilizations in various industrial and environmental contexts. The role of microorganisms in converting agro-industrial waste into valuable nanomaterials and the production of antimicrobial nanoparticles are discussed in depth, highlighting their relevance in waste management and healthcare.

This book also emphasizes the cutting-edge advancements in microbial nano biosensors and their potential applications in monitoring and diagnostics. Sustainable approaches to recycling electronic waste through nanobiotechnology and the use of genetically and metabolically engineered microbes for nanoparticle production illustrate the innovative strides made toward a circular economy. Environmental sustainability forms a core theme of this book, with chapters dedicated to the removal of toxic pollutants from water, hazardous pollutant adsorption, and bioremediation of industrial effluents, all facilitated by microbial nanotechnology. Additionally, the text explores interdisciplinary collaborations, emphasizing the importance of integrating microbial nanotechnology across fields to foster sustainable futures.

Emerging applications in medicine, healthcare, and energy sectors underscore the versatility of microbial nanotechnology. Furthermore, the ethical and environmental considerations of these applications are critically examined, ensuring a balanced perspective on their societal impacts.

Finally, this book addresses the critical aspects of technology transfer and commercialization, guiding readers on translating microbial nanotechnology innovations from research to real-world applications. This volume serves as a valuable resource for researchers, academicians, industry professionals, and policymakers interested in the intersection of microbiology, nanotechnology, and sustainability. By bridging knowledge gaps and fostering interdisciplinary collaboration, this book aims to inspire innovative solutions for a sustainable future.

We are confident that the insights and knowledge shared in this book will catalyze further research and development in this vital field.

Dr. Yugal Kishore Mohanta

Dr. Bishwambhar Mishra

Dr. Tamanna Bhuyan

Acknowledgments

The successful completion of this book, *Microbial Nanotechnology for Sustainable Future: Industrial and Environmental Perspectives*, would not have been possible without the Almighty God, whose infinite wisdom, boundless grace, and unwavering support have been the foundation of our perseverance and strength, making this achievement possible.

We express our profound gratitude to the contributing authors, whose expertise, dedication, and hard work have enriched this volume. Their commitment to advancing knowledge in the field of microbial nanotechnology and their willingness to share their cutting-edge research have been invaluable. We extend our sincere appreciation to the reviewers, whose critical insights and constructive feedback have significantly enhanced the quality and coherence of the chapters. Their expertise has ensured that the content meets the highest academic standards. We are also grateful to our publisher for their unwavering support and guidance throughout the publication process. Their professionalism and attention to detail have been instrumental in bringing this project to fruition. We are deeply grateful to CRC Press for selecting our book for publication. Our heartfelt thanks go to Ms. Renu Upadhyay, Commissioning Editor at CRC Press, and Ms. Jyotsna Jangra, Editorial Assistant, for their steadfast editorial support and insightful suggestions, which have been instrumental in shaping this book into its final form. Lastly, we express our heartfelt gratitude to our families, friends, and colleagues for their continuous encouragement and understanding during the preparation of this book. Their support has been a source of strength and motivation.

We hope this book will serve as a valuable resource for researchers, students, and professionals, inspiring further innovation and sustainable development in the field of microbial nanotechnology.

Dr. Yugal Kishore Mohanta

Dr. Bishwambhar Mishra

Dr. Tamanna Bhuyan

About the Editors

Dr. Yugal Kishore Mohanta is currently an Assistant Professor at the Department of Applied Biology, University of Science and Technology, Meghalaya (USTM), India. He has over six years of extensive post-PhD research experience in the fields of nanobiotechnology, microbiology, and Sustainable Development Goals (SDG). As a dedicated researcher and academician, he has honed the skills and knowledge in these areas. Furthermore, he has successfully published over 110 scientific articles (Q1/Q2/Q3) in SCIE-indexed journals, edited five books (Elsevier, CRC Press, Springer-Nature), 27 book chapters, and two Indian patents in his field of expertise. From the publication, Dr. Mohanta has garnered >4000 citations, an h-index of 36, and an i10-index of 76. Dr. Mohanta is also a reviewer for more than 30 reputed journals and a guest editor for many special issues of Springer-Nature, Frontiers, etc. Additionally, he established a research lab (NanoBioTraK_Lab) through the DST-ANRF grant (2022–2024), Government of India. In 2024–2025, XJENZA MALTA, the Government of Malta, sponsored an international research network project scheme grant for him. He is also the recipient of the International Leadership Award for "Education for Sustainable Development-2024" from Edufam International Academy, Al Bustan, Dubai. With his strong set of skills and qualities, he excels in the fields of science, education, and social work. Dr. Mohanta is particularly interested in prioritizing the concept of translational research in the field of sustainable development. In addition, he serves as the Chief Coordinator for the Microbiologists' Society India, promoting science and research in the northeastern states of India.

Dr. Bishwambhar Mishra brings over a decade of post-PhD experience in teaching and research and is currently an Assistant Professor in the Department of Biotechnology at Chaitanya Bharathi Institute of Technology in Hyderabad, India. His primary research areas include industrial microbiology and nanobiotechnology, with a focus on their applications. His work is aimed at harnessing the potential of microorganisms and nanotechnology to address pressing global challenges such as environmental sustainability, food security, and resource management. Dr. Mishra has made significant contributions to the scientific community, authoring 75 research publications in high-impact, SCIE-indexed journals, with his work garnering 1,443 citations, an h-index of 21, and an i10-index of 42. These publications reflect the high caliber of his research, spanning various interdisciplinary domains. In addition to his extensive publication record, he has authored one book and 20 book chapters that further advance knowledge in the field. His dedication to the academic and research community extends beyond his publications. Dr. Mishra has served as a reviewer for over 30 prestigious international journals, including those published by renowned publishers such as Springer, Elsevier, Wiley, Frontiers, and MDPI. His expertise is further recognized through his role as Topical Editor and Editorial Board Member of SCIE-indexed journals, where he plays a critical role in shaping scientific discourse and guiding future research.

Dr. Tamanna Bhuyan is currently an Assistant Professor at the Department of Applied Biology, University of Science and Technology Meghalaya, India. Her primary research areas include the fabrication of micro-/nanobots, nanoparticle synthesis, drug delivery studies, anticancer therapeutics, and environmental remediation. Her research focuses on exploring nanostructured and nanocomposite materials in the context of societal needs and all other areas of industrial applications. Dr. Bhuyan received the Best PhD Thesis Award from IIT Guwahati and has made significant contributions to the scientific community by publishing more than 20 research papers in high-impact journals, and holds two patents which gardening >1,197 citations, a h-index of 09 and an i10 index of 09. She has been awarded three projects sponsored by the Ministry of Electronics and Information Technology, the New Generation Innovation and Entrepreneurship Development Centre project, and DST-ANRF, funded by the Government of India. She also received the Women Scientist Award by the Microbiologists Society, India, and SITARE-GYTI (Gandhian Young Technological Innovation) Appreciation Award for her notable contributions in interdisciplinary domains.

Contributors

Ramzan Ahmad
Department of Transfusion Medicine and Clinical Microbiology, Faculty of Allied Sciences
Chulalongkorn University
Bangkok, Thailand

Pooja Aich
Department of Medicinal and Applied Chemistry
Kaohsiung Medical University
Kaohsiung City, Taiwan

Bhavna Alke
LAQV/Requimte, Department of Chemistry, NOVA School of Science and Technology, FCT NOVA, NOVA
University Lisbon
Caparica, Portugal

Abdullah G. Al-Sehemi
Department of Chemistry
King Khalid University
Abha, Saudi Arabia

G. Alwina
Department of Biotechnology
Chaitanya Bharathi Institute of Technology
Hyderabad, India

Anuranjeeta
Department of Vocational Studies and Skill Development, School of Life Long Learning,
Central University of Haryana
Mahendergarh, India

Manjunatha Bangeppagari
Department of Cell Biology and Molecular Genetics
Sri DevarajUrs Academy of Higher Education and Research (A Deemed to Be University)
Tamaka, India

Santosh Kumar Behera
National Institute of Pharmaceutical Education and Research (NIPER-A)
Ahmedabad, India

Smita Behera
Department of Biotechnology
Indian Institute of Management and Science
Bhubaneswar, India

Mriganka Shekhar Borah
Department of Agricultural Engineering
Assam University
Silchar, India

Setu Chakrabarty
School of Agriculture and Allied Sciences
Doon Business School, Group, Selaqui
Dehradun, India

Minisrang Daimary
Department of Botany
Bodoland University
Rangalikhata, India

Swayamprava Dalai
Centre for Nanoscience and Nanotechnology
Symbiosis International (Deemed) University
Pune, India

Debashree Debasmita
Centre for Nanotechnology
IIT Guwahati
North Guwahati, India

N.O. Gopal
Department of Physics
Vikrama Simhapuri University College
Kavali, India

Rutuja Gumathannavar
PhD Scholar at Symbiosis Centre for Nanoscience and Nanotechnology
Symbiosis International (Deemed) University
Pune, India

Rashmi Ranjan Guru
University Hospital and Research Centre (SUHRC)
Symbiosis International (Deemed) University
Pune, India

Mwalimu Raphael Jilani
Department of Biotechnology
Vel Tech Rangarajan Dr. Sagunthala R&D Institute of Science and Technology
Chennai, India

Onkar Kadam
Junior Research Fellow at Symbiosis Centre for Nanoscience and Nanotechnology
Symbiosis International (Deemed) University
Pune, India

Srinivasan Kameswaran
Department of Botany
Vikrama Simhapuri University College
Kavali, India

Tanisha Kishan
Department of Biotechnology
Vel Tech Rangarajan Dr. Sagunthala R&D Institute of Science and Technology
Chennai, India

N. Krishna Prasad
Department of Biotechnology
P. A. College of Engineering
Mangalore, India

Abinaya Krishnan
Department of Biotechnology
Vel Tech Rangarajan Dr. Sagunthala R&D Institute of Science and Technology
Chennai, India

Jay Kumar
Department of Ophthalmology, Visual and Anatomical Sciences
Wayne State University
Detroit, Michigan

Neelam Amit Kungwani
Marine Biotechnology Division, Department of Environmental Biotechnology
Gujarat Biotechnology University
Gujarat, India

Vennela Lakavath
Department of Biotechnology
Chaitanya Bharathi Institute of Technology
Hyderabad, India

D. Mahaboob Basha
Department of Biotechnology
Chaitanya Bharathi Institute of Technology
Hyderabad, India

Anita Mandal
Department of Environmental Biotechnology
Gujarat Biotechnology University
Gujarat, India

Sanjeeb Kumar Mandal
Department of Biotechnology
Chaitanya Bharathi Institute of Technology
Hyderabad, India

Thameridus B. Marak
Department of Agricultural Engineering
Assam University
Silchar, India

Pinku Chandra Nath
Department of Biotechnology,
Manav Rachna International Institute of Research and Studies,
Faridabad 121004, Haryana, India

Debasis Nayak
Department of Wildlife and Biodiversity Conservation
Maharaja Sriram Chandra Bhanja Deo University
Baripada, India

Amilia Nongbet
Department of Botany
University of Science and Technology Meghalaya
Baridua, India

Azhagu Saravana Babu Packirisamy
Department of Biotechnology
Vel Tech Rangarajan Dr. Sagunthala R&D Institute of Science and Technology
Chennai, India

Bibhu Prasad Panda
Environmental Sciences, Department of Chemistry
ITER, Siksha 'O' Anusandhan (A Deemed to Be University)
Bhubaneswar, India

Jibanjyoti Panda
Department of Wildlife and Biodiversity Conservation
Maharaja Sriram Chandra Bhanja Deo University
Baripada, India

M. Pooja
Department of Biotechnology
Vel Tech Rangarajan Dr. Sagunthala R&D Institute of Science and Technology
Chennai, India

Bharathi Prakash
Department of Microbiology
University College Mangalore
Mangalore, India

Bellamkonda Ramesh
Molecular Biology
University of Nebraska Medical Center
Omaha, Nebraska

M. K. Ramis
Department of Mechanical Engineering
P. A. College of Engineering
Mangalore, India

A. Sivashankar Reddy
Department of Physics
Vikrama Simhapuri University College
Kavali, India

Hemen Sarma
Department of Botany
Bodoland University
Rangalikhata, India

N. Seema Devi
Department of Botany
The Assam Royal Global University
Guwahati, India

S. Shanmugavani
Department of Biotechnology, School of Life Sciences
Vels Institute of Science, Technology and Advanced Studies
Chennai, India

Deboja Sharma
Department of Applied Biology
University of Science and Technology Meghalaya (USTM)
Baridua, India

Rohit Shendkar
Symbiosis Centre for Nanoscience and Nanotechnology
Symbiosis International (Deemed) University
Pune, India

Bodika Shynisha
Department of Biotechnology
Chaitanya Bharathi Institute of Technology
Hyderabad, India

Anitha T. Simon
Department of Mechanical Engineering
University of Minnesota
Minneapolis, Minnesota

S. Simran
Department of Biotechnology
Vel Tech Rangarajan Dr. Sagunthala R&D Institute of Science and Technology
Chennai, India

Niraj Singh
Department of Microbiology
The Assam Royal Global University
Guwahati, India

B. Sneha
Department of Biotechnology
Chaitanya Bharathi Institute of Technology
Hyderabad, India

Siddharth Solanki
Department of Environmental Biotechnology
Gujarat Biotechnology University
Gandhinagar, India

B. Sumithra
Department of Biotechnology
Chaitanya Bharathi Institute of Technology
Hyderabad, India

Akbar Husen Sunasara
Department of Environmental Biotechnology
Gujarat Biotechnology University
Gandhinagar, India

Usman Taqui Syed
Division of Biological and Environmental Science and Engineering
King Abdullah University of Science and Technology (KAUST)
Thuwal, Saudi Arabia

Qudsiya Y. Tamboli
Department of Basic and Applied Science
MGM University
Aurangabad, India

M. Thenmozhi
Department of Biotechnology, School of Life Sciences
Vels Institute of Science, Technology and Advanced Studies
Chennai, India

Ajita Tiwari
Department of Agricultural Engineering
Assam University
Silchar, India

Shareefraza J. Ukkund
Department of Biotechnology
P. A. College of Engineering
Mangalore, India

Ramya Sree Vasamsetty
Department of Biotechnology
Vel Tech Rangarajan Dr. Sagunthala R&D Institute of Science and Technology
Chennai, India

Kalyani Vikhe
Marine Biotechnology Division, Department of Environmental Biotechnology
Gujarat Biotechnology University
Gandhinagar, India

Maheswari Vinodkumar
Department of Biotechnology
Vel Tech Rangarajan Dr. Sagunthala R&D Institute of Science and Technology
Chennai, India

Shashwati Wankar
Symbiosis Centre for Nanoscience and Nanotechnology,
Symbiosis International (Deemed) University
Pune, India

1 Antimicrobial Efficacy of Microbial-Derived Nanoparticles

An Alternative, Ecofriendly, and Green Approach against Multidrug-Resistant Pathogens

Jibanjyoti Panda, Smita Behera, Ramzan Ahmad, Debasis Nayak, and Santosh Kumar Behera

1.1 INTRODUCTION

In recent years, nanotechnology has emerged as a transformational science that has a wide range of applications, ranging from healthcare to environmental remediation. These applications integrate both nanotechnology and biotechnology disciplines. In the realm of science, nanotechnology is an advanced field that investigates the distinct physical, chemical, and biological characteristics of nanostructures, as well as their applications in a variety of fields such as agriculture and medicine. In order to improve existing biotechnological procedures and solve their limitations, such as the harmful effects that are linked with conventional treatments, nano-based methodologies are continuously being developed. Given that nanotechnology may be applied to a wide range of fields, it is projected that techniques based on nanotechnology will play a significant role in the development of a future that is environmentally sustainable [1]. Integrating four highly functional scientific domains—nanotechnology, biotechnology, information technology, and cognitive sciences (NBIC)—has attracted a lot of attention in the past few decades; these fields are together called "convergent technologies" [2]. Green-synthesized nanoparticles (NPs) have been produced with nanostructure dimensions between 1 and 100 nm, utilizing biological methods in conjunction with techniques from chemistry, physics, engineering, and computational sciences [3].

Bacterial infections significantly influence chronic diseases and mortality rates. Antibiotics have emerged as the most favoured approach for addressing bacterial infections due to their economic viability and robust effectiveness. However, numerous investigations have demonstrated that the extensive use of antibiotics has resulted in the rise of multidrug-resistant bacterial strains. Recently, super-bacteria have evolved, showcasing resistance to nearly all antibiotics as a result of the overuse of these medications [4,5]. Nanomaterials are substances that either have at least one dimension ranging from 1 to 100 nm or have a basic unit in its three-dimensional space within this range. Nanoparticles demonstrate significant antibacterial effectiveness against both gram positive and gram negative bacteria [6]. For instance, ZnO nanoparticles were observed to inhibit *Staphylococcus aureus*, while silver nanoparticles demonstrate concentration-dependent antibacterial efficacy against *Escherichia coli* and *Pseudomonas aeruginosa* [7]. The antibacterial mechanism of action of nanoparticles is typically categorized into three models: oxidative stress induction, metal ion release, or non-oxidative processes [8].

DOI: 10.1201/9781003487142-1

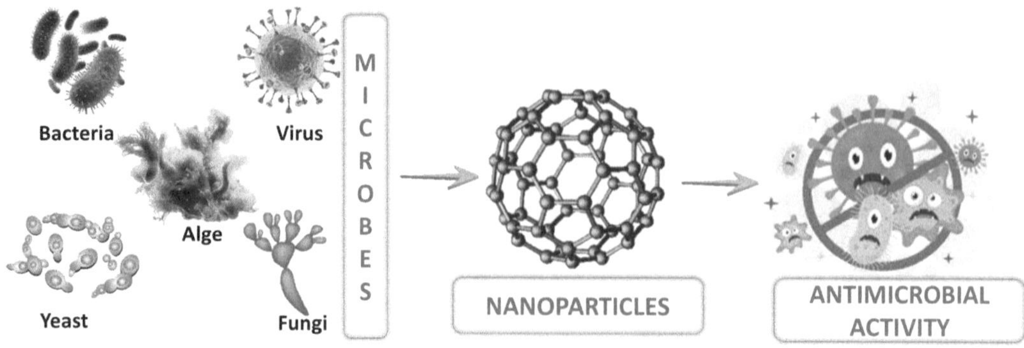

FIGURE 1.1 Antimicrobial activity of microbial-based nanoparticles.

The biosynthesis of nanoparticles is efficient, economical, and ecologically friendly. Moreover, the toxicity and dimensional characteristics of nanoparticles can be controlled. Numerous researchers have examined the synthesis of nanoparticles using bacteria, actinobacteria, fungi, yeast, microalgae, and viruses for the production of targeted nanoparticles [9]. Nanoparticles can be engineered to possess unique compositions and functionalities, providing novel tools and techniques that were previously inaccessible in biological study. Bacteria such as *Pseudomonas deceptionensis, Bacillus licheniformis, Pyrobaculum islandicum, P. aeruginosa*, and *Pyrococcus furiosus* have successfully synthesized nanoparticles of gold, silver, zinc, iron, nickel, and other significant elements [10,11]. Microorganisms, such as fungi, bacteria, and algae, exhibit rapid growth, are straightforward to manage, and have a natural ability to thrive in diverse environments. Microalgae and bacteria are capable of synthesizing distinctive nanomaterials, including exopolysaccharides, nanocellulose, and nanowires [12]. The current chapter presents various sources of microbial nanoparticles and outlines the key processes that contribute to their antibacterial properties (Figure 1.1). These processes include the disruption of bacterial cell membranes, the generation of reactive oxygen species (ROS), the infiltration of bacterial cell membranes, and the induction of intracellular antibacterial effects, which involve interactions with DNA and proteins.

1.2 MDR AND ITS SUSTAINABILITY

Bacteria are living organisms that evolve over time, primarily aiming to proliferate, survive, and spread rapidly. Bacteria adapt to their environment and formulate strategies that guarantee their survival. Any chemical, including an antibiotic, that impedes bacterial growth may induce genetic alterations, resulting in the bacteria's resistance to the treatment and thus enhancing their survival [13]. The development of MDR organisms, which are resistant to antimicrobial drugs or combination therapy, occurs when a species develops resistance to more than one antibiotic (Figure 1.2). This situation occasionally facilitates the spread of resistance by encouraging bacterial proliferation that is not swiftly eliminated [14–16]. The World Health Organization (WHO) has identified the rise of antibiotic resistance as one of the most pressing public health concerns of the modern era. The most prominent cause for concern is the ESKAPE group of bacteria, which includes the more dangerous *Enterococcus, Staphylococcus, Klebsiella, Actinobacter, Pseudomonas*, and *Enterobacter*. There is a significant mortality rate linked to all of them [17]. There is still a major threat to animal and human health from the spread of MDR and XDR bacteria, which are responsible for higher rates of illness and death as well as higher treatment costs. Various bacterial strains' abilities to form biofilms intensify the problem, as these microbes are linked to 65%–80% of human diseases [18]. The antibiotic susceptibility of biofilm cells is typically 100–1000 times less than that of planktonic cells. Drug resistance in biofilms develops due to various factors, including decreased drug absorption through the extracellular polymeric matrix, reduced intracellular drug levels, lowered bacterial metabolism,

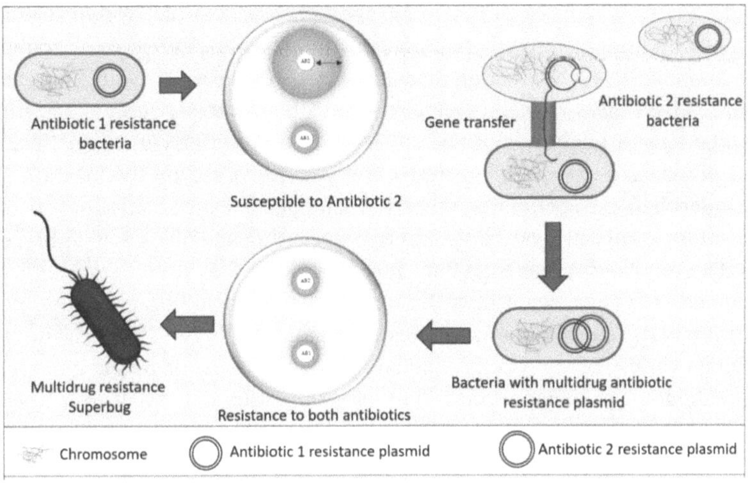

FIGURE 1.2 Development of multi-drug-resistance bacteria. (Figure obtained from open-access article by Imran et al. [21].)

and the transfer of resistance genes. MDR influences the management of diseases and diminishes the effectiveness of treatments, resulting in prolonged infections for patients. Individuals with compromised immune systems, including those with chronic conditions, undergoing cancer treatment, taking immunosuppressive medications, or who have had organ transplants, are at a heightened risk for infections caused by multidrug-resistant organisms [19,20]. Moreover, the expense of treatment has increased due to multidrug resistance (MDR), since microbes have developed resistance to existing antibiotics, necessitating the adoption of more expensive alternatives. The spread of multidrug-resistant (MDR) organisms is promoted by the increase in worldwide trade and travel [13].

At least 700,000 deaths globally are attributed to multidrug-resistant bacteria annually, with 23,000 occurring in the US and 25,000 in the EU. According to WHO reports, the improper use and excessive consumption of antibiotics account for over 80% of MDR and XDR bacteria, which are linked to considerable negative consequences [22,23]. By 2050, bacterial infections will kill 10 million people worldwide if we don't do anything to stop the spread of antibiotic-resistant bacteria or develop new drugs. The rise in multidrug-resistant infections and other resistant organisms has led to a limited number of treatment and preventive options currently available. The limitations of current antibiotics necessitate the investigation of alternative treatment options. Numerous bacterial infections can be effectively addressed using nanoparticles, especially those induced by multidrug-resistant pathogens. Antibiotics and nanoparticles can be utilized individually or in combination to produce noteworthy synergistic effects [4].

Antibiotic resistance is a result of metabolic pathway and target site changes by pathogenic microorganisms. Additionally, as a result of their ongoing exposure, these microorganisms become more pathogenic and resistant to medications [24]. Thus, it is critical to find new, less expensive, and safer ways to combat bacteria and other microbes. Biofabricated NPs are both smaller and have a greater surface area. These characteristics of nanoparticles (NPs) enable them to effectively penetrate microbial cell membranes, disrupt metabolic pathways, and impede DNA replication [25]. Research on the antibacterial properties of various NPs is abundant.

1.3 MICROBIAL-DERIVED NPs

The WHO has identified antimicrobial drug resistance as one of the foremost worldwide public health challenges, and antimicrobial nanoparticles represent a significant potential solution to mitigate this disaster. Multiple studies have underscored the significance of metallic nanoparticles (e.g., silver, gold, iron oxide, zinc oxide) as potent agents against various drug-resistant bacteria [26]. Microbes demonstrate a quick growth rate and possess a distinctive innate capability to synthesize inorganic nanomaterials by reduction mechanisms via intracellular and extracellular pathways, owing to their adaptive fitness in metal-rich environments. Microbes capture and metabolize environmental metallic ions through specialized enzymatic activity. Biosynthetic pathways can be classified into intracellular and extracellular production systems based on the location of nanoparticle formation. The microbial extracellular synthesis of nanoparticles presents a promising opportunity due to its potential for straightforward downstream processing and quick scalability, hence enabling large-scale production. Microorganisms can thus be regarded as prospective biofactories for the synthesis of biomedically significant metallic nanoparticles, including cadmium, gold, and silver. Consequently, they are preferred choices in green nanotechnology [27,28]. Nanoparticles can be produced through a variety of approaches, including physical, chemical, and biological techniques. Nonetheless, the creation of nanoparticles via biological systems is swift, effective, and environmentally sustainable. Furthermore, the toxicity and dimensional properties of nanoparticles can be regulated. A variety of studies have shown the synthesis of nanoparticles utilizing bacteria, actinobacteria, fungi, yeast, microalgae, and viruses for the production of specific nanoparticles [12,29].

Microbial diversity operates as small nanofactories, transforming metal ions into metal nanoparticles through the activity of enzymes and other biomolecular compounds produced or secreted by the microbes [30]. Table 1.1 depicts the existing nanoparticles with antibacterial properties obtained from various microorganisms. The ecologically benign nanoparticles synthesized by green methods can be created via a single-step reduction process that utilizes comparatively low energy, showcasing significant efficiency. In the process of NP synthesis, solvent solutions, which may include water or the natural extract of the component, are frequently more preferable. Ionic liquids, which are also referred to as room temperature ionic liquids, are ideal due to the fact that they contain ions and have a melting point that is lower than at least 100°C. The hydrophobic or hydrophilic nature of these ionic liquids is determined by the presence of particular anions or cations in the liquid formulation. These substances are essential for the synthesis of a variety of nanoparticles, including those of silver, gold, aluminium, and platinum, because they serve as reductants and protective agents [3,31].

1.3.1 Bacterial-Derived NPs

More and more research is focusing on the chemical, optical, and mechanical properties of nanoparticles and their potential applications. A relatively new idea in the field of therapeutics and diagnostics that has emerged within the past 20 years is drug delivery using nanoparticles. The ability of NPs to enter bacterial cells, despite their diminutive size, has piqued researchers' interest in them as a potential new weapon in the fight over drug susceptibility [59]. Nanoparticles can be biomineralized by certain bacteria using chemicals and enzymes found on their cell walls, which allows them to convert metal ions to metal elements. Emerging as a drug-replacement item in numerous biomedical fields, silver nanoparticles derived from bacteria are effective against microbes [60]. Among the many of microbial species, prokaryotic bacteria have been the subject of substantial investigation primarily due to their manageable nature, cost-effective production, and broad potential for genetic manipulation. Common bacteria such as *E. coli, Bacillus subtilis*, and *Lactobacillus*; filamentous cyanobacteria like *Plectonema boryanum*; mesophilic bacteria such as *Shewanella* algae; endophytic bacteria like *Bacillus cereus*; and extremophilic actinomycetes such as *Thermomonospora* species, among others, have been extensively utilized for the production of metal nanoparticles [61]. Biomolecules excreted by bacteria can serve as both capping and stabilizing agents in nanoparticle formation, eliminating the necessity for external additives. The medicinal application of silver dates back to antiquity. The antimicrobial properties of this noble metal have been recognized for centuries. Consequently, investigating the antibacterial properties of silver nanoparticles was a logical direction during the initial stages of nanotechnology development.

TABLE 1.1
Microbes-Mediated Synthesis of Nanoparticles with Antimicrobial Properties

Microbial Strains	Nanoparticles	Size (mm)	References
Acinetobacter pittii	Te	60–130	[32]
Aspergillus niger	Ag	13.2–646.8	[33]
Bacillus subtilis	Ag	3–20	[34]
Bacillus sp.	Se	141.6	[35]
Brevundimonas diminuta	Ag	5–50	[36]
Candida albicans	CdS	50–60	[37]
Chlorella ellipsoidea	Ag	220±31.3	[38]
Escherichia coli	CdS	40–80	[39]
Escherichia coli	Ag	5–50	[36]
Exiguobacterium aurantiacumm	Ag	5–50	[36]
Halomonas sp.	Ag	22	[35]
Lactobacillus plantarum TA4	ZnO	292.6±83.2	[40]
Lactobacillus acidophilus	CdO	40, 48, and 67	[41]
Marinobacter sp. 2C8	ZnO	10.23±2.48	[42]
Marinospirillum alkaliphilum	Ag	30–70	[43]
Morchella esculenta	Au	16.51	[44]
Padina sp.	Ag	25–60	[45]
Penicillium oxalicum	Ag	60–80	[46]
Proteus vulgaris	Fe_2O_3	19–30	[47]
Pseudoduganella eburnean	Ag	8–24	[48]
Pseudomonas kunmingensis ADR19	Pt	2–8	[49]
Sphingobium sp. MAH-11^T	Ag	7–22	[50]
Spirulina platensis	Au	15.60–77.13	[51]
Streptomyces kunmingensis MR03	CdS	4.62	[52]
Streptomyces sp. OSIP1	Ag	8	[53]
Streptomyces sp. OSNP14	Ag	15	[53]
Ulva armoricana	Ag	33	[54]
Vibrio sp. B4L	Ag	32.67–107.18	[55]
Vibrio sp. VLA	ZnO	20.26±4.44	[42]
Vibrio sp. VLC	CuO	8.83–8.77	[56]
Volvariella volvacea	Fe_2O_3	–	[57]
Yarrowia lipolytica DSM 3286	Ag	12.4±5.22	[58]

Silver nanoparticles have been identified as the most potent antibacterial agents among all nanomaterials, and their stability enhances their potency [26]. They are effective against a wide variety of bacteria, including Gram-negative and Gram-positive varieties, as well as MDR strains of bacteria, and they may even be able to combat fungal infections like *Candida albicans*. Moreover, when combined with antibiotics or antibacterial organic compounds such as nitrofurantoin, carbenicillin, vancomycin, rifampicin, and ciprofloxacin, they show a considerable synergistic effect against harmful bacteria, such as ESBL-producing Enterobacteriaceae (e.g., *E. coli* and *Klebsiella pneumoniae*), methicillin-resistant *S. aureus*, multi-drug-resistant *S. enterica*, *Vibrio* species, and *Bacillus anthracis*. The efficacy of the combination can increase by as much as 50% [27]. Various silver-tolerant bacterial species possess the inherent ability to efficiently reduce silver ions, making them prevalent standard candidates for the synthesis of silver nanoparticles with diverse sizes, shapes, and morphologies in laboratories around. Examples of bacterial strains that have been extensively utilized for this purpose include *E. coli, P. proteolytica, L. casei, Aeromonas* sp. SH10, *B. cereus, Bacillus cecembensis, B. licheniformis, Bacillus amyloliquefaciens, Bacillus indicus,*

Enterobacter cloacae, and *Arthrobacter gangotriensis*, among others [28]. Since antimicrobial resistance (AMR) is a worldwide concern, bacteria-mediated synthesis of silver nanoparticles provides a great environmentally friendly substitute because it can be used to safely and effectively prevent and treat infections, sterilize industrial products, decontaminate medical supplies, and create new bactericidal products for purification or disinfection. Their appeal is evidenced by the consistent expansion of their market, the rising demand in industrial sectors, and the augmented research investment associated with them [62]. Similarly, various bacterial species have been employed in the synthesis of gold nanoparticles, including *E. coli* DH5a, *Brevibacterium casei, P. aeruginosa, B. subtilis* 168, *Bacillus megaterium* D01, *P. boryanum* UTEX 485, *Desulfovibrio desulfuricans, Rhodopseudomonas capsulata,* and *Shewanella alga*, among others [63]. Gold nanoparticles have remarkable stability, superior biocompatibility, and notable antibacterial capabilities, and they can be readily changed for functional improvement. They are superior drug carriers capable of enhancing the efficacy of loaded antibacterial agents, hence serving as an exceptional nano-arsenal against ESBL-producing resistant infections [64]. The antibacterial capabilities of silver nanoparticles synthesized from *B. subtilis* against four strains of multidrug-resistant bacteria: *Acinetobacter baumannii, P. aeruginosa,* methicillin-resistant *Staphylococcus aureus* (MRSA), and *E. coli* [65]. The spherical nanoparticles, which had a diameter of around 80 ± 0.18 nm and an absorbance maximum at 426 nm, were also validated by Zeta Sizer Nano. The antibacterial efficacy of cadmium sulphide nanoparticles synthesized by *K. pneumoniae* K-6 and *E. coli* E-30 ranged from 3.2 to 44.9 nm, and these nanoparticles were found to be most effective against *Acinetobacter fumigatus, Geracilla candidum, B. subtilis, S. aureus*, and *E. coli* strains [66]. The colour of the gold nanoparticles (AuNPs) containing colloidal solution showed tremendous fluctuation when incubated with *Acinetobacter* sp. SW30 and varying concentrations of gold chloride. This variation in turn suggests variations in size and shape of AuNPs. Silver nanoparticles have been successfully synthesized and further stabilized using water-based extracts of the cyanobacterium *Oscillatoria limnetica*. Nanoparticles with a size range of 3.30–17.97 nm exhibited anti-cancer and antimicrobial properties [67]. Microscopically small silver nanoparticles produced by *Desertifilum* sp. (4.5–26 nm) were found to possess antibacterial properties and to exhibit cytotoxic effects when tested against HepG2, MCF-7, and Caco-2 cancer cells [68].

1.3.2 Algae-Derived NPs

Algae is an exceptional natural resource with diverse applications across different industries, including healthcare and pharmaceuticals. They are exceptionally adaptable and sustainable organisms. They possess the ability to perform photosynthesis, exhibit rapid growth in various settings, and generate important bioactive chemicals with a range of intriguing features [69]. Algae are presently categorized into ten principal phyla: Cyanobacteria, Bacillariophyta, Charophyta, Chlorophyta, Glaucophyta, Cryptophyta, Ochrophyta, Rhodophyta, Euglenozoa, and Miozoa. These can be further divided into macroalgae and microalgae based on notable variations in morphology and dimensions [70]. Both have significantly influenced the medical field and offer exceptional platforms for the biosynthesis of high-value products, including diverse nanoparticles, thereby fostering the cost-effective and environmentally sustainable discipline of phyco-nanotechnology, which shows considerable potential [71,72]. Their abundant secondary metabolites are particularly advantageous since they can function as capping and reducing agents in the synthesis process. Many marine algae species, including brown alga *Cystoseira myrica*, green alga *Ulva rigida*, and red alga *Gracilaria foliifera*, as well as freshwater microalgae *Scenedesmus* species, *Arthrospira platensis, Chlorella vulgaris,* cyanobacterium *Chroococcus minutus*, and many more have been used to synthesize antimicrobial metal nanoparticles [72,73]. Their antimicrobial properties make them useful against a wide range of bacteria and fungi, including *S. aureus, B. subtilis, P. aeruginosa, E. coli, Serratia marcescens,* and *K. pneumoniae* [73]. Additional research is necessary to fully understand how algal metabolites contribute to green synthesis, as this is a relatively young area of study. To get the most out of your crop in the least amount of time possible, this can be useful for adjusting growth parameters and manufacturing settings.

1.3.3 Fungal-Derived NPs

Synthesizing antimicrobial nanoparticles using fungal systems is a practical and efficient alternative. For instance, in the early 2000s, silver-tolerant fungi, such as yeast strain MKY3, were able to create 2–5 nm silver nanoparticles outside of their cells in response to a challenge of 1 mM soluble silver during the log phase of cell growth [74]. The extracellular synthesis of nanoparticles is economically advantageous and feasible for large-scale manufacturing and straightforward downstream processing. Indeed, numerous laboratories have achieved superior yields of nanoparticles with varying forms, sizes, and therapeutic efficacy when utilizing fungal systems, attributable to their greater biomass compared to bacterial systems [75]. Numerous fungal species have been utilized to generate antimicrobial nanoparticles, including *Saccharomyces cerevisiae, Aspergillus oryzae, Alternaria alternata, P. fellutanum, Fusarium oxysporum, Trichoderma viride, Verticillium luteoalbum, Colletotrichum* sp., and *Trichothecium* sp. Numerous specimens exhibit diverse levels of antimicrobial efficacy, encompassing antifungal activity against both fluconazole-susceptible and fluconazole-resistant *C. albicans* isolates, as well as against pathogenic bacteria [76,77].

1.3.4 Virus-Derived NPs

Virus-derived green metal nanoparticles are recognized for their broad-spectrum antiviral potential against several viruses, including hepatitis B virus, herpes simplex virus types I and II, HIV, respiratory syncytial virus, and monkeypox virus. They can function by inhibiting the processes of viral attachment to the host cell, viral penetration, or replication within the host. This facilitates numerous applications in the healthcare sector, bioremediation, food and textile industries, smart agriculture, air filtration, and wastewater treatment [78]. Conversely, other research groups have pursued virus-mediated nanoparticle manufacturing, particularly utilizing plant viruses and bacteriophages for the treatment of infections and disorders in people and animals [30]. Viral particles can be regarded as naturally occurring nanoparticles, with dimensions ranging from 20 to 500 nm, and have been investigated for applications in medication delivery, targeted vaccinations, and immunotherapy [79]. The primary problem of this approach is its safety, which must be addressed with the highest level of caution and priority.

1.3.5 Actinomycetes-Derived NPs

Actinomycetes attract specific attention within the realm of green nanotechnology. Many see them as the preeminent groupings among commercially significant microbial species due to their diverse bioactive components and extracellular enzymes, which are essential for metal nanoparticle formation. Species of *Streptomyces*, *Thermomonospora*, *Rhodococcus*, and *Nocardia* have been identified as the most effective actinomycetes for producing antimicrobial nanoparticles that exhibit significant efficacy against *E. coli, S. aureus, Enterococcus faecalis, P. aeruginosa, C. albicans, Candida glabrata, A. alternata*, and *F. oxysporum*, as established by standard protocols such as the agar well diffusion technique, minimum inhibitory concentration (MIC), and minimum bactericidal concentration (MBC) analysis [30,80]. Consequently, the utilization of microbial biofactories is a feasible environmentally sustainable approach for the manufacture of antimicrobial nanoparticles, but they possess certain restrictions [12]. Synthetic methods utilizing fungi and bacteria provide difficulties due to the necessity of strain separation, adequate preservation of pure cultures, and the maintenance of optimal growth conditions and culture media.

1.4 ANTIMICROBIAL INTERACTION OF NANOPARTICLES

Nanoparticles are characterized by their complex nature, which makes it challenging to determine the particular mechanisms of action that they employ. Nanomaterials have the ability to attack bacteria through a variety of bactericidal processes (Figure 1.3) [81]. ROS production, cell wall and membrane breakdown, drug transport by membrane fusion, and interaction with intracellular components including DNA and ribosomes are all examples of such mechanisms [82]. Microbes have adapted their cell membranes to physically withstand antimicrobials. Teichoic acids, found in Gram-positive bacteria's cell walls, and lipopolysaccharide, found in Gram-negative bacteria's outer membrane, are the two types of carbohydrates that give bacteria their negative charge. Hydrophobic antimicrobials are less effective against bacteria in this extremely polar environment because they cannot cross cell membranes [83]. The cell wall experiences depolarization, reduced permeability, and reduced fluidity as a result of electrostatic adsorption. This destroys cells by blocking their ability to absorb and use energy. Accumulating nanoparticles can also induce the formation of "pits" in the cell walls of bacteria. This allows them to infiltrate cells, change their membranes, and eventually damage or kill cells [84]. Positively charged ions produced by nanoparticles interact with the negatively charged surfaces of bacteria, particularly through carboxyl or phosphate groups, improving the bactericidal efficiency. However, nanoparticles (NPs) are unable to enter the cells of Gram-positive bacteria due to their thick peptidoglycan covering [85]. This means that NPs can only interact with surface microorganisms.

ROS are by-products of oxidative metabolism. They affect cellular development, signalling, survival, and apoptosis. Moreover, they have a robust positive redox potential. ROS encompass hydroxyl radicals, superoxide radicals, singlet oxygen, and hydrogen peroxide. Diverse nanoparticles produce distinct combinations of ROS, each exhibiting unique antibacterial properties [8]. For example, MgO and ZnO nanoparticles only generate superoxide radicals and a combination of hydroxyl and hydrogen peroxide radicals, respectively, whereas Ag and Cu nanoparticles produce all varieties of ROS. ROS are induced by nanoparticles or by the disruption of the respiratory chain. Typically, scavenger molecules maintain them at low levels as reduced glutathione [86]. In a standard environment, the generation and elimination of ROS are proportionate. Under acute

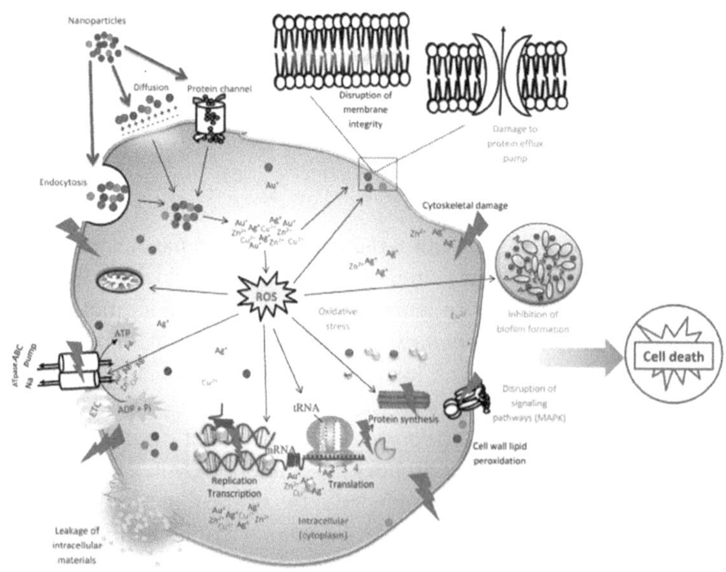

FIGURE 1.3 Antimicrobial mechanism of nanoparticles. (Figure obtained from open-access article by Sharmin et al. [95] with Copyright permission.)

stress, excessive ROS production alters cell membrane flexibility, resulting in bacterial damage. In addition to oxidative stress, ROS can damage cellular macromolecules by modifying proteins, inhibiting enzymes, lipid peroxidation, RNA/DNA damage, and electron transport chain disruption. Nanoparticles of iron, zinc oxide, silver, copper, and zinc oxide have been shown to provide significant antibacterial effects by way of ROS [87–90].

The maintenance of bacterial cellular homeostasis and the integration of intracellular signalling networks are both essential to the survival and activity of bacteria. The manipulation of nanomaterials in order to disrupt certain pathways could potentially result in the death of cells. Disruptions can take several forms, including changes in protein synthesis, DNA damage, and altered gene expression [91]. AuNPs were combined with 4,6-diamino-2-pyrimidinethiol (DAPT), which is a derivative of 2-pyrimidinethiol that is present in *E. coli*. This was done in order to produce pyrimidine-capped AuNPs, also known as Au-DAPT. It was found that multidrug-resistant strains of *E. coli* and *P. aeruginosa* were unable to grow when these nanoparticles were present [92].

The improved penetrating characteristics of nanoparticles make it possible for them to eliminate biofilm [93]. They destroy bacteria by interacting with them once they have penetrated the deep layers of the biofilm, which is the extracellular polysaccharide matrix. The charge of nanoparticles has a significant impact on their ability to interact with one another and penetrate [94]. A useful illustration of this would be the fact that cationic nanoparticles, which have a positive charge, have powerful interactions with anionic matrices.

1.5 POTENTIALITY OF NANOBIOTICS OVER TRADITIONAL ANTIBIOTICS AND EFFECTS ON MDR

Antibiotics are medications that hinder the growth of microorganisms, address particular illnesses, and save lives when used correctly. The primary pharmaceutical options for tackling biofilm and planktonic infections are detailed. Nonoantibiotics (NAb) refer to antibiotic molecules that possess a minimum average diameter of 100 nm in at least one dimension, or they are nanoparticles specifically designed to encapsulate these antibiotic molecules [84]. Antibiotics often lead to bacterial death by interfering with the proton motive force across the cell membrane. This disruption diminishes the microorganism's capacity to generate or store energy, hinders protein synthesis, and compromises cell wall integrity, among other mechanisms. Multifunctional "intelligent" antimicrobials encapsulated in nanoparticles respond to stimuli, enhancing interaction with bacterial cell walls and membranes, thereby facilitating improved membrane penetration and targeted drug delivery [96]. Neutralizing antibodies include a target-specific, controlled release mechanism enabling single-dose delivery, unlike most traditional antibiotics that require many doses for systematic release. Unlike traditional "free" antibiotics, the use of nanoparticle-based delivery systems for antibiotics offers numerous advantages such as reduced dosage and frequency, enhanced efficacy through extended drug release, and precise targeting of drug delivery to specific locations within bacterial cells [97]. Compared to molecular materials, nanoparticles have a high surface-to-volume ratio and a unique surface area, leading to an extraordinarily large contact area. Combining antibiotics with nanoparticles directs the antibiotics to specific areas, while the nanoparticles simultaneously activate multiple antibacterial processes (Figure 1.4). Consequently, NPs provide greater functionality than the effects of an individual medicine or the combination of several treatments for targeting MDR strains [98,99]. β-Lactam antibiotics kill microbes in various ways, including interfering with the transpeptidase enzyme that binds peptidoglycan in bacterial cell walls, blocking penicillin-binding proteins (PBPs), an enzyme crucial for peptidoglycan synthesis, and disrupting FtsI (PBP3), a protein complex that microbes rely on for survival [100]. Researchers have linked the production of ROS by nanoantibiotics to DNA degradation, protein denaturation, membrane malfunction, and impaired membrane integrity [101]. Antibiotic-conjugated nanoparticles have the potential to physically disrupt cellular structures through the dissolution and release of metal ions, resulting in biomechanical damage [102]. Signal transmission impedes ATPase activity

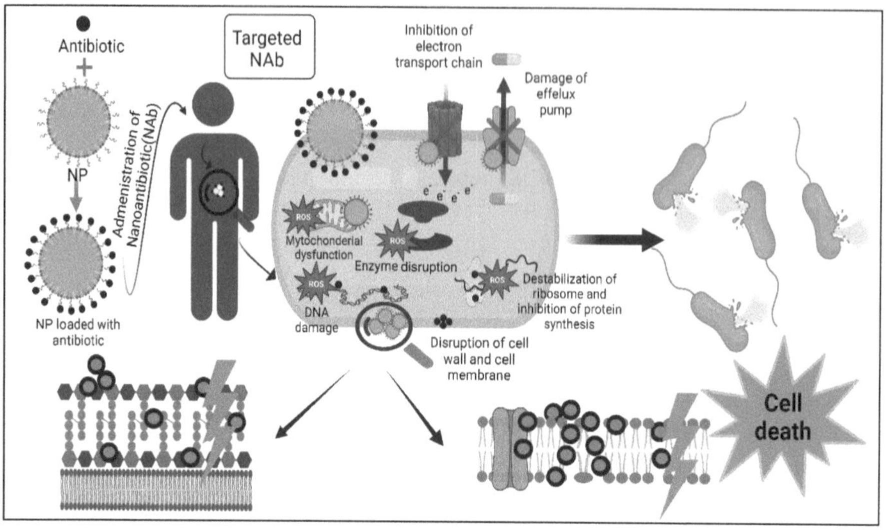

FIGURE 1.4 Mechanism of nanoantibiotics along with conventional antibiotics against MDR. (Figure obtained from open-access article by Hetta et al. [84].)

and obstructs the binding of ribosomal subunits to tRNA. This leads to changes in metabolic processes such as protein regulation, lipid and carbohydrate metabolism, energy production, and stopping bacterial growth, which ultimately leads to cell death [103]. Nanosized antibiotics must interact with parts inside bacteria cells in order to work properly in many situations, including killing microbes, moving drugs around, penetrating cell membranes, and stopping protein synthesis. Clinical conditions view nanobiotics as a promising alternative to current antimicrobial resistance and treatment approaches [101].

1.6 BIOCOMPATIBILITY AND SAFETY

Microbial-based nanoparticles demonstrate superior biocompatibility compared to chemically synthesized counterparts, making them safer for medical applications. Biocompatibility is the ability of a biomaterial to fulfil its intended role in medical treatment without causing negative local or systemic reactions in the recipient. It also fosters optimal beneficial cellular or tissue responses, thereby enhancing the clinically significant efficacy of the therapy [30]. The intrinsic interactions among microorganisms have the fundamental potential to enable nanoparticle manufacturing without reliance on detrimental and expensive chemicals [104]. However, attention has shifted to sustainable technology due to the urgent demand for a non-toxic, efficient, rapid, cost-effective, and environmentally friendly solution. Microbes are often referred to as tiny "nanofactories," and the formation of microbial nanoparticles has brought together nanotechnology, microbiology, and biotechnology, resulting in the intriguing domain of "nano-biotechnology." The synthesis of nanoparticles from biological sources is far safer than utilizing chemical and physical processes [9]. The ability to successfully cultivate bacteria in artificial conditions with appropriate growth rates makes them a preferred microbe for nanoparticle manufacturing. Researchers have recognized various fungi, such as *F. oxysporum, V. luteoalbum, Trichothecium* sp., *Colletotrichum* sp., and *A. oryzae*, as makers of nanoparticles with varying shapes and sizes [105]. However, the synthesis of nanoparticles from fungi faces various challenges, including extended production durations and expensive downstream processing, which require resolution to develop viable methods for large-scale nanoparticle production at an economical cost.

1.7 CHALLENGES

Although NPs offer substantial benefits over traditional antibiotics, their size and physical characteristics make their distribution a formidable obstacle. Their antibiotic delivery systems allow for tailored administration and prolonged release. Hence, less frequent dosing, less host cell toxicity, and less resistance can be achieved with smaller and lower dosages [106]. Still, there are a lot of other problems that need fixing before NPs can be systemically administered. These include finding the best dose, figuring out the best ways to administer NPs, dealing with toxicity after both short-term and long-term exposures, and figuring out the exact mechanism of cellular uptake [107]. In addition, there may be unforeseen health hazards associated with long-term exposure to NPs at work. Secondary exposure to NPs can also occur by inhalation of air contaminants that contain nanoparticles. It is possible for these inhaled nanoparticles to cause systemic health concerns if they manage to avoid the immune system. The spleen, lungs, and bone marrow are sites of NP accumulation following intravenous injection. In addition, because of their tiny size, huge surface area, and rapid absorption rate, inhaled NPs can travel via the cardiovascular system, liver, spleen, and brain [108]. There is a possibility of multiorgan nanotoxicity with therapeutic administration of NP. Metabolic changes, such as fatty acid oxidation, glycolysis, and reduced ketogenesis, which occur through ROS, are associated with NPs in all toxic cases and can lead to liver and lung toxicity and nephrotoxicity [91,108]. Due to the difficulties that have been encountered, the Food and Drug Administration has not yet granted approval for the use of nanobiotics in the systemic administration of human medicine. In light of this, it is imperative that future preclinical investigations encompass not only the safety, toxicity, pharmacokinetic, and pharmacodynamic characteristics of NP systems, but also the therapeutic efficacy metrics that are applied in clinical trials. For the final point, it is essential to evaluate the financial repercussions of these nurse practitioner clinical translations in connection to the efficacy of their therapy [109].

1.8 CONCLUSIONS AND RECOMMENDATION

The alarming increase in the number of germs that are resistant to several drugs calls for a response from the entire world. Treatment of multi-drug-resistant planktonic bacterial and biofilm infections can be accomplished by the utilization of nanomaterials, which offer a novel and unconventional method. It is essential to maximize their physical features, particularly their size and surface charge, in order to amplify their therapeutic potential while simultaneously reducing the risk to the host. There are now limitations placed on the utilization of nanoparticles in clinical practice due to concerns regarding the systemic safety of nanoparticles as well as the potential long-term consequences of nanoparticles. With regard to metabolic, toxic, stability, and mechanism-level problems, it is essential that future research provide a comprehensive understanding of the influence that NP has on gene expression. Understanding the precise and detailed method by which nanoparticles interact with biological systems is required in order to bring about the creation of nanomaterials that possess desirable physicochemical properties and are able to adapt to various biological contexts in order to provide therapeutic benefits without causing any adverse effects. Future experiments will benefit from having this information at their disposal. The field of nanoantibiotics will be the focus of the next generation of medication development when it comes to the fight against bacteria that are resistant to several agents.

REFERENCES

1. Pokrajac, L.; Abbas, A.; Chrzanowski, W.; Dias, G.M.; Eggleton, B.J.; Maguire, S.; Maine, E.; Malloy, T.; Nathwani, J.; Nazar, L.; et al. Nanotechnology for a Sustainable Future: Addressing global challenges with the international network4sustainable nanotechnology. *ACS Nano* **2021**, *15*, 18608–18623, doi:10.1021/acsnano.1c10919.

2. Shahcheraghi, N.; Golchin, H.; Sadri, Z.; Tabari, Y.; Borhanifar, F.; Makani, S. Nano-Biotechnology, an applicable approach for sustainable future. *3 Biotech* **2022**, *12*, 65, doi:10.1007/s13205-021-03108-9.
3. Mayegowda, S.B.; Sarma, G.; Gadilingappa, M.N.; Alghamdi, S.; Aslam, A.; Refaat, B.; Almehmadi, M.; Allahyani, M.; Alsaiari, A.A.; Aljuaid, A.; et al. Green-synthesized nanoparticles and their therapeutic applications: A review. *Green Process. Synth.* **2023**, *12*, 20230001, doi:10.1515/gps-2023-0001.
4. Naghavi, M.; Vollset, S.E.; Ikuta, K.S.; Swetschinski, L.R.; Gray, A.P.; Wool, E.E.; Robles Aguilar, G.; Mestrovic, T.; Smith, G.; Han, C.; et al. Global burden of bacterial antimicrobial resistance 1990–2021: A systematic analysis with forecasts to 2050. *Lancet* **2024**, *404*, 1199–1226, doi:10.1016/S0140-6736(24)01867-1.
5. Thirumalai, A.; Girigoswami, A.; Girigoswami, K. Bacterial nanocellulose-robust preparation and application—A literature review. *Appl. Chem. Eng.* **2023**, *6*, 2170, doi:10.24294/ace.v6i3.2170.
6. Joudeh, N.; Linke, D. Nanoparticle classification, physicochemical properties, characterization, and applications: A comprehensive review for biologists. *J. Nanobiotech.* **2022**, *20*, 262, doi:10.1186/s12951-022-01477-8.
7. Wang, L.; Hu, C.; Shao, L. The antimicrobial activity of nanoparticles: Present situation and prospects for the future. *Int. J. Nanomed.* **2017**, *12*, 1227–1249, doi:10.2147/IJN.S121956.
8. Mondal, S.K.; Chakraborty, S.; Manna, S.; Mandal, S.M. Antimicrobial nanoparticles: Current landscape and future challenges. *RSC Pharm.* **2024**, *1*, 388–402, doi:10.1039/D4PM00032C.
9. Pandit, C.; Roy, A.; Ghotekar, S.; Khusro, A.; Islam, M.N.; Emran, T. Bin; Lam, S.E.; Khandaker, M.U.; Bradley, D.A. Biological agents for synthesis of nanoparticles and their applications. *J. King Saud Univ. Sci.* **2022**, *34*, 101869, doi:10.1016/j.jksus.2022.101869.
10. Iravani, S. Bacteria in nanoparticle synthesis: Current status and future prospects. *Int. Sch. Res. Not.* **2014**, *2014*, 359316, doi:10.1155/2014/359316.
11. Mubeen, B.; Ansar, A.N.; Rasool, R.; Ullah, I.; Imam, S.S.; Alshehri, S.; Ghoneim, M.M.; Alzarea, S.I.; Nadeem, M.S.; Kazmi, I. Nanotechnology as a novel approach in combating microbes providing an alternative to antibiotics. *Antibiotics* **2021**, *10*, 1473, doi:10.3390/antibiotics10121473.
12. Singh, N.A.; Narang, J.; Garg, D.; Jain, V.; Payasi, D.; Suleman, S.; Swami, R.K. Nanoparticles synthesis via microorganisms and their prospective applications in agriculture. *Plant Nano Biol.* **2023**, *5*, 100047, doi:10.1016/j.plana.2023.100047.
13. Uddin, T.M.; Chakraborty, A.J.; Khusro, A.; Zidan, B.M.R.M.; Mitra, S.; Emran, T. Bin; Dhama, K.; Ripon, M.K.H.; Gajdács, M.; Sahibzada, M.U.K.; et al. Antibiotic resistance in microbes: History, mechanisms, therapeutic strategies and future prospects. *J. Infect. Public Health* **2021**, *14*, 1750–1766, doi:10.1016/j.jiph.2021.10.020.
14. Torky, H.A.; Saad, H.M.; Khaliel, S.A.; Kassih, A.T.; Sabatier, J.-M.; Batiha, G.E.; Hetta, H.F.; Elghazaly, E.M.; De Waard, M. Isolation and molecular characterization of corynebacterium pseudotuberculosis: Association with proinflammatory cytokines in caseous lymphadenitis pyogranulomas. *Animals* **2023**, *13*, 296.
15. Mabrok, M.; Algammal, A.M.; Sivaramasamy, E.; Hetta, H.F.; Atwah, B.; Alghamdi, S.; Fawzy, A.; Avendaño-Herrera, R.; Rodkhum, C. Tenacibaculosis caused by *Tenacibaculum maritimum*: Updated knowledge of this marine bacterial fish pathogen. *Front. Cell. Infect. Microbiol.* **2023**, *12*, 1068000, doi:10.3389/fcimb.2022.1068000.
16. Saraswathi, N.; Girigoswami, K.; Divya, K.C.; Kumar, S.G.; Girigoswami, A. Degree of gelatination on Ag-nanoparticles to inactivate multi-drug resistant bacterial biofilm isolated from sewage treatment plant. *Curr. Drug Deliv.* **2023**, *20*, 566–574, doi:10.2174/1567201819666220509160432.
17. Algammal, A.M.; Hetta, H.F.; Elkelish, A.; Alkhalifah, D.H.H.; Hozzein, W.N.; Batiha, G.E.-S.; El Nahhas, N.; Mabrok, M.A. Methicillin-Resistant *Staphylococcus aureus* (MRSA): One health perspective approach to the bacterium epidemiology, virulence factors, antibiotic-resistance, and zoonotic impact. *Infect. Drug Resist.* **2020**, *13*, 3255–3265, doi:10.2147/IDR.S272733.
18. Lebeaux, D.; Chauhan, A.; Rendueles, O.; Beloin, C. From in vitro to in vivo models of bacterial biofilm-related infections. *Pathogens* **2013**, *2*, 288–356.
19. Salah, M.; Badr, G.; Hetta, H.F.; Khalifa, W.A.; Shoreit, A.A. Fig latex inhibits the growth of pathogenic bacteria invading human diabetic wounds and accelerates wound closure in diabetic mice. *Sci. Rep.* **2022**, *12*, 21852, doi:10.1038/s41598-022-26338-0.
20. Kungwani, N.A.; Panda, J.; Mishra, A.K.; Chavda, N.; Shukla, S.; Vikhe, K.; Sharma, G.; Mohanta, Y.K.; Sharifi-Rad, M. Combating bacterial biofilms and related drug resistance: Role of phyto-derived adjuvant and nanomaterials. *Microb. Pathog.* **2024**, *195*, 106874, doi:10.1016/j.micpath.2024.106874.

21. Imran, M.; Jha, S.K.; Hasan, N.; Insaf, A.; Shrestha, J.; Shrestha, J.; Devkota, H.P.; Khan, S.; Panth, N.; Warkiani, M.E.; et al. Overcoming multidrug resistance of antibiotics via nanodelivery systems. *Pharmaceutics* **2022**, *14*, 586, doi:10.3390/pharmaceutics14030586.
22. Salam, M.A.; Al-Amin, M.Y.; Salam, M.T.; Pawar, J.S.; Akhter, N.; Rabaan, A.A.; Alqumber, M.A.A. Antimicrobial resistance: A growing serious threat for global public health. *Healthcare* **2023**, *11*, **1946**, doi:10.3390/healthcare11131946.
23. Pham, M.N.; Nishimura, F.; Lan, J.C.W.; Khoo, K.S. Recent advancement of eliminating antibiotic resistance bacteria and antibiotic resistance genes in livestock waste: A review. *Environ. Technol. Innov.* **2024**, *36*, 103751, doi:10.1016/j.eti.2024.103751.
24. Larsson, D.G.J.; Flach, C.-F. Antibiotic resistance in the environment. *Nat. Rev. Microbiol.* **2022**, *20*, 257–269, doi:10.1038/s41579-021-00649-x.
25. Singh, A.; Gautam, P.K.; Verma, A.; Singh, V.; Shivapriya, P.M.; Shivalkar, S.; Sahoo, A.K.; Samanta, S.K. Green synthesis of metallic nanoparticles as effective alternatives to treat antibiotics resistant bacterial infections: A review. *Biotechnol. Rep.* **2020**, *25*, e00427, doi:10.1016/j.btre.2020.e00427.
26. Skłodowski, K.; Chmielewska-Deptuła, S.J.; Piktel, E.; Wolak, P.; Wollny, T.; Bucki, R. Metallic nanosystems in the development of antimicrobial strategies with high antimicrobial activity and high biocompatibility. *Int. J. Mol. Sci.* **2023**, *24*, 2104, doi:10.3390/ijms24032104.
27. Bhardwaj, B.; Singh, P.; Kumar, A.; Kumar, S.; Budhwar, V. Eco-friendly greener synthesis of nanoparticles. *Adv. Pharm. Bull.* **2020**, *10*, 566–576, doi:10.34172/apb.2020.067.
28. Singh, J.; Dutta, T.; Kim, K.H.; Rawat, M.; Samddar, P.; Kumar, P. "Green" synthesis of metals and their oxide nanoparticles: applications for environmental remediation. *J. Nanobiotech.* **2018**, *16*, 84, doi:10.1186/s12951-018-0408-4.
29. Kulkarni, D.; Sherkar, R.; Shirsathe, C.; Sonwane, R.; Varpe, N.; Shelke, S.; More, M.P.; Pardeshi, S.R.; Dhaneshwar, G.; Junnuthula, V.; et al. Biofabrication of nanoparticles: Sources, synthesis, and biomedical applications. *Front. Bioeng. Biotechnol.* **2023**, *11*, 1159193, doi:10.3389/fbioe.2023.1159193.
30. Ghosh, S.; Ahmad, R.; Zeyaullah, M.; Khare, S.K. Microbial nano-factories: Synthesis and biomedical applications. *Front. Chem.* **2021**, *9*, 626834, doi:10.3389/fchem.2021.626834.
31. Roy, A.; Bulut, O.; Some, S.; Mandal, A.K.; Yilmaz, M.D. Green synthesis of silver nanoparticles: Biomolecule-nanoparticle organizations targeting antimicrobial activity. *RSC Adv.* **2019**, *9*, 2673–2702, doi:10.1039/c8ra08982e.
32. Tang, A.; Ren, Q.; Wu, Y.; Wu, C.; Cheng, Y. Investigation into the antibacterial mechanism of biogenic tellurium nanoparticles and precursor tellurite. *Int. J. Mol. Sci.* **2022**, *23*, 11697.
33. Gursoy, N. Fungus-mediated synthesis of silver nanoparticles (Agnp) and inhibitory effect on aspergillus Spp. in combination with antifungal agent. *Cumhur. Sci. J.* **2020**, *41*, 311–318, doi:10.17776/csj.653627.
34. Alsamhary, K.I. Eco-friendly synthesis of silver nanoparticles by *Bacillus subtilis* and their antibacterial activity. *Saudi J. Biol. Sci.* **2020**, *27*, 2185–2191, doi:10.1016/j.sjbs.2020.04.026.
35. Abdollahnia, M.; Makhdoumi, A.; Mashreghi, M.; Eshghi, H. Exploring the potentials of halophilic prokaryotes from a solar saltern for synthesizing nanoparticles: The case of silver and selenium. *PLoS One* **2020**, *15*, e0229886, doi:10.1371/journal.pone.0229886.
36. Saeed, S.; Iqbal, A.; Ashraf, M.A. Bacterial-mediated synthesis of silver nanoparticles and their significant effect against pathogens. *Environ. Sci. Pollut. Res. Int.* **2020**, *27*, 37347–37356, doi:10.1007/s11356-020-07610-0.
37. Piragash Kumar, R.M.; Venkatesh, A.; Moorthy, V.H.S. Nanopits based novel hybrid plasmonic nanosensor fabricated by a facile nanofabrication technique for biosensing. *Mater. Res. Express* **2019**, *6*, 1150b6, doi:10.1088/2053-1591/ab33b9.
38. Borah, D.; Das, N.; Das, N.; Bhattacharjee, A.; Sarmah, P.; Ghosh, K.; Chandel, M.; Rout, J.; Pandey, P.; Ghosh, N.N.; et al. Alga-mediated facile green synthesis of silver nanoparticles: Photophysical, catalytic and antibacterial activity. *Appl. Organomet. Chem.* **2020**, *34*, e5597, doi:10.1002/aoc.5597.
39. Shivashankarappa, A.; Sanjay, K.R. Escherichia coli-based synthesis of cadmium sulfide nanoparticles, characterization, antimicrobial and cytotoxicity studies. *Brazilian J. Microbiol.* **2020**, *51*, 939–948, doi:10.1007/s42770-020-00238-9.
40. Mohd Yusof, H.; Abdul Rahman, N.; Mohamad, R.; Zaidan, U.H.; Samsudin, A.A. Biosynthesis of zinc oxide nanoparticles by cell-biomass and supernatant of lactobacillus plantarum TA4 and its antibacterial and biocompatibility properties. *Sci. Rep.* **2020**, *10*, 19996, doi:10.1038/s41598-020-76402-w.
41. Rajeshkumar, S.; Jayakodi, S.; Tharani, M.; Alharbi, N.S.; Thiruvengadam, M. Antimicrobial activity of probiotic bacteria-mediated cadmium oxide nanoparticles against fish pathogens. *Microb. Pathog.* **2024**, *189*, 106602, doi:10.1016/j.micpath.2024.106602.

42. Barani, M.; Masoudi, M.; Mashreghi, M.; Makhdoumi, A.; Eshghi, H. Cell-free extract assisted synthesis of ZnO nanoparticles using aquatic bacterial strains: Biological activities and toxicological evaluation. *Int. J. Pharm.* **2021**, *606*, 120878, doi:10.1016/j.ijpharm.2021.120878.
43. Nazari, N.; Jookar Kashi, F. A novel microbial synthesis of silver nanoparticles: Its bioactivity, Ag/Ca-Alg beads as an effective catalyst for decolorization disperse blue 183 from textile industry effluent. *Sep. Purif. Technol.* **2021**, *259*, 118117, doi:10.1016/j.seppur.2020.118117.
44. Acay, H. Utilization of morchella esculenta-mediated green synthesis golden nanoparticles in biomedicine applications. *Prep. Biochem. Biotechnol.* **2021**, *51*, 127–136, doi:10.1080/10826068.2020.1799390.
45. Bhuyar, P.; Rahim, M.H.A.; Sundararaju, S.; Ramaraj, R.; Maniam, G.P.; Govindan, N. Synthesis of silver nanoparticles using marine macroalgae padina Sp. and its antibacterial activity towards pathogenic bacteria. *Beni-Suef Univ. J. Basic Appl. Sci.* **2020**, *9*, 3, doi:10.1186/s43088-019-0031-y.
46. Feroze, N.; Arshad, B.; Younas, M.; Afridi, M.I.; Saqib, S.; Ayaz, A. Fungal mediated synthesis of silver nanoparticles and evaluation of antibacterial activity. *Microsc. Res. Tech.* **2020**, *83*, 72–80, doi:10.1002/jemt.23390.
47. Majeed, S.; Danish, M.; Mohamad Ibrahim, M.N.; Sekeri, S.H.; Ansari, M.T.; Nanda, A.; Ahmad, G. Bacteria mediated synthesis of iron oxide nanoparticles and their antibacterial, antioxidant, cytocompatibility properties. *J. Clust. Sci.* **2021**, *32*, 1083–1094, doi:10.1007/s10876-020-01876-7.
48. Huq, M.A. Green synthesis of silver nanoparticles using Pseudoduganella eburnea MAHUQ-39 and their antimicrobial mechanisms investigation against drug resistant human pathogens. *Int. J. Mol. Sci.* **2020**, *21*, 510, doi:10.3390/ijms21041510.
49. Eramabadi, P.; Masoudi, M.; Makhdoumi, A.; Mashreghi, M. Microbial cell lysate supernatant (CLS) alteration impact on platinum nanoparticles fabrication, characterization, antioxidant and antibacterial activity. *Mater. Sci. Eng. C* **2020**, *117*, 111292, doi:10.1016/j.msec.2020.111292.
50. Akter, S.; Huq, M.A. Biologically rapid synthesis of silver nanoparticles by *Sphingobium* sp. MAH-11(T) and their antibacterial activity and mechanisms investigation against drug-resistant pathogenic microbes. *Artif. Cells, Nanomed. Biotechnol.* **2020**, *48*, 672–682, doi:10.1080/21691401.2020.1730390.
51. El-Sheekh, M.M.; Shabaan, M.T.; Hassan, L.; Morsi, H.H. Antiviral activity of algae biosynthesized silver and gold nanoparticles against herps simplex (HSV-1) virus in vitro using cell-line culture technique. *Int. J. Environ. Health Res.* **2022**, *32*, 616–627, https://doi.org/10.1080/09603123.2020.1789946.
52. Elumalai, L.; Palaniyandi, S.; Anbazhagan, G.K.; Mohanam, N.; Munusamy, S.; Sri Ragavi, G.K.; Pudukadu Munusamy, A.; Chinnasamy, M.; Ramasamy, B. Synthesis of biogenic cadmium sulfide nanoparticles (MR03-CdSNPs) using marine *Streptomyces kunmingensis* – MR03 for in-vitro biological determinations and in silico analysis on biofilm virulence proteins: A novel approach. *Environ. Res.* **2023**, *235*, 116698, doi:10.1016/j.envres.2023.116698.
53. Bakhtiari-Sardari, A.; Mashreghi, M.; Eshghi, H.; Behnam-Rasouli, F.; Lashani, E.; Shahnavaz, B. Comparative evaluation of silver nanoparticles biosynthesis by two cold-tolerant streptomyces strains and their biological activities. *Biotechnol. Lett.* **2020**, *42*, 1985–1999, doi:10.1007/s10529-020-02921-1.
54. Massironi, A.; Morelli, A.; Grassi, L.; Puppi, D.; Braccini, S.; Maisetta, G.; Esin, S.; Batoni, G.; Della Pina, C.; Chiellini, F. Ulvan as novel reducing and stabilizing agent from renewable algal biomass: Application to green synthesis of silver nanoparticles. *Carbohydr. Polym.* **2019**, *203*, 310–321, doi:10.1016/j.carbpol.2018.09.066.
55. Zamanpour, N.; Mohammad Esmaeily, A.; Mashreghi, M.; Shahnavaz, B.; Reza Sharifmoghadam, M.; Kompany, A. Application of a marine luminescent *Vibrio* sp. B4L for biosynthesis of silver nanoparticles with unique characteristics, biochemical properties, antibacterial and antibiofilm activities. *Bioorg. Chem.* **2021**, *114*, 105102, https://doi.org/10.1016/j.bioorg.2021.105102
56. Nakhaeepour, Z.; Mashreghi, M.; Matin, M.M.; NakhaeiPour, A.; Housaindokht, M.R. Multifunctional CuO nanoparticles with cytotoxic effects on KYSE30 esophageal cancer cells, antimicrobial and heavy metal sensing activities. *Life Sci.* **2019**, *234*, 116758, doi:10.1016/j.lfs.2019.116758.
57. Nayak, D.; Mishra, A.K.; Malla, S.; Swain, U.; Panigrahi, K.; Panda, J.; Rustagi, S.; Rauta, P.R.; Mohanta, Y.K. Eco-friendly mycogenic synthesis of iron oxide nanoparticles using Volvariella volvacea and unveiling its potential as antibiofilm and anticancer agents. *ECS J. Solid State Sci. Technol.* **2024**, *13*, 087003, doi:10.1149/2162-8777/ad658b.
58. Bolbanabad, E.M.; Ashengroph, M.; Darvishi, F. Development and evaluation of different strategies for the clean synthesis of silver nanoparticles using Yarrowia lipolytica and their antibacterial activity. *Process Biochem.* **2020**, *94*, 319–328, doi:10.1016/j.procbio.2020.03.024.
59. Altammar, K.A. A review on nanoparticles: Characteristics, synthesis, applications, and challenges. *Front. Microbiol.* **2023**, *14*, 1155622, doi:10.3389/fmicb.2023.1155622.

60. Arora, A.; Lashani, E.; Turner, R.J. Bacterial synthesis of metal nanoparticles as antimicrobials. *Microb. Biotechnol.* **2024**, *17*, e14549, doi:10.1111/1751-7915.14549.
61. Velusamy, P.; Kumar, G.V.; Jeyanthi, V.; Das, J.; Pachaiappan, R. Bio-inspired green nanoparticles: Synthesis, mechanism, and antibacterial application. *Toxicol. Res.* **2016**, *32*, 95–102, doi:10.5487/TR.2016.32.2.095.
62. Bruna, T.; Maldonado-Bravo, F.; Jara, P.; Caro, N. Silver nanoparticles and their antibacterial applications. *Int. J. Mol. Sci.* **2021**, *22*, 7202, doi:10.3390/ijms22137202.
63. Biswas, A. Potential of honey in sustainable health management. *Ecol. Environ. Conserv.* **2023**, *29*, S81–S87, doi:10.53550/eec.2023.v29i01s.013.
64. Rizvi, S.M.D.; Lila, A.S.A.; Moin, A.; Hussain, T.; Kamal, M.A.; Sonbol, H.; Khafagy, E.S. Antibiotic-loaded gold nanoparticles: A nano-arsenal against ESBL producer-resistant pathogens. *Pharmaceutics* **2023**, *15*, 430, doi:10.3390/pharmaceutics15020430.
65. Tariq, F.; Ahmed, N.; Afzal, M.; Khan, M.A.U.; Zeshan, B. Synthesis, characterization and antimicrobial activity of *Bacillus subtilis*-derived silver nanoparticles against multidrug-resistant bacteria. *Jundishapur J. Microbiol.* **2020**, *13*, e91934, doi:10.5812/jjm.91934.
66. Abd Elsalam, S.S.; Taha, R.H.; Tawfeik, A.M.; Abd El-Monem, M.O.; Mahmoud, H.A. Antimicrobial activity of bio and chemical synthesized cadmium sulfide nanoparticles. *Egypt. J. Hosp. Med.* **2018**, *70*, 1494–1507.
67. Hamouda, R.A.; Hussein, M.H.; Abo-elmagd, R.A.; Bawazir, S.S. Synthesis and biological characterization of silver nanoparticles derived from the cyanobacterium *Oscillatoria limnetica. Sci. Rep.* **2019**, *9*, 13071, doi:10.1038/s41598-019-49444-y.
68. Hamida, R.S.; Abdelmeguid, N.E.; Ali, M.A.; Bin-Meferij, M.M.; Khalil, M.I. Synthesis of silver nanoparticles using a novel cyanobacteria Desertifilum sp. extract: Their antibacterial and cytotoxicity effects. *Int. J. Nanomed.* **2020**, *15*, 49–63, doi:10.2147/IJN.S238575.
69. Parveen, A.; Bhatnagar, P.; Bisht, B.; Kumar, S.; Joshi, S.; Gautam, P.; Kumar, V.; Adarchenko Irina, A.; Kurbatova Anna, K. Algal nanobionics to enhance value added products – A review. *Curr. Res. Biotechnol.* **2023**, *5*, 100124, doi:10.1016/j.crbiot.2023.100124.
70. Dai, N.; Wang, Q.; Xu, B.; Chen, H. Remarkable natural biological resource of algae for medical applications. *Front. Mar. Sci.* **2022**, *9*, 912924, doi:10.3389/fmars.2022.912924.
71. Mandal, A.K.; Nayak, R.; Pradhan, B.; Behera, C.; Behera, A.K.; Parida, S.; Patra, S.; Hembram, P.; Jena, M. Algal-derived nanoparticles and their antibacterial potential: Current evidence and future prospectives. *J. Microbiol. Methods* **2023**, *211*, 106790, doi:10.1016/j.mimet.2023.106790.
72. El-Sheekh, M.M.; Morsi, H.H.; Hassan, L.H.S.; Ali, S.S. The efficient role of algae as green factories for nanotechnology and their vital applications. *Microbiol. Res.* **2022**, *263*, 127111, doi:10.1016/j.micres.2022.127111.
73. Sidorowicz, A.; Fais, G.; Casula, M.; Borselli, M.; Giannaccare, G.; Locci, A.M.; Lai, N.; Orrù, R.; Cao, G.; Concas, A. Nanoparticles from microalgae and their biomedical applications. *Mar. Drugs* **2023**, *21*, 352, doi:10.3390/md21060352.
74. Kowshik, M.; Ashtaputre, S.; Kharrazi, S.; Vogel, W.; Urban, J.; Kulkarni, S.K.; Paknikar, K.M. Extracellular synthesis of silver nanoparticles by a silver-tolerant yeast strain MKY3. *Nanotechnology* **2003**, *14*, 95–100, doi:10.1088/0957-4484/14/1/321.
75. Adeleke, B.S.; Olowe, O.M.; Ayilara, M.S.; Fasusi, O.A.; Omotayo, O.P.; Fadiji, A.E.; Onwudiwe, D.C.; Babalola, O.O. Biosynthesis of nanoparticles using microorganisms: A focus on endophytic fungi. *Heliyon* **2024**, *10*, e39636, doi:10.1016/j.heliyon.2024.e39636.
76. Abdel-Hadi, A.; Iqbal, D.; Alharbi, R.; Jahan, S.; Darwish, O.; Alshehri, B.; Banawas, S.; Palanisamy, M.; Ismail, A.; Aldosari, S.; et al. Myco-synthesis of silver nanoparticles and their bioactive role against pathogenic microbes. *Biology (Basel)* **2023**, *12*, 661, doi:10.3390/biology12050661.
77. Elshafei, A.M.; Othman, A.M.; Elsayed, M.A.; Al-Balakocy, N.G.; Hassan, M.M. Green Synthesis of silver nanoparticles using *Aspergillus oryzae* NRRL447 exogenous proteins optimization via central composite design, characterization and biological applications. *Environ. Nanotechnology, Monit. Manag.* **2021**, *16*, 100553, doi:10.1016/j.enmm.2021.100553.
78. Luceri, A.; Francese, R.; Lembo, D.; Ferraris, M.; Balagna, C. Silver nanoparticles: Review of antiviral properties, mechanism of action and applications. *Microorganisms* **2023**, *11*, 629, doi:10.3390/microorganisms11030629.
79. Chung, Y.H.; Cai, H.; Steinmetz, N.F. Viral nanoparticles for drug delivery, imaging, immunotherapy, and theranostic applications. *Adv. Drug Deliv. Rev.* **2020**, *156*, 214–235, doi:10.1016/j.addr.2020.06.024.

80. Dayma, P.; Choudhary, N.; Ali, D.; Alarifi, S.; Dudhagara, P.; Luhana, K.; Yadav, V.K.; Patel, A. Patel, R. Exploring the potential of halotolerant actinomycetes from Rann of Kutch, India: A study on the synthesis, characterization, and biomedical applications of silver nanoparticles. *Pharmaceuticals* **2024**, *17*, 743, doi:10.3390/ph17060743.
81. Hochvaldová, L.; Večeřová, R.; Kolář, M.; Prucek, R.; Kvítek, L.; Lapčík, L.; Panáček, A. Antibacterial nanomaterials: Upcoming hope to overcome antibiotic resistance crisis. *Nanotechnol. Rev.* **2022**, *11*, 1115–1142, doi:10.1515/ntrev-2022-0059.
82. Cheeseman, S.; Christofferson, A.J.; Kariuki, R.; Cozzolino, D.; Daeneke, T.; Crawford, R.J.; Truong, V.K.; Chapman, J.; Elbourne, A. Antimicrobial metal nanomaterials: From passive to stimuli-activated applications. *Adv. Sci.* **2020**, *7*, 1902913, doi:10.1002/advs.201902913.
83. Riu, F.; Ruda, A.; Ibba, R.; Sestito, S.; Lupinu, I.; Piras, S.; Widmalm, G.; Carta, A. Antibiotics and carbohydrate-containing drugs targeting bacterial cell envelopes: An overview. *Pharmaceuticals (Basel)*. **2022**, *15*, 942, doi:10.3390/ph15080942.
84. Hetta, H.F.; Ramadan, Y.N.; Al-Harbi, A.I.; A. Ahmed, E.; Battah, B.; Abd Ellah, N.H.; Zanetti, S.; Donadu, M.G. Nanotechnology as a promising approach to combat multidrug resistant bacteria: A comprehensive review and future perspectives. *Biomedicines* **2023**, *11*, 413.
85. Girma, A. Alternative mechanisms of action of metallic nanoparticles to mitigate the global spread of antibiotic-resistant bacteria. *Cell Surf.* **2023**, *10*, 100112, doi:10.1016/j.tcsw.2023.100112.
86. Rosli, N.A.; Teow, Y.H.; Mahmoudi, E. Current approaches for the exploration of antimicrobial activities of nanoparticles. *Sci. Technol. Adv. Mater.* **2021**, *22*, 885–907, doi:10.1080/14686996.2021.1978801.
87. Hailan, W.A.; Al-Anazi, K.M.; Farah, M.A.; Ali, M.A.; Al-Kawmani, A.A.; Abou-Tarboush, F.M. Reactive oxygen species-mediated cytotoxicity in liver carcinoma cells induced by silver nanoparticles biosynthesized using Schinus molle extract. *Nanomaterials* **2022**, *12*, 161.
88. Banerjee, S.; Vishakha, K.; Das, S.; Sangma, P.D.; Mondal, S.; Ganguli, A. Oxidative stress, DNA, and membranes targets as modes of antibacterial and antibiofilm activity of facile synthesized biocompatible keratin-copper nanoparticles against multidrug resistant uro-pathogens. *World J. Microbiol. Biotechnol.* **2022**, *38*, 20, doi:10.1007/s11274-021-03187-z.
89. Fahimmunisha, B.A.; Ishwarya, R.; AlSalhi, M.S.; Devanesan, S.; Govindarajan, M.; Vaseeharan, B. Green fabrication, characterization and antibacterial potential of zinc oxide nanoparticles using aloe socotrina leaf extract: A novel drug delivery approach. *J. Drug Deliv. Sci. Technol.* **2020**, *55*, 101465, doi:10.1016/j.jddst.2019.101465.
90. Bardestani, A.; Ebrahimpour, S.; Esmaeili, A.; Esmaeili, A. Quercetin attenuates neurotoxicity induced by iron oxide nanoparticles. *J. Nanobiotechnology* **2021**, *19*, 327, doi:10.1186/s12951-021-01059-0.
91. Xuan, L.; Ju, Z.; Skonieczna, M.; Zhou, P.-K.; Huang, R. Nanoparticles-induced potential toxicity on human health: Applications, toxicity mechanisms, and evaluation models. *MedComm* **2023**, *4*, e327, doi:10.1002/mco2.327.
92. Saed, M.; Ayivi, R.D.; Wei, J.; Obare, S.O. Gold nanoparticles antibacterial activity: Does the surface matter? *Colloid Interface Sci. Commun.* **2024**, *62*, 100804, doi:10.1016/j.colcom.2024.100804.
93. Sahli, C.; Moya, S.E.; Lomas, J.S.; Gravier-Pelletier, C.; Briandet, R.; Hémadi, M. Recent advances in nanotechnology for eradicating bacterial biofilm. *Theranostics* **2022**, *12*, 2383–2405, doi:10.7150/thno.67296.
94. Öztürk, K.; Kaplan, M.; Çalış, S. Effects of nanoparticle size, shape, and zeta potential on drug delivery. *Int. J. Pharm.* **2024**, *666*, 124799, doi:10.1016/j.ijpharm.2024.124799.
95. Sharmin, S.; Rahaman, M.M.; Sarkar, C.; Atolani, O.; Islam, M.T.; Adeyemi, O.S. Nanoparticles as antimicrobial and antiviral agents: A literature-based perspective study. *Heliyon* **2021**, *7*, e06456, doi:10.1016/j.heliyon.2021.e06456.
96. Muteeb, G.; Rehman, M.T.; Shahwan, M.; Aatif, M. Origin of antibiotics and antibiotic resistance, and their impacts on drug development: A narrative review. *Pharmaceuticals* **2023**, *16*, 1615.
97. Razei, A.; Javanbakht, M.; Hajizade, A.; Heiat, M.; Zhao, S.; Aghamollaei, H.; Saadati, M.; Khafaei, M.; Asadi, M.; Cegolon, L.; et al. Nano and microparticle drug delivery systems for the treatment of brucella infections. *Biomed. Pharmacother.* **2023**, *169*, 115875, doi:10.1016/j.biopha.2023.115875.
98. Moradi, F.; Ghaedi, A.; Fooladfar, Z.; Bazrgar, A. Recent advance on nanoparticles or nanomaterials with anti-multidrug resistant bacteria and anti-bacterial biofilm properties: A systematic review. *Heliyon* **2023**, *9*, e22105, doi:10.1016/j.heliyon.2023.e22105.
99. Sathyakamala, R.; Peace, A.R.; Shanmugam, P. A comparative study on bacterial co-infections and prevalence of multidrug resistant organisms among patients in COVID and non-COVID intensive care units. *J. Prev. Med. Hyg.* **2022**, *63*, E19–E26, doi:10.15167/2421-4248/jpmh2022.63.1.2175.

100. Wong, F.; Wilson, S.; Helbig, R.; Hegde, S.; Aftenieva, O.; Zheng, H.; Liu, C.; Pilizota, T.; Garner, E.C.; Amir, A.; et al. Understanding beta-lactam-induced lysis at the single-cell level. *Front. Microbiol.* **2021**, *12*, 712007, doi:10.3389/fmicb.2021.712007.
101. Mamun, M.M.; Sorinolu, A.J.; Munir, M.; Vejerano, E.P. Nanoantibiotics: Functions and properties at the nanoscale to combat antibiotic resistance. *Front. Chem.* **2021**, *9*, 687660, doi:10.3389/fchem.2021.687660.
102. Mohanta, Y.K.; Mishra, A.K.; Panda, J.; Chakrabartty, I.; Sarma, B.; Panda, S.K.; Chopra, H.; Zengin, G.; Moloney, M.G.; Sharifi-Rad, M. Promising applications of phyto-fabricated silver nanoparticles: Recent trends in biomedicine. *Biochem. Biophys. Res. Commun.* **2023**, *688*, 149126, doi:10.1016/j.bbrc.2023.149126.
103. Modi, S.K.; Gaur, S.; Sengupta, M.; Singh, M.S. Mechanistic insights into nanoparticle surface-bacterial membrane interactions in overcoming antibiotic resistance. *Front. Microbiol.* **2023**, *14*, doi:10.3389/fmicb.2023.1135579.
104. Lahiri, D.; Nag, M.; Sheikh, H.I.; Sarkar, T.; Edinur, H.A.; Pati, S.; Ray, R.R. Microbiologically-synthesized nanoparticles and their role in silencing the biofilm signaling cascade. *Front. Microbiol.* **2021**, *12*, 636588, doi:10.3389/fmicb.2021.636588.
105. Adebayo, E.A.; Azeez, M.A.; Alao, M.B.; Oke, A.M.; Aina, D.A. Fungi as veritable tool in current advances in nanobiotechnology. *Heliyon* **2021**, *7*, e08480, doi:10.1016/j.heliyon.2021.e08480.
106. Varier, K.M.; Gudeppu, M.; Chinnasamy, A.; Thangarajan, S.; Balasubramanian, J.; Li, Y.; Gajendran, B. Nanoparticles: Antimicrobial applications and its prospects. *Adv. Nanostructured Mater. Environ. Remediat.* **2019**, *25*, 321–355.
107. Mitchell, M.J.; Billingsley, M.M.; Haley, R.M.; Wechsler, M.E.; Peppas, N.A.; Langer, R. Engineering precision nanoparticles for drug delivery. *Nat. Rev. Drug Discov.* **2021**, *20*, 101–124, doi:10.1038/s41573-020-0090-8.
108. Egbuna, C.; Parmar, V.K.; Jeevanandam, J.; Ezzat, S.M.; Patrick-Iwuanyanwu, K.C.; Adetunji, C.O.; Khan, J.; Onyeike, E.N.; Uche, C.Z.; Akram, M.; et al. Toxicity of nanoparticles in biomedical application: Nanotoxicology. *J. Toxicol.* **2021**, *2021*, 9954443, doi:10.1155/2021/9954443.
109. Yusuf, A.; Almotairy, A.R.Z.; Henidi, H.; Alshehri, O.Y.; Aldughaim, M.S. Nanoparticles as drug delivery systems: A review of the implication of nanoparticles' physicochemical properties on responses in biological systems. *Polymers (Basel)* **2023**, *15*, 1596.

2 Green Power
Photosynthetic Microbes Pioneering Nanotechnology

Niraj Singh, N. Seema Devi, Setu Chakrabarty, Jibanjyoti Panda, and Jay Kumar

2.1 INTRODUCTION

Photosynthetic microbes constitute a diverse array of organisms that hold pivotal roles in Earth's ecosystems and the sustenance of life. This group includes bacteria, algae, and specific types of protists [1]. By utilising sunlight, these organisms convert carbon dioxide and water into organic compounds, simultaneously releasing oxygen as a by-product. This process, termed photosynthesis, forms a fundamental pillar of the food chain and oxygen production on our planet. Among these microbes, cyanobacteria stand out as one of the most well-known examples, often mistakenly referred to as blue-green algae. Despite the common name, cyanobacteria are actually bacteria capable of photosynthesis. These microbes played a crucial role in shaping Earth's atmosphere billions of years ago by generating oxygen through photosynthesis, leading to the oxygenation of the atmosphere and the subsequent evolution of aerobic life forms [2,3]. Algae represent another vital group of photosynthetic microbes essential to both marine and freshwater ecosystems. They encompass a wide variety of organisms ranging from single-celled to multicellular forms. Certain algae, such as diatoms and dinoflagellates, serve as primary producers at the base of the food chain, providing sustenance for a diverse range of aquatic organisms. Other types, such as seaweeds, create complex marine ecosystems and offer habitats for numerous marine species [4]. Additionally, photosynthetic protists, including euglenoids and specific types of protozoa, contribute to aquatic ecosystems. These organisms are frequently found in freshwater environments and aid in the food chain through photosynthesis. Beyond their ecological significance, photosynthetic microbes hold practical value across various industries. For instance, algae are being investigated as a potential source of biofuels due to their rapid growth rate and capacity to produce oils convertible into biodiesel. Similarly, cyanobacteria are under scrutiny for their potential to generate biofuels and other valuable compounds. In agriculture, photosynthetic microbes find applications as biofertilisers and biopesticides. Some cyanobacteria can fix nitrogen from the atmosphere, thereby making it accessible to plants and reducing the need for synthetic fertilisers [5]. Moreover, other photosynthetic microbes are being researched for their ability to degrade pollutants and remediate contaminated environments.

Nanotechnology has rapidly advanced in recent years due to its versatile applications in electronics, medicine, agriculture, textiles, paints, and environmental protection. Nanomaterials offer an alternative to traditional technologies, addressing existing limitations. Due to their unique physical and chemical properties, nanoparticles (NPs) are utilised in various therapeutic applications, including diagnosis, imaging, drug delivery, triggered release, and wound dressings [6]. While NPs are typically synthesised using physical and chemical methods like laser ablation, sol-gel, and electro-explosion, these techniques pose health and environmental risks [7]. Given their applications in biomedicine and agriculture, it's crucial to adopt safer NP fabrication methods [8]. The concept of "green nanoparticles" has emerged, involving the environmentally benign synthesis of

 DOI: 10.1201/9781003487142-2

NPs using bacteria, fungi, algae, and plants [9]. These biological entities' extracts contain proteins and functional groups that act as reducing and capping agents, respectively, for NP synthesis and stabilisation [10,11].

Green synthesis results in more stable NPs, with properties influenced by physical factors such as pH, temperature, and contact time. Morphological and chemical characterisation of these NPs is conducted using techniques like scanning electron microscopy (SEM), transmission electron microscopy (TEM), X-ray diffraction (XRD), energy dispersive spectroscopy (EDS), and Fourier transform infrared (FTIR) spectroscopy [12,13].

Phyconanotechnology, the creation of metal NPs using algae, is an emerging field. The process involves three main steps: isolation and preparation of algal extracts, preparation of a precursor metal solution, and incubation of the metal precursor with the algal extract. A change in colour indicates the initiation of nucleation and subsequent NP synthesis. Algae are efficient accumulators of metal and can produce biomass quickly and at scale. This biomass can be harvested easily, making algae cost-effective nanobiofactories compared to other natural extracts [14,15].

The scope of this book chapter is to explore the diverse applications of photosynthetic microbes in the field of nanotechnology. It aims to provide a comprehensive overview of how these microbes, including bacteria, algae, and certain types of protists, can be utilised in various nanotechnological applications.

2.2 PHOTOSYNTHETIC MICROBES AS NANOPARTICLE SYNTHESISERS

Microbes create metallic, nonmetallic, and metal-oxide NPs through an enzymatic reduction process, either intracellularly or extracellularly. Metallic NPs can be synthesised by marine bacteria, fungus, and microalgae using intracellular or extracellular routes. Cyanobacteria (blue-green algae) are Gram-negative photosynthetic eubacteria that are ubiquitous, and they also are the largest group of photosynthetic prokaryotes rich in novel biologically active compounds. They are known to produce silver nanoparticles (Ag NPs) with applications in the treatment of infectious diseases [16]. Among cyanobacteria, species such as *Oscillatoria*, *Spirulina*, *Anabaena*, and *Calothrix* form Au, Ag, platinum (Pt), and palladium (Pd) NPs of different sizes [17]. *Nostoc linckia* forms amorphous selenium (Sp) NPs (10–80 nm) by selenite reduction involving thiol groups of cyanobacterial proteins [18]. Ag NPs made from cyanobacteria such as *Phormidium* sp., *Synechocystis* sp., and *Arthrospira sp.* polysaccharides have also demonstrated wound-healing applications with positive results [19].

The location of NP synthesis varies depending on the species and might occur intracellularly underneath the surface of the cell wall, as in *Verticillium* sp. or on the surface of mycelia/plasma membrane (*Rhodococcus* sp., *Tetraselmis kochinensis*) [20]. Similar to this, surface proteins or enzyme secretion plays a role in the extracellular creation of metallic nanoclusters. Negatively charged proteins that aid in the formation of nanostructures reduce positively charged metal ions that are trapped in the cytoplasm or on the cell wall to tiny nuclei as part of the mechanism of microbial production of NPs [21]. When exposed to metal salts in the external media, a number of fungal and algal species have been shown to create gold (Au) and silver (Ag) NPs. Chaudhary et al. [15] report that around 38 species of brown algae, 23 species of red algae, 49 species of green algae, and 21 species of blue-green algae are known to produce NPs with a wide variety of applications.

There are also reports of microalgae using polysaccharides found in their cell walls to produce metal-oxide NPs. *Sargassum muticum*, *Chlamydomonas reinhardtii*, and *Gracilaria gracilis* are only a few of the many species of algae that can generate ZnO NPs [23].

Phytoplanktons are the photosynthetic elements of the aquatic grazer food chain. Grazer populations of zooplanktons, like copepods, are significantly shaped by chemicals secreted by several phytoplankton species as a means of self-defence [24]. They have the capacity to absorb significant amounts of heavy metals and transform them into useful nanostructures. This skill has proven to be quite useful in the field of bioremediation, which cleans contaminated places using a variety of techniques. Nevertheless, depending on the concentration, size, form, and type of the metal involved, these NPs can also seriously harm phytoplanktons. Thus, appropriate research methodologies are required to use algae for the goal of bioremediation while keeping them intact and viable [25].

2.3 MECHANISMS OF NANOPARTICLE GREEN SYNTHESIS

The biosynthetic mechanism of Mt NPs in microorganisms can be carried out by trapping target metal ions from the surrounding environment and enzymatically converting them into elemental form, following a reduction mechanism [26,27]. Not all microorganisms are able to produce Mt NPs because they are produced through metabolic pathways and through cellular enzymes that may not be present in some organisms. The synthesis of Mt NPs is also dependent on the capacity of microorganisms for tolerating heavy metals. High metal stresses can affect various microbial activities, and some microorganisms are able to reduce metal ions to the respective metals under stress condition. In general, microorganisms that live in metal-rich habitats are highly resistant to those metals due to their uptake and chelation by intracellular and extracellular proteins.

Intracellular biosynthesis involves unique transport systems in microorganisms, in which the cell wall plays an important role due to its negative charge: positively charged metal ions are deposited in negatively charged cell walls through electrostatic interactions. After transport into the cells of microorganisms, ions are reduced using metabolic reactions mediated by enzymes such as nitrate reductase to form Mt NPs [22]. Mt NPs accumulated in the periplasmic space can then be passed through the cell wall [28,29]. The *extracellular* biosynthesis of Mt NPs is also mediated by nitrate reductase, where the Mt NPs are produced by reductase enzymes which are either located in the cell wall or secreted from the cell to the growth medium. In this process, nitrate reductase reduces metal ions to metallic forms [29,30]. A general schematic of intra- and extracellular biosyntheses of NPs is given in Figure 2.1.

Cyanobacteria are increasingly used in green synthesis (Table 2.1) due to their abundance of secondary metabolites, pigments, proteins, and peptides, which operate as bio-nanofactories [31]. Metabolites such as proteins, enzymes, and other biomolecules serve as reducing, capping, and stabilising agents. The varied shape, size, dispersity, and stability of NPs are all connected to secreted metabolites [30]. The reductase enzymes released by cyanobacteria in certain growth conditions can carry out extracellular ZnO NP generation. NADH-dependent reductase enzymes act as electron carriers, reducing Zn^{2+} to ZnO and then forming ZnO NPs [32]. The reductase enzymes acquired electrons from NADH, which were then oxidised to NAD+, while metal ions were reduced to nanoscales [33].

Green NP synthesis is a non-toxic, cost-effective, and ecologically friendly option. Velusamy et al. [44] and Nowruzi et al. [45] found that plants, fungi, bacteria, and algae contribute to the green synthesis of pure NPs. These biological systems serve as both regenerative and inhibitory agents in the synthesis of NPs, either intracellularly or externally. Functional groups (hydroxyl, carbonyl, amine, and thiol), proteins, and peptides in biological organisms convert dissolved metal salts to NPs in a straightforward, repeatable, and consistent process.

In NP biosynthesis, organisms convert the metallic ions produced into metallic components of the required shape and size using enzymes created by cell activities. First, metal ions are trapped on the surface or inside microbial cells, which is the initiation phase. These metal ions are then reduced by enzymes, followed by nucleation of these metal atoms, called the activation phase. This is sometimes followed by the growth phase, where small adjacent NPs spontaneously coalesce into larger aggregates of NPs.

Intracellular mechanism involves ion transport via the microbial cell to generate NPs using enzymes, pigments, and action of reductase and nitrogenase enzymes [37,46]. Extracellular synthesis of NPs occurs when metallic ions are trapped on the cell surface, and secreted enzymes reduce the ions to NP forms, which can then be extracted from the growth media. The capacity of

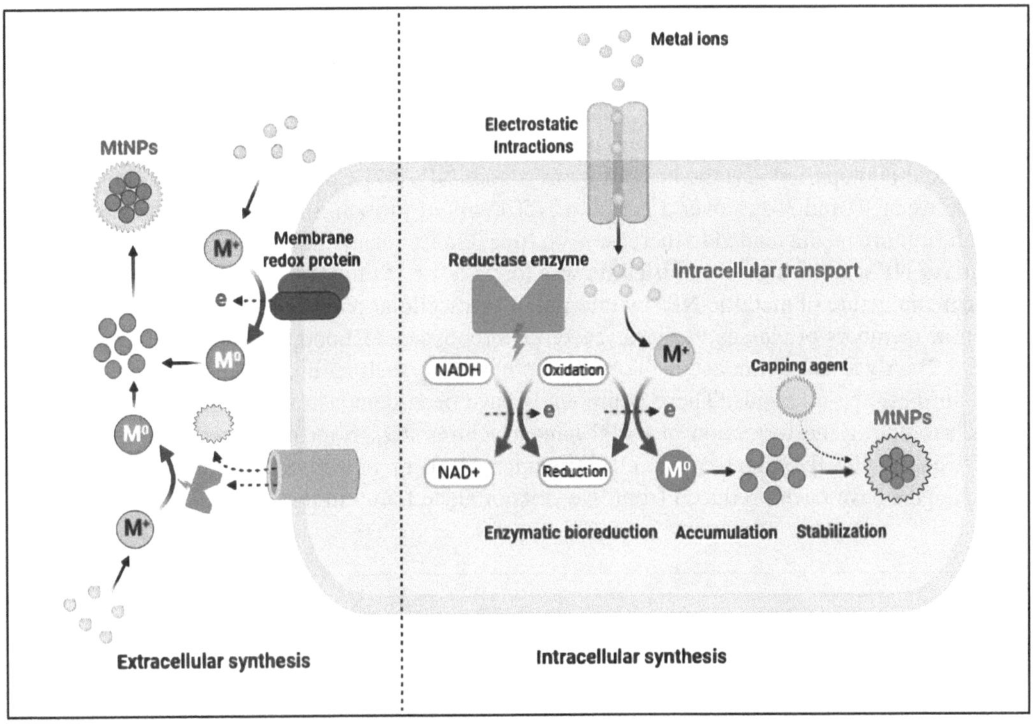

FIGURE 2.1 Schematic representation of the mechanisms of extracellular and intracellular biosyntheses of Mt NPs. Extracellular biosynthesis of Mt NPs carried out by trapping metal ions on the cell wall and reducing them in the presence of secreted enzymes or metabolites. In the intracellular biosynthesis of Mt NPs, after the transfer of metal ions into cell cytoplasm, the metal ions are reduced as a result of metabolic reactions with enzymes, such as nitrate reductase. (Figure obtained from open-access article [22].)

TABLE 2.1
Cyanobacteria Species Commonly Used for Green Synthesis of Nanoparticles

Cyanobacteria Species	Nanoparticles	Mechanism	Morphology	References
Anabaena doliolum	Ag NPs	Extract	Spherical (10–50 nm)	[34]
Nostoc sp. EA03	ZnO NPs	Extract	Star-shaped (50–80 nm)	[35]
Limnothrix sp.	Ag NPs	Cell-free culture	Elongated (~31.86 nm)	[36]
Limnothrix sp.	Ag NPs	Cell-free culture	Spherical, elongated (~25.65 nm)	[36]
Phormidium ambiguum	Ag NPs	Extract	6.46–12.2 nm	[37]
Oscillatoria sp.	AgO NPs	Whole-cell cultivation, intracellular	Spherical (14.42–48.97 nm)	[38]
Spirulina platensis	Au NPs	Whole-cell cultivation, extracellular	Octahedral, pentagonal, triangular (15.60–77.13 nm)	[38]
Oscillatoria limnetica	Ag NPs	Extract	Quasi-spherical (3.30–17.97 nm)	[39]
Anabaena variabilis	Ag NPs	Extract	Spherical (11–15 nm)	[40]
Lyngbya majuscula	Au NPs	Whole-cell cultivation, intracellular/extracellular	Spherical (~41.7 nm)	[41]
Microchaete NCCU 342	Ag NPs	Extract	Spherical (60–80 nm)	[42]
Spirulina platensis	Fe_3O_4/Ag	Nanocomposite extract	Spherical (30–68 nm)	[43]
Nostoc sp.	ZnO NPs	Extract	Star like (50–80 nm)	[35]

cyanobacteria to absorb ions at the surface depends on their cell wall. Glycoprotein layers cover the outer and inner surface of the cell walls, creating extracellular polymers and negatively charged groups that promote adherence and a reducing environment for the formation of NPs. Some studies have shown that the size and shape of NPs depend on the organism used, the pH, and the light conditions provided. A pH in the range of 6–7 resulted in smaller NPs (80–150nm), while a 12 hour/12 hour photoperiod was the best for homogenous NPs [47]. Some other studies indicate a size range between 10 and 30nm over a period of 150hours of growth. In this study, interestingly, the pH of the culture media tended to increase with time. On frequently adjusting the pH down to 6, the median Ag NP size during culture lifetime was found to be 15nm [48].

The manufacture of metallic NPs by internal or extracellular pathways has also been extensively studied in members of classes Phaeophyceae, Chlorophyceae, Rhodophyceae, and Cyanophyceae (Table 2.2). Algae are abundant in bioactive compounds, such as antioxidants, pigments, terpenoids, amines, and alkaloids. These compounds have been demonstrated to function as reducing agents, promoting the formation of metal nanostructures [69]. Nanostructures have been assembled by employing live algal cells, cell-free supernatant, or bioactive compounds isolated from cells. Ag NPs have been produced from live cells of algae found in the Chlorophyta, Ochrophyta,

TABLE 2.2
Algae Species Commonly Used for Green Synthesis of Nanoparticles

Algae Species	Nanoparticles	Mechanism	Morphology	References
		Macro Brown Algae		
Cystoseira crinita	MgO NPs	Extract	Spherical shape (3–18nm)	[49]
Sargassum latifolium	Cu_2O NPs	Extract	Irregular shapes (12 and 26nm)	[50]
Fucus gardeneri	Ag NPs	Extract	Different sizes and shapes	[51]
Fucus evanescens	Ag NPs	Extract	Spherical (~50nm)	[52]
Padina boryana	Pd NPs	Extract	Face-cantered cubic (~11.16nm)	[53]
Petalonia fascia	Fe_3O_4 NPs	Extract	Spherical (6.54–13.46nm)	[54]
Colpomenia sinuosa	Fe_3O_4 NPs	Extract	Cubic (10.56–19.91nm)	[54]
Padina pavonica	Fe_3O_4 NPs	Extract	Spherical (26.60–34.01nm)	[54]
		Macro Red Algae		
Lemanea fluviatilis	Au NPs	Dried biomass	Cubic (5–15nm)	[55]
Acanthophora spicifera	Ag NPs	Extract	Cubic (33–81nm)	[56]
Laurencia catarinensis	Ag NPs	Powder and extract	Polymorphic shapes (39–78nm)	[57]
Portieria hornemannii	Ag NPs	Extract	Spherical (70–75nm)	[58]
		Macro Green Algae		
Chaetomorpha linum	Ag NPs	Extract	Spherical shaped (70–80nm)	[59]
Chaetomorpha ligustica	Ag NPs	Extract	Spherical shaped (2–12nm)	[60]
Caulerpa serrulata	Ag NPs	Extract	Spherical (~10nm)	[61]
Ulva lactuca	ZnO NPs	Extract	Hexagonal (10–50nm)	[62]
Cladophora glomerata	ZnO NPs	Extract	Spherical (14.39–37.85nm)	[63]
Spirogyra varians	Ag NPs	Extract	Quasi-spherical (~35nm)	[64]
		Micro Green Algae		
Chlorella vulgaris	Ag NPs	Extract	Spherical, nanorods (~10nm)	[65]
Botryococcus braunii	Cu NPs	Extract	Cubical and spherical with an elongated shape (10–70nm)	[66]
Nannochloropsis oculata	Ag NPs	Whole-cell cultivation, intracellular	Spherical (11.6–26.1nm)	[67]
Chlorella vulgaris	Pd NPs	Extract	Spherical, truncated triangular (~70nm)	[68]

Haptophyta, and Spirulina families [70–73]. Dahoumane et al. [74] used this approach to synthesise bimetal Au–Ag alloy NPs. In another investigation, cell-free extracts of *Euglena* sp. were used to accumulate Ag NPs [75]. After being exposed to Ag salts for about 4 hours, *Chlorella vulgaris* accumulated 1,200–3,300 µg/g dry weight of Ag NPs. Using a similar approach, *Raphidocelis subcapitata* accumulated 45.0 µg/g dry weight of Ag NPs after 24 hours of exposure. The difference in absorption ability appears correlated to exposure period, Ag ion concentration, reducing conditions, and the algal species involved [76].

Factors affecting NP accumulation in algae depend largely on the initial concentration of NPs in water. NPs have been seen to be absorbed at faster rates with a strongly acidic culture media [77], and in the initial phases of the culture, reducing gradually as saturation is approached [78]. Algal size, algal biomass, physiochemical properties of algae, and dosage of NPs are some of the other factors that influence the absorption, varying with the algae species used [79].

2.4 APPLICATIONS OF BIOGENIC NANOPARTICLES

2.4.1 Energy Productions

Nanoparticles synthesised from photosynthetic microbes offer a sustainable and eco-friendly approach to energy production. These NPs, often referred to as biogenic NPs, are produced through the metabolic processes of photosynthetic microbes, such as algae and cyanobacteria [80]. Here's how these NPs are being used for energy production.

2.4.1.1 Catalysts for Energy Conversion

Biogenic NPs derived from photosynthetic microbes offer several advantages as catalysts in energy conversion processes. These NPs can be synthesised using sustainable and eco-friendly methods, making them an attractive alternative to traditional catalysts, which often involve harsh chemicals and high energy inputs [81,82]. In fuel cells, biogenic NPs can serve as catalysts to improve the efficiency of electrochemical reactions. For example, in proton exchange membrane fuel cells (PEMFCs), biogenic NPs can be used as catalysts at the cathode to facilitate the reduction of oxygen, a key reaction in generating electricity. The high surface area and unique chemical properties of these NPs can enhance the electrocatalytic activity, leading to improved fuel cell performance [83,84].

Furthermore, biogenic NPs can be engineered to exhibit specific catalytic properties, such as high catalytic activity, selectivity, and stability [55]. These properties are crucial for optimising energy conversion processes and increasing the overall efficiency of energy conversion devices. Overall, biogenic NPs derived from photosynthetic microbes show great promise as catalysts for energy conversion processes. Their environmentally friendly nature, combined with their unique catalytic properties, makes them a sustainable and efficient option for enhancing energy conversion technologies (Figure 2.2) [85].

2.4.1.2 Enhancing Solar Energy Harvesting

Enhancing the efficiency of solar energy harvesting is crucial for advancing solar cell technology and increasing the adoption of renewable energy sources. Biogenic NPs, synthesised from photosynthetic microbes, offer a sustainable and promising approach to achieve this goal [86,87].

One key advantage of biogenic NPs is their ability to enhance light absorption in solar cells. These NPs can be engineered to efficiently capture a broad spectrum of sunlight, including visible and infrared wavelengths [88]. By incorporating these NPs into the design of solar panels, researchers can increase the amount of light absorbed by the solar cells, leading to higher energy conversion efficiencies [89,90].

Additionally, biogenic NPs can improve electron transport within the solar cell. When sunlight is absorbed by NPs, it generates electron-hole pairs. These electrons can then be efficiently transported

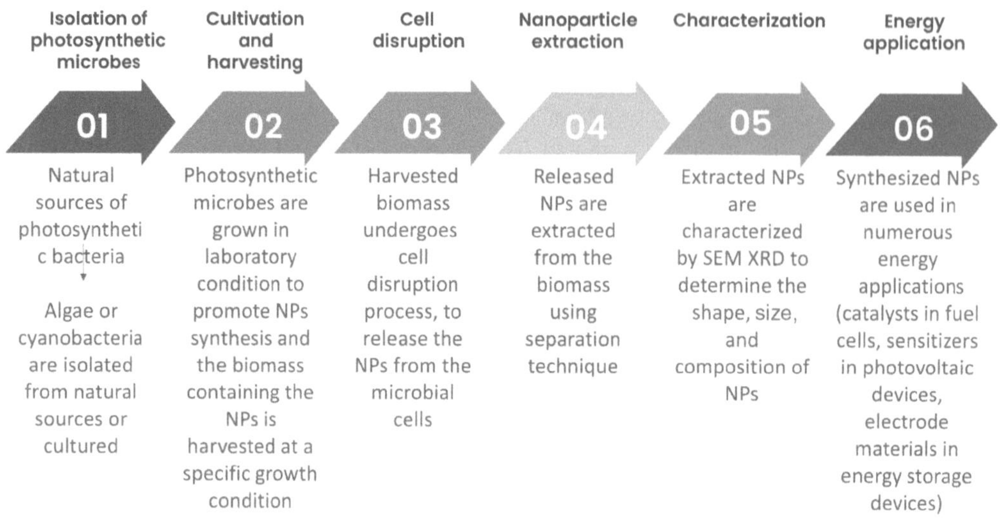

FIGURE 2.2 Process of NP synthesis for energy production from photosynthetic microbes.

through the solar cell material, contributing to the generation of electrical current. By enhancing electron transport, biogenic NPs can further improve the overall efficiency of solar cells [91].

Furthermore, the unique properties of biogenic NPs, such as their high surface area and catalytic activity, make them ideal candidates for enhancing the performance of solar cells. These NPs can be tailored to exhibit specific optical and electronic properties, allowing for fine-tuning of their performance in solar energy–harvesting applications. Biogenic NPs derived from photosynthetic microbes offer several advantages for enhancing solar energy harvesting in solar cells [92]:

i. **Improved Light Absorption:** Biogenic NPs can be engineered to have unique optical properties, such as plasmonic resonances, that enhance light absorption. By incorporating these NPs into the design of solar panels, researchers can increase the absorption of sunlight, especially in the visible and near-infrared regions, where traditional solar cells may have lower efficiency.
ii. **Enhanced Charge Separation:** When sunlight is absorbed by NPs, it generates electron-hole pairs. These electrons and holes need to be efficiently separated to generate an electrical current. Biogenic NPs can be functionalised to facilitate charge separation, improving the overall efficiency of the solar cell.
iii. **Reduced Reflectance:** Biogenic NPs can reduce the amount of light that is reflected off the surface of the solar cell. By minimising reflectance losses, more light can be absorbed and converted into electricity, increasing the overall efficiency of the solar cell.
iv. **Improved Electron Transport:** The unique structure of biogenic NPs can facilitate the transport of electrons within the solar cell material. This improved electron transport can enhance the overall efficiency of the solar cell by ensuring that more of the generated charge carriers contribute to the electrical current.
v. **Compatibility with Green Technology**: Biogenic NPs are synthesised using environmentally friendly methods, making them compatible with green technology initiatives. Their sustainable nature aligns with the goal of developing renewable energy sources that have minimal environmental impact.
vi. **Cost-Effectiveness:** The production of biogenic NPs can be cost-effective compared to traditional NPs synthesised using chemical methods. This cost-effectiveness makes them an attractive option for large-scale solar energy–harvesting applications.

2.4.2 Bioremediation and Biomonitoring

Bioremediation, the use of biological agents to remove or neutralise contaminants from the environment, is an increasingly important field in environmental science [93]. One emerging area within bioremediation is the use of NPs derived from photosynthesisers, such as algae and cyanobacteria, for environmental cleanup and monitoring [94]. These NPs offer unique properties that make them effective tools for tackling pollution and monitoring environmental health.

2.4.2.1 Wastewater Treatment

The use of NPs in wastewater treatment is growing because of their high reactivity, huge surface area, and degree of functionalisation. Because of their synergistic metabolic action, pure or mixed microbial cultures can be used to remediate wastewater. Dye degradation can be achieved with the application of periphyton biofilm [95]. It has been documented that using genetically modified bacteria may remove dyes [96]. Dong et al. [97] discovered that the azoreductase enzyme broke down azo dyes in *Enterococcus gallinarum* and *Streptomyces* S27. For numerous colours, laccase has shown considerable degradation potential. According to Chivukula and Renganathan [98], laccase breaks down dye by a general free radical process, resulting in the creation of phenolic compounds without the production of aromatic amines.

A study explores using iron oxide NPs and photosynthetic bacteria (PSB) for wastewater treatment. Iron oxide NPs are known for their adsorption capacity and inertness, while PSB can degrade contaminants. Fe_3O_4/biochar nanocomposites loaded with PSB showed enhanced nutrient removal capabilities in wastewater, removing a significant percentage of chemical oxygen demand, ammonia, and phosphate. The composites were also recyclable, maintaining effectiveness over multiple cycles, making them promising for bioremediation in wastewater treatment [99]. The use of microbes such as *Aspergillus niger* (Vassilev et al., 1997), *Rhodopseudomonas sphaeroides* [75], and *Pseudomonas aeruginosa* [100] in wastewater treatment has been studied.

Another study explores the biological synthesis of cadmium sulphide (CdS) NPs using blue-green algae, a common food supplement. Unlike traditional methods, this synthesis occurs within the algal medium without the need for an external sulphur precursor. The CdS NPs produced exhibit photoluminescence and act as efficient photocatalysts for degrading the dye pollutant malachite green. This process is in line with wastewater treatment objectives, as it both adsorbs and degrades pollutants. Additionally, the study underscores the method's potential for bioremediation, as it converts toxic Cd(II) ions into less harmful CdS NPs within the algal framework [101].

2.4.2.2 Removal of Heavy Metals from Environment

Due to their carcinogenic and teratogenic properties, heavy metals have a serious negative impact on human and animal health. Thus, finding a mechanism to remove heavy metals from the environment in a sustainable manner will become essential. Heavy metals like arsenic and copper are frequently found in industrial wastewaters from the mining and metallurgical processing industries. Arsenic, copper, zinc, and chromium can be adsorbed from wastewater by iron NPs derived from natural sources [102,103]. Biogenic precipitates from crops included siderite ($FeCO_3$), magnetite (Fe_3O_4), and vivianite ($Fe_3(PO_4)_2 8H_2O$). When compared to chromate, these biogenic adsorbents exhibit higher affinity and adsorption capacity for arsenate. Organic NP elements had also been adsorbed onto copper and zinc. High concentrations of arsenic, Fe(II), and Mn(II) in surface and groundwater have a negative impact on human health as well as the aquatic ecosystem [104] The manganese (II)–oxidising bacterium *Pseudomonas* sp. QJX-1, isolated from the soil of manganese mines, was utilised in the manufacture of biogenic Fe–Mn oxides (BFMO). Arsenic has been found to be oxidised and adsorbed by those oxides. The suggested mechanism involves the first precipitation of BFMO, on the surface of which Mn(II) was biologically oxidised to generate MnO_2 and Fe(II) was chemically oxidised to FeOOH. These two processes co-precipitated to adsorb and oxidise equivalent to Fe–Mn oxide. In addition to Biogenic Mn Oxide (BMO), which has a higher oxidation efficiency, FeOOH has a higher adsorption potential and a lower oxidation potential. Consistently, BFMO is produced by FeOOH and BMO, facilitating the appropriate adsorption and oxidation of As (III) and As (V). Remediating three pollutants at the same time, BMO also succeeds in oxidising Fe (II) to FeOOH, which in turn adsorbs any leftover As (III) and As(V) [105].

2.4.3 Photosynthetic Microbes in the Agriculture Industry

Blue-green algae (cyanobacteria), brown algae, red algae, and green algae are photosynthetic microbes utilised extensively for various commercial and industrial purposes, including the agriculture and food industries. These are mostly used as biofertilisers to promote plant growth and health as well as to promote soil nutrient status and microbiome. In recent times, however, there has been a shift in emphasis to nanotechnology. Nanotechnology in this regard promises to increase agricultural productivity through the formulation and use of agricultural components such as nanopesticides, nanofertilisers, nanofungicides, nanoherbicides, and nanosensors. These can be used in monitoring plant and animal health, identification of diseases and pest and their remedial measures, and even crop harvest and production management [106], thereby increasing quality and yield of agricultural products while reducing chemical pollution and dependence on chemical pesticides, herbicides, and fertilisers.

2.4.3.1 Use of Photosynthetic Microbes in Agricultural Nanotechnology

In the current world situation, population growth and the associated food security issues are burning issues requiring urgent and innovative solution for the survival of humans. Rapid population expansion is already creating a scarcity of various resources due to high demand from the population, including land that is otherwise needed for habitation. There is an emphasis on the increase in agricultural production on a steadily decreasing amount of agricultural land. Nanoformulations of pesticides, herbicides, fungicides, and fertilisers have significant advantages over the use of traditional versions of these products. Primarily, nanotechnology affords benefits such as increased efficiency and durability, and reduced amount of the active ingredient used. These nanoformulations have demonstrated a stronger effect against phytopathogens, pest, and other insects that damage crops; controlled release; longer half-lives; and protection from degradation by environmental factors [22], enabling the concept of smart agriculture and precision farming (Figure 2.3).

2.4.3.1.1 Nanofertilisers

Conventional fertilisers are synthesised via chemical processes, and applications in agricultural fields are subject to a wide range of environmental and biotic factors that can reduce uptake and also accelerate the degradation of fertilisers. In addition, excessive and long-term use of chemically synthesised fertilisers can reduce soil fertility, requiring more of the fertiliser being used in successive crop cycles. This can and does have a myriad of downstream adverse effects on soil fertility, and cause environmental pollution, including eutrophication.

Nanofertilisers and biofertilisers are significantly more effective and environment friendly, compared to the conventional methods of fertilisation. Nanofertilisers, owing to their size and structures, can provide advantages such as high surface area, encapsulation, and controlled and gradual release in an efficient manner in response to abiotic cues. This greatly enables the protection and preservation of nutrients and efficient delivery of nourishment to plants, and therefore reduces the amounts of fertilisers used. In many cases, the availability of nutrients to the plants and nutrient uptake are also increased, resulting in improvements in crop growth, health, biomass production, and yield and quality of crops [96,107,108]. Encapsulated nanoformulations provide the ability to release nutrients in a controlled and gradual manner directly to target sites, preventing the exposure to soil, water, air, and other microorganisms, and help in minimising the risk of environmental leaching and degradation of fertilisers. They can also play important roles in providing protection

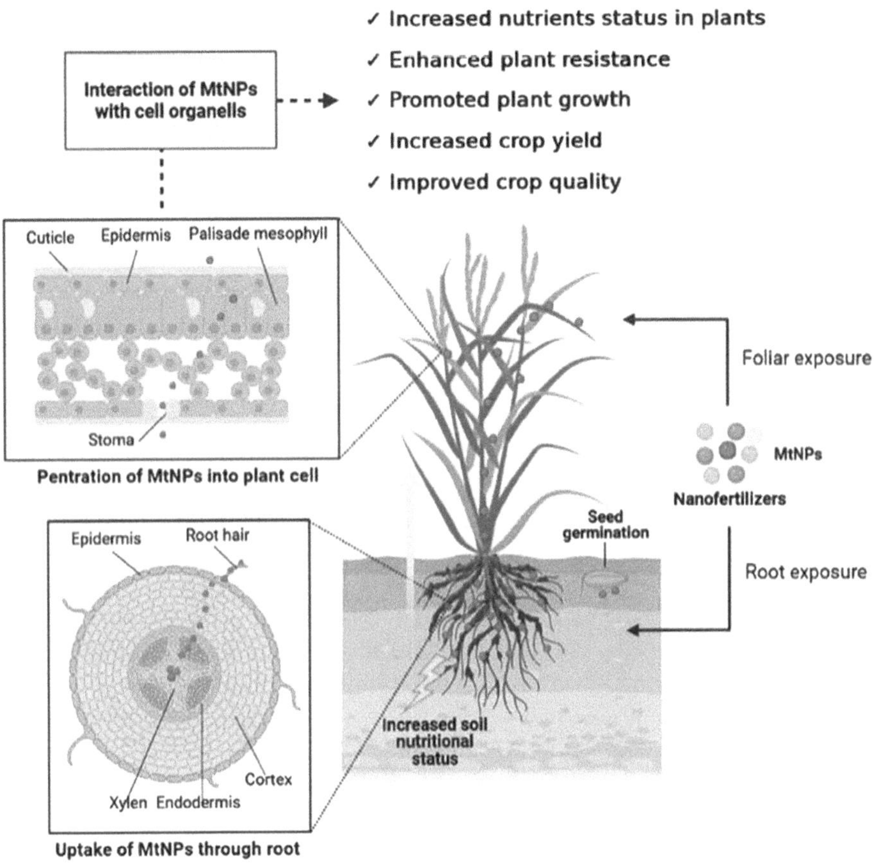

FIGURE 2.3 Schematic representation of the entry of NPs into plants through soil and roots or through extra-soil parts of plants as nanofertilisers and their uptake, translocation, and potential effects on plants. (Figure obtained from open-access article [22].)

to the plant from various abiotic stresses such as drought, salt stress, temperature, heavy-metal tolerance, and UV radiation [109].

Nanofertilisers can be of three types based on their usage. Nanoscale fertilisers (NPs that contain nutrients) are the nano-sized particles of the nutrients, viz. nano-urea, nano-zinc, and nano-copper that can be applied directly to the roots or via foliar means owing to their small size. Nanoscale additives refer to the additives formulated on a nanoscale and are added as supplements to conventional fertilisers as and when required. Nanoscale coating/encapsulation (traditional fertilisers coated with or loaded into NPs) allows the application of small amounts of fertilisers in a way that prevents degradation by the environment while enabling gradual and controlled release of the nutrients depending on the soil moisture, alkalinity, and salinity and maintaining conditions favourable to crop growth and yield.

2.4.3.1.2 Nanoherbicides

Herbicides are generally used to prevent the growth of undesirable weeds that interfere and compete with the crop species for nutrients. Some herbicides may also affect the growth of crops in addition to eliminating the weeds. Conventional herbicides are used both as pre-emergence and post-emergence control methods. Herbicides that are more stable in the soil have a delayed and gradual release of active ingredients, and have a long half-life and therefore preferred for usage for long-term benefits.

Nanoherbicide formulations that are encapsulated or coated versions of the herbicides, using nanopolymers and nanoshells, provide an exciting potential platform for the delivery of bioactive ingredients to the soil and plants. Copper oxide NPs have been utilised for controlling weeds in a targeted delivery approach [109]. Nanoherbicides can also be a better, eco-friendly alternative for weed control, leaving minimal toxic residue in the soil [110]. Poly (epsilon-caprolactone) (PCL) nanocapsules have been used as a carrier system for the conventional herbicide atrazine, and have shown higher potency as a post-emergence herbicide on mustard, compared to commercial atrazine formulations [111].

2.4.3.1.3 Nanopesticides

Pesticides enhance and improve crop yield and efficiency by protecting plants from damaging factors, including plant diseases and insects. However, a lot of pesticides have been found to be toxic and damaging to the environment as well as to human and animal health. Significant and indiscriminate use of pesticides harms the environment and results in the bioaccumulation of toxins by disrupting the soil microbiota and harming the health of end users. Nanostructured alumina and copper NPs are some well-studied nanopesticides that have been shown promising activities in the cultivation of crops such as cotton [112]. Photochemically active TiO_2 and ZnO NPs have shown promising results in the control of spot disease in tomato plants [113].

Encapsulating or packaging of conventional pesticides with NPs or polymers is a research area in high demand as it is expected to address the issues of enhancing pesticide efficacy, reduction of run-off, and long-term activity, by slow and targeted release of bioactive ingredients while protecting from the elements. Nanoemulsions have also found favour in recent years as pest control methods. These are oil-in-water emulsions into which the pesticides are dispersed as nano-sized droplets. The main advantage in this case is the increased solubility and efficacy [114] due to increase in uptake [115]. Studies have reported that the nanoemulsion of emamectin-benzoate is more effective against the Asian rice borer than its normal formulation [116].

2.4.3.1.4 Nanofungicides

Fungal diseases account for up to 70% of major damages to crop plants such as rice, wheat, maize, and barley [117]. Conventional fungicides work to prevent losses due to these diseases, but they can be non-specific in their targets, leading to problems such as biodiversity imbalances and bioaccumulation in higher trophic levels of end users. Alternative approaches to achieve better precision in the management of fungal diseases are being researched. Development of NPs as an effective strategy to combat fungal pathogens is a very good approach in this regard, owing to the antimicrobial properties of NPs such as Ag NP, Cu NPs, Zn NPs, and Mg NPs. In the pathogenic fungi *Bipolaris sorokiniana* and *Magnaporthe grisea*, Ag NPs were found to significantly reduce the disease severity when applied for 3 hours [118]. Cu NPs and Ag NPs were also found to be effective against *Alternaria* and *Botrytis* species (Ouda, 2014) [119]. Similarly, ZnO NPs and MgO NPs show antifungal activities against *Alternaria*, *Rhizopus*, *Fusarium*, and *Mucor* species [120]. Encapsulated conventional pesticides zineb and mancozeb were also confirmed to be potent fungicide against *Alternaria alternate* [121].

2.4.4 Biomedical and Health Care Applications

Green NPs exhibit notable antibacterial efficacy against a broad spectrum of both gram-negative and gram-positive bacteria. Stan et al. [122] conducted a study comparing the synthesis of zinc oxide NPs through both chemical and green methods. The experimental findings demonstrated a superior antibacterial effect of green zinc oxide compared to chemically prepared zinc oxide [122].

Beyond their antimicrobial properties, biogenic NPs present compelling advantages over conventional counterparts as drug delivery agents due to their stability, biocompatibility, enhanced bioavailability, controlled drug release kinetics, targeted delivery capabilities, and non-toxic nature [123–125]. Magnetotactic bacteria can convert magnetic materials such as greigite Fe_3S_4 and magnetite Fe_3O_4 into bilayer membrane-bound structures known as magnetosomes. These magnetosomes can serve as carriers for encapsulating and transporting drugs [126]. Studies have demonstrated the efficacy of bacterial magnetosomes loaded with doxorubicin in suppressing tumours in H22 tumour-bearing mice, surpassing the effects of doxorubicin alone [127]. Additionally, magnetosomes from *Magnetospirillum gryphiswaldense* loaded with anti-4-1BB agonistic antibody have been employed for cancer immunotherapy in TC-1 mouse models [128]. Gold NPs obtained from the fungus *Humicola* sp., conjugated with Taxol, have been utilised for anti-tumour drug delivery applications [129]. Biogenic gold NPs functionalised with molecules like transferrin possess the ability to traverse the blood–brain barrier for targeted drug delivery to the brain [130]. Platinum NPs from *Saccharomyces boulardii* have exhibited effectiveness against A431 epidermoid carcinoma and MCF-7 breast cancer cell lines [131]. Similarly, gold NPs biosynthesised from *Streptomyces cyaneus* demonstrated anticancer activity against HEPG-2 human liver cancer cells and MCF-7 breast cancer cells *in vitro*. Silver NPs derived from the water extract of the endophytic fungi *Cladosporium perangustum* have been observed to reduce MCF-7 cell viability by enhancing the expression levels of caspase-3, caspase-7, caspase-8, and caspase-9 [132].

Nanoparticles are increasingly finding applications in diagnostics and biosensors, often conjugated with diagnostic enzymes [133,134]. Recent research has also explored biogenic NPs in biosensors and imaging modalities like MRI. Bacteriogenic metal NPs, including copper, palladium, and gold, have been investigated for their biosensing potential [133]. In an intriguing study, gold NPs synthesised from *Candida albicans* were conjugated with liver cancer cell surface-specific antibodies. When used to probe liver cancer cells, they exhibited unique binding to liver cancer-specific surface antigens, enabling differentiation from normal cells. Such NPs hold promise for diagnostic applications or as carriers for anticancer drugs [135].

2.5 CONCLUSION AND PROSPECT

Microbially synthesised NPs exhibit significant promise across various domains such as agriculture, bioremediation, and biomedical applications due to their distinctive characteristics. The adoption of green synthesis methods has gained traction as a safe, eco-friendly, and economical approach. This innovative technique generates composite NPs that harness the advantages of multiple metals while minimising chemical inputs, highlighting the capacity of microorganisms to facilitate synthesis. These naturally occurring compounds, serving as both capping and reducing agents, have unveiled a wide array of applications across different sectors, driving transformative changes.

Despite their potential, biogenic NPs encounter certain challenges that must be addressed for scalable deployment. Issues such as lack of uniformity, time-consuming production processes, low output rates, and variations between batches have hindered their commercial utilisation thus far. Addressing these challenges requires a meticulous approach to synthesis, considering factors like organism selection, genetic attributes, optimal growth conditions, and catalyst state. Ensuring the stability and characterisation of NPs calls for biological protocols that encompass these critical aspects.

A comprehensive evaluation of toxicity and the optimisation of genetically modified microorganisms are essential steps towards fully capitalising on the potential of green-synthesized NPs. Unlocking their full capabilities necessitates thorough toxicological assessments and the refinement of genetically engineered microbial strains. Moreover, while many biomedical studies have demonstrated the efficacy of microbial NPs *in vitro*, large-scale clinical trials and safety assessments are imperative to validate their *in vivo* effects. Through further research and exploration, it is anticipated that microbial NPs will emerge as indispensable assets across diverse fields including agriculture, environmental remediation, and healthcare.

REFERENCES

1. Corliss, J.O. Biodiversity and biocomplexity of the protists and an overview of their significant roles in maintenance of our biosphere. *Acta Protozool.* 2002, *41*, 199–220.
2. Blankenship, R.E. Origin and early evolution of photosynthesis. *Photosynth. Res.* 1992, *33*, 91–111.
3. Dallakyan, S.; Olson, A.J. Small-molecule library screening by docking with PyRx. In *Chemical Biology. Methods in Molecular Biology*; Hempe, l J., Williams, C., Hong, C., Eds.; Humana Press, New York, 2015; pp. 243–250. ISBN 978-1-4939-2269-7.
4. Geider, R. *Algal Photosynthesis*; Springer Science & Business Media, Berlin, 2013; Vol. 2; ISBN 1475721536.
5. Ray, M.K.; Mishra, A.K.; Mohanta, Y.K.; Mahanta, S.; Chakrabartty, I.; Kungwani, N.A.; Avula, S.K.; Panda, J.; Pudake, R.N. Nanotechnology as a promising tool against phytopathogens: A futuristic approach to agriculture. *Agriculture* 2023, *13*, 1856.
6. Ranpariya, B.; Salunke, G.; Karmakar, S.; Babiya, K.; Sutar, S.; Kadoo, N.; Kumbhakar, P.; Ghosh, S. Antimicrobial synergy of silver-platinum nanohybrids with antibiotics. *Front. Microbiol.* 2021, *11*, 610968.
7. Bloch, K.; Pardesi, K.; Satriano, C.; Ghosh, S. Bacteriogenic platinum nanoparticles for application in nanomedicine. *Front. Chem.* 2021, *9*, 624344.
8. Ghosh, S.; Ghosh, A.; Pramanik, S.; Kuiri, P.K.; Sen, R.; Neogi, S.K. Synthesis of ZnO nanoparticles by co-precipitation technique and characterize the structural and optical properties of these nanoparticles. *J. Phys. Conf. Ser.* 2022, *2349*, doi:10.1088/1742-6596/2349/1/012014.
9. Orshiso, T.A.; Zereffa, E.A.; Murthy, H.C.A.; Demissie, T.B.; Pardeshi, O.; Avhad, L.S.; Ghotekar, S. Biosynthesis of artemisia abyssinica leaf extract-mediated bimetallic ZnO–CuO nanoparticles: Antioxidant, anticancer, and molecular docking studies. *ACS Omega* 2023, *8*, 41039–41053, doi:10.1021/acsomega.3c01814.
10. Nayak, D.; Mishra, A.K.; Malla, S.; Swain, U.; Panigrahi, K.; Panda, J.; Rustagi, S.; Rauta, P.R.; Mohanta, Y.K. Eco-friendly mycogenic synthesis of iron oxide nanoparticles using volvariella volvacea and unveiling its potential as antibiofilm and anticancer agents. *ECS J. Solid State Sci. Technol.* 2024, *13*, 087003, doi:10.1149/2162-8777/ad658b.
11. Mohanta, Y.K.; Mishra, A.K.; Panda, J.; Chakrabartty, I.; Sarma, B.; Panda, S.K.; Chopra, H.; Zengin, G.; Moloney, M.G.; Sharifi-Rad, M. Promising applications of phyto-fabricated silver nanoparticles: Recent trends in biomedicine. *Biochem. Biophys. Res. Commun.* 2023, *688*, 149126, doi:10.1016/j.bbrc.2023.149126.
12. Mohanta, Y.K.; Biswas, K.; Mishra, A.K.; Patra, B.; Mishra, B.; Panda, J.; Avula, S.K.; Varma, R.S.; Panda, B.P.; Nayak, D. Amelioration of gold nanoparticles mediated through ocimum oil extracts induces reactive oxygen species and mitochondrial instability against MCF-7 breast carcinoma. *RSC Adv.* 2024, *14*, 27816–27830, doi:10.1039/D4RA04807E.
13. Kimta, N.; Chauhan, A.; Puri, S.; Kumari, A.; Sharma, R.; Kumar, A.; Kapoor, D. Evaluating the biomedical potential of phytomediated silver doped zinc oxide nanoparticles derived from N-butanol fraction of adiantum venustum D. Don. *J. Indian Chem. Soc.* 2024, *101*, 101364, doi:10.1016/j.jics.2024.101364.
14. Khanna, P.; Kaur, A.; Goyal, D. Algae-based metallic nanoparticles: Synthesis, characterization and applications. *J. Microbiol. Methods* 2019, *163*, 105656, doi:10.1016/j.mimet.2019.105656.
15. Chaudhary, R.; Nawaz, K.; Khan, A.K.; Hano, C.; Abbasi, B.H.; Anjum, S. An overview of the algae-mediated biosynthesis of nanoparticles and their biomedical applications. *Biomolecules* 2020, *10*, 1498.
16. Bishoyi, A.K.; Sahoo, C.R.; Padhy, R.N. Recent progression of cyanobacteria and their pharmaceutical utility: An update. *J. Biomol. Struct. Dyn.* 2023, *41*, 4219–4252, doi:10.1080/07391102.2022.2062051.
17. Ali, D.M.; Sasikala, M.; Gunasekaran, M.; Thajuddin, N. Biosynthesis and characterization of silver nanoparticles using marine cyanobacterium, Oscillatoria willei NTDM01. *Dig. J. Nanomater Biostruct.* 2011, *6*, 385–390.
18. Rai, S.; Wenjing, W.; Shrivastava, A.K.; Singh, P.K. 11: Cyanobacteria as a source of nanoparticles and their applications. In *Role of Plant Growth Promoting Microorganisms in Sustainable Agriculture and Nanotechnology*; Kumar, A., Singh, A.K., Choudhary, K.K., Eds.; Woodhead Publishing, Sawston, 2019; pp. 183–198. ISBN 978-0-12-817004-5.
19. Priambodo, G.; Samekto, W.; Susanto, H.; Dwiantoro, L.; Subagio, A. Green synthesis of silver nanoparticles as a wound healing agent: A systematic review of the biological activities. *Indones. J. Med.* 2023, *08*, 100–113.

20. Ahmad, A.; Mukherjee, P.; Senapati, S.; Mandal, D.; Khan, M.I.; Kumar, R.; Sastry, M. Extracellular biosynthesis of silver nanoparticles using the fungus Fusarium oxysporum. *Colloids Surf. B Biointerfaces* 2003, *28*, 313–318, doi:10.1016/S0927-7765(02)00174-1.
21. Golinska, P.; Wypij, M.; Ingle, A.P.; Gupta, I.; Dahm, H.; Rai, M. Biogenic synthesis of metal nanoparticles from actinomycetes: Biomedical applications and cytotoxicity. *Appl. Microbiol. Biotechnol.* 2014, *98*, 8083–8097, doi:10.1007/s00253-014-5953-7.
22. Bahrulolum, H.; Nooraei, S.; Javanshir, N.; Tarrahimofrad, H.; Mirbagheri, V.S.; Easton, A.J.; Ahmadian, G. Green synthesis of metal nanoparticles using microorganisms and their application in the agrifood sector. *J. Nanobiotechnol.* 2021, *19*, 86, doi:10.1186/s12951-021-00834-3.
23. Mahdavi, M.; Namvar, F.; Ahmad, M.B.; Mohamad, R. Green biosynthesis and characterization of magnetic iron oxide ($_{Fe3O4}$) nanoparticles using seaweed (*Sargassum muticum*) aqueous extract. *Molecules* 2013, *18*, 5954–5964, doi:10.3390/molecules18055954.
24. Dhanker, R.; Molinero, J.C.; Kumar, R.; Tseng, L.-C.; Ianora, A.; Hwang, J.-S. Responses of the estuarine copepod Pseudodiaptomus annandalei to diatom polyunsaturated aldehydes: Reproduction, survival and postembryonic development. *Harmful Algae* 2015, *43*, 74–81, doi:10.1016/j.hal.2015.02.002.
25. Dhanker, R.; Hussain, T.; Tyagi, P.; Singh, K.J.; Kamble, S.S. The emerging trend of bio-engineering approaches for microbial nanomaterial synthesis and its applications. *Front. Microbiol.* 2021, *12*, doi:10.3389/fmicb.2021.638003.
26. Kumari, A.; Naveen, T.; Dhatwalia, J.; Thakur, S.; Radhakrishnan, A.; Chauhan, A.; Chandan, G.; Choi, B.H.; Kimta, N.; Bhardwaj, N. Antioxidant, antimicrobial, and cytotoxic potential of Euphorbia royleana extract-mediated silver and copper oxide nanoparticles. *Chem. Pap.* 2023, *77*, 4643–4657, doi:10.1007/s11696-023-02814-3.
27. Prabakaran, L.; Sathyaraj, W.V.; Yesudhason, B.V.; Subbaraj, G.K.; Atchudan, R. Green synthesis of multifunctional silver nanoparticles using Plectranthus amboinicus for sensitive detection of triethylamine, with potential in vitro antibacterial and anticancer activities. *Chemosensors* 2023, *11*, 373, doi:10.3390/chemosensors11070373.
28. Tiquia-Arashiro, S.; Rodrigues, D. Nanoparticles synthesized by microorganisms. In *Extremophiles: Applications in Nanotechnology*; Tiquia-Arashiro, S., Rodrigues, D.F., Eds.; Springer International Publishing, Cham, 2016; pp. 1–51. ISBN 978-3-319-45215-9.
29. Hulkoti, N.I.; Taranath, T.C. Biosynthesis of nanoparticles using microbes—A review. *Colloids Surf. B Biointerfaces* 2014, *121*, 474–483, doi:10.1016/j.colsurfb.2014.05.027.
30. Mohd Yusof, H.; Mohamad, R.; Zaidan, U.H.; Abdul Rahman, N.A. Microbial synthesis of zinc oxide nanoparticles and their potential application as an antimicrobial agent and a feed supplement in animal industry: A review. *J. Anim. Sci. Biotechnol.* 2019, *10*, 57, doi:10.1186/s40104-019-0368-z.
31. Sahoo, G.; Wani, A.M.; Satpathy, B.; Rout, S. Traditional medicinal plants of Odisha. *Res. Rev. A J. Pharmacogn.* 2020, *7*, 7–10.
32. Soliman, M.; Qari, S.H.; Abu-Elsaoud, A.; El-Esawi, M.; Alhaithloul, H.; Elkelish, A. Rapid green synthesis of silver nanoparticles from blue gum augment growth and performance of maize, fenugreek, and onion by modulating plants cellular antioxidant machinery and genes expression. *Acta Physiol. Plant.* 2020, *42*, doi:10.1007/s11738-020-03131-y.
33. El-Belely, E.F.; Farag, M.M.S.; Said, H.A.; Amin, A.S.; Azab, E.; Gobouri, A.A.; Fouda, A. Green synthesis of zinc oxide nanoparticles (ZnO-NPs) using Arthrospira platensis (Class: Cyanophyceae) and evaluation of their biomedical activities. *Nanomaterials* 2021, *11*, 95, doi:10.3390/nano11010095.
34. Singh, G.; Babele, P.K.; Shahi, S.K.; Sinha, R.P.; Tyagi, M.B.; Kumar, A. Green synthesis of silver nanoparticles using cell extracts of anabaena doliolum and screening of its antibacterial and antitumor activity. *J. Microbiol. Biotechnol.* 2014, *24*, 1354–1367, doi:10.4014/jmb.1405.05003.
35. Ebadi, M.; Zolfaghari, M.R.; Aghaei, S.S.; Zargar, M.; Shafiei, M.; Zahiri, H.S.; Noghabi, K.A. A bio-inspired strategy for the synthesis of zinc oxide nanoparticles (ZnO NPs) using the cell extract of cyanobacterium Nostoc Sp. EA03: From biological function to toxicity evaluation. *RSC Adv.* 2019, *9*, 23508–23525, doi:10.1039/C9RA03962G.
36. Patel, V.; Berthold, D.; Puranik, P.; Gantar, M. Screening of cyanobacteria and microalgae for their ability to synthesize silver nanoparticles with antibacterial activity. *Biotechnol. Rep.* 2015, *5*, 112–119, doi:10.1016/j.btre.2014.12.001.
37. Hanna, A.L.; Hamouda, H.M.; Goda, H.A.; Sadik, M.W.; Moghanm, F.S.; Ghoneim, A.M.; Alenezi, M.A.; Alnomasy, S.F.; Alam, P.; Elsayed, T.R. Biosynthesis and characterization of silver nanoparticles produced by Phormidium ambiguum and Desertifilum tharense cyanobacteria. *Bioinorg. Chem. Appl.* 2022, 2022, 9072508, doi:10.1155/2022/9072508.

38. El-sheekh, M.; Hassan, L.H.S.; Morsi, H.H. Evaluation of antimicrobial activities of blue-green algae-mediated silver and gold nanoparticles. *Rend. Lincei. Sci. Fis. e Nat.* 2021, *32*, 747–759.
39. Hamouda, R.A.; Hussein, M.H.; Abo-elmagd, R.A.; Bawazir, S.S. Synthesis and biological characterization of silver nanoparticles derived from the cyanobacterium oscillatoria limnetica. *Sci. Rep.* 2019, *9*, 13071, doi:10.1038/s41598-019-49444-y.
40. Ahamad, I.; Aziz, N.; Zaki, A.; Fatma, T. Synthesis and characterization of silver nanoparticles using anabaena variabilis as a potential antimicrobial agent. *J. Appl. Phycol.* 2021, *33*, 829–841, doi:10.1007/s10811-020-02323-w.
41. Bakir, E.M.; Younis, N.S.; Mohamed, M.E.; El Semary, N.A. Cyanobacteria as nanogold factories: Chemical and anti-myocardial infarction properties of gold nanoparticles synthesized by *Lyngbya majuscula*. *Mar. Drugs* 2018, *16*, 217, doi:10.3390/md16060217.
42. Husain, S.; Afreen, S.; Hemlata; Yasin, D.; Afzal, B.; Fatma, T. Cyanobacteria as a bioreactor for synthesis of silver nanoparticles-an effect of different reaction conditions on the size of nanoparticles and their dye decolorization ability. *J. Microbiol. Methods* 2019, *162*, 77–82, doi:10.1016/j.mimet.2019.05.011.
43. Shokoofeh, N.; Moradi-Shoeili, Z.; Naeemi, A.S.; Jalali, A.; Hedayati, M.; Salehzadeh, A. Biosynthesis of Fe_3O_4@Ag nanocomposite and evaluation of its performance on expression of norA and norB efflux pump genes in ciprofloxacin-resistant Staphylococcus aureus. *Biol. Trace Elem. Res.* 2019, *191*, 522–530, doi:10.1007/s12011-019-1632-y.
44. Velusamy, P.; Kumar, G.V.; Jeyanthi, V.; Das, J.; Pachaiappan, R. Bio-inspired green nanoparticles: Synthesis, mechanism, and antibacterial application. *Toxicol. Res.* 2016, *32*, 95–102, doi:10.5487/TR.2016.32.2.095.
45. Nowruzi, B.; Jalil, B.S.; Metcalf, J.S. Antifungal screening of selenium nanoparticles biosynthesized by microcystin-producing Desmonostoc alborizicum. *BMC Biotechnol.* 2023, *23*, 41, doi:10.1186/s12896-023-00807-4.
46. Rajeshkumar, S.; Tharani, M.; Rajeswari, V.D.; Alharbi, N.S.; Kadaikunnan, S.; Khaled, J.M.; Gopinath, K.; Vijayakumar, N.; Govindarajan, M. Synthesis of greener silver nanoparticle-based chitosan nanocomposites and their potential antimicrobial activity against oral pathogens. *Green Process. Synth.* 2021, *10*, 658–665.
47. Alipour, S.; Kalari, S.; Morowvat, M.H.; Sabahi, Z.; Dehshahri, A. Green synthesis of selenium nanoparticles by cyanobacterium Spirulina platensis (Abdf2224): Cultivation condition quality controls. *Biomed Res. Int.* 2021, *2021*, 6635297.
48. Cicci, A.; Sed, G.; Tirillò, J.; Stoller, M.; Bravi, M. Production and characterization of silver nanoparticles in cultures of the cyanobacterium A. platensis (Spirulina). *Chem. Eng. Trans.* 2017, *57*, 1405–1410.
49. Fouda, A.; Eid, A.M.; Abdel-Rahman, M.A.; EL- Belely, E.F.; Awad, M.A.; Hassan, S.E.-D.; AL- Faifi, Z.E.; Hamza, M.F. Enhanced antimicrobial, cytotoxicity, larvicidal, and repellence activities of brown algae, *Cystoseira Crinita*-mediated green synthesis of magnesium oxide nanoparticles. *Front. Bioeng. Biotechnol.* 2022, *10*, 849921, doi:10.3389/fbioe.2022.849921.
50. Taherzadeh Soureshjani, P.; Shadi, A.; Mohammadsaleh, F. Algae-mediated route to biogenic cuprous oxide nanoparticles and spindle-like $CaCO_3$: A comparative study, facile synthesis, and biological properties. *RSC Adv.* 2021, *11*, 10599–10609, doi:10.1039/D1RA00187F.
51. Princy, K.F.; Gopinath, A. Green synthesis of silver nanoparticles using polar seaweed fucus gardeneri and its catalytic efficacy in the reduction of nitrophenol. *Polar Sci.* 2021, *30*, 100692, doi:10.1016/j.polar.2021.100692.
52. Yugay, Y.A.; Usoltseva, R. V; Silant'ev, V.E.; Egorova, A.E.; Karabtsov, A.A.; Kumeiko, V. V; Ermakova, S.P.; Bulgakov, V.P.; Shkryl, Y.N. Synthesis of bioactive silver nanoparticles using alginate, fucoidan and laminaran from brown algae as a reducing and stabilizing agent. *Carbohydr. Polym.* 2020, *245*, 116547, doi:10.1016/j.carbpol.2020.116547.
53. Sonbol, H.; Ameen, F.; AlYahya, S.; Almansob, A.; Alwakeel, S. Padina boryana mediated green synthesis of crystalline palladium nanoparticles as potential nanodrug against multidrug resistant bacteria and cancer cells. *Sci. Rep.* 2021, *11*, 5444, doi:10.1038/s41598-021-84794-6.
54. El-Sheekh, M.M.; El-Kassas, H.Y.; Shams El-Din, N.G.; Eissa, D.I.; El-Sherbiny, B.A. Green synthesis, characterization applications of iron oxide nanoparticles for antialgal and wastewater bioremediation using three brown algae. *Int. J. Phytoremediation* 2021, *23*, 1538–1552, doi:10.1080/15226514.2021.1915957.
55. Sharma, D.; Kanchi, S.; Bisetty, K. Biogenic synthesis of nanoparticles: A review. *Arab. J. Chem.* 2019, *12*, 3576–3600, doi:10.1016/j.arabjc.2015.11.002.

56. Ibraheem, I.; Saad, Ae. B.; WF, S.; WAWA, F. Green biosynthesis of silver nanoparticles using marine red algae Acanthophora specifera and its antibacterial activity. *J. Nanomed. Nanotechnol.* 2016, *07*, 409, doi:10.4172/2157-7439.1000409.
57. AbdelRahim, K.; Mahmoud, S.Y.; Ali, A.M.; Almaary, K.S.; Mustafa, A.E.-Z.M.A.; Husseiny, S.M. Extracellular biosynthesis of silver nanoparticles using Rhizopus stolonifer. *Saudi J. Biol. Sci.* 2017, *24*, 208–216, doi:10.1016/j.sjbs.2016.02.025.
58. Fatima, R.; Priya, M.; Indurthi, L.; Radhakrishnan, V.; Sudhakaran, R. Biosynthesis of silver nanoparticles using red algae Portieria hornemannii and its antibacterial activity against fish pathogens. *Microb. Pathog.* 2020, *138*, 103780, doi:10.1016/j.micpath.2019.103780.
59. Acharya, D.; Satapathy, S.; Somu, P.; Parida, U.K.; Mishra, G. Apoptotic effect and anticancer activity of biosynthesized silver nanoparticles from marine algae Chaetomorpha linum extract against human colon cancer cell HCT-116. *Biol. Trace Elem. Res.* 2021, *199*, 1812–1822, doi:10.1007/s12011-020-02304-7.
60. Al-Zahrani, S.A.; Bhat, R.S.; Rashed, S.A. Al; Mahmood, A.; Fahad, A. Al; Alamro, G.; Almusallam, J.; Subki, R. Al; Orfali, R.; Daihan, S. Al green-synthesized silver nanoparticles with aqueous extract of green algae Chaetomorpha ligustica and its anticancer potential. *Green Process. Synth.* 2021, *10*, 711–721, doi:10.1515/gps-2021-0067.
61. Aboelfetoh, E.F.; El-Shenody, R.A.; Ghobara, M.M. Eco-friendly synthesis of silver nanoparticles using green algae (Caulerpa serrulata): Reaction optimization, catalytic and antibacterial activities. *Environ. Monit. Assess.* 2017, *189*, 349, doi:10.1007/s10661-017-6033-0.
62. Ishwarya, R.; Vaseeharan, B.; Kalyani, S.; Banumathi, B.; Govindarajan, M.; Alharbi, N.S.; Kadaikunnan, S.; Al-anbr, M.N.; Khaled, J.M.; Benelli, G. Facile green synthesis of zinc oxide nanoparticles using ulva lactuca seaweed extract and evaluation of their photocatalytic, antibiofilm and insecticidal activity. *J. Photochem. Photobiol. B Biol.* 2018, *178*, 249–258, doi:10.1016/j.jphotobiol.2017.11.006.
63. Abdulwahid, K.E.; Dwaish, A.S.; Dakhil, O.A. Green synthesis and characterization of zinc oxide nanoparticles from Cladophora glomerata and its antifungal activity against some fungal isolates. *Plant Arch.* 2019, *19*, 3527–3532.
64. Salari, Z.; Danafar, F.; Dabaghi, S.; Ataei, S.A. Sustainable synthesis of silver nanoparticles using macroalgae Spirogyra varians and analysis of their antibacterial activity. *J. Saudi Chem. Soc.* 2016, *20*, 459–464, doi:10.1016/j.jscs.2014.10.004.
65. Soleimani, M.; Habibi-Pirkoohi, M. Biosynthesis of silver nanoparticles using Chlorella vulgaris and evaluation of the antibacterial efficacy against Staphylococcus aureus. *Avicenna J. Med. Biotechnol.* 2017, *9*, 120–125.
66. Arya, A.; Gupta, K.; Chundawat, T.S.; Vaya, D. Biogenic synthesis of copper and silver nanoparticles using green alga Botryococcus braunii and its antimicrobial activity. *Bioinorg. Chem. Appl.* 2018, *2018*, 7879403, doi:10.1155/2018/7879403.
67. El-Kassas, H.Y.; Ghobrial, M.G. Biosynthesis of metal nanoparticles using three marine plant species: Anti-algal efficiencies against "Oscillatoria simplicissima." *Environ. Sci. Pollut. Res.* 2017, *24*, 7837–7849, doi:10.1007/s11356-017-8362-5.
68. Mishra, V.; Arya, A.; Chundawat, T.S. High catalytic activity of Pd nanoparticles synthesized from green alga chlorella vulgaris in buchwald-hartwig synthesis of N-aryl piperazines. *Curr. Organocatal.* 2019, *7*, 23–33, doi:10.2174/2213337206666190515091945.
69. Asmathunisha, N.; Kathiresan, K. A review on biosynthesis of nanoparticles by marine organisms. *Colloids Surf. B Biointerfaces* 2013, *103*, 283–287, doi:10.1016/j.colsurfb.2012.10.030.
70. Merin, D.D.; Prakash, S.; Bhimba, B.V. Antibacterial screening of silver nanoparticles synthesized by marine micro algae. *Asian Pac. J. Trop. Med.* 2010, *3*, 797–799.
71. Mohseniazar, M.; Barin, M.; Zarredar, H.; Alizadeh, S.; Shanehbandi, D. Potential of microalgae and lactobacilli in biosynthesis of silver nanoparticles. *BioImpacts BI* 2011, *1*, 149.
72. Dahoumane, S.A.; Djediat, C.; Yéprémian, C.; Couté, A.; Fiévet, F.; Coradin, T.; Brayner, R. Species selection for the design of gold nanobioreactor by photosynthetic organisms. *J. Nanoparticle Res.* 2012, *14*, 1–17.
73. Rösken, L.M.; Cappel, F.; Körsten, S.; Fischer, C.B.; Schönleber, A.; van Smaalen, S.; Geimer, S.; Beresko, C.; Ankerhold, G.; Wehner, S. Time-dependent growth of crystalline Au^0-nanoparticles in cyanobacteria as self-reproducing bioreactors: 2. Anabaena cylindrica. *Beilstein J. Nanotechnol.* 2016, *7*, 312–327.
74. Dahoumane, S.A.; Wijesekera, K.; Filipe, C.D.M.; Brennan, J.D. Stoichiometrically controlled production of bimetallic gold-silver alloy colloids using micro-alga cultures. *J. Colloid Interface Sci.* 2014, *416*, 67–72.

75. Liu, S.; Li, X.; Zhang, G.; Zhang, J. Optimization of influencing factors on biomass accumulation and 5-aminolevulinic acid (ALA) yield in Rhodobacter sphaeroides wastewater treatment. *J. Microbiol. Biotechnol.* 2015, *25*, 1920–1927.
76. Ribeiro, F.; Gallego-Urrea, J.A.; Goodhead, R.M.; Van Gestel, C.A.M.; Moger, J.; Soares, A.M.V.M.; Loureiro, S. Uptake and elimination kinetics of silver nanoparticles and silver nitrate by Raphidocelis subcapitata: The influence of silver behaviour in solution. *Nanotoxicology* 2015, *9*, 686–695.
77. Vijayaraghavan, K.; Mahadevan, A.; Sathishkumar, M.; Pavagadhi, S.; Balasubramanian, R. Biosynthesis of Au (0) from Au (III) via biosorption and bioreduction using brown marine alga Turbinaria conoides. *Chem. Eng. J.* 2011, *167*, 223–227.
78. Harja, M.; Buema, G.; Bulgariu, L.; Bulgariu, D.; Sutiman, D.M.; Ciobanu, G. Removal of cadmium (II) from aqueous solution by adsorption onto modified algae and ash. *Korean J. Chem. Eng.* 2015, *32*, 1804–1811.
79. Esmaeili, A.; Beni, A.A. Novel membrane reactor design for heavy-metal removal by alginate nanoparticles. *J. Ind. Eng. Chem.* 2015, *26*, 122–128.
80. Nitnavare, R.; Bhattacharya, J.; Thongmee, S.; Ghosh, S. Photosynthetic microbes in nanobiotechnology: Applications and perspectives. *Sci. Total Environ.* 2022, *841*, 156457.
81. Rajeshkumar, S.; Bharath, L.V. Mechanism of plant-mediated synthesis of silver nanoparticles – A review on biomolecules involved, characterisation and antibacterial activity. *Chem. Biol. Interact.* 2017, *273*, 219–227, doi:10.1016/j.cbi.2017.06.019.
82. Singh, P.; Kim, Y.-J.; Zhang, D.; Yang, D.-C. Biological synthesis of nanoparticles from plants and microorganisms. *Trends Biotechnol.* 2016, *34*, 588–599, doi:10.1016/j.tibtech.2016.02.006.
83. Elkhrachy, I.; Singh, V.; Kumar, A.; Roy, A.; Abbas, M.; Gacem, A.; Alam, M.W.; Yadav, K.K.; Verma, D.; Jeon, B.-H. Use of biogenic silver nanoparticles on the cathode to improve bioelectricity production in microbial fuel cells. *Front. Chem.* 2023, *11*, 1273161.
84. Frattini, D.; Karunakaran, G.; Cho, E.-B.; Kwon, Y. Sustainable syntheses and sources of nanomaterials for microbial fuel/electrolysis cell applications: An overview of recent progress. *Processes* 2021, *9*, 1221.
85. Sekar, N.; Ramasamy, R.P. Recent advances in photosynthetic energy conversion. *J. Photochem. Photobiol. C Photochem. Rev.* 2015, *22*, 19–33.
86. Tanvir, R.U.; Zhang, J.; Canter, T.; Chen, D.; Lu, J.; Hu, Z. Harnessing solar energy using phototrophic microorganisms: A sustainable pathway to bioenergy, biomaterials, and environmental solutions. *Renew. Sustain. Energy Rev.* 2021, *146*, 111181.
87. Kim, Y.J.; Hong, H.; Yun, J.; Il Kim, S.; Jung, H.Y.; Ryu, W. Photosynthetic nanomaterial hybrids for bioelectricity and renewable energy systems. *Adv. Mater.* 2021, *33*, 2005919.
88. Wang, Z.; Gao, D.; Zhan, Y.; Xing, C. Enhancing the light coverage of photosynthetic bacteria to augment photosynthesis by conjugated polymer nanoparticles. *ACS Appl. Bio Mater.* 2020, *3*, 3423–3429.
89. Talabani, R.F.; Hamad, S.M.; Barzinjy, A.A.; Demir, U. Biosynthesis of silver nanoparticles and their applications in harvesting sunlight for solar thermal generation. *Nanomaterials* 2021, *11*, 2421.
90. Danladi, E.; Ichoja, A.; Onoja, E.D.; Adepehin, D.S.; Onwoke, E.E.; Ekwu, O.M.; Alfred, D.O. Broad-band-enhanced and minimal hysteresis perovskite solar cells with interfacial coating of biogenic plasmonic light trapping silver nanoparticles. *Mater. Res. Innov.* 2023, *27*, 521–536.
91. Murugadoss, G. Bio-mediated synthesis of nanomaterials for dye-sensitized solar cells. *Mater. Res. Found.* 2022, *121*, 175–210.
92. Liu, L.; Choi, S. Enhanced biophotoelectricity generation in cyanobacterial biophotovoltaics with intracellularly biosynthesized gold nanoparticles. *J. Power Sources* 2021, *506*, 230251.
93. Pande, V.; Pandey, S.C.; Sati, D.; Pande, V.; Samant, M. Bioremediation: An emerging effective approach towards environment restoration. *Environ. Sustain.* 2020, *3*, 91–103.
94. Chaudhary, P.; Ahamad, L.; Chaudhary, A.; Kumar, G.; Chen, W.-J.; Chen, S. Nanoparticle-mediated bioremediation as a powerful weapon in the removal of environmental pollutants. *J. Environ. Chem. Eng.* 2023, *11*, 109591.
95. Shabbir, S.; Faheem, M.; Dar, A.A.; Ali, N.; Kerr, P.G.; Yu, Z.-G.; Li, Y.; Frei, S.; Albasher, G.; Gilfedder, B.S. Enhanced periphyton biodegradation of endocrine disrupting hormones and microplastic: Intrinsic reaction mechanism, influential humic acid and microbial community structure elucidation. *Chemosphere* 2022, *293*, 133515, doi:10.1016/j.chemosphere.2022.133515.
96. Kumar, N.; Balamurugan, A.; Mohiraa Shafreen, M.; Rahim, A.; Vats, S.; Vishwakarma, K. Nanomaterials: Emerging trends and future prospects for economical agricultural system. In *Biogenic Nano-Particles and Their Use in Agro-Ecosystems*; Ghorbanpour, M., Bhargava, P., Varma, A., Choudhary, D., Eds.; Springer, Singapore, 2020, pp. 281–305.

97. Dong, H.; Guo, T.; Zhang, W.; Ying, H.; Wang, P.; Wang, Y.; Chen, Y. Biochemical characterization of a novel azoreductase from Streptomyces Sp.: Application in eco-friendly decolorization of azo dye wastewater. *Int. J. Biol. Macromol.* 2019, *140*, 1037–1046, doi:10.1016/j.ijbiomac.2019.08.196.
98. Chivukula, M.; Renganathan, V. Phenolic azo dye oxidation by laccase from Pyricularia oryzae. *Appl. Environ. Microbiol.* 1996, *61*, 4374–4377, doi:10.1128/AEM.61.12.4374-4377.1995.
99. He, S.; Zhong, L.; Duan, J.; Feng, Y.; Yang, B.; Yang, L. Bioremediation of wastewater by iron oxide-biochar nanocomposites loaded with photosynthetic bacteria. *Front. Microbiol.* 2017, *8*, 823.
100. Shukla, V.Y.; Tipre, D.R.; Dave, S.R. Optimization of chromium(VI) detoxification by Pseudomonas aeruginosa and its application for treatment of industrial waste and contaminated soil. *Bioremediat. J.* 2014, *18*, 128–135, doi:10.1080/10889868.2013.834872.
101. Mandal, R.P.; Sekh, S.; Sarkar, N.S.; Chattopadhyay, D.; De, S. Algae mediated synthesis of cadmium sulphide nanoparticles and their application in bioremediation. *Mater. Res. Express* 2016, *3*, 55007.
102. Ray, M.K.; Panda, J.; Panda, B.P.; Mohanta, T.K.; Mohanta, Y.K. Mycoremediation of heavy metals and/or metalloids in soil. In *Land Remediation and Management: Bioengineering Strategies*; Sarma, H., Joshi, S., Eds.; Springer Nature, Singapore, 2023; pp. 161–190. ISBN 978-981-99-4221-3.
103. Poonia, K.; Singh, P.; Ahamad, T.; Le, Q. Van; Phan Quang, H.H.; Thakur, S.; Mishra, A.K.; Selvasembian, R.; Hussain, C.M.; Nguyen, V.-H.; et al. Sustainability, performance, and production perspectives of waste-derived functional carbon nanomaterials towards a sustainable environment: A review. *Chemosphere* 2024, *352*, 141419, doi:10.1016/j.chemosphere.2024.141419.
104. Davolos, D.; Pietrangeli, B. A molecular study on bacterial resistance to arsenic-toxicity in surface and underground waters of latium (Italy). *Ecotoxicol. Environ. Saf.* 2013, *96*, 1–9, doi:10.1016/j.ecoenv.2013.05.039.
105. Bai, Y.; Yang, T.; Liang, J.; Qu, J. The role of biogenic Fe-Mn oxides formed in situ for arsenic oxidation and adsorption in aquatic ecosystems. *Water Res.* 2016, *98*, 119–127, doi:10.1016/j.watres.2016.03.068.
106. Mukhopadhyay, S.S. Nanotechnology in agriculture: Prospects and constraints. *Nanotechnol. Sci. Appl.* 2014, *7*, 63–71, doi:10.2147/NSA.S39409.
107. Knowd, I.; Mason, D.; Docking, A. Urban agriculture: The new frontier. In *Proceedings of the Planning for Food Seminar*; Vancouver, 2006; Vol. 21.
108. Itodo, H.U.; Nnamonu, L.A.; Wuana, R.A. Green synthesis of copper chitosan nanoparticles for controlled release of pendimethalin. *Asian J. Chem. Sci.* 2017, *2*, 1–10.
109. Mittal, D.; Kaur, G.; Singh, P.; Yadav, K.; Ali, S.A. Nanoparticle-based sustainable agriculture and food science: Recent advances and future outlook. *Front. Nanotechnol.* 2020, 2, 579954, doi:10.3389/fnano.2020.579954.
110. Pérez-de-Luque, A. Interaction of nanomaterials with plants: What do we need for real applications in agriculture? *Front. Environ. Sci.* 2017, *5*, doi:10.3389/fenvs.2017.00012.
111. Oliveira, H.C.; Stolf-Moreira, R.; Martinez, C.B.R.; Grillo, R.; de Jesus, M.B.; Fraceto, L.F. Nanoencapsulation enhances the post-emergence herbicidal activity of atrazine against mustard plants. *PLoS One* 2015, *10*, e0132971.
112. Zhao, L.; Lu, L.; Wang, A.; Zhang, H.; Huang, M.; Wu, H.; Xing, B.; Wang, Z.; Ji, R. Nano-biotechnology in agriculture: Use of nanomaterials to promote plant growth and stress tolerance. *J. Agric. Food Chem.* 2020, *68*, 1935–1947, doi:10.1021/acs.jafc.9b06615.
113. Paret, M.L.; Vallad, G.E.; Averett, D.R.; Jones, J.B.; Olson, S.M. Photocatalysis: Effect of light-activated nanoscale formulations of TiO_2 on xanthomonas perforans and control of bacterial spot of tomato. *Phytopathology* 2013, *103*, 228–236.
114. Anjali, C.H.; Sharma, Y.; Mukherjee, A.; Chandrasekaran, N. Neem oil (Azadirachta indica) nanoemulsion—A potent larvicidal agent against Culex quinquefasciatus. *Pest Manag. Sci.* 2012, *68*, 158–163.
115. Suresh Kumar, R.S.; Shiny, P.J.; Anjali, C.H.; Jerobin, J.; Goshen, K.M.; Magdassi, S.; Mukherjee, A.; Chandrasekaran, N. Distinctive effects of nano-sized permethrin in the environment. *Environ. Sci. Pollut. Res.* 2013, *20*, 2593–2602.
116. Fan, P.; Gu, Z.; Xu, D.; Xu, X.; Xu, G. Action analysis of drops of emamection-benzoate microemulsion on rice leaf. *Chine. J. Rice Sci.* 2010, *24*, 503, doi:10.3969/j.issn.1001-7216.2010.05.010.
117. Godfray, H.C.J.; Mason-D'Croz, D.; Robinson, S. Food system consequences of a fungal disease epidemic in a major crop. *Philos. Trans. R. Soc. B Biol. Sci.* 2016, *371*, 20150467.
118. Jo, Y.-K.; Kim, B.H.; Jung, G. Antifungal activity of silver ions and nanoparticles on phytopathogenic fungi. *Plant Dis.* 2009, *93*, 1037–1043, doi:10.1094/PDIS-93-10-1037.
119. Ouda, S.M., 2014. Antifungal activity of silver and copper nanoparticles on two plant pathogens, Alternaria alternata and Botrytis cinerea. *R. J. Microbiology.* 2014, *9*, 34.

120. Wani, A.H.; Shah, M.A. A unique and profound effect of MgO and ZnO nanoparticles on some plant pathogenic fungi. *J. Appl. Pharm. Sci.* 2012, *2*, 40–44.
121. Sarlak, N.; Taherifar, A.; Salehi, F. Synthesis of nanopesticides by encapsulating pesticide nanoparticles using functionalized carbon nanotubes and application of new nanocomposite for plant disease treatment. *J. Agric. Food Chem.* 2014, *62*, 4833–4838.
122. Stan, M.; Popa, A.; Toloman, D.; Silipas, T.-D.; Vodnar, D.C. Antibacterial and antioxidant activities of ZnO nanoparticles synthesized using extracts of Allium sativum, Rosmarinus officinalis and Ocimum basilicum. *Acta Metall. Sin. (English Lett.)* 2016, *29*, 228–236.
123. Ghotekar, S.; Ravikumar, C.R.; Chauhan, A.; Hikku, G.S.; Lin, K.-Y.A.; Rahdar, A.; Hitler, L.; Jabir, M.S.; Marzban, A.; Oza, R. Eco-friendly fabrication of CdO nanoparticles using Polyalthia longifolia leaves extract for antibacterial and electrochemical sensing studies. *J. Sol-Gel Sci. Technol.* 2024, *110*, 221–232, doi:10.1007/s10971-024-06352-6.
124. Sharma, D.; Chauhan, A.; Verma, R.; kumari, S.; Thakur, P.; Mahajan, A.K.; Kumar, V.; Sharma, M. Structural, morphological, optical and biomedical applications of Berberis aristata mediated ZnO and Ag-ZnO nanoparticles. *Nano Express* 2023, *4*, 45003, doi:10.1088/2632-959X/ad05e1.
125. Rao, Y.B.S.; Parajuli, D.; Someswara Rao, M.V; Ramakrishna, A.; Ramanjaneyulu, K.; Mammo, T.W.; Shanmukhi, P.S.V; Prasad, M.S.N.A.; Rao, G.M.; Suryanarayana, B.; et al. Effect of Fe doped and capping agent – Structural, optical, luminescence, and antibacterial activity of ZnO nanoparticles. *Chem. Phys. Impact* 2023, *7*, 100270, doi:10.1016/j.chphi.2023.100270.
126. Ahmad, F.; Ashraf, N.; Ashraf, T.; Zhou, R.-B.; Yin, D.-C. Biological synthesis of metallic nanoparticles (MNPs) by plants and microbes: Their cellular uptake, biocompatibility, and biomedical applications. *Appl. Microbiol. Biotechnol.* 2019, *103*, 2913–2935.
127. Sun, J.-B.; Wang, Z.-L.; Duan, J.-H.; Ren, J.; Yang, X.-D.; Dai, S.-L.; Li, Y. Targeted distribution of bacterial magnetosomes isolated from Magnetospirillum gryphiswaldense MSR-1 in healthy sprague-dawley rats. *J. Nanosci. Nanotechnol.* 2009, *9*, 1881–1885, doi:10.1166/jnn.2009.410.
128. Tang, Y.; Wang, D.; Zhou, C.; Zhang, S. Preparation and anti-tumor efficiency evaluation of bacterial magnetosome–anti-4-1BB antibody complex: Bacterial magnetosome as antibody carriers isolated from Magnetospirillum gryphiswaldense. *Biotechnol. Appl. Biochem.* 2019, *66*, 290–297.
129. Khan, S.A.; Gambhir, S.; Ahmad, A. Extracellular biosynthesis of gadolinium oxide (Gd_2O_3) nanoparticles, their biodistribution and bioconjugation with the chemically modified anticancer drug Taxol. *Beilstein J. Nanotechnol.* 2014, *5*, 249–257, doi:10.3762/bjnano.5.27.
130. Tripathi, R.M.; Shrivastav, A.; Shrivastav, B.R. Biogenic gold nanoparticles: As a potential candidate for brain tumor directed drug delivery. *Artif. Cells Nanomed. Biotechnol.* 2015, *43*, 311–317.
131. Borse, V.; Kaler, A.; Banerjee, U.C. Microbial synthesis of platinum nanoparticles and evaluation of their anticancer activity. *Int. J. Emerg. Trends Electr. Electron* 2015, *11*, 2320–9569.
132. Govindappa, M.; Lavanya, M.; Aishwarya, P.; Pai, K.; Lunked, P.; Hemashekhar, B.; Arpitha, B.M.; Ramachandra, Y.L.; Raghavendra, V.B. Synthesis and characterization of endophytic fungi, Cladosporium perangustum mediated silver nanoparticles and their antioxidant, anticancer and nano-toxicological study. *Bionanoscience* 2020, *10*, 928–941.
133. Ghosh, S. Copper and palladium nanostructures: A bacteriogenic approach. *Appl. Microbiol. Biotechnol.* 2018, *102*, 7693–7701.
134. Rossi, L.M.; Quach, A.D.; Rosenzweig, Z. Glucose oxidase–magnetite nanoparticle bioconjugate for glucose sensing. *Anal. Bioanal. Chem.* 2004, *380*, 606–613.
135. Chauhan, A.; Zubair, S.; Tufail, S.; Sherwani, A.; Sajid, M.; Raman, S.C.; Azam, A.; Owais, M. Fungus-mediated biological synthesis of gold nanoparticles: Potential in detection of liver cancer. *Int. J. Nanomed.* 2011, *6*, 2305–2319.

3 Deciphering Bacterial Nanoparticle Generation and Its Implementation in Diversified Applications

Debashree Debasmita, Qudsiya Y. Tamboli, and Abdullah G. Al-Sehemi

3.1 INTRODUCTION

Nanomaterials have emerged as a savior in various research fields. This 1–100 nm sized entities are different from their bulk counterparts and also bear properties that are superior to them. The electric, magnetic and surface, optical and chemical properties of nanomaterials are utilized in the area of bio-imaging such as positron emission tomography (PET) scan, combination therapy against cancer, drug delivery vehicle, biosensing, semiconductor industry and bioremediation. Nanoparticles are formed in various sizes, shapes and dimensions. They can be classified into different forms based on their origin: natural or artificial. Their porosity regulates their various applications, and they can be microporous, mesoporous or macroporous structures. The materials with pore sizes less than 2 nm are microporous nanoparticles, and they can only hold gases or small molecules. They are helpful in preparing membrane filters. The mesoporous materials such as SBA15 or MCM41 have pore sizes ranging from 2 to 50 nm and can be used in adsorbing other cargoes. Macroporous nanoparticles have larger pores, often more than 50 nm, and can accommodate large quantity of cargoes. Nanoparticles can be carbon based, organic, inorganic or composite type. They are classified based on the dimensions such as 0-dimensional (0D) or 1–3-dimensional (1D, 2D and 3D) structures. For example, nanorods and quantum dots are 0D; films of nanomaterials are 1D; carbon nanotubes, whether single layered or multi-layered, can be of 2D; and 3D nanoparticles are dendrimers [1,2]. Inorganic nanoparticles are usually metal and metal oxide, semiconductor or ceramic based. Organic nanoparticles are carbon based such as fullerene, carbon nanotubes or graphene based.[3] These abovementioned nanomaterials can be synthesized using various physicochemical methods and under different parameters. These regulate the type of nanomaterials to be obtained. These methods have been discussed in detail in the subsequent paragraphs.

3.1.1 Nanoparticle Synthesis Methodology

There exists a wide variety of nanomaterials out of which the most commonly used are nanoparticles (NP). NPs can be synthesized following two methodologies: top-down and bottom-up approaches. The first method includes breaking bulk materials into NPs. The benefit of top-down approach is to achieve NPs in a shorter time. In the second approach, the atoms are arranged into nuclei and further develop into nanosized particles. The benefit of bottom-up approach is to make NPs with required crystallization, surface to volume ratio and well-defined shapes (Figure 3.1). Some common methods that aid these two approaches are physical methods such as milling, grinding, pyrolysis, vapor pressure and thermal ablation. Moreover, chemical and photochemical reduction,

DOI: 10.1201/9781003487142-3

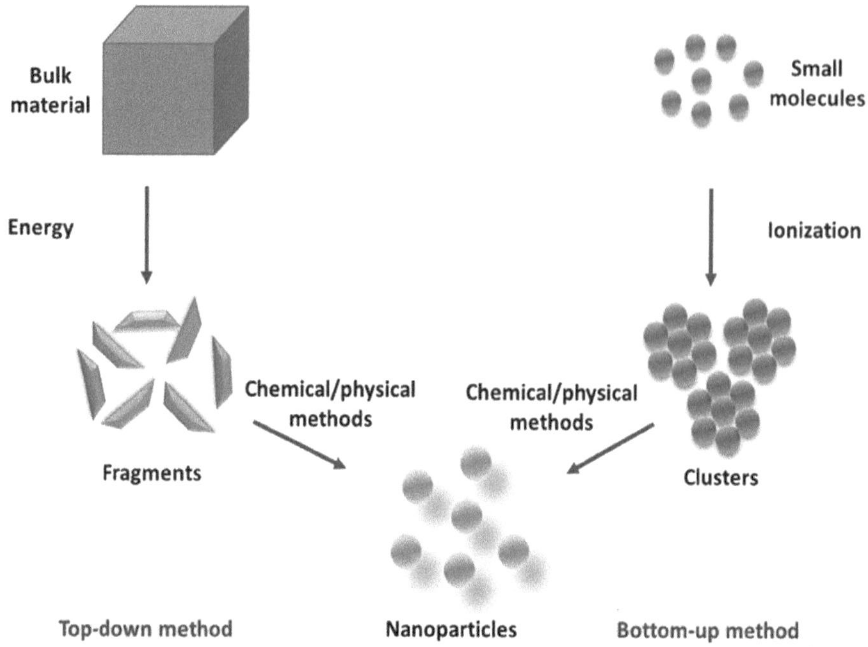

FIGURE 3.1 Schematic representation of top-down and bottom-up approach methods.

irradiation and electrochemistry are also the most used methods as these do not require any energy for production, and their yield is higher than that of physical methods. However, chemical wastes thus generated are hazardous to the environment and can cause side effects to living organisms. This drives synthesis methodologies toward more ecofriendly and sustainable strategies, and thus, the concept of green synthesis came in to light. The biogenic agents used for green synthesis are bacteria, plants, virus, algae, fungi and yeast [3,4].

3.1.2 Green Synthesis of Nanoparticles

Green synthesis of NPs is an emerging method that is meant to establish more sustainable, yield-making, safer and cost-effective processes. This is one of the bottom-up approaches of NP synthesis. The nanomaterial synthesis has a major drawback related to the degradation of the components used that are usually solvents, reducing agents that could be hazardous to the environment. Thus, in order to deal with the issue reusing the utilized raw materials or their recovery could be considered as a safe option. There is necessity of a sustainable and environmentally friendly methodology for the synthesis of nanomaterials [5]. The concept of green synthesis of nanoparticles is an outcome of this necessity. Biogenic nanomaterials are mostly prepared from plants, bacteria, fungus and algae. The extracts of biogenic agents, the medium in which they are grown also has capability to act as reducing agents for their substrates for preparing nanomaterials. Apart from biological agents, pH, temperature and pressure also play vital roles. The chemical synthesis of materials involves toxic and harmful chemicals as reducing and stabilizing agents. On the other hand, the green synthesis method serves as a natural and ecofriendly alternative. The ecofriendly nature of the green synthesis of nanoparticles is also dependent on the type of solvents that are used [6].

The green synthesis mainly involves the usage of water as a solvent, and an additional advantage of water is the presence of oxygen that reduces metals partially. The subsequent reduction can be carried out by the biogenic agent (Figure 3.2). Ionic liquids are thermally stable, and they have both hydrophobic and hydrophilic chemical groups. They do not evaporate into the environment like volatile solvents, but they need to be modified to overcome their degradability issue [7]. The most

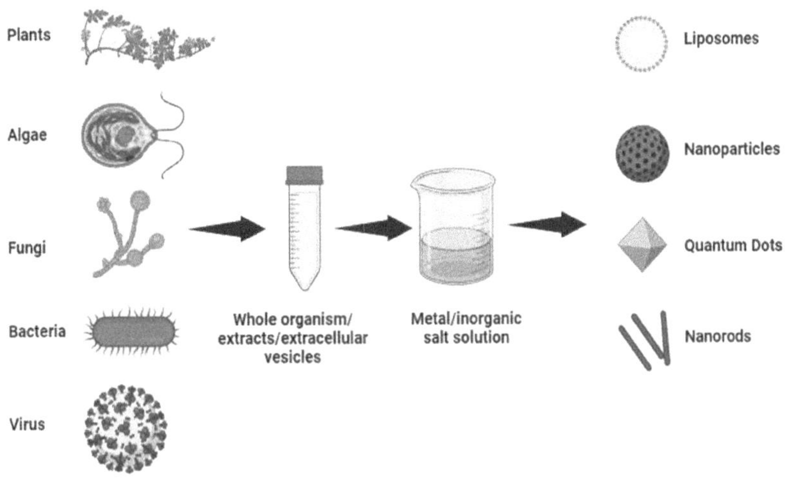

FIGURE 3.2 Graphical representation of biogenic agent–mediated nanomaterial synthesis.

common metal NPs synthesized via biogenic agents are gold, platinum, silver, zinc oxide, iron, copper, palladium, titanium oxide and palladium oxide [8]. AgNPs synthesized on the extracellular surface of *Bacillus cereus* are used for various applications such as antibacterial and antibiofilm. Moreover, they are used in the antitumour therapy against breast cancer. AgNPs formed using *Bacillus brevis* were efficiently able to eradicate the biofilms of *Salmonella typhi* and drug-resistant *Staphylococcus aureus*. The ultra-small palladium NPs synthesized by *Shewanella loihica* were used for catalytic activities. Often AgNPs lead to color change of the medium containing the bacterium from yellow to brown [9–12].

The bacterium that are isolated from the gold mines are tolerant to the inherent cytotoxicity effect of the gold and thus they act as biocatalysts for AuNP synthesis. *Pseudomonas stutzeri* AG259 isolated from silver mines is capable to form AgNPs. Marine bacterial strains are also utilized for these purposes such as *Stenotrophomonas*, *Kocuria flava* and *Alteromonas macleodii*. Magnetotactic bacteria form Fe_2O_3 or Fe_2S NPs intracellularly. The common species are *Magnetotactic cocci*, *Magnetospirillum magnetotacticum* and *Magnetospirillum magneticum*. It is believed that the nanoparticle synthesis by biogenic materials is the result of the defense mechanism of these organisms against metal ions. Various types of nanomaterials can be synthesized. For example, *Pseudomonas* sp. is capable of producing fluorescent quantum dots of CdSe. Synthesis of nanofibers of bacterial chitin and cellulose was carried out by addition of labeled chitin into the medium of *Acetobacter* [13–15]. In the next section, we will discuss about the various mechanisms of action used by biogenic agents for the nanomaterial formation.

3.1.3 Mechanism of Bacterial Nanoparticle Generation

Bacteria are nanofactories of nanoparticle synthesis. The cell wall of bacteria can allow the entry of metal ions and also facilitate the release of NPs into the extracellular space. The chemical groups present on the cell walls can be manipulated by chemical treatments to upsurge the metal ion percolation inside the cells. For example, a surge of 20,000–40,000-fold was seen in *Bacillus subtilis*. Bacteria can uptake metal ions from the medium they are grown in. For example, *Klebsiella pneumoniae* was able to synthesize CdS NPs when the medium was supplemented with Cd ions [16].

The initiation of metal nanoparticles formation by microbes happens by the incarceration of metal ions on their surface via electrostatic interactions and sometimes by the internalization of these metals that are reduced further by the action of the enzymes produced by microbes such as NADH-dependent nitrate reductase. In general, bacteria are grown under ambient conditions to

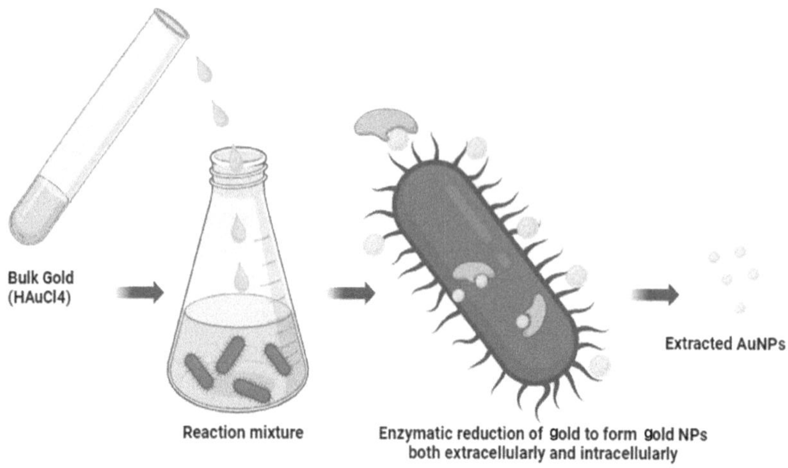

FIGURE 3.3 Graphical representation of bacteria-mediated gold nanoparticle synthesis.

generate the required amount of biomass by centrifugation. Furthermore, metal salt solutions are added to the biomass for a defined incubation time. The reduction of metal ions is often marked with the change in color (Figure 3.3).

The various mechanisms of NPs generation are dissimilatory oxidation to prevent cytotoxicity from metals, precipitation of metal ions, biochemical reduction of metal ions and complex formation. For example, *Streptomyces* sp. forms AgNPs that are stable without addition of any capping agents by simply using nitrate reductase enzymes that could convert nitrates to nitrites and further to nitrogenous gases. The reduction of silver ion is an endothermic reaction, and their further aggregation is an exothermic reaction that is mediated by surface proteins and enzymes. It has been seen that the metal ion reduction by a bacterial strain is a matter of the strain's defense mechanism [17]. The metal ion can reach intracellular compartments and intercalate with DNA causing a damage leading to cell death [18]. Strains that have a reducing enzyme system can reduce metal ions and thrive. Not all the strains have similar quantity or enzymatic activity against metal ions. Thus, all the bacterial strains are not capable of NP synthesis [19].

Redox mediators and cytochromes also play vital roles in extracellular production of metal NPs. Cytochromes mostly transfer the electron from the inner membrane to the outer membrane and cause the reduction of metal ions via enzymatic reduction. Exopolysaccharides (EPS) of the cell membrane, which in other ways responsible for cell-to-cell interaction and protection, have hemiacetal and aldehyde groups which reduce metal ions. Moreover, the various organic groups that form EPS are mannose, glucose, galactose or N-acetyl glucosamine or N-acetyl galactosamine [20].

The negative charge is attained due to the phosphate, hydroxyl, carboxyl and pyruvate groups. Metal ions are chelated and reduced, and capped and stabilized by the various groups. Magnetotactic bacteria have a unique mechanism of NP formation that is yet to be revealed completely. The initial step is the formation of a vesicle that allows invagination, leading to biomineralization of iron. Furthermore, these vesicles align themselves in the form of long chains following the cytoskeletal arrangement, and iron ions are internalized using iron transporters into these vesicles. The associated proteins assist the magnetite nucleation and formation of NPs of a particular morphology. The proteins released by the biomass mark the end of a stabilized NPs generation by acting as stabilizing and capping agents. *P. stutzeri* follows intracellular reduction of metal ions to form NPs [21].

3.1.4 Parameters Affecting Bacterial Nanoparticle Generation

The biogenic nanomaterial synthesis requires careful optimization of various factors affecting their synthesis. The parameters affecting the synthesis are varying. Some of them are as follows:

Technique: Physicochemical properties of nanomaterials are affected by the method used to synthesize them. Chemical reduction and biogenic synthesis methods provide different outcomes of the synthesis including the associated hazards. The biogenic synthesis is an ecofriendly and safer option.

Type of Species: The biogenic material itself is a controlling parameter. The concentration of the bacterial culture holds primary roles to synthesize NPs as it is directly associated with the number of NPs formed. The functional groups present on their cell membrane determine the types of NPs that are formed [22].

pH: The pH of the medium used and temperatures also control the outcome [23]. AgNPs formed using *Morganella* were of homogenous size when formed at a specific temperature of 20°C and their shapes changed on varying temperatures. Sizes also fluctuated with either increasing or decreasing the temperature beyond 20°C. A few studies have proven that AgNPs prepared from *Arthrobacter* showed varying sizes on changing the pH between 5 and 8. Lower pH produced smaller and spherical AuNPs, whereas higher pH produced irregular and larger NPs. The change in pH will result in NPs with different sizes and properties, but this is depended on the type of bioagent being used. Lactic acid bacteria produce smaller NPs in acidic pH, whereas few other bacterial strains produce NPs at a faster rate at higher pH. At higher pH, the catalytic activity of enzyme reduces and larger NPs are formed. Reductase enzymes can work efficiently in higher pH especially for synthesizing AgNPs. Thiol-containing proteins that reduce metal ions mostly work under alkaline condition. The acidic environment also supports the synthesis of metal ions; however, it takes days to synthesize as compared to alkaline pH that does the same work within few hours.

Temperature: Similar to the other parameters, temperature is vital. During synthesis, a higher temperature attains a more number of nuclei and small-sized NPs as compared to a lower temperature. The physical methods usually require higher temperatures, more than 300°C, and the chemical synthesis can be carried out in lesser temperatures. However, the biogenic synthesis is mostly set up in temperatures below 100°C.

Time and Pressure: The synthesis of nanoparticles greatly depends on the time of incubation and duration of the method used. Agitation speed, time, light sources and storage of nanoparticles for a certain time point affect their aggregation and sizes. The pressure of the synthesis method determines the size of nanoparticles. The physical methods mostly rely on the pressure parameters.

Enzyme Activity and Other Parameters: The higher activity of enzymes leads to the smaller-sized NPs. The extracellular accessories of bacteria called pili are made up of pilin protein. Studies have shown that the presence and absence of this protein led to the formation of NPs at either extracellular or intracellular space. Over- or underexpression of pilin protein controlled the periplasmic localization of metal ions and also affected their biomineralization [24]. Moreover, this protein reduces the metal ions extracellularly. The time of reaction holds importance in determining the size of NPs as the tie increases the aggregation increases, leading to larger NPs [25]. The presence of supplements in the growth medium regulates the amount, property and quality of NPs present. The presence of H_2 as an electron donor generated Pd NPs at a faster rate [26].

The concentration of precursor materials also matters as it has been seen that the higher concentration of metal ions generates smaller NPs as they form a layer on the bioagent and doesn't allow further accumulation of metal NPs. In a particular bacterium, the amount of enzymes available are different than others and depends on the age of that bacterium at which the enzymes are produced. Thus, the stage of the life cycle of bacteria is also important in determining the size of NPs. The prolonged incubation of bacteria with metals yields irregular-shaped NPs [8].

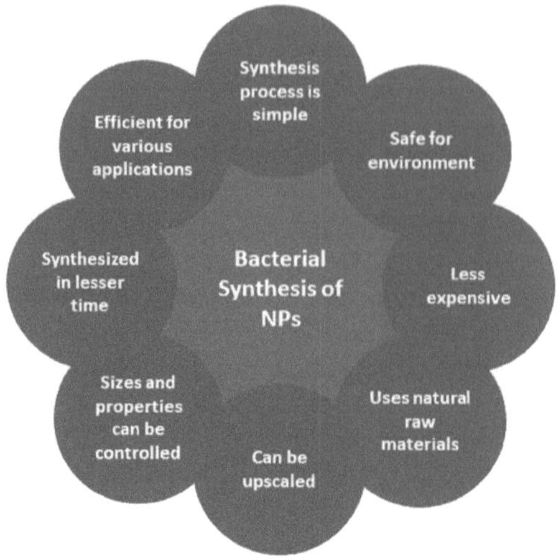

FIGURE 3.4 Advantages of bacterial synthesis of nanoparticles.

3.2 BACTERIAL NANOPARTICLE GENERATION: BOON OR BANE?

3.2.1 Advantages of Bacterial Nanoparticle Generation over Existing Methods

The NPs produced by bacteria are biocompatible, less expensive and biodegradable. The use of an aqueous environment for the synthesis is a non-toxic method and ecofriendly alternative to the toxic chemical reduction methods. Moreover, bacteria can grow at a fast rate and produce a large amount of nanomaterials since they can be easily grown and genetically manipulated as per the required nanoparticles. The biosynthesis method is cheaper and faster when compared with the existing methods as the raw materials for their growth are cheaper than the chemicals and solvents required for the chemical synthesis methods. The NPs formed using biofilms of bacteria are highly productive as the biomass involved is larger and also the biofilms are stoic toward the metal toxicity as compared to their planktonic versions. The synthesis process itself is less energy consuming, and it decreases the overall cost of production. The NPs produced have more surface to volume ratio, and the bacterial intermediate helps in their attachment to enzymes or other functional groups (Figure 3.4). As discussed in the previous sections, NPs can be synthesized extracellularly by the bacterium avoiding the requirement of expensive down-streaming processes [27].

3.2.2 Bacterial Strains Utilized for Nanoparticle Generation

Bacteria are a great source for producing metal nanoparticles of silver, gold, iron and copper. The common species exhibiting these capabilities are *Escherichia coli, Bacillus cecembensis, Bacillus amyloliquefaciens, Bacillus indicus, Lactobacillus casei* [28], *Pseudomonas proteolytica, Enterobacter cloacae* and *Corynebacterium* sp. The nanoparticles prepared by bacteria can be spherical, octahedral, or even triangular in shape (Table 3.1).

3.2.3 Limitations and Challenges of Bacterial Nanoparticle Generation

Although advantages of the green synthesis of nanomaterials are far beyond any other method, limitations also exist such as low yield, scaling-related issues and optimization of various parameters for large-scale production. Above these, there is a lack of studies related to exact mechanisms that lead to the synthesis of a particular type or shape of NPs. Extracellularly synthesized NPs are of varied

TABLE 3.1
Bacterial Species Used for Nanoparticle (NP) Synthesis

NPs	Bacteria	Size (nm)	Application	Mechanism	References
Ag	*Pseudomonas* sp.	20–70	Antimicrobial	Incubation of metal salt with bacteria	[29]
Ag	*Bacillus thuringiensis*	40–50	Antimicrobial	Heating and centrifugation	[30]
Ag	*Escherichia coli*	5–50	Antimicrobial	Agitation of metal salt with biomass	[31]
Ag	*Bacillus brevis*	40–70	Antimicrobial	Incubation for 1 week in dark condition	[32]t
Au	*Micrococcus* sp.	53	Antimicrobial and anticancer	Microwave-mediated heating	[33]
Au	*Mycobacterium* sp.	5–55	Antimicrobial and anticancer	Incubation and heating with metal salt	[34]
TiO_2 and ZnO_2	*Aeromonas hydrophila*	28–54	Antibacterial	Centrifugation	[35]
Cu	*Shewanella loihica*	10–16	Antimicrobial	Agitation in anoxic condition	[36]
ZnO_2	*Staphylococcus aureus*	10–15	Antimicrobial	Refluxing	[37]
ZnO_2	*Sphingobacterium* sp.	40	Antimicrobial	Incubation of metal salt with bacteria	[38]
ZnO_2	*Lactobacillus plantarum*	7	Wound healing	Agitation of metal salt with biomass	[39]
Au	*Rhodopseudomonas* sp.	50–60	Bioremediation	Incubation of metal salt with bacteria	[40]

sizes in comparison to intracellularly synthesized ones. However, the extraction of extracellular NPs is easier, and thus, the synthesis methods need to be standardized. The osmotic shock induction, and heating or freeze-thaw methods of NPs extraction mostly result in the breakage of NPs or disintegration of the bioagent and alteration in the properties of NPs. Therefore, methods need to be developed in order to smoothly extract NPs [41]. Although enzymatic extraction or surfactants are a better method, it is more expensive than the other ways. Thus, centrifugation methods need to be improved as an alternative to them [42]. The monodispersed NPs are less achievable with the current methods of green synthesis, so more efforts should be made to obtain monodispersed NPs and reduce their aggregation into larger units.

3.3 APPLICATION OF BACTERIAL-GENERATED NANOPARTICLES

The biogenic agent–mediated nanoparticle synthesis has various industrial and biomedical applications (Figure 3.5).

3.3.1 Anticancer Therapeutics and Drug Delivery

AuNPs are known for their low toxicity and high surface to volume ratio, which make them great carriers [42] of targeted molecules [43]. Silver nanoparticles showed anti-angiogenic property when used to treat bovine retinal endothelial cells. They reduce the IC_{50} concentrations of various drugs when used as a carrier or in combination, thereby lowering the adversities associated with heavy drug doses otherwise. These NPs can be fabricated to attain targeted therapies and also in diagnostics [44]. Iron NPs can be specially used in hyperthermia-mediated cancer therapies. Iron oxide NPs have been seen to induce apoptosis in breast cancer cell lines. Biosynthesized AgNPs and platinum NPs from *Saccharomyces boulardii* cause toxicity and reduce viability in epidermal and breast cancer cell lines. *Streptomyces*-originated selenium nanorods treat triple negative breast cancer and liver cancer cell lines [45]. ZnO NPs synthesized from *Rhodococcus* sp. and encapsulated with drugs induce apoptosis by increasing toxicity against colon cancer cell lines. Moreover, they act as carrier systems for anticancer drugs. In another study, ZnO NPs have been loaded with anthraquinone to work against colon cancer cells. The Superparamagnetic Iron Oxide Nanoparticles (SPION)

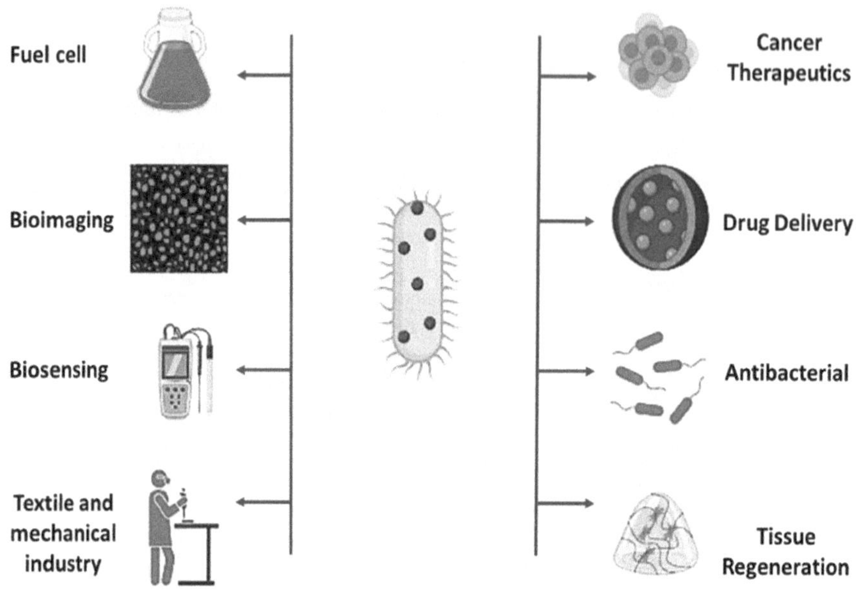

FIGURE 3.5 Schematic representation of applications of bacteria-mediated synthesized nanoparticles.

are used to carry out anticancer therapeutics and used against drug-resistant cancer cells with lower doses as compared to only drugs. Metal NPs are used for protein, antibody and gene delivery due to their high surface to volume ratio and tunability [46]. Due to their small size, they can easily slide into the tight junction gaps between the cells where only drugs are difficult to target. The biocompatibility of these NPs made them suitable to be assigned as a nanocarrier system that escapes from immunogenic reactions. Magnetosomes synthesized by magnetotactic bacteria [47] are used for the delivery of drugs as the thiol group present in them creates easy passage into cells and also for immunotherapy [48]. They can also be used for photodynamic or photothermal therapy [49].

3.3.2 Antimicrobial Agent

The initiation of microbial knockdown happens with the electrostatic interaction of NPs and the cell membrane of bacteria. The interaction leads to the disruption of the membrane by the changes in the wall structure or even by depolarization leading to eventual cell death. Various mechanisms of actions are associated with antimicrobial properties of bacterial NPs. Often, they enter into the cells and then interfere with the DNA and protein synthesis pathways, conformation changes and enzymatic inactivation of various proteins. Antibacterial effects are mostly reactive oxygen species (ROS) mediated, leading to membrane deformation and DNA mutations also by dis-functioning the electron transport chain. Iron NPs act on *E. coli* and *S. aureus* by disrupting their respiration. AgNPs are designed using various bacterial species to be used in healing materials as antibacterial agents [50]. Silver NPs [51] produced by microbes can affect the other target microbes in various ways depending on their shape, size and other physicochemical properties [52]. The shapes supporting the high atomic density within NPs exhibit better efficacy against microbes. Some of them can affect the microbial membrane through the sulfhydryl groups of the enzymes produced, while the others denature their membranes and induce ROS generation. Zn ions released on the surface destabilize the cell membrane causing cell death. Gold NPs produced via green synthesis are non-toxic, and at the same time, they can be functionalized for antibacterial effects [53]. TiO_2 [35], Fe_2O_3, CuO [36] and ZnO NPs are known for their antibacterial properties against gram-positive bacterial strains such as *S. aureus* and *B. subtilis* and gram-negative strains such as *E. coli* and *P. aeruginosa.*

Iron oxide nanoparticles show the lowest antibacterial efficacy as compared to zinc oxide NPs that have the highest efficacy. It is assumed that the NPs synthesized thorough green synthesis methods also bear the properties of biogenic entity it is made up of. Hence, their efficacies are superior to that of the chemically synthesized NPs [54].

3.3.3 Wound Healing and Regeneration

Silver NPs are mostly incorporated with different treatment options such as wound-healing nanofibers, scaffolds and polymers. The antibacterial effects of silver NPs provide elevated effects of the standard treatment options. Sometimes, AgNPs are used with ZnO [55] or TiO_2 for burn wound or skin irritation or catheter cleaning. Biomaterials are fabricated to induce repair and regrow the wounded and damaged tissues. They are mostly implanted within the area of injury to release the loaded drugs/growth factors or act as a scaffold to support cell growth and repair. Often, these biomaterials are overpowered by microbes and cause secondary infections. NPs are incorporated into biomaterials to provide additional mechanical support, antimicrobial properties, drug encapsulation and release, and targeted drug delivery. For example, TiO_2 NPs are used to coat the implants to increase the cell growth and infiltrations. ZnO NPs slowly release into cells and contribute to the cascade of cell signaling for cell growth [56].

3.3.4 Bioimaging

The silver and gold NPs are used in creating contrast images for diagnostic purposes [57]. They are incorporated for MRI scans and fluorescence-based imaging, as they can be loaded with dye for particular cells, targeted molecules, radioactive materials and even antibodies. ZnO NPs are biocompatible and get readily absorbed by cells. Therefore, they are ideal for imaging purposes. These NPs are better than the conventional dyes as they do not undergo instant photobleaching and also have low luminescence and intensity. ZnO NPs can be easily fabricated; as per the requirement, they can be doped and manipulated [48,58].

3.3.5 Biosensor

Selenium NPs formed out of *B. subtilis* are highly efficient when it comes to biosensing ability as they have high surface to volume ratio, adsorption efficacy, biocompatibility and sensitivity toward various functional groups. Silver and gold NPs or their alloy NPs are used for the glucose detection in sensors and also for detecting vanillin from vanilla beans. Fe_3O_4 NPs are synthesized in cube shapes to detect glucose through rapid test devices [48]. These iron oxide NPs are also used to detect peroxide ions [59]. Bacterial nanocellulose is extracted and formed into biofilms or hydrogels by the bottom-up method. These cellulose fibers have diameter less than 100 nm. They are often formed into paper cellulose for biosensing. The nanocellulose obtained is used for colorimetric analysis, SERS-mediated sensing, and bioluminescence- and fluorescence-based assays. NPs can be incorporated into these hydrogels that can be further used for detection of various chemical groups. They are helpful in detecting analytes in gaseous phase. These fibers can be modified to induce conductivity and have shown electric properties when incorporated into electro-biosensing devices [60]. Furthermore, bioagent-mediated alloy NPs of Ag and Au were used to sense vanillin with five-fold increased precision and a lower detection limit of 40 nM [18,64].

3.3.6 Bioremediation

Discharging untreated industrial and domestic wastes in to the environment, specifically to water bodies or soils, cause pollutions affecting the health. NPs can be modified to chelate harmful heavy metals or inorganic chemicals; for example, carbon particles are used to remove oil spills in ocean and other water bodies. Similarly, metal oxide NPs are used to remove impurities from water bodies. Zeolites or carbon nanotubes are used for heavy metal chelating. Apart from fabrication, the conjugation of iron NPs with other metal ions is also helpful in removing the effluents by enhancing the rate of chelating. The ground water pollution is a great threat to the health, and it has been seen that the water filter membranes incorporated with Cu NPs showed reduction in the population of *E. coli*. NPs break pollutants into lesser toxic entities and sometimes to recyclable forms. The high surface area of biogenic NPs bears a greater number of active sites and show catalytic properties. They have reduced dyes like methylene blue, erichrome black T, etc. Dyes like malachite green and Congo red were reduced by Cu NPs formed using *E. coli*. These pollutants sometimes directly affect our nervous system. Au and Pd NPs synthesized by *Bacillus* sp. degrade these pollutants such as 4-nitrophenol or carbon tetrachloride by Fe_3O_4 NPs. Pd NPs are also pivotal in the removal of water pollutants such as heavy metals like Cr [61].

3.3.7 Microbial Fuel Cell

NPs are also used for energy storage and also carry out catalysis of petroleum products to generate energy. This is a much cost-effective method compared to electrolysis, chemical pyrolysis and thermolysis. In the drilling fluids used for petroleum recovering, NPs are added to enhance the productivity. NPs give more stability and act as lubricating agents and bore-hole cleaners due to their large surface to volume ratio. Iron oxide NPs were used in the Xanthan gum drilling fluid. This addition helped in friction reduction and improved filtration in the fluid. Titanium and silicon oxide NPs are also chosen for this purpose [62].

3.3.8 Insecticide and Nanofertilizers

Arthropods, especially the mosquitoes, are of great threat to humans. They are responsible for the epidemics such as malaria, filariasis, zika virus, dengue and Japanese encephalitis. They are periodically controlled using insecticides and pesticides that also impact the health. These arboviruses need to be checked. Therefore, a carefully thought measure must be developed to impart the desired outcome without hampering the health. Here come nanomaterials as rescuers. NPs produced using various bioagents are used as insecticides or pesticides such that they are specifically toxic against the larvae of vectors like mosquitoes and not to affect the other non-targeted species. Bacteria are common candidates for the synthesis of metal NPs as the metabolites released from them can reduce metal ions to form NPs. The cell wall, as discussed earlier, is negatively charged and causes the reduction of positively charged metal ions. Studies suggest that *Streptomyces* and *Bacillus safensis* isolated from soils of the western ghats in India especially in the states of Gujarat, Maharashtra, Goa, Karnataka, Kerala, and Tamil Nadu and duping yard, respectively, are bioagents for silver NPs synthesis. These NPs elevate oxidative stress in the larvae, and via exoskeletons, they enter the larvae and further bind to the thiol of proteins and phosphate of DNA causing disruption of cellular activity and fatality of the larvae [63]. Another advantage of amalgamation of NPs with insecticides is their slow and sustained release. This not only enables targeted application of insecticides but also reduces wastage and promotes lower dosage of insecticide usage. Silica NPs, when mixed with soils, are up-taken by plant roots and greatly improve the soil quality and thus act as nanofertilizers. ZnO and TiO_2 NPs have prevented silkworms from diseases and greatly affected the silk cocoon yield [64].

3.3.9 Catalytic Activity

NPs provide high surface to volume ratio for the reactions to happen and check the activation energy–related barriers. The green synthesis of these NPs made them ecofriendly and degradable after the catalytic activity is over, thus reducing the usage of harmful catalytic agents like

4-nitrophenol. The addition of Ag NPs (formed from biogenic agents) accelerates the reactions. The industrial effluents from textile industries contain harmful dyes that persist in the environment for years and undergoes a slow degradation process. The addition of these NPs can greatly enhance the adsorption of these dyes from water bodies and waste treatment plants, and can begin the catalysis at a faster rate. The rate of removal will increase, and the quality of the water bodies will be upgraded. Apart from this, NPs, being cost-effective, nanosized and tunable in nature, also provide opportunity to sense the presence of heavy metals, such as Hg and Cd, by colorimetric analysis. Magnetic NPs formed out of *Pseudomonas delafieldii* are used to desulfurize dibenzothiophene due to their high surface energies that resulted in easy adsorption [65].

3.3.10 Textile Industry

NPs are added to the textiles while preparation so that they provide additional ballistic protection, avoid crease formation of cloths, and also make them antibacterial. The specialized fabricated fabrics that are antibacterial and do not require repeated washing and cleaning are also helpful for the creation of uniforms for soldiers who work in harsh climates and conditions. ZnO NPs in the form of nanoflowers formed by *Serratia* sp. were incorporated into fabrics to make them antibacterial. The new-age fabric is the smart fabric that is designed to bear sensors that help in tracking and monitoring physiological conditions, and trapping solar energy for harvesting solar energy through body movements [48].

3.3.11 Mechanical Applications

As discussed in the previous sections, the properties of NPs elevate the overall efficiency of any material. Thus, by virtue of their stress and strain tolerance, they can be used as coating materials to enhance shine and as lubricants of machineries. Moreover, they provide resistance to the metals due to their delamination properties, and they also reduce friction and related damages to polymer and metal scaffolds or matrices. Mostly, NPs of aluminum, titanium and carbon are used in the mechanical industry as the additional resistance, and reduction in wearing is provided by NPs [18,64,66].

3.4 CONCLUSION

In conclusion, the advent of synthesis of NPs via biogenic methods is a boon for scientific research areas. These methods are based on simple principles to produce higher amounts of NPs without forming by-products that are hazardous to the environment. Biogenic agents have the perquisites in them to reduce the metal ions to their NP form, either extracellularly or intracellularly. These are either enzymes or proteins that help a biogenic agent to downregulate the effect of the presence of metal ions in the growth medium by reducing them by virtue of a defense mechanism. This property is being harnessed to produce NPs. Plants, fungi, bacteria, algae, etc. are used to synthesize NPs. However, bacteria are more advantageous in terms of their fast and high growth rate, robust tolerance and easy-to-manipulate nature. Nanomaterial synthesis is a critical method that involves various parameters that determine the quality and quantity of the product. Therefore, careful designing of a synthesis process is required. However, this in turn is depended on the in-depth knowledge about the mechanism of NP synthesis. The scaling up of these methods also need careful attention.

3.5 FUTURE PROSPECTIVE

As discussed in the previous sections, the biogenic synthesis of NPs is superior to the chemically synthesized NPs. Therefore, to overcome the currently existing challenges in producing biogenic NPs, few measures can be taken such as developing methods for elaborate studies on the mechanical and physicochemical properties of these NPs for better understanding of the mechanism of synthesis. These detailed studies can be helpful in standardizing the parameters for NPs synthesis. In chemical extraction of NPs, the quality often gets hampered. Down-streaming processes need to be upgraded or replaced with modern techniques that keep the properties of NPs intact. For establishing sustainable biogenic NPs generation systems, large-scale production needs to be set up. Recycling methods for raw materials for the biogenic synthesis can be established for sustainable systems that reduce environmental hazards [34].

REFERENCES

1. Paramasivam, G.; Palem, V.V.; Sundaram, T.; Sundaram, V.; Kishore, S.C.; Bellucci, S. Nanomaterials: Synthesis and applications in theranostics. *Nanomaterials* **2021**, 11, 3228.
2. Khan, M.R.; Fromm, K.M.; Rizvi, T.F.; Giese, B.; Ahamad, F.; Turner, R.J.; Füeg, M.; Marsili, E. Metal nanoparticle–microbe interactions: Synthesis and antimicrobial effects. *Part. Part. Syst. Charact.* **2020**, 37(5), 1900419.
3. Mekuye, B.; Abera, B. Nanomaterials: An overview of synthesis, classification, characterization, and applications. *NanoSelect.* **2023**, 4, 486. https://doi.org/10.1002/nano.202300038.
4. Nam, N.H.; Luong, N.H. Nanoparticles: Synthesis and applications. *Mater. Biomed. Eng.* **2019**, 211–40. https://doi.org/10.1016/B978-0-08-102814-8.00008-1.
5. Sharma, S.; Kumar, K.; Thakur, N. Green synthesis of silver nanoparticles and evaluation of their anti-bacterial activities: Use of *Aloe Barbadensis Miller* and *Ocimum Tenuiflorum* leaf extracts. *Nanofabrication* **2021**, 6(1), 52–67. https://doi.org/10.1515/nanofab-2020-0102.
6. Bakir, E.; Younis, N.; Mohamed, M.; El Semary, N. Cyanobacteria as nanogold factories: Chemical and anti-myocardial infarction properties of gold nanoparticles synthesized by *Lyngbya majuscula. Marine Drugs.* **2018**, 16(6), 217.
7. Singh, J.; Dutta, T.; Kim, K.-H.; Rawat, M.; Samddar, P.; Kumar, P. 'Green' synthesis of metals and their oxide nanoparticles: Applications for environmental remediation. *J. Nanobiotechnol.* **2018**, 16(1), 84. https://doi.org/10.1186/s12951-018-0408-4.
8. Mohd Yusof, H.; Mohamad, R.; Zaidan, U.H.; et al. Microbial synthesis of zinc oxide nanoparticles and their potential application as an antimicrobial agent and a feed supplement in animal industry: A review. *J. Animal Sci. Biotechnol.* **2019**, 10, 57. https://doi.org/10.1186/s40104-019-0368-z.
9. Banerjee, K.; Rai, V.R. A review on mycosynthesis, mechanism, and characterization of silver and gold nanoparticles. *BioNanoScience* **2018,** 8(1), 17–31.
10. Buszewski, B.; Railean-Plugaru, V.; Pomastowski, P.; Rafińska, K., Szultka-Mlynska, M.; Golinska, P.; Wypij, M.; Laskowski, D.; Dahm, H. Antimicrobial activity of biosilver nanoparticles produced by a novel *Streptacidiphilus durhamensis* strain. *J. Microbiol. Immunol. Infect.* **2018**, 51(1), 45–54.
11. Das, M.; Chatterjee, S. Green synthesis of metal/metal oxide nanoparticles toward biomedical applications: Boon or bane. In: Shukla, A.K., & Iravani, S. (Eds.). *Green Synthesis, Characterization and Applications of Nanoparticles.* Elsevier. **2019**, 265–301.
12. Fang, X.; Wang, Y.; Wang, Z.; Jiang, Z.; Dong, M. Microorganism assisted synthesized nanoparticles for catalytic applications. *Energies* **2019**, 12(1), 190.
13. Lahiri, D.; Nag, M.; Sheikh, H.I.; Sarkar, T.; Edinur, H.A.; Pati, S.; Ray, R.R. Microbiologically-synthesized nanoparticles and their role in silencing the biofilm signaling cascade. *Front. Microbiol.* **2021**, 12, 636588. https://doi.org/10.3389/fmicb.2021.636588.
14. Yadav, R.; Kumar, M.; Tomar, R.S. Revisiting the microbial biosynthesis of metal nanoparticles and their applications. *J. Appl. Pharm. Sci.* **2023**, 13(07), 013–031. https://doi.org/10.7324/JAPS.2023.89282.
15. Ayesha, A. Bacterial synthesis and applications of nanoparticles. *Nano Sci. Nano Technol.* **2017**, 11(2), 119.
16. Wypij, M.; Golińska, P. Mechanism of microbial synthesis of nanoparticles. *Microbial Nanotechnology.* **2020**. eBook ISBN9780429276330.
17. Singh, R.; Shedbalkar, U.U.; Wadhwani, S.A.; et al. Bacteriagenic silver nanoparticles: Synthesis, mechanism, and applications. *Appl. Microbiol. Biotechnol.* **2015**, 99, 4579–4593. https://doi.org/10.1007/s00253-015-6622-1
18. Karunakaran, G.; Sudha, K.G.; Ali, S.; Cho, E.-B. Biosynthesis of nanoparticles from various biological sources and its biomedical applications. *Molecules* **2023**, 28(11), 4527. https://doi.org/10.3390/molecules28114527.
19. Alsamhary, K.I. Eco-friendly synthesis of silver nanoparticles by *Bacillus subtilis* and their antibacterial activity. *Saudi J. Biol. Sci.* **2020**, 27, 2185–2191.

20. Hamouda, R.A., Hussein, M.H., Abo-elmagd, R.A.; et al. Synthesis and biological characterization of silver nanoparticles derived from the cyanobacterium *Oscillatoria limnetica*. *Sci. Rep.* **2019**, 9, 13071. https://doi.org/10.1038/s41598-019-49444-y.
21. John, M.S.; Nagoth, J.A.; Ramasamy, K.P.; Mancini, A.; Giuli, G.; Natalello, A.; Ballarini, P.; Miceli, C.; Pucciarelli, S. Synthesis of bioactive silver nanoparticles by a pseudomonas strain associated with the antarctic psychrophilic protozoon euplotes focardii. *Mar. Drugs* **2020**, 18(1), 38. https://doi.org/10.3390/md18010038.
22. Ibrahim, S.; Ahmad, Z.; Manzoor, M.Z.; et al. Optimization for biogenic microbial synthesis of silver nanoparticles through response surface methodology, characterization, their antimicrobial, antioxidant, and catalytic potential. *Sci. Rep.* **2021**, 11, 770. https://doi.org/10.1038/s41598-020-80805-0.
23. Campaña, A.L.; Saragliadis, A.; Mikheenko, P.; Linke, D. Insights into the bacterial synthesis of metal nanoparticles. *Front. Nanotechnol.* **2023**, 5, 1216921. https://doi.org/10.3389/fnano.2023.1216921.
24. Gahlawat, G.; Choudhury, A.R. A review on the biosynthesis of metal and metal salt nanoparticles by microbes. *RSC Adv.* **2019**, 9(23), 12944–12967. https://doi.org/10.1039/C8RA10483B.
25. Patil, S.; Chandrasekaran, R. Biogenic nanoparticles: A comprehensive perspective in synthesis, characterization, application and its challenges. *J. Genet. Eng. Biotechnol.* **2020**, 18(1), 67. https://doi.org/10.1186/s43141-020-00081-3.
26. Patra, J.K.; Baek, K.H. Green nanobiotechnology: Factors affecting synthesis andcharacterization techniques. *J. Nanomater.* **2014**, 12, Article ID 417305. https://doi.org/10.1155/2014/417305.
27. Sabri, M.A.; Umer, A.; Awan, G.H.; Hassan, M.F.; Hasnain, A. Selection of suitable biological method for the synthesis of silver nanoparticles. *Nanomater. Nanotechnol.* **2016**, 6, 29. https://doi.org/10.5772/62644.
28. Tukenmez, U.; Aktas, B.; Aslim, B.; Yavuz, S. The relationship between the structural characteristics of lactobacilli-EPS and its ability to induce apoptosis in colon cancer cells in vitro. *Sci. Rep.* **2019**, 9(1), 8268. https://doi.org/10.1038/s41598-019-44753-8.
29. Syed, B.; Prasad, N.; Dhananjaya, B.L.; Yallappa, S.; Satish, S. Synthesis of silver nanoparticles by endosymbiont *Pseudomonas fluorescens* CA 417 and their bactericidal activity. *Enzym. Microb. Technol.* **2016**, 95, 128–136.
30. Ge, J; Hu, J.; Cui, S.; Wang, Y.; Xu, C.; Liu, W. Biosynthesis of Bt-Ag_2O nanoparticles using *Bacillus thuringiensis* and their pesticidal and antimicrobial activities. *Appl. Microbiol. Biotechnol.* **2024**, 108, 157. https://doi.org/10.1007/s00253-023-12859-9.
31. Kathiresan, K.; Alikunhi, N.M.; Pathmanaban, S.M.; Nabikhan, A.; Kandasamy, S. Analysis of antimicrobial silver nanoparticles synthesized by coastal strains of *Escherichia coli* and *aspergillus niger*. *Can. J. Microbiol.* **2010**, 56, 1050–1059.
32. Pugazhenthiran, N.; Anandan, S.; Kathiravan, G.; Udaya Prakash, N.K.; Crawford, S.; Ashokkumar, M. Microbial synthesis of silver nanoparticles by *Bacillus* sp. *J. Nanoparticle Res.* **2009**, 11, 1811–1815.
33. Kumari S, Shivakrishna P, Sreenivasulu K. Synthesis and characterization of gold nanoparticles from marine *Micrococcus* sp. OUS9. *Bioinformation* **2020**, 16(11), 849–855. https://doi.org/10.6026/97320630016849.
34. Menon, S.; Rajeshkumar, S.; Venkat Kumar, S. A review on biogenic synthesis of gold nanoparticles, characterization, and its applications. *Resour. Efficient Technol.* **2017**, 3(4), 516–527. https://doi.org/10.1016/j.reffit.2017.08.002.
35. Jayaseelan, C.; Rahuman, A.A.; Roopan, S.M.; Kirthi, A.V.; Venkatesan, J.; Kim, S.K.; Iyappan, M.; Siva, C. Biological approach to synthesize TiO_2 nanoparticles using Aeromonas hydrophila and its antibacterial activity. *Spectrochim. Acta A Mol. Biomol. Spectrosc.* **2013**, 107, 82–89. https://doi.org/10.1016/j.saa.2012.12.083.
36. Lv, Q.; Zhang, B.; Xing, X.; Zhao, Y.; Cai, R.; Wang, W.; Gu, Q. Biosynthesis of copper nanoparticles using Shewanella loihica PV-4 with antibacterial activity: Novel approach and mechanisms investigation. *J. Hazard Mater.* **2018**, 347, 141–149. https://doi.org/10.1016/j.jhazmat.2017.12.070.
37. Mahamuni, P.P.; Patil, P.M.; Dhanavade, M.J.; Badiger, M.V.; Shadija, P.G.; Lokhande, A.C.; Bohara, R.A. Synthesis and characterization of zinc oxide nanoparticles by using polyol chemistry for their antimicrobial and antibiofilm activity. *Biochem. Biophys. Rep.* **2018**, 17, 71–80. https://doi.org/10.1016/j.bbrep.2018.11.007.
38. Murali, M.; Gowtham, H.G.; Shilpa, N.; Singh, S.B.; Aiyaz, M.; Sayyed, R.Z.; Shivamallu, C.; Achar, R.R.; Silina, E.; Stupin, V.; Manturova, N.; Shati, A.A.; Alfaifi, M.Y.; Elbehairi, S.E.I.; Kollur, S.P. Zinc oxide nanoparticles prepared through microbial mediated synthesis for therapeutic applications: A possible alternative for plants. *Front. Microbiol.* **2023**, 14, 1227951. https://doi.org/10.3389/fmicb.2023.1227951.

39. Mohd Yusof, H.; Abdul Rahman, N.; Mohamad, R.; et al. Biosynthesis of zinc oxide nanoparticles by cell-biomass and supernatant of *Lactobacillus plantarum* TA4 and its antibacterial and biocompatibility properties. *Sci. Rep.* **2020**, 10, 19996. https://doi.org/10.1038/s41598-020-76402-w.
40. Liu, B.; Jin, Y.; Wang, Z.; Xing, D.; Ma, C.; Ding, J.; Ren, N. Enhanced photo-fermentative hydrogen production of *Rhodopseudomonas* sp. nov. strain A7 by the addition of TiO_2, ZnO and SiC nanoparticles. *Int. J. Hydrogen Energy* **2017**, 42(29), 18297–18287. https://doi.org/10.1016/j.ijhydene.2017.04.147.
41. Capuzzo, A.M. *Bacterial Synthesis of Nanoparticles: Current Trends in Biotechnology and Biomedical Fields*; preprint; LIFE SCIENCES, 2021. https://doi.org/10.20944/preprints202103.0358.v1.
42. Park, B.-W.; Zhuang, J.; Yasa, O.; Sitti, M. Multifunctional bacteria-driven microswimmers for targeted active drug delivery. *ACS Nano* **2017**, 11(9), 8910–8923. https://doi.org/10.1021/acsnano.7b03207.
43. Debasmita, D.; Ghosh, S.S.; Chattopadhyay, A. Living gut bacteria functionalized with gold nanoclusters and drug for facile cancer theranostics. *ACS Appl. Bio Mater.* **2023**, 6(2), 628–639. https://doi.org/10.1021/acsabm.2c00911.
44. Debasmita, D.; Ghosh, S.S.; Chattopadhyay, A. Hierarchical passage of gold nanoclusters in living bacteria. *ACS Appl. Bio Mater.* **2022**, 5(6), 2543–2548. https://doi.org/10.1021/acsabm.2c00315.
45. Bisht, N.; Phalswal, P.; Khanna, P.K. Selenium nanoparticles: A review on synthesis and biomedical applications. *Mater. Adv.* **2022**, 3(3), 1415–1431. https://doi.org/10.1039/D1MA00639H.
46. Holay, M.; Guo, Z.; Pihl, J.; Heo, J.; Park, J.H.; Fang, R.H.; Zhang, L. Bacteria-inspired nanomedicine. *ACS Appl. Bio Mater.* **2021**, 4(5), 3830–3848. https://doi.org/10.1021/acsabm.0c01072.
47. Stanton, M.M.; Park, B.-W.; Vilela, D.; Bente, K.; Faivre, D.; Sitti, M.; Sánchez, S. Magnetotactic bacteria powered biohybrids target *E. coli* biofilms. *ACS Nano* **2017**, 11(10), 9968–9978. https://doi.org/10.1021/acsnano.7b04128.
48. Ghosh, S.; Ahmad, R.; Zeyaullah, Md.; Khare, S.K. Microbial nano-factories: Synthesis and biomedical applications. *Front. Chem.* **2021**, 9, 626834. https://doi.org/10.3389/fchem.2021.626834.
49. Wu, M.; Wu, W.; Duan, Y.; Li, X.; Qi, G.; Liu, B. Photosensitizer-bacteria biohybrids promote photodynamic cancer cell ablation and intracellular protein delivery. *Chem. Mater.* **2019**, 31(18), 7212–7220. https://doi.org/10.1021/acs.chemmater.9b01518.
50. Shumi, G.; Demissie, T.B.; Eswaramoorthy, R.; Bogale, R.F.; Kenasa, G.; Desalegn, T. Biosynthesis of silver nanoparticles functionalized with histidine and phenylalanine amino acids for potential antioxidant and antibacterial activities. *ACS Omega* **2023**, 8(27), 24371–24386. https://doi.org/10.1021/acsomega.3c01910.
51. Gan, L.; Zhang, S.; Zhang, Y.; He, S.; Tian, Y. Biosynthesis, characterization and antimicrobial activity of silver nanoparticles by a halotolerant *Bacillus Endophyticus* SCU-L. *Prep. Biochem. Biotechnol.* **2018**, 48(7), 582–588. https://doi.org/10.1080/10826068.2018.1476880.
52. Saeed, S.; Iqbal, A.; Ashraf, M.A. Bacterial-mediated synthesis of silver nanoparticles and their significant effect against pathogens. *Environ. Sci. Pollut. Res.* **2020**, 27(30), 37347–37356. https://doi.org/10.1007/s11356-020-07610-0.
53. Jafari, M.; Rokhbakhsh-Zamin, F.; Shakibaie, M.; Moshafi, M.H.; Ameri, A.; Rahimi, H.R.; Forootanfar, H. Cytotoxic and antibacterial activities of biologically synthesized gold nanoparticles assisted by micrococcus yunnanensis strain J2. *Biocatal. Agric. Biotechnol.* **2018**, 15, 245–253. https://doi.org/10.1016/j.bcab.2018.06.014.
54. Li, X.; Xu, H.; Chen, Z.-S.; Chen, G. Biosynthesis of nanoparticles by microorganisms and their applications. *J. Nanomater.* **2011**, 2011, 1–16. https://doi.org/10.1155/2011/270974.
55. Rauf, M.A.; Owais, M.; Rajpoot, R.; Ahmad, F.; Khan, N.; Zubair, S. Biomimetically synthesized ZnO nanoparticles attain potent antibacterial activity against less susceptible S. aureus skin infection in experimental animals. *RSC Adv.* **2017**, 7(58), 36361–36373. https://doi.org/10.1039/C7RA05040B.
56. Pandit, C.; Roy, A.; Ghotekar, S.; Khusro, A.; Islam, M.N.; Emran, T.B.; Lam, S.E.; Khandaker, M.U.; Bradley, D.A. Biological agents for synthesis of nanoparticles and their applications. *J. King Saud Univ. Sci.* **2022**, 34(3), 101869. https://doi.org/10.1016/j.jksus.2022.101869.
57. Camas, M.; Sazak Camas, A.; Kyeremeh, K. Extracellular synthesis and characterization of gold nanoparticles using *Mycobacterium* Sp. BRS2A-AR2 isolated from the aerial roots of the ghanaian mangrove plant, *Rhizophora racemosa. Indian J. Microbiol.* **2018**, 58(2), 214–221. https://doi.org/10.1007/s12088-018-0710-8.
58. Rajabairavi, N.; Raju, C.S.; Karthikeyan, C.; Varutharaju, K.; Nethaji, S.; Hameed, A.S.H.; Shajahan, A. Biosynthesis of novel zinc oxide nanoparticles (ZnO NPs) using endophytic bacteria sphingobacterium thalpophilum. In *Recent Trends in Materials Science and Applications*; Ebenezar, J., Ed.; Springer Proceedings in Physics; Springer International Publishing: Cham, **2017**; Vol. 189, pp. 245–254. https://doi.org/10.1007/978-3-319-44890-9_23.

59. Ghosh, S.; Ahmad, R.; Banerjee, K.; AlAjmi, M.F.; Rahman, S. Mechanistic aspects of microbe-mediated nanoparticle synthesis. *Front. Microbiol.* **2021**, 12, 638068. https://doi.org/10.3389/fmicb.2021.638068.
60. Golmohammadi, H.; Morales-Narváez, E.; Naghdi, T.; Merkoçi, A. Nanocellulose in sensing and biosensing. *Chem. Mater.* **2017**, 29(13), 5426–5446.
61. Kapoor, R.T.; Salvadori, M.R.; Rafatullah, M.; Siddiqui, M.R.; Khan, M.A.; Alshareef, S.A. Exploration of microbial factories for synthesis of nanoparticles – A sustainable approach for bioremediation of environmental contaminants. *Front. Microbiol.* **2021**, 12, 658294. https://doi.org/10.3389/fmicb.2021.658294.
62. Mughal, B.; Zaidi, S.Z.J.; Zhang, X.; Hassan, S.U. Biogenic nanoparticles: Synthesis, characterisation and applications. *Appl. Sci.* **2021**, 11(6), 2598. https://doi.org/10.3390/app11062598.
63. John Wilson, J.; Harimuralikrishnaa, T.; Ponmanickam, P.; Ponseetha Lakshmi, M. Bacterial silver nanoparticles: Method, mechanism of synthesis and application in mosquito control. In *Mosquito Research – Recent Advances in Pathogen Interactions, Immunity, and Vector Control Strategies*; Puerta-Guardo, H., Manrique-Saide, P., Eds.; IntechOpen, **2023**. https://doi.org/10.5772/intechopen.104144.
64. Sandhu, A.; Goel, A. Biosynthesis of nanoparticles by micro-organisms and its applications. *J. Young Pharm.* **2023**, 15(3), 430–440. https://doi.org/10.5530/jyp.2023.15.58.
65. Das, R.K.; Pachapur, V.L.; Lonappan, L.; Naghdi, M.; Pulicharla, R.; Maiti, S.; Cledon, M.; Dalila, L.M.A.; Sarma, S.J.; Brar, S.K. Biological synthesis of metallic nanoparticles: Plants, animals and microbial aspects. *Nanotechnol. Environ. Eng.* **2017**, 2(1), 18. https://doi.org/10.1007/s41204-017-0029-4.
66. Capuzzo, A. Bacterial synthesis of nanoparticles: Current trends in biotechnology and biomedical fields. *Ann. Adv. Biomed. Sci.* **2021**, 4(1), 000161. https://doi.org/10.20944/preprints202103.0358.v1.

4 Utilization of Agro-Industrial Waste as a Green Resource for Production of Nanomaterials

Mriganka Shekhar Borah, Pinku Chandra Nath, Jibanjyoti Panda, Deboja Sharma, Thameridus B. Marak, and Ajita Tiwari

4.1 INTRODUCTION

The swift expansion of the world population has spurred a surge in their requirements for a variety of food and energy resources, leading to exponential growth [1]. This rising demand has helped industrialize and modernize agro and food-related sectors, improving production diversity and quality and quantity. This progress has led to a large generation of agro-industrial food waste, which is commonly dumped in landfills due to the high cost of treating it to reduce the organic load from industry and household sources [2–4]. Globally, the quantity of food waste generated has been documented to reach staggering levels, ranging from 1.3 to 1.4 billion tonnes, with projections indicating a potential increase to 2.6 billion tonnes by 2025 [5]. Reports from the Food and Agriculture Organization (FAO) of the United Nations (UN) have underscored the significant impact of vegetable residues on the carbon footprint and fruit wastes on freshwater resources [6]. Economic analyses have revealed a substantial financial loss associated with food waste, estimated at approximately 400 billion USD, according to the United Nations Environment Programme (UNEP) [7,8]. To address these challenges, it is necessary to use sustainable and eco-friendly methods that focus on optimizing the use of agro-industrial and food waste as valuable resources [9]. These resources have the potential to be converted into bio-products, such as biofuels, organic acids, biopolymers, enzymes, and other valuable items. Instead of being carelessly disposed of in the environment, where they can cause substantial pollution, they can be utilized responsibly [10,11].

In recent decades, there has been a major increase in the popularity of the circular bio-economy concept. This concept focuses on maximizing the use of waste from primary production processes as cost-effective resources for making secondary products [12]. Nevertheless, the diverse and intricate biochemical makeup of food waste frequently presents difficulties in finding a single production method that can fully utilize it. This difficulty has stimulated the advancement of integrated manufacturing methods, enabling the creation of numerous goods by utilizing a focused waste valorization approach [13,14]. The increasing popularity of utilizing agro-industrial food waste for product development requires careful optimization of pre-treatment procedures, fermentation processes, and downstream processing. Furthermore, it is crucial to implement regulatory reforms at the governmental level and increase public knowledge regarding the significance of adopting circular processes to guarantee the long-term viability of biorefinery methods based on food waste [15,16]. This requires a primary change in perception, viewing waste not as a burden but as a valuable bio-resource to be reused, recycled, refurbished, and remanufactured. The complex biochemical composition of food waste makes it highly suitable for producing a diverse range of products through fermentation approaches. However, achieving optimal results often necessitates pre-treatment methods involving thermo-chemical and enzymatic processes before fermentation can

 DOI: 10.1201/9781003487142-4

begin [17,18]. Effective collection and segregation of waste before pre-treatment are crucial steps for optimizing the overall process. Unfortunately, this requirement represents a significant challenge and serves as a major obstacle in the commercialization of waste valorization methods, as it increases labor and capital requirements [19,20]. Moreover, the affordability and adaptability of bio-economy concepts vary from one location to another due to the absence of a standardized policy or strategy for their operation and preservation. This lack of uniformity further complicates the widespread adoption and sustainability of these methods.

Researchers have focused a lot of attention on the many uses of nanotechnology in agriculture, especially nano-fertilizers and nano-pesticides and their functions in managing phytopathogens, regulating the release of agrochemicals, and improving nutrient utilization [21–23]. All sorts of methods fall under the umbrella of these applications, such as using nano-applications to clean up polluted areas [24], managing agricultural waste with nano-adsorbents [25], making supercapacitors out of garlic peels [26], using nanocellulose as an ingredient in pharmaceuticals [27], making nano-silica [28], and turning waste lignin into nano-biochar. The purpose of this chapter is to provide light on the numerous sources of agro-industrial wastes, their classification, and the technique of green synthesis of valuable NMs from agro-industrial waste by utilizing microorganisms inside the waste environment. In addition to that, it also explores the role that microorganisms play in the process of synthesizing NMs from agro-industrial waste.

4.2 SOURCES OF AGRO-INDUSTRIAL WASTE

Presently, there exists a variety of agro-industrial wastes. Figure 4.1 illustrates the different types of waste and their conversion into value-added products.

4.2.1 Fruit and Vegetable Wastes (FVWs)

FVWs can be transformed into bio-fertilizers through anaerobic processes, thereby indirectly mitigating environmental pollution and enhancing soil nutrition [29,30]. The high C/N ratio, caused by the FVW's high carbon content and low nitrogen content, poses a substantial barrier during the bio-processing of 500 g of FVW. This ratio undermines soil protection and prevents soil bacteria from releasing nutrients to crops. This problem was tackled by co-digesting FVW with slaughterhouse wastewater (SWW), another type of eco-waste. Optimal conditions for the use of *Trichoderma reesei* in anaerobic

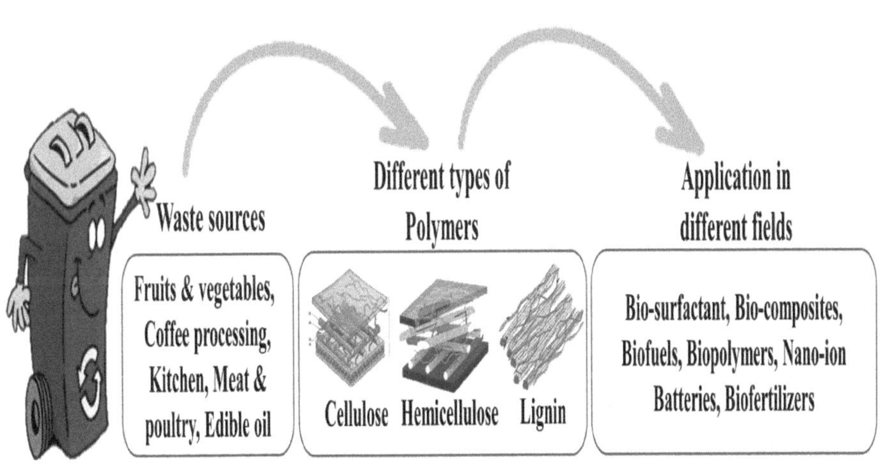

FIGURE 4.1 An illustrative depiction of utilizing agricultural and food waste to generate products with increased economic worth.

digestion and subsequent aerobic digestion improve the production of bio-fertilizers. After the second step, the valuable bio-fertilizer is formed with a reduced C/N ratio from 12 to 10.8, mitigating the impact of FVW. Moreover, FVW, mainly comprising grapes, can serve as bio-adsorbents for metals, facilitating the entire management of waste [31,32]. These results underscore the practical utility of FVW management at both household and industrial levels.

To address the current feed shortages, animal feed made from fruit waste (papaya, pineapple, and orange) and vegetable waste (mostly okra, spinach, and cabbage) can be used in developing nations. The manufacture of animal feed has the potential to tackle waste management head-on and is highly relevant in satisfying the feed needs of the growing number of livestock. In order to determine whether FVWs were suitable as additives to animal feed, they were tested for water capacity, in vitro digestibility, nutritional and microbiological contents, and more [33,34]. Greenhouse gas (GHG) emissions were reduced by 10%–12%, and nutrient utilization was increased in sheep fed a FVW diet (10%). As a result, food wastewater can be effectively recycled and used as a supplement in animal feed, taking the place of more costly ingredients in traditional diets [35]. This helps with food security, promotes safe waste management, and reduces pollution. In addition, by applying the life cycle assessment (LCA) approach, FVW can be organically transformed into an earthworm meal, which can be used as both food and feed [36,37].

4.2.2 Edible-Oil Wastes

It is possible to make toilet (bath) soap from used cooking oil that is thrown from homes and restaurants after frying. Utilizing waste cooking oil (WCO), a more environmentally friendly and secure pest control product has been created for use in agriculture. Inert components in traditional neem emulsifiable concentrate (EC) formulations include organic solvents (60%–70%) and emulsifying agents (8%–10%). Many of these formulations contain volatile organic compounds (VOCs) derived from petroleum, which can permeate plant tissues and negatively impact a range of physiological responses. In order to reduce these adverse impacts, WCO, which has a linolenic acid content of around 3%, can be included as a substitute solvent in the formulation. This method safely enhances insect control by as much as 90% while simultaneously aiding with trash management [38]. Toxic effects on the skin, lungs, and plants (such as phytotoxicity) are not present in WCO-derived formulations. An eco-friendly, biodegradable, see-through thermoplastic biopolymer, poly-(lactic acid) (PLA) is a great choice. Additives such as citrate esters, poly(butyl acrylate), and poly(propylene glycol) have been mixed into the PLA matrix to make it more flexible [39].

Microbial lipids can be produced by co-fermenting WCO [40,41]. The presence of triglycerides and fatty acids in WCO increases the generation of lipids from waste food by employing *Rhodosporidium* species. Nevertheless, an abundance of WCO can impede the synthesis of lipids by disrupting the availability of oxygen. The World Customs Organization is also utilized in the textile industry. It can substitute virgin oil in oil–water systems for reactive dyeing without affecting the quality of dyeing [42,43]. The WCO–water dyeing technique demonstrates superior dye fixing, uniform dye distribution, outstanding color durability, and notably decreased wastewater production. Performing alkaline refinement of WCO after each dyeing cycle can restore its optimal functioning, allowing it to be efficiently reused in reactive dyeing procedures. Moreover, green emulsion liquid membrane (GELM) can be produced using WCO, providing a substitute for conventional organic solvents. Emulsions made from waste oil can be used to create liquid membranes that are capable of extracting and recovering lead (97%) from water solutions [44].

4.2.3 Coffee Processing Wastes (CPWs)

CPWs present an opportunity to reuse and recycle nitrogen-containing carbon at a reasonable cost, because of the massive global production of coffee beans [42]. The nitrogen-doped carbon content of CPW was utilized in the synthesis of a composite electrode material by mixing $NaCrO_2$

powder with CPW and then carbonizing the mixture. By analyzing the nitrogen-doped carbon from CPW on $NaCrO_2$ particles quantitatively and qualitatively, we found that NC-$NaCrO_2$ had better electrochemical performance than bare $NaCrO_2$ in terms of cycle stability and rate capability. Additionally, after a pre-adsorption and post-calcination method, biochar produced from discarded coffee grounds was immobilized with $NiCoO_2$-NPs. The adsorption of target molecules was greatly improved by the highly dispersed active sites on the charcoal substrate [42]. This demonstrates the possibility of using coffee grounds for making absorbent materials. To create urea resin, mixed CPW can be added [42]. The use of renewable coffee waste-activated carbons (CWACs) as anode materials in microbial fuel cells (MFCs) is effective [45]. MFCs that are based on modified CWAC create a power density of 3,928 mW/m^2, which is much higher than the power density of activated carbon that is available for commercial use (976 mW/m^2). The removal of copper, zinc, and lead from soil wastewater can be accomplished with the use of spent coffee grounds (SCG), which are classified as effective adsorbents [46,47]. Additionally, a unique and effective bio-fungicide known as *Trichoderma harzianum*, which is used to combat pathogen assaults on pepper and ginger crops, may be generated in a laboratory-scale fermentation process by employing ten grams of coffee husk [48].

4.2.4 Kitchen-Based Wastes

In order to optimize the utilization of kitchen refuse, a 45 days investigation was conducted into vermicomposting at temperatures ranging from 20°C to 25°C and a moisture content of 60%. Vermicomposting considerably increased phosphorus (31.35%–55.80%) and potassium (33.45%–63.20%) production, according to the research. The growth of earthworms was enhanced by kitchen refuse, whereas an overabundance of waste impeded their development [49]. An additional investigation involved the synthesis of fluorescent carbon dots (CDs) using tea powder, a frequently discarded kitchen waste that is carbon-rich owing to the inclusion of sugar and milk proteins [50]. The water-soluble carbon dots (WTR-CDs) were generated through the controlled pyrolysis of tea residue. These dots demonstrated emission at a wavelength of 415 nm. These particles exhibited remarkable stability in ionic media and remained stable across a broad pH range. In contrast to other prevalent interfering ions, WTR-CDs exhibited exceptional fluorescence quenching owing to the oxidative characteristics of ClO-ions, which conferred high sensitivity and selectivity. In comparison to alternative CDs, they demonstrated a low detection limit, swift response time, high reliability, and excellent accuracy. To bioconvert the organic components in kitchen wastewater into electricity, microorganisms can be used as biocathodes. Two photosynthetic bacteria, *Synechococcus* and *Chlorococcum* sp., act as cathode catalysts in a MFC configuration, which treats wastewater from kitchens in the anode chamber. Under 1,600 lux lighting, *Synechococcus* sp. outperforms *Chlorococcum* sp. in terms of biomass growth rate and power density (42.0 ± 0.6 mW/m^2), according to research. *Synechococcus* sp. is responsible for increasing the concentration of dissolved oxygen, which in turn increases the generation of bioelectricity and, consequently, the power density. Bioelectricity, CO_2 sequestration, and chemical oxygen demand (COD) removal from wastewater are all made possible by biocathodes used in MFCs [51,52].

4.2.5 Meat and Poultry Industry Wastes

Bone, meal, and feathers are three common types of chicken waste that contain both organic and inorganic materials that could be used as polymer fillers. It is possible to add these things to the base polymer polypropylene at up to 40% (wt) for bone and meal and up to 60% (wt) for feathers [53]. Poultry meal has natural parts like proteins and fats, as well as calcium sulfate (with small amounts of sodium), silicon oxide, and magnesium oxide, all of which affect how biodegradable polymers are. A big part of chicken feathers is keratin, which can be used as filler in many different types of polymers. Its high energy and carbon content lowers costs for both the polymer and chicken businesses. Traditional fake wood binders are often made from formaldehyde, which is bad for people's health. Hydrolysates made from chicken feathers, on the other hand (20 g), can be used as bio-adhesives to make wood goods like particle boards. The results of experiments show that chicken feather waste produces 70% keratin hydrolysate protein extracts. This makes it a potentially useful feedstock for making bio-binders for wood composites that are good for the environment [54]. Furthermore, poultry abattoir sludge (PAS) can be used to grow walnuts. Mineral fertilizers and four different amounts of PAS were put side by side. PAS has a lot of nutrients and organic matter, like nitrogen, phosphorus, potassium, and other vitamins that make the soil more fertile. It was found that 400 gN/tree was the best amount of PAS to use. This amount acts as a fertilizer storage and releases nutrients continuously [55].

4.3 GREEN SYNTHESIS OF NMS FROM AGRO-INDUSTRIAL WASTES BY MICROORGANISMS

Microorganisms offer significant promise in synthesizing NPs, offering cost-effectiveness and environmental friendliness by sidestepping harmful chemicals and minimizing energy consumption compared to physiochemical methods. Various reductase enzymes play a crucial role in both accumulating and detoxifying metals, facilitating the reduction of metal salts into NPs with minimal variability in size and distribution [56,57]. Extracellular synthesis has gained significant traction and interest due to its ability to bypass downstream processing steps required in intracellular methods for NP recovery. This method eliminates the need for cell wall breakdown via sonication and centrifugation, along with subsequent washing stages necessary for NP purification. Additionally, various elements such as metal resistance genes, enzymes, peptides, proteins, organic materials, and reducing cofactors are pivotal in green synthesis [58,59]. These components act as reducing agents, aiding in NP synthesis, while also serving as natural capping agents. They prevent aggregation and impart prolonged stability to NPs [60,61]. To achieve sustainable stabilization and surface functionalization of NPs, biocompatible stabilizing agents such as microbes are utilized. These microbes can produce a wide range of stable NPs with various shapes like flower-shaped, spherical, and rod-shaped, offering diverse potential applications. Genetically engineered organisms are particularly advantageous for environmentally friendly bio-reduction in NP synthesis due to their ease of genetic manipulation and resistance to toxicity [62,63]. Metal and metal oxide NPs have been extensively researched for their efficacy against multidrug-resistant bacteria. Numerous studies have highlighted the antimicrobial properties of these NPs against a broad spectrum of bacterial species [64]. Resistance to such bacteria has also been observed in NPs synthesized using *Lactobacillus* sp.

Various types of NPs, including Ag, Au, ZnO, TiO_2, Se, Pd, Pt, and various others, have been synthesized using microbes as both reducing and capping agents [65,66]. These NPs found applications across a wide range of areas are shown in Figure 4.2.

4.4 MECHANISM OF GREEN SYNTHESIS OF NMs

The creation of NPs by bacteria takes place through either intracellular or extracellular mechanisms. During intracellular production, the microbial cell facilitates ion movement, with the cell wall of the bacterium playing a vital function. The electrostatic attraction between the cations of the metal ions and the anions of the cell wall is crucial. Microorganisms' cell walls include enzymes that convert these ions into NPs, which subsequently diffuse out of the cell wall [67]. The process of NP synthesis has three primary stages: entrapment, bio-reduction, and encapsulation [68]. An example of intracellular synthesis of Ag-NPs has been accomplished utilizing *Rhodococcus* sp. The procedure entailed cultivating the microbe aerobically in an M9 medium supplemented with salts for a duration of 24 hours at a temperature of 30°C while subjecting it

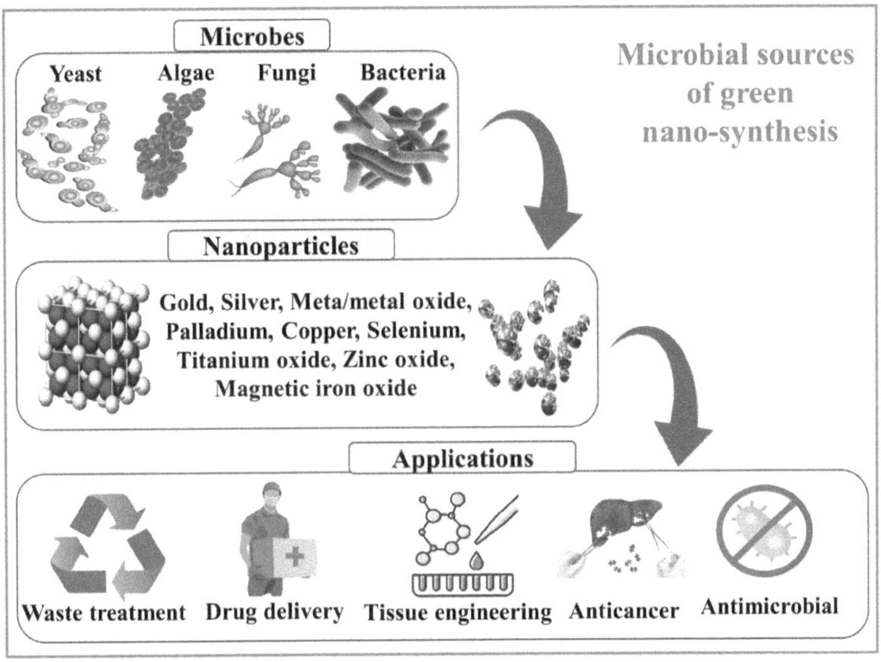

FIGURE 4.2 Microbial synthesis of various metal/metal oxide NPs including Ag (silver), Au (gold), Pd (palladium), Pt (platinum), Cu (copper), Se (selenium), TiO_2 (titanium dioxide), ZnO (zinc oxide), and Fe_2O_3 (iron oxide) NPs and their applications in environmental remediation and biomedical fields.

to agitation at a speed of 130 rpm. After 24 hours, the resulting culture was used as the synthesis medium, combined with silver nitrate ($AgNO_3$) salt at a pH of 7.0. During this period, the color changed from white to brown, indicating the progress of the synthesis process. Incubation was carried out for more than 24 hours, and the samples were collected afterward. Electrostatic interaction ensues when metal ions make contact with the cell surface, resulting in the entrapment of ions. The enzymes within the cellular wall are accountable for the conversion of metal ions into their metallic state [69].

During extracellular synthesis, bacteria are first cultivated in a revolving shaker for a period of 1–2 days under ideal circumstances. Subsequently, the culture is subjected to centrifugation to separate the biomass, and the resulting liquid containing microbial metabolites is gathered for the synthesis of NPs. The filter-sterilized metal salt solution is subsequently introduced into the supernatant, followed by another incubation period [70]. The synthesis progress can be tracked visually through changes in the color of the culture medium. For instance, Au-NPs cause the medium to turn deep purple from ruby red, while Ag-NPs turn it dark brown. Centrifugation of the reaction mixture can eliminate large particles or medium components. Ultimately, NPs can be additionally refined using centrifugation using a density gradient or at a high velocity, subsequently followed by meticulous washing with a mixture of ethanol and water before collection [71]. Shu and colleagues have clarified the process of synthesizing Ag-NPs using yeast extract. The process entails the creation of clusters of yeast particles in the yeast extract by the spontaneous arrangement of biological molecules. This is then followed by the reduction of silver ions (Ag^+) by a procedure that occurs inside the system itself, resulting in the stability of the NPs. The coating of the surface of Ag-NPs enhances their affinity to the bacterial membrane, leading to increased permeability of the cell wall. When Ag-NPs interact with peptidoglycan, it triggers a change in the configuration of peptidoglycan molecules, ultimately causing damage to the bacteria through the apoptosis process [72]. The mechanism of green synthesis of NMs is shown in Figure 4.3.

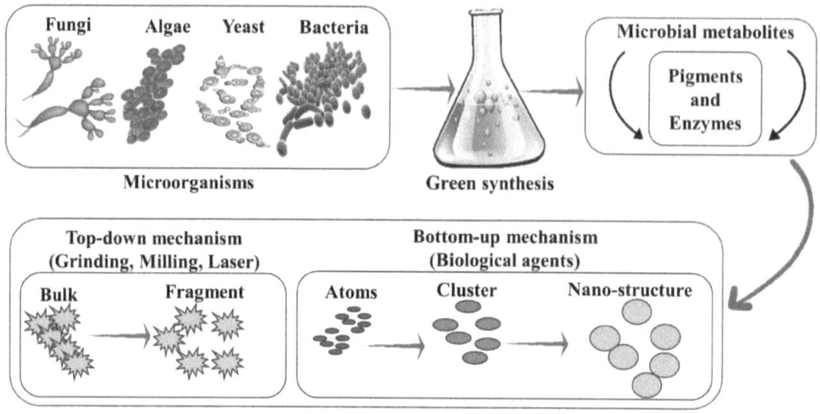

FIGURE 4.3 Mechanism of green synthesis of NMs.

4.5 ROLE OF MICROORGANISMS

4.5.1 Algae

Several authors have extensively studied the environmentally friendly production of different NPs utilizing algae, as outlined in Table 4.1. An investigation was conducted on the green synthesis of Ag-NPs using freshwater algae *Chara vulgaris* [73,74]. The procedure entailed dehydrating and pulverizing the algae to create a 100 ppm solution, which was then combined with $AgNO_3$ in different proportions. The most efficient production of Ag-NPs was accomplished using a ratio of 3:1 of *C. vulgaris* to $AgNO_3$, which yielded NPs with a size of 16.98±0.2 nm, measured using scanning electron microscope (SEM) analysis. These Ag-NPs exhibited significant against bacteria such as *Pseudomonas aeruginosa*, *Klebsiella pneumonia*, and *Escherichia coli*. Additionally, various other algae species have been utilized for NP synthesis, including *Bifurcaria bifurcate*, *Caulerpa peltate*, *Hypnea valencia*, *Sargassum myriocystum*, and *Sargassum muticum* [75].

4.5.2 Fungi

The utilization of fungus for the manufacture of NPs presents numerous benefits in comparison to other microorganisms [101,102]. Fungal *mycelial* mesh has the unique ability to endure agitation, flow pressure, and various conditions within chambers and bioreactors, unlike bacteria and plants. Fungi are readily cultivable, manageable, and amenable to manipulation. They exhibit remarkable metal bioaccumulation ability and strong binding capability, as well as the ability to tolerate diverse circumstances, including intracellular absorption, which is advantageous for research applications [46,103]. Currently, several fungal species are in insist for NPs production, including *Volvariella volvaceae* [104], *Trichothecium harzianum* [105], *Aspergillus fumigatus* [106], *Penicillium brevicompactum* [107], *Aspergillus niger* [108], *Colletotrichum* sp. [109], *Fusarium semitectum* [110], *Phoma glomerata* [111], *Penicillium fellutanum* [112], *Phaenerochaete chrysosporium* [113], *Fusarium oxysporum* [114], and *Cladosporium cladosporioides* [115]. Fungi-mediated biosynthesis can occur via either extracellular or intracellular processes. In the process of extracellular synthesis, the liquid that contains fungal bio-molecules is mixed with a metal precursor, resulting in the creation of NPs in the liquid phase. This technology is preferred as it eliminates the need to release NPs from cells, simplifying the process. On the other hand, intracellular synthesis entails introducing the metal precursor into the fungal *mycelial* culture and subsequently incorporating it inside the biomass. This approach necessitates the isolation of NPs by a range of methods, including centrifugation, chemical treatments, or filtering, to separate NPs from the biomass [116,117].

TABLE 4.1
Green and Sustainable Production of Nanoparticles Utilizing Algae

Nanoparticles	Algal Species	Size (nm)	Morphology	Applications	References
Silver	*Amphiroa rigida*	25	Globular	Anticancer and antibacterial properties	[76]
	Chara vulgaris	16.98±0.2	–	Antibacterial	[77]
	Sargassum myriocystum	18–22	Crystalline	Antibacterial and anticancer activities	[78]
	Padina sp.	25–60	Globular	Antibacterial properties	[79]
	Portieria hornemannii	–	Globular	Antimicrobials for the treatment of infections that affect fish	[80]
	Corymbia ellipsoidea	220.9±31.2	Globular	Breakdown of methyl-orange and antimicrobial	[81]
	Noctiluca scintillans	4.6	Globular	Antimicrobial and anticancer activity	[82]
	Botryococcus braunii	40–90	Globular, cubical, and truncated triangular	The use of catalysts in the production of benzimidazoles	[83]
	Ulva armoricana	12.4	Globular	Antibacterial	[84]
	Laminaria japonica	20–30	Ellipsoidal	Seed treatment and pharmacy	[85]
	Sargassum wightii	18.47–41.58	Globular	Medicinal agent	[86]
	C. vulgaris	40–90	Globular, cubical, and truncated triangular	The catalyst for the synthesis of quinolines	[87]
	Spyridia fusiformis	5–50	Globular and rounded	Antibacterial	[88]
Copper oxide	*Macrocystis pyrifera*	2–50	Globular	–	[89]
Gold	*Gelidiella acerosa*	5.81–117.59	Globular	α-Glucosidase possesses antibacterial and antioxidant properties	[90]
	Sargassum crassifolium	5–300	Varied	Pharmaceuticals	[91]
	Cystoseira baccata	8.4±2.2	Globular	Cancer-fighting agents	[92]
	Sargassum cymosum	7–20	Globular	Commercial application	[93]
Zinc oxide	*Cladophora glomerata*	14.38–37.87	Globular	Pharmaceuticals	[94]
	Anabaena cylindrica	40–60	Rod	Antimicrobial activity	[95]
	Tetraselmis indica	20–40	Globular	Cosmetics, textiles, biomedicine, and food industries	[96]
Palladium	*C. vulgaris*	15	Globular	–	[97]
	Padina boryana	11.15	Crystal	Nanomedicine for combating bacterial infections and cancer	[98]
Copper and silver	*B. braunii*	Cu (10–70) and Ag (40–100)	Truncated, cuboidal, and spheri-cubical in shape	Antimicrobial	[99]
Cadmium sulfide	*Chlamydomonas reinhardtii*	2–7	Globular	Organic dyes that are photodegradable	[100]

4.5.3 Bacteria

Bacteria are highly effective candidates for converting metal ions to NPs due to their exceptional capabilities, which are helped by their rapid growth rate and ease of manipulation. Bacteria exhibit a higher degree of genetic manipulability for metal ion bio-mineralization in comparison to other microorganisms [118]. Bacteria frequently reside in environments characterized by elevated levels of heavy metal ions, which are known to be poisonous and harsh. To endure such harsh environments, bacteria have evolved innate defense mechanisms such as external precipitation, internal sequestration, control of metal ion levels, and efflux pumps. These methods allow bacteria to flourish under challenging conditions and have been successfully utilized for the creation of NPs.

When specific bacteria, like *Sulfolobus acidocaldarius*, *Acidithiobacillus thiooxidans*, and *Thiobacillus ferrooxidans*, are cultivated using elemental sulfur as a source of energy, they possess the potential to convert ferric ions into the ferrous form. *T. thiooxidans* is capable of reducing ferrous ions under aerobic circumstances with low pH. However, it cannot re-oxidize these ions, resulting in their stability and preventing auto-oxidation. Unlike the aerobic process of bacterial re-oxidation of ferrous ions, the bio-reduction of ferric ions by *T. ferrooxidans* does not require the presence of oxygen [119]. Various bacterial species are employed in the manufacture of NPs, including *Pseudomonas rhodesiae*, *Bacillus megaterium*, *Actinomycetes* sp., *Bacillus haynesii*, and *Streptomyces* sp. [120,121]. Table 4.2 contains further information on the manufacture of NPs by different bacteria, including their properties and applications.

TABLE 4.2
Green and Sustainable Production of Nanoparticles Utilizing Bacteria

Nanoparticles	Bacteria Employed	Size (nm)	Morphology	Applications	References
Zirconium oxide	*Enterobacter* sp.	33–75	Globular	Antifungal action against twig blight disease	[122]
Silver	*Bacillus subtilis*	2–26	Globular	Antibacterial properties	[123]
Titanium oxide	*Achrmobacter* sp.	5–10	Improper	Antimicrobial activity	[124]
Silver	*Dietzia maris*	40–50	Globular	Wound healing properties	[125]
Silver	*Pseudomonas* sp. and *Enterobacter*	63.50 and 45.81		Antibacterial activity	[126]
Silver	*Bacillus amyloliquefaciens* (MH046937)	1.23–10.80	Globular	Promising antimicrobial effects against plant disease agents	[127]
Silver	*Pseudoduganella eburnea* (MAHUQ-39)	8–24	Globular	Antimicrobial agents	[128]
Gold	*Bacillus licheniformis*	40	Improper	Antimicrobial activity	[129]
Silver	*Pseudomonas rhodesiae*	20–100	Globular	Antibacterial properties	[130]
Zinc oxide	*Bacillus haynesii*	20–100	Globular	Environmental and health care industries	[131]
Silver	*Nostoc linckia*	5–60	Globular	Antimicrobial activity	[132]
Zinc oxide	*Bacillus megaterium* (NCIM2326)	45–95	Rod and cube shapes	Antibacterial and therapeutic agents	[133]
Gold	*Streptomyces griseoruber*	5–50	Globular and hexagonal	Degradation of methylene blue	[134]
Silver	*Streptomyces* genus	160	Globular	Nanomedicine and cosmetology industries	[135]
Silver	*Actinomycetes* sp.	–	–	Antibacterial activity	[136]
Silver	*Streptomyces* sp.	2.3–85	Globular	Antimicrobial activity	[137]

4.5.4 Yeast

Yeast extract has a double function as both a reducing and capping agent. It contains carbohydrates, vitamins, and amino acids. Meanwhile, metal ions serve as electron acceptors. The organic capping agents provide stability to the produced monodispersed NPs, allowing them to be stored for more than a year without forming precipitates [138]. Yeast production can be readily regulated under controlled laboratory conditions. The exponential proliferation of yeast strains and their uncomplicated nutritional demands provide numerous benefits for the generation of NPs. For example, researchers have created TiO_2-NPs by utilizing *Lactobacillus* sp. and *Saccharomyces cerevisae* yeast species. The pH and redox potential of the culture fluid are important factors in the synthesis process [139]. Several researchers have investigated the synthesis of NPs utilizing yeast extracts [140–142].

4.6 NMS PRODUCED USING MICROORGANISMS

4.6.1 Silver Nanoparticles (Ag-NPs)

Silver and silver salts were frequently employed in ancient human civilizations. Nevertheless, in recent years, there has been an increasing need for the production of Ag-NPs because of their exceptional use in the biological domain. These applications encompass the ability to combat fungal and bacterial infections, protect against oxidative damage, and contribute to improving agricultural practices and environmental cleanup efforts. Ag-NPs are widely studied in the biomedical field and are considered important compared to other metals. They have been extensively researched, especially using environmentally friendly production methods [143]. Gevorgyan et al. examine the effectiveness of Ag-NPs as strong inhibitors of both gram-positive *Staphylococcus aureus* and gram-negative *Salmonella typhimurium* [144].

Kashyap et al. used a method called intracellular green synthesis, which involved using a type of microalgae called *Scenedesmus* sp., to create nanohybrids made of Ag/AgCl. They employed quantities of 0.5 and 1 mM of $AgNO_3$ precursor, in addition to the extract of *Scenedesmus* sp., as a reducing agent for the synthesis [145]. Spherical NPs with dimensions ranging from 10–20 nm to 10–50 nm were obtained, respectively. The mixture of $AgNO_3$ and bacterial strain solution transformed transparency to a deep brown color, indicating the creation of hexagonal-shaped Ag-NPs by the use of *B. anthracis* (PAFB2). The NPs had a zeta potential (ZP) of –15.5 mV and a polydispersity index (PDI) of 0.428, demonstrating their favorable colloidal properties and long-term stability. The NPs had a size of around 84 nm, as determined by atomic force microscope (AFM) analysis [146]. In addition, spherical NPs with sizes ranging from 13 to 27 nm were created utilizing the *Paenarthrobacter nicotinovorans* (MAHUQ-43) bacterial strain [147].

4.6.2 Gold Nanoparticles (Au-NPs)

The process of creating Au-NPs involves combining a 10 mL extract of *Gelidiella acerosa* marine algae with a 90 mL aqueous solution of $HAuCl_4{\cdot}3H_2O$ (gold chloride) at a concentration of 1 mM. The mixture was then kept at a constant temperature of 37°C without any movement. The Au-NPs obtained were isolated using a process of washing and centrifugation at a speed of 10,000 rpm for a duration of 15 minutes. Subsequently, they were dried at a temperature of 50°C and stored at 4°C for subsequent analysis and utilization. The crystalline NPs had a size range of 5.81–117.59 nm, with both spherical and hexagonal forms detected. Au-NPs exhibited significant inhibition against α-glucosidase and α-amylase enzymes, with IC50 values of 2.8 ± 0.02 and 4.1 ± 0.01 µg/mL, respectively [90]. The Au-NPs showed remarkable efficacy in suppressing the growth of colon cancer cells, as evidenced by their dose-dependent inhibition. The synthesis of these NPs was undertaken as a chemotherapeutic option to reduce the toxicity and adverse effects commonly associated with traditional medications. The IC50 value of the NPs was determined to be 15 µg/mL, indicating their efficacy compared to the standard treatment. Morphological analysis revealed cytotoxic influences on the human colon cancer cell line (HCA-7), with distinct damage observed in treated cells compared to untreated cells. The biogenic Au-NPs induced significant cell damage, leading to the release of cell debris and suggesting their potential for combating human colon cancer [148]. Furthermore, Au-NPs synthesized with *Ecklonia cava*, a marine alga, exhibited antimicrobial properties against various pathogenic organisms. The zone of inhibition for different microbial strains ranged from 16.6 to 31.8 mm, with the highest activity observed against *A. niger* (ATCC-1015) and *E. coli* (ATCC-10,536). Additionally, *Sargassum incisifolium*–assisted Au-NPs were investigated for their anticancer and antimicrobial properties [149].

4.6.3 Copper Nanoparticles (Cu-NPs)

The process of synthesizing Cu-NPs using *Escherichia* sp. involved cultivating and incubating the bacteria in nutritional broth for 24 hours at a speed of 150 rpm. Subsequently, a solution of $CuSO_4$ with a concentration of 5 mM was introduced, and the incubation process was prolonged. An evident shift in color from bluish-green to a more concentrated green hue was detected, providing phenotypic evidence of Cu-NP synthesis. A peak at 325.89 nm was discovered in the ultraviolet (UV) region, providing additional evidence of the production of NPs [150]. Cu-NPs were produced by utilizing *S. loihica* (PV-4) through the process of extracellular bio-reduction of Cu(II). The transmission electron microscopy (TEM) examination determined that the size of the NPs was between 10 and 16 nm. Additionally, the energy-dispersive X-ray (EDX) analysis detected strong copper signals, confirming the synthesis process [151].

Textile wastewater is recognized for its presence of diverse pollutants including hardness, elevated turbidity, pH fluctuations, Total Suspended Solids (TSS), Total Dissolved Solids (TDS), COD, sulfates, chlorides, and other impurities, which have harmful impacts on aquatic species. The use of biogenic Cu-NPs derived from *Escherichia* sp., as shown in [150], showed significant efficacy in removing color from azo dyes (25 mg/L). The decolorization rates achieved were 83.62% ± 1.94%, 88.43% ± 2.87%, 90.54% ± 2.05%, and 97.06% ± 1.23% for RB-5, DB-1, MG, and CR, correspondingly, within 5 hours of sunshine exposure. Copper nanoparticles (Cu-NPs) produced by *Streptomyces* sp., a type of endophytic actinomycetes, have shown promise in a range of biotechnological applications [152]. Cu-NPs supported by *A. niger* (STA9) exhibited antidiabetic, antibacterial, and anticancer properties [153].

4.6.4 TiO_2 Nanoparticles (TiO_2-NPs)

The enhanced photocatalytic activities, along with the excellent antioxidant properties of TiO_2-NPs, have led to their considerable study and use [154]. Implant biomaterials, sunscreen items, self-cleaning sanitary ceramics, cement, sugar, newspaper, plastic, healthcare ceramics, printer ink, colorants, and antimicrobial plastic packaging films are just a few of the many numerous sectors that find extensive use for them [155].

The use of *Streptomyces* sp. (HC1) in the synthesis of TiO_2-NPs was studied by Ağçeli and colleagues [156]. The bacterial colony was grown in 50 mL of sterile nutrient broth and incubated in a shaker at 150 rpm at 37°C for 24 hours. Once turbidity was seen, the bacterial culture was introduced to a $TiO(OH)_2$ solution (20 mL, 0.025 M). The mixture was then incubated in a steam bath for 30 minutes at 60°C. Centrifugation was used to collect the white clusters that had formed at the base of the flask after incubation; distilled water was used to wash the precipitate such that its pH remained neutral [156]. Another work used *Bacillus* sp. bacteria to produce spherical TiO_2-NPs with a size of 50 nm. The Taguchi method, which optimizes process parameters to maximize production dependability, was used to synthesize these NPs that were mediated by bacteria. TEM and SEM micrographs verified that the NPs that were obtained had extremely dense spherical forms. Weight loss up to 670°C was shown by thermogravimetric analysis (TGA). The first loss, below

150°C, was due to water evaporation. The second loss, between 250°C and 670°C, was due to the disintegration of TiO_2-NPs and organic compounds [157].

TiO_2-NPs are perfect for use in biomedical applications and the pharmaceutical industry because to their non-toxicity, biocompatibility, and outstanding contributions to photo-electrochemical energy production [158]. The inhibitory effects of TiO_2-NPs produced by *Aspergillus flavus* were seen in a range of bacterial species. For example, *K. pneumonia* (18 mm), *P. aeruginosa* (27 mm), *S. aureus* (25 mm), *Bacillus subtilis* (22 mm), and *E. coli* (35 mm) all had their own zone of inhibition. The tetracycline antibiotics were used as a control in the tests, and the cell diffusion method and minimum inhibitory concentration (MIC) assays were used to evaluate the zone of inhibition. The presence of inhibitory zones against gram-positive and gram-negative bacteria should not be overlooked. The MIC values for *E. coli* (MTCC-1721), *S. aureus* (MTCC-3160), *B. subtilis* (MTCC-1427), *K. pneumoniae* (MTCC-4030), and *P. aeruginosa* (MTCC-1034) were found to be 40, 45, 70, and 80 μg/mL, respectively. The most effective result was noted against *E. coli* [159]. Additionally, TiO_2 has demonstrated antibacterial activity against *B. megaterium* [160] and *E. coli* [161] under environmental light.

4.6.5 Iron Oxide Nanoparticles (Fe_2O_3-NPs)

For the synthesis of Fe_2O_3-NPs using three fungal strains – *Fusarium incarnatum*, *Phialemoniopsis ocularis*, and *Trichoderma asperellum* – initially, fungal cell filtrate (10 mL) from each fungal strain was mixed with a salt solution containing $FeCl_3$ and $FeCl_2$ in a 2:1 mM ratio. The mixture was agitated at room temperature for 5 minutes, and the change in color of the reaction mixture indicated the formation of NPs with the selected fungal strain. Subsequently, the synthesized Fe_2O_3-NPs were centrifuged for 20 minutes at 12,000 rpm and washed thoroughly with de-ionized water. The entire synthesis process is described in the flowchart provided by [162].

In another method, the synthesis of Fe_2O_3-NPs was carried out using an aqueous extract of *Aegle marmelos*. Initially, 5 g of *A. marmelos* extract was dissolved in 100 mL of distilled water and stirred with heating for 1 hour. The resulting extract was then filtered using Whatman filter paper. Subsequently, 10 mL of this filtered extract was mixed with 90 mL of ferric nitrate solution and stirred for 1 hour. The mixture was then heated in a hot air oven at 150°C. Finally, the obtained powder was calcinated for 5 hours at 400°C to produce Fe_2O_3-NPs [163]. Furthermore, these Fe_2O_3-NPs exhibited remarkable photocatalytic activity for the degradation of Brilliant Green dye, achieving a degradation efficiency of 95.89% in 90 minutes under UV light. The degradation efficiency and kinetics followed a pseudo-first-order kinetic model with a constant (K) value of 0.04058/minute [163].

4.7 ADVANTAGES AND DISADVANTAGES OF NM SYNTHESIS USING MICROORGANISMS

Synthesizing NPs using microorganisms offers several advantages and disadvantages, which are summarized in the following.

4.7.1 Advantages

a. **Eco-friendly:** Microorganism-mediated synthesis typically occurs under mild conditions without the need for harsh chemicals, making it environmentally friendly.
b. **Cost-Effective:** Microorganisms are readily available and can be cultured inexpensively in laboratory settings, reducing production costs.
c. **Biocompatibility:** NPs synthesized using microorganisms often exhibit biocompatibility, making them suitable for biomedical applications such as drug delivery and tissue engineering.

d. **Controlled Synthesis:** Microorganisms provide a natural mechanism for controlling the synthesis process, leading to uniform particle size and shape distribution.
e. **Versatility:** Microorganisms can synthesize a wide range of NPs with diverse properties, enabling customization for specific applications.
f. **Scalability:** Microbial synthesis processes can be easily scaled up for large-scale production, making them suitable for industrial applications.

4.7.2 Disadvantages

a. **Process Variability:** Microbial synthesis processes may exhibit variability due to factors such as changes in culture conditions, leading to inconsistencies in NP properties.
b. **Contamination Risk:** Microbial cultures are susceptible to contamination by other microorganisms, which can affect the purity and quality of synthesized NPs.
c. **Limited Control:** Despite efforts to optimize culture conditions, microorganisms may produce NPs with unpredictable properties, limiting control over the synthesis process.
d. **Longer Synthesis Times:** Microorganism-mediated synthesis processes may take longer compared to chemical methods, leading to slower production rates.
e. **Scale-Up Challenges:** Scaling up microbial synthesis processes for industrial production may present challenges related to maintaining culture purity, optimizing growth conditions, and ensuring consistent NP quality.
f. **Regulatory Considerations:** Regulatory approval for biomedical applications of microorganism-synthesized NPs may be more complex due to concerns about safety, quality control, and standardization.

Overall, while microorganism-mediated synthesis offers numerous advantages such as eco-friendliness, biocompatibility, and versatility, careful consideration of potential disadvantages is necessary to optimize the process for specific applications and ensure the production of high-quality NPs.

4.8 CONCLUSIONS AND FUTURE TRENDS

In recent times, there has been significant attention toward sustainable or green synthesis methods, which hold economic importance by offering a clean, straightforward, effective, non-toxic, and environmentally friendly approach to producing NPs. Utilizing microbes in the synthesis of metal/metal oxide NPs has shown promise, with their well-ordered structures proving beneficial for various applications in waste treatment, as well as biological and therapeutic fields. Microbes offer advantages such as biocompatibility, controlled morphology, and other useful properties when in nano-form, making them suitable candidates for NP synthesis. Key features of microbial cells, including easy maintenance, rapid growth, and safety in use, contribute to their effectiveness as reducing and capping agents in this process. Green synthesis addresses the limitations of conventional synthetic methods by coating NPs with bio-molecules or lipid layers, enhancing their physiological solubility, processability, and stability. This also allows for surface functionalization, making the NPs well-suited for biomedical applications.

REFERENCES

1. Choe, C., et al., Critical aspect of renewable syngas production for power-to-fuel via solid oxide electrolysis: Integrative assessment for potential renewable energy source. *Renewable and Sustainable Energy Reviews* 2022. **161**: p. 112398.
2. Awasthi, M.K., et al., Recent trends and developments on integrated biochemical conversion process for valorization of dairy waste to value added bioproducts: A review. *Bioresource Technology* 2022. **344**: p. 126193.

3. Panda, J., et al., Exploring biopolymer for food and pharmaceuticals application in the circular bioeconomy: An agro-food waste-to-wealth approach. *Waste and Biomass Valorization* 2024. **15**: p. 5607.
4. Ramandani, A.A., et al., Upcycling nutrients derived from food waste via microalgae cultivation: A review on impacts on cellular compounds, economy and environment analyses for achieving circular bioeconomy. *Biochemical Engineering Journal* 2024. **211**: p. 109454.
5. Sinha, S. and P. Tripathi, Trends and challenges in valorisation of food waste in developing economies: A case study of India. *Case Studies in Chemical and Environmental Engineering* 2021. **4**: p. 100162.
6. Ben-Othman, S., I. Jõudu, and R.J.M. Bhat, Bioactives from agri-food wastes: Present insights and future challenges. *Molecules* 2020. **25**(3): p. 510.
7. Theagarajan, R., et al., Valorisation of grape pomace (cv. Muscat) for development of functional cookies. *International Journal of Food Science & Technology* 2019. **54**(4): pp. 1299–1305.
8. Rawindran, H., et al., Bio-refinery of palm kernel expeller waste into attached microalgal biodiesel, the life cycle analysis of waste biomass-to-energy. *Process Safety and Environmental Protection* 2024. **189**: p. 134.
9. Poonia, K., et al., Sustainability, performance, and production perspectives of waste-derived functional carbon nanomaterials towards a sustainable environment: A review. *Chemosphere* 2024. **352**: p. 141419.
10. Ashokkumar, V., et al., Advanced technologies on the sustainable approaches for conversion of organic waste to valuable bioproducts: Emerging circular bioeconomy perspective. *Fuel* 2022. **324**: p. 124313.
11. Nath, P.C., et al., Biogeneration of valuable nanomaterials from agro-wastes: A comprehensive review. *Agronomy* 2023. **13**(2): p. 561.
12. Chakraborty, A., et al., Jackfruit waste for the production of sustainable biorefinery products: A step towards bio-economic advancement. *Biomass Conversion and Biorefinery* 2024. https://doi.org/10.1007/s13399-024-05954-y.
13. Zabaniotou, A. and P. Kamaterou, Food waste valorization advocating circular bioeconomy – A critical review of potentialities and perspectives of spent coffee grounds biorefinery. *Journal of Cleaner Production* 2019. **211**: pp. 1553–1566.
14. Htoo, N.Y.M., et al., Advancements in valorization of food waste resources by microalgae for sustainable biorefinery production. *BioEnergy Research* 2024. **17**: pp. 1–20.
15. Nath, P.C., et al., Fundamentals of enzyme-based biorefinery for conversion of waste to value-added products, in *Enzymes in the Valorization of Waste*. 2022, CRC Press. pp. 1–24. https://doi.org/10.1201/9781003187721-1.
16. Upadhyay, S.K., et al., Transforming bio-waste into value-added products mediated microbes for enhancing soil health and crop production: Perspective views on circular economy. *Environmental Technology & Innovation* 2024. **34**: p. 103573.
17. Wagh, M.S., et al., Valorisation of agro-industrial wastes: Circular bioeconomy and biorefinery process–a sustainable symphony. *Process Safety and Environmental Protection* 2024. **183**: p. 708.
18. Ray, M.K., et al., Mycoremediation of heavy metals and/or metalloids in soil, in *Land Remediation and Management: Bioengineering Strategies*, H. Sarma and S. Joshi, Editors. 2023, Springer Nature: Singapore. pp. 161–190.
19. Ghotekar, S., et al., A bio-waste-mediated novel fabrication of $ZnCr_2O_4$ nanoparticles: A heterogeneous catalyst in Biginelli reaction. *Research on Chemical Intermediates* 2024. **50**(5): pp. 2199–2219.
20. Singh, K., et al., Waste water treatment using piezoelectric materials: A review on piezo-photocatalysis. *Topics in Catalysis* 2024. **67**: pp. 1–28.
21. Pramanik, P., et al., *Application of Nanotechnology in Agriculture*. 2020. pp. 317–348. https://doi.org/10.5772/intechopen.88390.
22. Mohanta, Y.K., et al., Promising applications of phyto-fabricated silver nanoparticles: Recent trends in biomedicine. *Biochemical and Biophysical Research Communications* 2023. **688**: p. 149126.
23. Ray, M.K., et al. Nanotechnology as a promising tool against phytopathogens: A futuristic approach to agriculture. *Agriculture* 2023. **13**: p. 1856. https://doi.org/10.3390/agriculture13091856.
24. Nath, P.C., et al., Recent advances in valorization of pineapple (*Ananas comosus*) processing waste by-products: A step towards circular bioeconomy. *Trends in Food Science & Technology* 2023. **136**, 100–111.
25. Sharma, R., et al., Recent advances in 3-D printing properties of natural food gels: Application of innovative food additives. *Food Chemistry* 2023. **432**: p. 137196.
26. Patra, D.C., N. Deka, and P.C. Nath, Ceramic multilayers and films for high-performance supercapacitors, in *Electroceramics for High Performance Supercapacitors*. 2023,Wiley. pp. 73–94. https://doi.org/10.1002/9781394167166.ch4.

27. Ahmed, S., et al., Development of pullulan/chitosan/salvianolic acid ternary fibrous membranes and their potential for chemotherapeutic applications. *International Journal of Biological Macromolecules* 2023. **250**: p. 126187.
28. Nath, P.C., et al., Recent advances in production of sustainable and biodegradable polymers from agro-food waste: Applications in tissue engineering and regenerative medicines. *International Journal of Biological Macromolecules* 2024. **259**: p. 129129.
29. Chakravarty, I. and S.A. Mandavgane, Valorization of fruit and vegetable waste for biofertilizer and biogas. *Journal of Food Process Engineering* 2021. **44**(2): p. e13512.
30. Lee, A., et al., Upcycling fruit waste into microalgae biotechnology: Perspective views and way forward. *Food Chemistry: Molecular Sciences* 2024. **8**: p. 100203.
31. Gupta, N., et al., Fruit waste management by pigment production and utilization of residual as bioadsorbent. *Journal of Environmental Management* 2019. **244**: pp. 138–143.
32. Ardo, F.M., et al., Modelling photoperiod in enhancing hydrogen production from *Chlorella vulgaris* sp. while bioremediating ammonium and organic pollutants in municipal wastewater. *Environmental Pollution* 2024. **346**: p. 123648.
33. Sahoo, A., et al., Utilization of fruit and vegetable waste as an alternative feed resource for sustainable and eco-friendly sheep farming. *Waste Management* 2021. **128**: pp. 232–242.
34. Sharma, R., et al., Rapid and sensitive approaches for detecting food fraud: A review on prospects and challenges. *Food Chemistry* 2024. **454**: p. 139817.
35. François, M., et al., Utilization of carbon-based nanomaterials for wastewater treatment and biogas enhancement: A state-of-the-art review. *Chemosphere* 2023. **350**: p. 141008.
36. Tedesco, D.E., et al., Bioconversion of fruit and vegetable waste into earthworms as a new protein source: The environmental impact of earthworm meal production. *Science of the Total Environment* 2019. **683**: pp. 690–698.
37. Sharma, R., et al., Microbial exopolysaccharides: Production, properties, and food applications, in *Food Process Engineering and Technology: Safety, Packaging, Nanotechnologies and Human Health*. 2024, Springer. pp. 441–469. https://doi.org/10.1007/978-981-99-6831-2_23.
38. Iqbal, N., et al., Successful utilization of waste cooking oil in Neem oil based fungicide formulation as an economic and eco-friendly green solvent for sustainable waste management. *Journal of Cleaner Production* 2021. **288**: p. 125631.
39. Thomas, J., V. Singh, and R. Jain, Synthesis and characterization of solvent free acrylic copolymer for polyurethane coatings. *Progress in Organic Coatings* 2020. **145**: p. 105677.
40. Nath, P.C., et al., Recent advances in artificial intelligence towards the sustainable future of agri-food industry. *Food Chemistry* 2024. **447**: p. 138945.
41. Gao, Z., et al., Waste cooking oil used as carbon source for microbial lipid production: Promoter or inhibitor. *Environmental Research* 2022. **203**: p. 111881.
42. Liu, L., et al., Clean cotton dyeing in circulated dyebath of waste cooking oil: A feasible industrialization strategy for pollution minimization. *Journal of Cleaner Production* 2021. **278**: p. 123799.
43. Lakshmayya, N.S.V., et al., Essential oils-based nano-emulsion system for food safety and preservation: Current status and future prospects. *Biocatalysis and Agricultural Biotechnology* 2023. **53**: p. 102897.
44. Sharma, R., et al., Exopolysaccharides for bioremediation, in *Microbial Exopolysaccharides*. CRC Press. pp. 275–303. https://doi.org/10.1201/9781003342687.
45. Hung, Y.-H., et al., Renewable coffee waste-derived porous carbons as anode materials for high-performance sustainable microbial fuel cells. *ACS Sustainable Chemistry & Engineering* 2019. **7**(20): pp. 16991–16999.
46. Sharma, R., et al., Recent advances in cellulose-based sustainable materials for wastewater treatment: An overview. *International Journal of Biological Macromolecules* 2023. **256**: p. 128517.
47. Futalan, C.M., et al., Adsorptive treatment via simultaneous removal of copper, lead and zinc from soil washing wastewater using spent coffee grounds. *Water Science & Technology* 2019. **79**(6): pp. 1029–1041.
48. Das, M.M., et al., Production of bio-fungicide, *Trichoderma harzianum* CH1 under solid-state fermentation using coffee husk. *Bioresource Technology Reports* 2021. **15**: p. 100708.
49. Zhi-Wei, S., et al., Investigation of rice straw and kitchen waste degradation through vermicomposting. *Journal of Environmental Management* 2019. **243**: pp. 269–272.
50. Gunjal, D.B., et al., Sustainable carbon nanodots synthesised from kitchen derived waste tea residue for highly selective fluorimetric recognition of free chlorine in acidic water: A waste utilization approach. *Journal of the Taiwan Institute of Chemical Engineers* 2019. **95**: pp. 147–154.

51. Mohamed, S.N., et al., Bioelectricity production from kitchen wastewater using microbial fuel cell with photosynthetic algal cathode. *Bioresource Technology* 2020. **295**: p. 122226.
52. Sharma, R., et al., Production of *Oscillatoria* sp. BTA-170 biomass in photobioreactor: Analysis of composition, drying behavior, sorption isotherm, and powder flow characteristics. *Journal of Food Process Engineering* 2022. **45**(10): p. e14044.
53. McGauran, T., et al., Feasibility of the use of poultry waste as polymer additives and implications for energy, cost and carbon. *Journal of Cleaner Production* 2021. **291**: p. 125948.
54. Fagbemi, O.D. and B. Sithole, Evaluation of waste chicken feather protein hydrolysate as a bio-based binder for particleboard production. *Current Research in Green and Sustainable Chemistry* 2021. **4**: p. 100168.
55. Yang, H., et al., Growth and physiological characteristics of four blueberry cultivars under different high soil pH treatments. *Environmental and Experimental Botany* 2022. **197**: p. 104842.
56. Roy, B., et al., Water pollution by heavy metals and their impact on human health, in *Handbook of Water Pollution*. 2024, Wiley. pp. 333–352. https://doi.org/10.1002/9781119904991.ch10.
57. Roy, B., et al., Application of genetically modified microorganisms for effective removal of heavy metals, in *Genetically Engineered Organisms in Bioremediation*, Inamuddin, C.O. Adetunji, M.I. Ahamed, T. Altalhi, Editors. 2024, CRC Press: Boca Raton, FL. pp. 136–150.
58. Nath, P.C., et al., Smart packaging for commercial food products, in *Green Sustainable Process for Chemical and Environmental Engineering and Science*. 2023, Elsevier. pp. 197–209. https://doi.org/10.1016/B978-0-323-95644-4.00016-4.
59. Ng, Z.Y., et al., Conversion of lignocellulosic biomass waste into mycoprotein: Current status and future directions for sustainable protein production. *Biomass Conversion and Biorefinery* 2024: pp. 1–27. https://doi.org/10.1007/s13399-024-06130-y.
60. Rana, A., K. Yadav, and S. Jagadevan, A comprehensive review on green synthesis of nature-inspired metal nanoparticles: Mechanism, application and toxicity. *Journal of Cleaner Production* 2020. **272**: p. 122880.
61. Messaoudi, O. and M. Bendahou, Biological synthesis of nanoparticles using endophytic microorganisms: Current development, in *Nanotechnology and the Environment*. 2020, InTech Open. pp. 1–19. https://doi.org/10.5772/intechopen.93734.
62. Nath, P.C., et al., Smart packaging to preserve fruit quality, in *Green Sustainable Process for Chemical and Environmental Engineering and Science*. 2023, Elsevier. pp. 267–281. https://doi.org/10.1016/B978-0-323-95644-4.00002-4.
63. Iravani, S. and R.S. Varma, Biofactories: Engineered nanoparticles via genetically engineered organisms. *Green Chemistry* 2019. **21**(17): pp. 4583–4603.
64. Niño-Martínez, N., et al., Molecular mechanisms of bacterial resistance to metal and metal oxide nanoparticles. *International Journal of Molecular Sciences* 2019. **20**(11): p. 2808.
65. Kour, D., et al., Microbial nanotechnology for agriculture, food, and environmental sustainability: Current status and future perspective. *Folia Microbiologica* 2024. **69**: pp. 1–30.
66. Kaushal, P., et al., Nano-green: Harnessing the potential of plant extracts for sustainable antimicrobial metallic nanoparticles. *Journal of Drug Delivery Science and Technology* 2024. **94**: p. 105488.
67. Zahera, M., K. Mishra, and F. Fatima, Exploiting the bioprospecting potential of nanoparticles with micro-organisms. In *The Nanotechnology Driven Agriculture*. Roy, S., Hossain, A., Eds 2024, CRC Press: Boca Raton, FL. pp. 245–263.
68. Fariq, A., T. Khan, and A. Yasmin, Microbial synthesis of nanoparticles and their potential applications in biomedicine. *Journal of Applied Biomedicine* 2017. **15**(4): pp. 241–248.
69. Otari, S., et al., Intracellular synthesis of silver nanoparticle by actinobacteria and its antimicrobial activity. *Spectrochimica Acta Part A: Molecular and Biomolecular Spectroscopy* 2015. **136**: pp. 1175–1180.
70. Ammar, H.A., et al., Extracellular myco-synthesis of nano-silver using the fermentable yeasts Pichia kudriavzevii HA-NY_2 and Saccharomyces uvarum HA-NY_3, and their effective biomedical applications. *Bioprocess and Biosystems Engineering* 2021. **44**: pp. 841–854.
71. Singh, P., et al., Biological synthesis of nanoparticles from plants and microorganisms. *Trends in Biotechnology* 2016. **34**(7): pp. 588–599.
72. Shu, M., et al., Biosynthesis and antibacterial activity of silver nanoparticles using yeast extract as reducing and capping agents. *Nanoscale Research Letters* 2020. **15**: pp. 1–9.
73. Rani, M., et al., An updated review on synthetic approaches of green nanomaterials and their application for removal of water pollutants: Current challenges, assessment and future perspectives. *Journal of Environmental Chemical Engineering* 2021. **9**(6): p. 106763.

74. Nath, P.C., et al., Optimization and kinetic study of ultrasonic-mediated phycoerythrin extraction from *Anabaena* sp. *Biomass Conversion and Biorefinery* 2022. **14**: pp. 1–12.
75. AlNadhari, S., et al., A review on biogenic synthesis of metal nanoparticles using marine algae and its applications. *Environmental Research* 2021. **194**: p. 110672.
76. Gopu, M., et al., Green biomimetic silver nanoparticles utilizing the red algae *Amphiroa rigida* and its potent antibacterial, cytotoxicity and larvicidal efficiency. *Bioprocess and Biosystems Engineering* 2021. **44**: pp. 217–223.
77. Hassan, K.T., et al., Facile green synthesis of Ag/AgCl nanoparticles derived from *Chara* algae extract and evaluating their antibacterial activity and synergistic effect with antibiotics. *Journal of Environmental Chemical Engineering* 2021. **9**(4): p. 105359.
78. Balaraman, P., et al., Phyco-synthesis of silver nanoparticles mediated from marine algae *Sargassum myriocystum* and its potential biological and environmental applications. *Waste and Biomass Valorization* 2020. **11**: pp. 5255–5271.
79. Bhuyar, P., et al., Synthesis of silver nanoparticles using marine macroalgae *Padina* sp. and its antibacterial activity towards pathogenic bacteria. *Beni-Suef University Journal of Basic and Applied Sciences* 2020. **9**: pp. 1–15.
80. Fatima, R., et al., Biosynthesis of silver nanoparticles using red algae *Portieria hornemannii* and its antibacterial activity against fish pathogens. *Microbial Pathogenesis* 2020. **138**: p. 103780.
81. Borah, D., et al., Alga-mediated facile green synthesis of silver nanoparticles: Photophysical, catalytic and antibacterial activity. *Applied Organometallic Chemistry* 2020. **34**(5): p. e5597.
82. Elgamouz, A., et al., Green synthesis, characterization, antimicrobial, anti-cancer, and optimization of colorimetric sensing of hydrogen peroxide of algae extract capped silver nanoparticles. *Nanomaterials* 2020. **10**(9): p. 1861.
83. Arya, A., V. Mishra, and T.S. Chundawat, Green synthesis of silver nanoparticles from green algae (*Botryococcus braunii*) and its catalytic behavior for the synthesis of benzimidazoles. *Chemical Data Collections* 2019. **20**: p. 100190.
84. Massironi, A., et al., Ulvan as novel reducing and stabilizing agent from renewable algal biomass: Application to green synthesis of silver nanoparticles. *Carbohydrate Polymers* 2019. **203**: pp. 310–321.
85. Kim, D.-Y., et al., Green synthesis of silver nanoparticles using *Laminaria japonica* extract: Characterization and seedling growth assessment. *Journal of Cleaner Production* 2018. **172**: pp. 2910–2918.
86. Deepak, P., et al., Facile green synthesis of nanoparticles from brown seaweed *Sargassum wightii* and its biological application potential. *Advances in Natural Sciences Nanoscience and Nanotechnology* 2018. **9**(3): p. 035019.
87. Mahajan, A., A. Arya, and T.S. Chundawat, Green synthesis of silver nanoparticles using green alga (*Chlorella vulgaris*) and its application for synthesis of quinolines derivatives. *Synthetic Communications* 2019. **49**(15): pp. 1926–1937.
88. Murugesan, S., S. Bhuvaneswari, and V. Sivamurugan, Green synthesis, characterization of silver nanoparticles of a marine red alga Spyridia fusiformis and their antibacterial activity. *International Journal of Pharmacy and Pharmaceutical Sciences* 2017. **9**(5): pp. 192–197.
89. Araya-Castro, K., et al., Green synthesis of copper oxide nanoparticles using protein fractions from an aqueous extract of brown algae *Macrocystis pyrifera*. *Processes* 2020. **9**(1): p. 78.
90. Senthilkumar, P., et al., Facile green synthesis of gold nanoparticles from marine algae *Gelidiella acerosa* and evaluation of its biological potential. *SN Applied Sciences* 2019. **1**: pp. 1–12.
91. Maceda, A.F., et al., Controlling the absorption of gold nanoparticles via green synthesis using *Sargassum crassifolium* extract. *Key Engineering Materials* 2018. **765**: pp. 44–48.
92. González-Ballesteros, N., et al., Green synthesis of gold nanoparticles using brown algae *Cystoseira baccata*: Its activity in colon cancer cells. *Colloids and Surfaces B: Biointerfaces* 2017. **153**: pp. 190–198.
93. Costa, L., et al., Green synthesis of gold nanoparticles obtained from algae *Sargassum cymosum*: Optimization, characterization and stability. *BioNanoScience* 2020. **10**: pp. 1049–1062.
94. Abdulwahid, K.E., A.S. Dwaish, and O.A. Dakhil, Green synthesis and characterization of zinc oxide nanoparticles from Cladophora glomerata and its antifungal activity against some fungal isolates. *Plant Archives* 2019. **19**: 3527–3532.
95. Bhattacharya, P., et al., Green synthesis of zinc oxide nanoparticles via algal route and its action on cancerous cells and pathogenic microbes. *Advanced Nano Research* 2020. **3**(1): pp. 15–27.
96. Thirumoorthy, G.S., et al., *Tetraselmis indica* mediated green synthesis of zinc oxide (ZnO) nanoparticles and evaluating its antibacterial, antioxidant, and hemolytic activity. *BioNanoScience* 2021. **11**(1): pp. 172–181.

97. Arsiya, F., M.H. Sayadi, and S.J.M.L. Sobhani, Green synthesis of palladium nanoparticles using *Chlorella vulgaris*. *Materials Letters* 2017. **186**: pp. 113–115.
98. Sonbol, H., et al., Padina boryana mediated green synthesis of crystalline palladium nanoparticles as potential nanodrug against multidrug resistant bacteria and cancer cells. *Scientific Reports* 2021. **11**(1): p. 5444.
99. Arya, A., et al., Biogenic synthesis of copper and silver nanoparticles using green alga *Botryococcus braunii* and its antimicrobial activity. *Bioinorganic Chemistry and Applications* 2018. **2018**: pp. 1–9.
100. Rao, M.D. and G. Pennathur, Green synthesis and characterization of cadmium sulphide nanoparticles from *Chlamydomonas reinhardtii* and their application as photocatalysts. *Materials Research Bulletin* 2017. **85**: pp. 64–73.
101. Singh, G. and S. Chandra, Insights on microbes-mediated greener synthesis of nanoparticles: Advantages and challenges, in *Applications of Nanotechnology in Microbiology*. 2024, Springer. pp. 41–57. https://doi.org/10.1007/978-3-031-49933-3_2.
102. Nath, P.C., et al., C-phycoerythrin production from *Anabaena* sp. BTA 903: optimization, production kinetics, thermodynamic, and stability analysis. *Biomass Conversion and Biorefinery* 2023. **14**: pp. 1–13.
103. Yadav, K.K., et al., Mechanistic understanding and holistic approach of phytoremediation: A review on application and future prospects. *Ecological Engineering* 2018. **120**: pp. 274–298.
104. Vijayakumar, G., et al., Macrofungal mediated biosynthesis of silver nanoparticles and evaluation of its antibacterial and wound-healing efficacy. *International Journal of Molecular Sciences (IJMS)* 2024. **25**(2): p. 861.
105. Moradi, N., et al., Biosynthesized silver nanoparticles using *Trichoderma harzianum* reduce charcoal rot disease in bean. *Rhizosphere* 2024. **29**: p. 100828.
106. Bathrinarayanan, P.V., et al., Biological synthesis and characterization of intracellular gold nanoparticles using biomass of *Aspergillus fumigatus*. *Bulletin of Materials Science* 2013. **36**: pp. 1201–1205.
107. Shaligram, N.S., et al., Biosynthesis of silver nanoparticles using aqueous extract from the compactin producing fungal strain. *Process Biochemistry* 2009. **44**(8): pp. 939–943.
108. Soni, N. and S. Prakash, Synthesis of gold nanoparticles by the fungus *Aspergillus niger* and its efficacy against mosquito larvae. *Reports in Parasitology* 2012. **2**: pp. 1–7.
109. Suryavanshi, P., et al., *Colletotrichum* sp.-mediated synthesis of sulphur and aluminium oxide nanoparticles and its *in vitro* activity against selected food-borne pathogens. *LWT - Food Science and Technology* 2017. **81**: pp. 188–194.
110. Basavaraja, S., et al., Extracellular biosynthesis of silver nanoparticles using the fungus *Fusarium semitectum*. *Materials Research Bulletin* 2008. **43**(5): pp. 1164–1170.
111. Gade, A., et al., Green synthesis of silver nanoparticles by *Phoma glomerata*. *Micron* 2014. **59**: pp. 52–59.
112. Chandrappa, C., et al., Endophytic synthesis of silver chloride nanoparticles from *Penicillium* sp. of *Calophyllum apetalum*. *Advances in Natural Sciences Nanoscience and Nanotechnology* 2016. **7**(2): p. 025016.
113. Sharma, J.L., V. Dhayal, and R.K. Sharma, Antibacterial effect of glycerol assisted ZnO nanoparticles synthesized by white rot fungus *Phanerochaete chrysosporium*. *Materials Today Proceedings* 2021. **43**: pp. 2855–2860.
114. Gupta, K. and T.S. Chundawat, Zinc oxide nanoparticles synthesized using Fusarium oxysporum to enhance bioethanol production from rice-straw. *Biomass and Bioenergy* 2020. **143**: p. 105840.
115. Hulikere, M.M. and C.G. Joshi, Characterization, antioxidant and antimicrobial activity of silver nanoparticles synthesized using marine endophytic fungus-*Cladosporium cladosporioides*. *Process Biochemistry* 2019. **82**: pp. 199–204.
116. Sadaf, A., J.S. Afolayan, and C.C. Perry, Developing gold nanoparticle mycelial composites: Effect of nanoparticle surface functionality on *Aspergillus niger* viability and cell wall biochemistry. *Current Research in Biotechnology* 2024. **7**: p. 100185.
117. Nath, P.C., et al., Sustainable production of cellulosic biopolymers for enhanced smart food packaging: An up-to-date review. *International Journal of Biological Macromolecules* 2024. **273**: p. 133090.
118. Vaseghi, Z., A. Nematollahzadeh, and O. Tavakoli, Green methods for the synthesis of metal nanoparticles using biogenic reducing agents: A review. *Reviews in Chemical Engineering* 2018. **34**(4): pp. 529–559.
119. Brock, T.D. and J. Gustafson, Ferric iron reduction by sulfur-and iron-oxidizing bacteria. *Applied and Environmental Microbiology* 1976. **32**(4): pp. 567–571.
120. Costa, J.S.D., et al., Whole genome sequence of the multi-resistant plant growth-promoting bacteria *Streptomyces* sp. Z38 with potential application in agroindustry and bio-nanotechnology. *Genomics* 2020. **112**(6): pp. 4684–4689.

121. Hashem, A.H., et al., *Bacillus megaterium*-mediated synthesis of selenium nanoparticles and their antifungal activity against *Rhizoctonia solani* in faba bean plants. *Journal of Fungi* 2021. **7**(3): p. 195.
122. Ahmed, T., et al., Green synthesis and characterization of zirconium oxide nanoparticles by using a native *Enterobacter* sp. and its antifungal activity against bayberry twig blight disease pathogen *Pestalotiopsis versicolor. NanoImpact* 2021. **21**: p. 100281.
123. Alsamhary, K.I., Eco-friendly synthesis of silver nanoparticles by *Bacillus subtilis* and their antibacterial activity. *Saudi Journal of Biological Sciences* 2020. **27**(8): pp. 2185–2191.
124. Farag, S., et al., Green synthesis of titanium dioxide nanoparticles via bacterial cellulose (BC) produced from agricultural wastes. *Cellulose* 2021. **28**(12): pp. 7619–7632.
125. Venil, C., et al., Green synthesis of silver nanoparticles using canthaxanthin from *Dietzia maris* AURCCBT01 and their cytotoxic properties against human keratinocyte cell line. *Journal of Applied Microbiology* 2021. **130**(5): pp. 1730–1744.
126. Saleh, G.M., Green synthesis concept of nanoparticles from environmental bacteria and their effects on pathogenic bacteria. *Iraqi Journal of Science* 2020. **61**: pp. 1289–1297. https://doi.org/10.24996/ijs.2020.61.6.6.
127. Abd El Aty, A.A. and M.M. Zohair, Green-synthesis and optimization of an eco-friendly nanobiofungicide from *Bacillus amyloliquefaciens* MH046937 with antimicrobial potential against phytopathogens. *Environmental Nanotechnology, Monitoring & Management* 2020. **14**: p. 100309.
128. Huq, M.A., Green synthesis of silver nanoparticles using *Pseudoduganella eburnea* MAHUQ-39 and their antimicrobial mechanisms investigation against drug resistant human pathogens. *International Journal of Molecular Sciences (IJMS)* 2020. **21**(4): p. 1510.
129. Scala, A., et al., Marine bacterial exopolymers-mediated green synthesis of noble metal nanoparticles with antimicrobial properties. *Polymers* 2019. **11**(7): p. 1157.
130. Hossain, A., et al., Green synthesis of silver nanoparticles with culture supernatant of a bacterium *Pseudomonas rhodesiae* and their antibacterial activity against soft rot pathogen *Dickeya dadantii. Molecules* 2019. **24**(12): p. 2303.
131. Rehman, S., et al., Isolation and characterization of a novel thermophile; *Bacillus haynesii*, applied for the green synthesis of ZnO nanoparticles. *Artificial Cells, Nanomedicine, and Biotechnology* 2019. **47**(1): pp. 2072–2082.
132. Vanlalveni, C., et al., Green synthesis of silver nanoparticles using *Nostoc linckia* and its antimicrobial activity: A novel biological approach. *BioNanoScience* 2018. **8**: pp. 624–631.
133. Saravanan, M., et al., Green synthesis of anisotropic zinc oxide nanoparticles with antibacterial and cytofriendly properties. *Microbial Pathogenesis* 2018. **115**: pp. 57–63.
134. Ranjitha, V. and V.R. Rai, Actinomycetes mediated synthesis of gold nanoparticles from the culture supernatant of *Streptomyces griseoruber* with special reference to catalytic activity. *3 Biotech* 2017. **7**: pp. 1–7.
135. de Amorim Silva, T., et al., Extracellular biogenic synthesis of silver nanoparticles by Actinomycetes from amazonic biome and its antimicrobial efficiency. *African Journal of Biotechnology* 2017. **16**(43): pp. 2072–2082.
136. Thomas, P.D., Actinomycetes synthesized nanoparticles and their antibacterial activity. *Research Journal of Science and Technology* 2017. **9**(2): pp. 219–223.
137. El-Gamal, M.S., S.S. Salem, and A.M. Abdo, Biosynthesis, characterization, and antimicrobial activities of silver nanoparticles synthesized by endophytic *Streptomyces* sp. *Egyptian Journal of Biotechnology* 2018. **56**: pp. 69–85.
138. Skalickova, S., M. Baron, and J. Sochor, Nanoparticles biosynthesized by yeast: A review of their application. *Kvasny Prumysl* 2017. **63**(6): pp. 290–292.
139. Jha, A.K., et al., Synthesis of TiO_2 nanoparticles using microorganisms. *Colloids and Surfaces B: Biointerfaces* 2009. **71**(2): pp. 226–229.
140. Kowshik, M., et al., Extracellular synthesis of silver nanoparticles by a silver-tolerant yeast strain MKY3. *Nanotechnology* 2002. **14**(1): p. 95.
141. Korbekandi, H., et al., Biosynthesis of silver nanoparticles using *Saccharomyces cerevisiae. Artificial Cells, Nanomedicine, and Biotechnology* 2016. **44**(1): pp. 235–239.
142. Kim, D.-Y., et al., Extracellular synthesis of silver nanoparticle using yeast extracts: Antibacterial and seed priming applications. *Applied Microbiology and Biotechnology* 2024. **108**(1): p. 150.
143. Lee, S.H. and B.-H. Jun, Silver nanoparticles: Synthesis and application for nanomedicine. *International Journal of Molecular Sciences* 2019. **20**(4): p. 865.
144. Gevorgyan, S., et al., Antibacterial activity of royal jelly-mediated green synthesized silver nanoparticles. *AMB Express* 2021. **11**: pp. 1–8.

145. Kashyap, M., et al., Synthesis, characterization and application of intracellular Ag/AgCl nanohybrids biosynthesized in *Scenedesmus* sp. as neutral lipid inducer and antibacterial agent. *Environmental Research* 2021. **201**: p. 111499.
146. Banerjee, S.L., et al., Ag NPs incorporated self-healable thermoresponsive hydrogel using precise structural "Interlocking" complex of polyelectrolyte BCPs: A potential new wound healing material. *Chemical Engineering Journal* 2021. **405**: p. 126436.
147. Huq, M.A. and S. Akter, Bacterial mediated rapid and facile synthesis of silver nanoparticles and their antimicrobial efficacy against pathogenic microorganisms. *Materials (Basel)* 2021. **14**(10): p. 2615.
148. Shunmugam, R., et al., Biosynthesis of gold nanoparticles using marine microbe (*Vibrio alginolyticus*) and its anticancer and antioxidant analysis. *Journal of King Saud University – Science* 2021. **33**(1): p. 101260.
149. Mmola, M., et al., Enhanced antimicrobial and anticancer activity of silver and gold nanoparticles synthesised using *Sargassum incisifolium* aqueous extracts. *Molecules* 2016. **21**(12): p. 1633.
150. Noman, M., et al., Use of biogenic copper nanoparticles synthesized from a native *Escherichia* sp. as photocatalysts for azo dye degradation and treatment of textile effluents. *Environmental Pollution* 2020. **257**: p. 113514.
151. Lv, Q., et al., Biosynthesis of copper nanoparticles using *Shewanella loihica* PV-4 with antibacterial activity: Novel approach and mechanisms investigation. *Journal of Hazardous Materials* 2018. **347**: pp. 141–149.
152. Hassan, S.E.-D., et al., Endophytic actinomycetes *Streptomyces* spp mediated biosynthesis of copper oxide nanoparticles as a promising tool for biotechnological applications. *Journal of Biological Inorganic Chemistry* 2019. **24**: pp. 377–393.
153. Noor, S., et al., A fungal based synthesis method for copper nanoparticles with the determination of anticancer, antidiabetic and antibacterial activities. *Journal of Microbiological Methods* 2020. **174**: p. 105966.
154. Boutillier, S., S. Fourmentin, and B. Laperche, History of titanium dioxide regulation as a food additive: A review. *Environmental Chemistry Letters* 2022. **20**: pp. 1–17.
155. Jangid, N.K., et al., Polyaniline-TiO_2-based photocatalysts for dyes degradation. *Polymer Bulletin* 2021. **78**: pp. 4743–4777.
156. Agceli, G.K., et al., A novel approach to synthesize TiO_2 nanoparticles: Biosynthesis by using *Streptomyces* sp. HC1. *Journal of Inorganic and Organometallic Polymers and Materials* 2020. **30**(8): pp. 3221–3229.
157. Moradpoor, H., et al., Green synthesis and antifungal effect of titanium dioxide nanoparticles on oral *Candida albicans* pathogen. *Inorganic Chemistry Communications* 2021. **130**: p. 108748.
158. Ahn, T.-K., et al., Modification of titanium implant and titanium dioxide for bone tissue engineering, in *Novel Biomaterials for Regenerative Medicine. Advances in Experimental Medicine and Biology*. 2018, Springer. pp. 355–368. https://doi.org/10.1007/978-981-13-0947-2_19.
159. Rajakumar, G., et al., Fungus-mediated biosynthesis and characterization of TiO_2 nanoparticles and their activity against pathogenic bacteria. *Spectrochimica Acta Part A: Molecular and Biomolecular Spectroscopy* 2012. **91**: pp. 23–29.
160. Karunakaran, G., et al., Impact of nano and bulk ZrO_2, TiO_2 particles on soil nutrient contents and PGPR. *Journal of Nanoscience and Nanotechnology* 2013. **13**(1): pp. 678–685.
161. Hong, X., et al., Enhanced hydrophilicity and antibacterial activity of PVDF ultrafiltration membrane using Ag_3PO_4/TiO_2 nanocomposite against E. coli. *Desalination and Water Treatment* 2017. **75**: pp. 26–33.
162. Mahanty, S., et al., Green synthesis of iron oxide nanoparticles mediated by filamentous fungi isolated from Sundarban mangrove ecosystem, India. *BioNanoScience* 2019. **9**: pp. 637–651.
163. Sriramulu, M., et al., Photo catalytic, antimicrobial and antifungal activity of biogenic iron oxide nanoparticles synthesised using Aegle marmelos extracts. *Journal of Inorganic and Organometallic Polymers and Materials* 2021. **31**: pp. 1738–1744.

5 Microbe-based Nano-warriors
Engineering Antimicrobial Nanoparticles

Shareefraza J. Ukkund, Bharathi Prakash, Bhavna Alke, Usman Taqui Syed, Krishna Prasad N., and Ramis M. K.

5.1 INTRODUCTION

5.1.1 NANOTECHNOLOGY

Nanotechnology, often dubbed as the science of the small, involves manipulating and engineering materials at the nanoscale, typically ranging from 1 to 100 nm [1,2]. At this scale, unique physical, chemical, and biological properties emerge, setting the stage for groundbreaking advancements across various disciplines [3]. Nanotechnology encompasses a multidisciplinary approach, integrating principles from physics, chemistry, biology, and engineering to design, create, and utilize materials with tailored properties [4].

The ability to control matter at the nanoscale has opened up new possibilities for the development of innovative materials and devices, offering unprecedented precision and functionality [5]. Nanotechnology has found applications in diverse fields such as medicine, electronics, energy, and materials science [6,7]. The manipulation of materials at the nanoscale enables scientists and engineers to design materials with enhanced properties, paving the way for the creation of novel technologies and solutions [8]. Breakthroughs in nanoscience have paved the way for a novel class of nanostructures, each characterized by distinct properties that underlie their remarkable range of applications. These advancements have ushered in a new era of possibilities, where the tailored characteristics of these nanostructures enable unprecedented uses across various fields [9]. Figure 5.1 depicts the multitudinous applications and different forms in which nanotechnology has taken precedence in various fields globally [10].

5.1.2 NANOPARTICLES AND NANOMATERIALS

Nanoparticles (NPs), as fundamental building blocks of nanotechnology, exhibit unique characteristics due to their small size and increased surface area-to-volume ratio [11,12]. These properties often lead to alterations in optical, electronic, and chemical behaviors, setting NPs apart from their macroscopic counterparts [13]. Nanomaterials, a broader category encompassing various nanoscale structures, including NPs, nanotubes, and nanocomposites, are engineered to capitalize on these distinctive properties [14,15].

The manipulation of NPs and nanomaterials have far-reaching implications in industries such as medicine, where drug delivery systems can be precisely controlled, and in electronics, where nanomaterials enable the development of smaller and more efficient devices [16,17]. As we explore the synthesis and applications of antimicrobial NPs, understanding the fundamentals of NPs and nanomaterials becomes crucial for appreciating their unique role in addressing microbial challenges [18,19].

 DOI: 10.1201/9781003487142-5

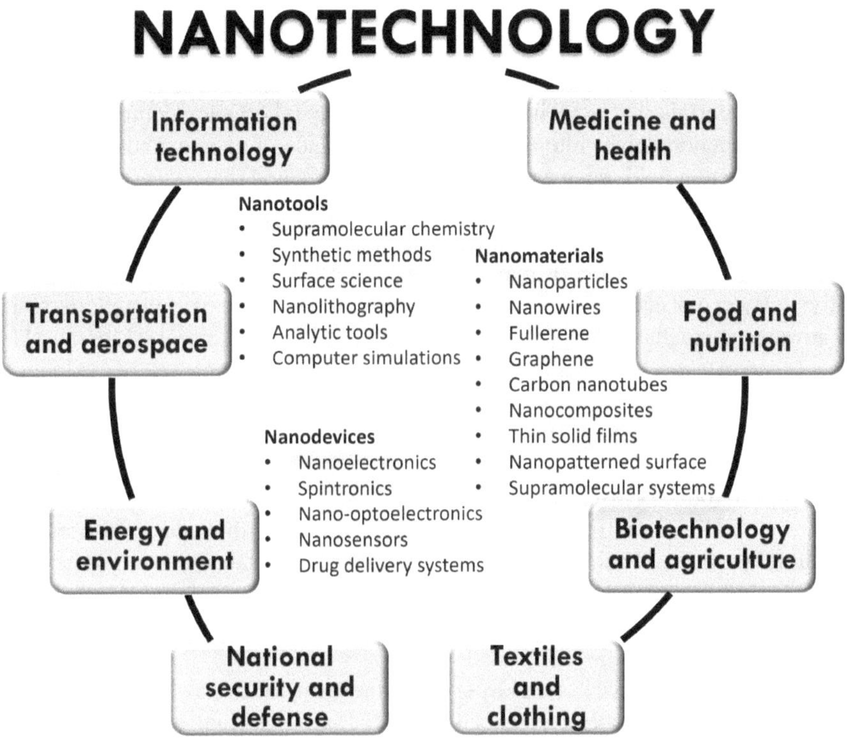

FIGURE 5.1 Nanotechnology applications in today's world. (Adapted from article [10].)

5.1.3 Antimicrobial Nanoparticles

The emergence of antimicrobial NPs represents a significant stride in the application of nanotechnology to combat microbial threats. Traditional antimicrobial agents face challenges such as resistance and limited effectiveness [20]. Antimicrobial NPs, particularly those composed of metals like silver (Ag NPs), gold (Au NPs), copper, and other materials such as titanium dioxide (TiO_2), exhibit potent antimicrobial properties. These NPs present a promising avenue for addressing microbial challenges in various domains, from healthcare to consumer products [11,12].

Antimicrobial NPs represent a specialized application of nanotechnology with the potential to revolutionize approaches to infection control. Their antimicrobial properties arise from specific interactions at the nanoscale, where unique physicochemical characteristics dictate their behavior [21,22]. This section focuses on the specific attributes that make antimicrobial NPs distinct from traditional antimicrobial agents. Understanding these properties is essential for evaluating their efficacy and potential applications in various settings [20].

The targeted nature of antimicrobial NPs allows for precision in inhibiting microbial growth, minimizing collateral damage to surrounding tissues or materials. Their adaptability to different environments and surfaces enhances their versatility for diverse applications. By elucidating the distinctive features of antimicrobial NPs, we lay the foundation for a detailed exploration of their synthesis methods and applications [21,22].

In this chapter, we embark on a comprehensive exploration of the synthesis methods, applications, and comparative studies of antimicrobial NPs. By understanding the foundations of nanotechnology, the unique attributes of NPs, and the potential of antimicrobial NPs, we set the stage for a detailed investigation into the multifaceted world of nanomaterials in the context of microbial control.

Nanotechnology holds a pivotal role in driving innovation across diverse industries. Its importance lies in the ability to manipulate materials at the nanoscale, enabling the creation of structures with unique properties [21,22]. In the context of antimicrobial NPs, nanotechnology provides a platform to engineer materials that exhibit enhanced antimicrobial efficacy compared to conventional counterparts. This section delves into the broader significance of nanotechnology, emphasizing its transformative impact on fields such as medicine, electronics, and materials science [6,7].

The precision offered by nanotechnology allows for the design and tailoring of materials at the molecular level. This precision is particularly crucial in the development of antimicrobial agents, where the controlled manipulation of materials enables the creation of NPs with targeted antimicrobial activity. By exploring the importance of nanotechnology, we gain insight into its role as an enabler of advancements that address pressing challenges related to microbial threats [11,12].

NPs, as a broad category, find applications in various fields, but their relevance in antimicrobial applications is particularly noteworthy. This section examines the specific role of NPs in addressing microbial challenges. By exploring the diverse range of NPs that exhibit antimicrobial properties, including metals like silver and gold, we gain insight into the rich landscape of materials that can be harnessed for microbial control [8].

The versatility of NPs for antimicrobial applications extends to different industries, from healthcare to consumer products. Their unique properties make them valuable candidates for the development of advanced antimicrobial solutions. As we delve into the role of NPs for antimicrobial applications, we set the stage for a more detailed exploration of specific types of antimicrobial NPs and their synthesis methods [20].

Microbial antibiotic resistance has rapidly evolved due to the life cycle of microbes. Nanomaterials possess the capability to concurrently execute various antimicrobial actions, including cellular membrane disintegration, reactive oxygen species (ROS) formation, ATP depletion, and damage to intracellular biomolecules like nucleic acids, proteins, ribosomes, and enzymes (Figure 5.2) [23].

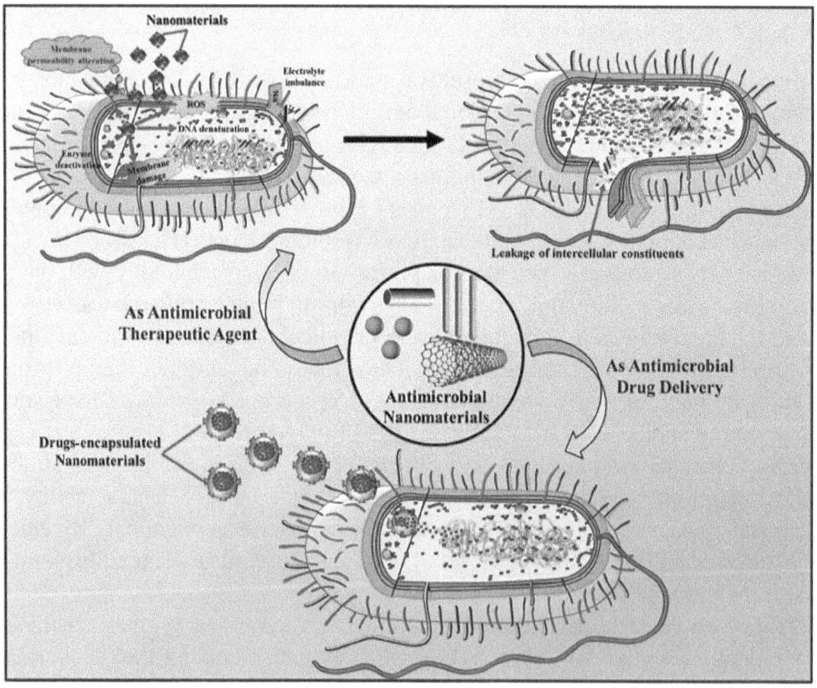

FIGURE 5.2 Antimicrobial nanomaterials hold potential applications as both therapeutic agents and carriers for drug delivery, showcasing their versatility in medical treatments. (Reprinted with permission from article [23].)

5.1.4 Synthesis of Nanoparticles

The synthesis of NPs is a critical step in harnessing their unique properties for specific applications. The general methods employed in the synthesis of NPs may be physical, chemical or biological methods [24,25]. The goal is to control the size, shape, and surface properties of NPs to tailor them for antimicrobial applications. The synthesis process plays a pivotal role in determining the characteristics of the resulting NPs, influencing factors such as stability, reactivity, and efficacy [26,27]. In this chapter, we will focus on the synthesis by biological methods, specifically microbes.

The biosynthesis of NPs harnesses the potential of microorganisms, including bacteria, fungi, actinomycetes, and even viruses, to produce nanomaterials in an eco-friendly and sustainable manner. These microorganisms act as natural nano factories, leveraging their metabolic pathways to reduce metal ions into NPs. Bacteria such as *Escherichia coli*, *Pseudomonas aeruginosa*, and *Bacillus subtilis* produce NPs through enzymatic reduction mechanisms. Fungi like *Aspergillus niger* and *Fusarium oxysporum* offer additional advantages, such as higher biomass yield and extracellular secretion of enzymes, making downstream processing easier. Actinomycetes such as *Streptomyces* species and viruses are also known for their precision in shaping NPs due to their unique structural and enzymatic properties, enabling them to produce uniform and stable NPs. This biosynthetic approach eliminates the need for toxic chemicals, aligning with the principles of green chemistry.

Microbial synthesis of NPs occurs through two main pathways: extracellular and intracellular mechanisms. In extracellular synthesis, microorganisms secrete enzymes or metabolites that reduce metal ions present in the surrounding medium into NPs. Conversely, intracellular synthesis occurs when metal ions penetrate microbial cells and are reduced to NPs within the cellular environment. Both pathways rely on reducing agents such as enzymes, proteins, or metabolites like NADH. The general approach for microbial synthesis of NPs can be summarized in the flowchart in Figure 5.3.

Biological synthesis, or biosynthesis, involves the use of living organisms or their components in the production of NPs. Microorganisms, plants, and enzymes can serve as biological entities for NP synthesis. The environmentally friendly nature of biological methods and the ability to control NP size and shape make this approach increasingly popular in the field [1,2].

Among the three methods employed for synthesizing antimicrobial NPs, the biological approach holds greater advantages, given its entirely natural synthesis of NPs. A more detailed explanation is discussed in Section 5.2 of this chapter.

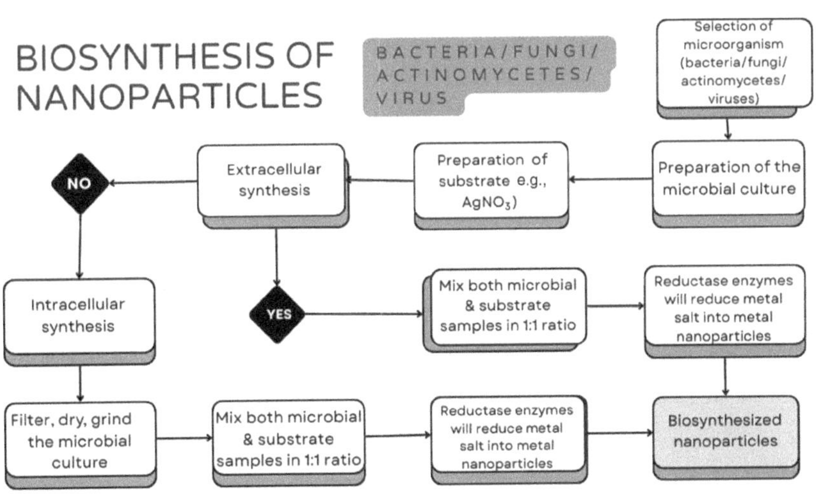

FIGURE 5.3 General steps involved in the biosynthesis of nanoparticles.

5.1.5 Synthesis of Antimicrobial Nanoparticles Derived from Different Types of Microbes

Microbial synthesis of antimicrobial NPs is an exciting and environmentally friendly approach, utilizing the natural abilities of various microorganisms to produce NPs with antimicrobial properties. For instance, certain bacteria, including *E. coli* and *P. aeruginosa*, are well known for their ability to produce silver nanoparticles (Ag NPs), which are highly effective in killing harmful microbes [28]. These bacteria use enzymes, such as nitrate reductase, to convert metal ions into NPs. Likewise, *B. subtilis* and *Staphylococcus aureus* are often employed to synthesize gold nanoparticles (Au NPs), which not only fight infections but also inhibit biofilm formation [29,30]. Other bacteria, like *Lactobacillus casei*, have shown the ability to produce zinc oxide nanoparticles (ZnO NPs), effective against a wide range of bacteria [31].

Fungi are another powerhouse in NP production, thanks to their ability to secrete enzymes and other biomolecules. Fungi such as *A. niger* and *F. oxysporum* are commonly used to produce silver NPs, which are highly stable and uniform in size [32]. Similarly, *Penicillium chrysogenum* can synthesize gold NPs, while *Trichoderma reesei* has been shown to generate titanium dioxide nanoparticles (TiO_2 NPs) that are effective antimicrobials [33]. Yeasts like *Saccharomyces cerevisiae* also contribute to this process by utilizing their cell wall components to reduce and stabilize metal ions, resulting in the formation of NPs with antimicrobial activity [34].

Actinomycetes, a unique group of filamentous bacteria, have proven to be excellent sources of NPs. *Streptomyces coelicolor* is widely studied for producing silver NPs that combat multidrug-resistant pathogens [35]. Additionally, *Rhodococcus erythropolis* has been shown to generate cadmium sulfide nanoparticles (CdS NPs), while *Nocardia farcinica* is used to produce copper nanoparticles (Cu NPs), which are particularly effective in tackling fungal infections and bacterial contaminants [36,37]. These microbes offer a precise control mechanism to produce NPs with specific shapes and sizes, which enhances their antimicrobial effectiveness.

Microalgae and viruses have also been explored for their potential in NP synthesis. Algae like *Chlorella vulgaris* and *Spirulina platensis* are used to produce NPs, including silver and gold, that are highly reactive and ideal for antimicrobial applications [38]. Even bacteriophages, such as *T4*, have shown promise in assembling metallic NPs by utilizing their structured protein capsids [39]. Archaea, such as *Halobacterium salinarum*, have been studied for their ability to produce selenium nanoparticles (Se NPs), which exhibit strong antimicrobial and antioxidant activities [40].

Some lesser-known microbes bring additional diversity to this field. Predatory bacteria like *Myxococcus xanthus* are capable of synthesizing silver and gold NPs that can effectively disrupt biofilms [41]. Similarly, *Corynebacterium glutamicum* has been used to produce magnetic iron oxide NPs (Fe_3O_4), which can be combined with antimicrobial agents for enhanced efficacy [30]. These examples highlight the vast potential of microbial synthesis, offering sustainable and innovative solutions for the development of next-generation antimicrobial technologies.

5.2 MICROBIAL SYNTHESIS OF NANOPARTICLES

The history of microbial-assisted synthesis of NPs dates back to the late 20th century, when researchers began exploring microorganisms' potential for environmentally friendly and sustainable NP production. Initial interest stemmed from observations that microbes naturally mediate metal transformations through bio-reduction and biomineralization processes in the environment. For example, studies on *Thiobacillus ferrooxidans* and *Leptospirillum ferrooxidans* in the 1980s highlighted their role in metal precipitation in natural ecosystems [42].

The direct application of microorganisms for NP synthesis gained momentum in the early 2000s. A landmark study by Klaus et al. demonstrated the ability of *Pseudomonas stutzeri* to biosynthesize silver NPs, sparking widespread interest in microbial NP synthesis [43]. Around the same time, fungi like *F. oxysporum* were explored for their ability to produce gold and silver NPs, facilitated by extracellular enzyme-mediated reduction processes [44]. These early findings underscored the potential of microbial systems to produce stable, bio-functional NPs under mild conditions.

In subsequent years, the range of microorganisms used for NP synthesis expanded significantly. Bacteria, fungi, actinomycetes, algae, and even viruses were investigated for producing NPs such as silver (Ag NPs), gold (Au NPs), zinc oxide (ZnO NPs), and titanium dioxide (TiO_2 NPs) [11]. For instance, *B. subtilis* and *A. niger* demonstrated exceptional capabilities in reducing metal ions into NPs [45]. Advances in molecular biology further clarified the roles of microbial enzymes, such as nitrate reductase and biomolecules like proteins and polysaccharides, in stabilizing NPs during synthesis [46].

By the 2010s, microbial synthesis became a cornerstone of green nanotechnology, recognized for its eco-friendly approach and the ability to tailor NP properties by selecting specific microbes and optimizing environmental conditions [47]. This approach has since been applied across diverse fields, including antimicrobial therapy, environmental remediation, and nanomedicine, paving the way for sustainable advancements in nanotechnology [48].

5.2.1 Mechanism of Action

Microbial synthesis of NPs involves a complex yet efficient mechanism where microorganisms reduce metal ions into NPs through their biochemical processes. This synthesis can occur either extracellularly, where microbial enzymes and metabolites reduce metal ions in the surrounding environment, or intracellularly, where ions are absorbed into the cell and reduced within the cytoplasm [49]. Key enzymes, such as nitrate reductase, play a pivotal role by transferring electrons to metal ions, facilitating their reduction to NPs [43]. Additionally, microbial biomolecules like proteins, polysaccharides, and secondary metabolites act as stabilizing agents, capping NPs to prevent aggregation and ensure stability. For instance, cell wall components in fungi and bacteria, such as peptides and polysaccharides, help nucleate and stabilize NPs during synthesis. Factors such as pH, temperature, and ion concentration significantly influence the size, shape, and functional properties of NPs. This green synthesis process is not only environmentally sustainable but also generates biocompatible NPs with applications in antimicrobial therapy, drug delivery, and environmental cleanup [47,50].

Almost all metal NPs carry a positive charge, which interacts with the negatively charged surfaces of microbes. This electrostatic interaction causes the metal NPs to disrupt the cell walls of the microbes, ultimately leading to their destruction. The mechanisms of various metallic NPs exhibiting antimicrobial activity are depicted in Figure 5.4.

5.2.2 Biosynthesis of Nanoparticles by Bacteria

Various NPs can be synthesized by using bacteria such as silver, gold, ZnO, CuS, and TiO_2 NPs. Biosynthesis of NPs are widely used for producing silver and gold NPs due to their excellent antibacterial properties over other NPs. In this section, we will explore the various steps involved in the synthesis of silver and gold NPs.

Silver NPs can be synthesized by various microbes including *Pseudomonas* and *Bacillus* [1,52,53]. The substrate used for the synthesis of silver NPs is silver nitrate ($AgNO_3$). Silver, known for its inherent bactericidal properties, exhibits an enhanced antimicrobial effect when reduced to nanoscale, increasing its efficacy by three to five times. Bacteria possess nitrate reductase enzymes, and when treated with $AgNO_3$, silver ions disrupt the bacterial cell wall, allowing them to interact with the nitrate reductase enzyme inside the bacteria. This enzyme reduces silver nitrate to silver ions, which are then further reduced to form silver nanoparticles (Ag NPs). The general steps involved in the biosynthesis of silver NPs from silver nitrate ($AgNO_3$) by microbes are illustrated in Figure 5.5. Similarly, even gold NPs can also be biosynthesized by bacteria with the gold chloride solution ($HAuCl_4$). The general approach is depicted in Figure 5.6.

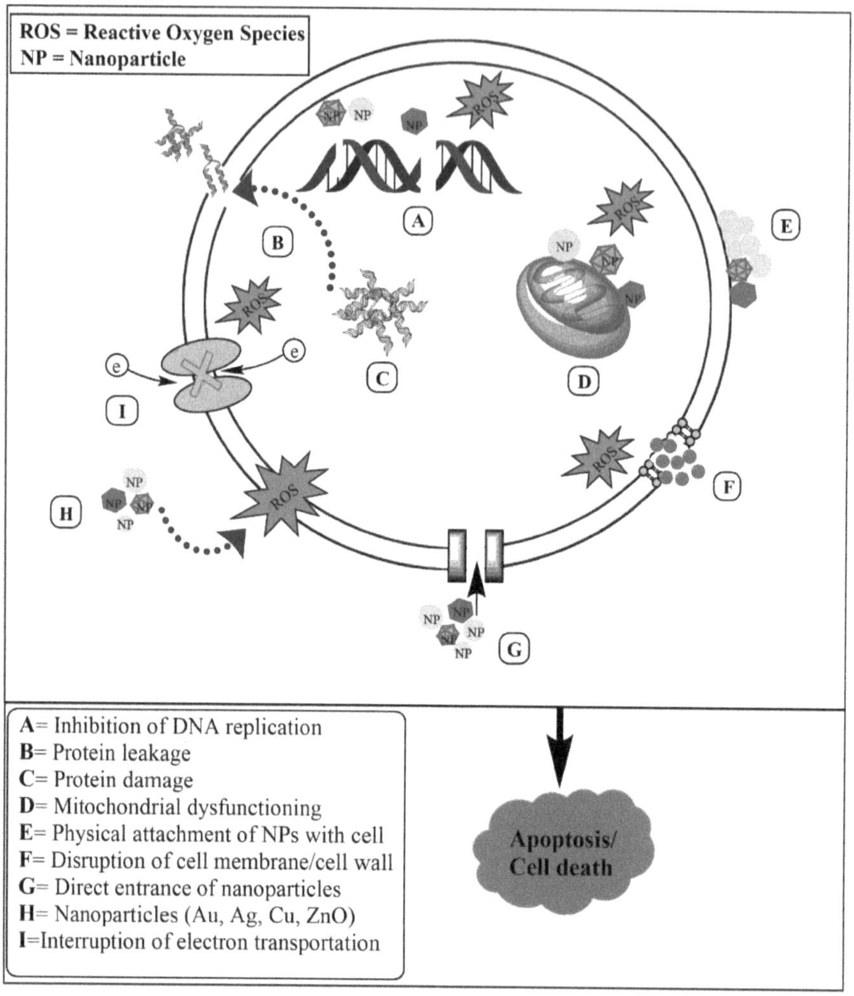

FIGURE 5.4 Mechanism of action of metallic nanoparticles exhibiting antimicrobial activities. (Reprinted with permission from article [51].)

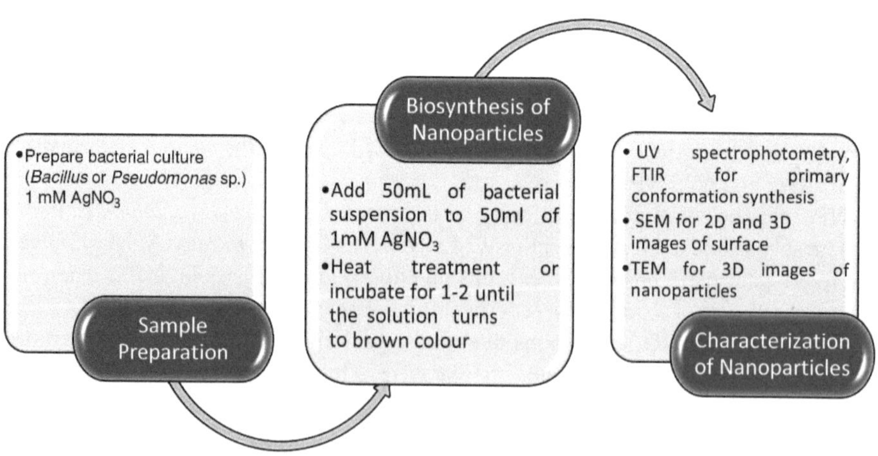

FIGURE 5.5 Biosynthesis of silver nanoparticles (Ag NPs) by bacteria from $AgNO_3$ [54].

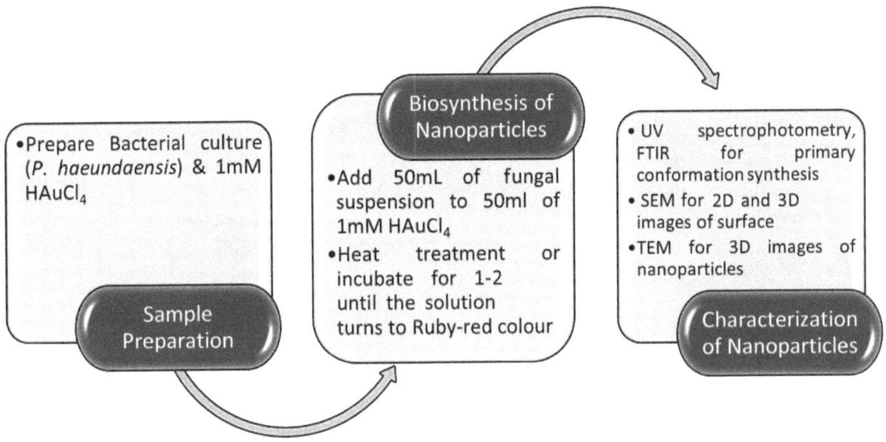

FIGURE 5.6 Biosynthesis synthesis of gold nanoparticles by *Paracoccus haeundaensis* [55].

In addition to bacteria and fungi, other microbes such as algae, actinomycetes, and viruses have also been explored for synthesizing antimicrobial NPs. Algae, for instance, have been used for the synthesis of gold NPs (*C. vulgaris*) through their phytochemicals acting as reducing and stabilizing agents [56]. Actinomycetes, such as *Streptomyces* sp., have demonstrated the ability to synthesize titanium dioxide (TiO_2) NPs, which exhibit notable antimicrobial properties [57]. Viruses, although less commonly studied, have been utilized for the synthesis of magnetic NPs by acting as biological templates, as demonstrated with Cowpea mosaic virus [58]. These diverse microbial systems expand the potential of NP synthesis, providing eco-friendly and sustainable approaches for developing NPs with enhanced antimicrobial properties.

Metallic NPs have been successfully synthesized using various microorganisms, exhibiting diverse sizes and shapes through both intracellular and extracellular processes. For instance, silver NPs, primarily spherical and measuring up to 50 nm, have been synthesized by *Aspergillus flavus*, *Aspergillus fumigatus*, *Bacillus cereus*, *Bacillus licheniformis*, and *F. oxysporum* [59]. Additionally, pyramid-shaped silver NPs reaching up to 200 nm have been reported in *Phaenerochaete chrysosporium* [60]. Other metallic NPs, such as those of mercury, palladium, uranium, cadmium telluride, and selenium, have been synthesized by microorganisms including *Enterobacter* species, *Desulfovibrio desulfuricans*, *Pyrobaculum islandicum*, *E. coli*, and *Shewanella* species, typically in spherical forms [59].

Furthermore, metal oxide NPs, such as Fe_3O_4, $FePO_4$ nanopowders, and spherical to tetragonal $BaTiO_3$, have been synthesized by various bacterial, fungal, and yeast strains through both intracellular and extracellular pathways [61]. Sulfide NPs like CdS, FeS, and ZnS have been produced using *E. coli* [62], *Lactobacillus* [63], and *Rhodobacter sphaeroides* [64], as well as fungi like *F. oxysporum* [65] and yeasts such as *Schizosaccharomyces pombe* and *Candida glabrata* [66]. The data summarizing NP production from bacteria are presented in Table 5.1, while Table 5.2 highlights the contributions of fungi, and Table 5.3 covers yeasts, molds, and algae.

5.2.3 Biosynthesis of Nanoparticles by Fungi

NPs can also be synthesized using fungi in a manner similar to bacterial synthesis, as discussed in Section 5.2.2. However, unlike bacterial synthesis, fungi-mediated biosynthesis of NPs can be classified into two types: intracellular and extracellular syntheses. In extracellular synthesis, the fungal culture medium is treated directly with substrates such as $AgNO_3$ for silver NP synthesis or $HAuCl_4$ for gold NP synthesis. By contrast, intracellular synthesis involves an additional step: the fungal cells grown in nutrient broth are first harvested through filtration, dried, and then ground to release cellular components, including the enzymes responsible for NP synthesis.

TABLE 5.1
Biosynthesis of Nanoparticles by Bacteria

Nanoparticles	Bacteria	Size (nm)	Morphology	Reference
Silver	*Escherichia coli*	118–200	Nanocrystalline	[29]
	Pseudomonas stutzeri AG259	200	Nanocrystalline	[43]
	Acetobacter xylinum	2–4	Ribbon shaped	[59]
	Aeromonas sp. SH10	6–7	Not determined	
	Bacillus licheniformis	50	Not determined	
	Bacillus sp.	5–15	Spherical	
Gold	*Escherichia coli*	9.5±4.5	Polydispersed, mostly spherical	[67]
	Paracoccus haeundaensis BC74171	20.93±3.46	Spherical	[55]
	Bacillus megatherium D01	1.9–2.6	Spherical	[59]
	Escherichia coli DH5α	70	Spherical	
	Lactobacillus sp.	>100	Not determined	
	Plectonema boryanum UTEX485	10	Cubic	
Titanium	*Lactobacillus* sp.	40–60	Spherical	
Fe_3O_4	*Aquaspirillum magnetotacticum*	40–50	Octahedral prism	
	Magnetotactic bacterium MV-1	40×40×60	Parallel	
	Magnetospirillum magnetotacticum	47	Cubo-octohedrons	
	Magnetospirillum magnetotacticum (MS-1)	50	Cubo-octahedral	

TABLE 5.2
Nanoparticles Produced by Fungi

Nanoparticles	Fungi	Size (nm)	Morphology	Reference
Silver	*Penicillium* sp.	50–75	Spherical	[1]
	Fusarium oxysporum	30–45	Nanocrystalline, mostly spherical	[4]
	Candida albicans	30–70	Spherical	[6]
	Penicillium camemberti MTCC 418	55.2±15	Spherical	[52]
	Alternaria alternata	20–60	Polydispersed/spherical	[59]
	Amylomyces rouxii	5–27	Spherical	
	Aspergillus niger	20	Spherical	
	Aspergillus terreus	1–20	Spherical	
	Fusarium oxysporum	5–50	Spherical	
	Fusarium semitectum	10–80	Spherical	
	Macrophomina phaseolina	5–40	Spherical	
	Penicillium nagiovense AJ12	25±2.8	Spherical	
	Penicillium fellutanum	5–25	Spherical	
	Trichoderma asperellum	13–18	Nanocrystalline	
Gold	*Alternaria alternate*	12±5	Spherical, triangular, hexagonal	
	Aspergillus clavatus	24.4±11	Triangular, spherical, and hexagonal	
	Aspergillus niger	10–20	Polydispersed	
	Aspergillus oryzae var. *viridis*	10–60	Various shapes, mostly spherical	
	Aspergillus sydowii	8.7–15.6	Spherical	
	Candida albicans	60–80	Nonspherical	
	Penicillium brevicompactum	10–60	Spherical, triangular, and hexagonal	
Platinum	*Fusarium oxysporum*	10–100	Rectangular, triangular, spherical hexagonal, pentagonal, and squares	
Zinc	*Fusarium* spp.	20.5±1.82	Irregular, some spherical	
Zinc oxide	*Aspergillus terreus*	54.8–82.6	Spherical	
Fe_3O_4	*Fusarium oxysporum*	20–50	Irregular, quasi-spherical	

TABLE 5.3
Nanoparticles Synthesized by Yeasts, Molds, and Algae

Nanoparticles	Fungi	Size (nm)	Morphology	Reference
Silver	*Fusarium oxysporum*	5–13	Spherical shape, face-centered cubic crystals	[59]
	Humicola sp.	5–25	Spherical shape, face-centered cubic crystals	
	Scenedesmus sp. (IMMTCC-25)	5–35	Spherical shape, highly crystalline cluster	
Gold	*Penicillium brevicompactum* KCCM 60390	25–60	Spherical	
	Tetraselmis kochinensis	5–35	Spherical and triangular shape	
Zirconium	Yeast	80–120	Irregular mesoporous	
Cadmium telluride	*Saccharomyces cerevisiae*	2–3.6	Cubic	
ZnS	*Aspergillus flavus*	12–24	Spherical	
PbS	*Aspergillus flavus*	35–100	Cubic crystal	

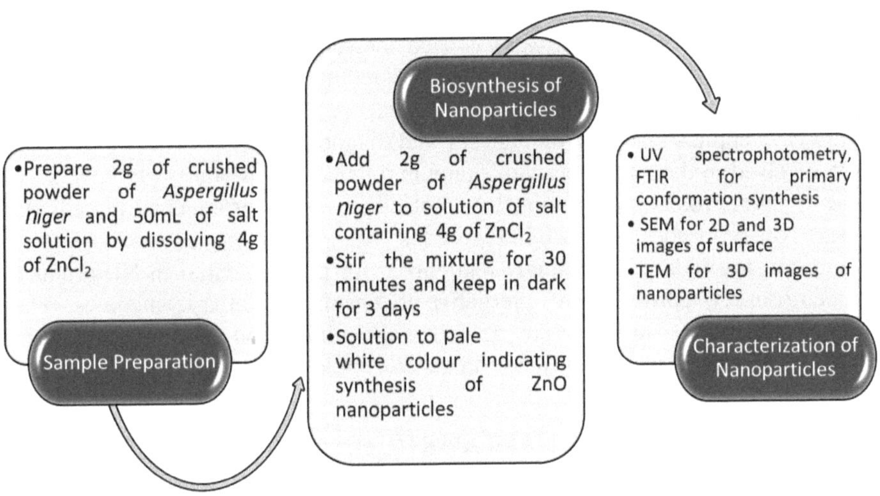

FIGURE 5.7 Biosynthesis of ZnO nanoparticles by *Aspergillus niger.*

Various species of fungi, such as *Aspergillus*, *Penicillium*, and *Candida*, can be employed for NP synthesis. Some of the most commonly used fungi include *A. niger*, *F. oxysporum*, *P. chrysogenum*, *Trichoderma resei*, and the yeast *S. cerevisiae.*

Fungal-mediated synthesis can be utilized for the production of a variety of NPs, including those of silver (Ag), gold (Au), copper (Cu), zinc oxide (ZnO), titanium dioxide (TiO_2), and copper sulfide (CuS). For instance, the intracellular synthesis of ZnO NPs using *A. niger* was demonstrated by Shamim et al. [68], as illustrated in Figure 5.7. Similarly, the extracellular synthesis of silver NPs using *F. oxysporum*, was also reported by Ukkund et al. [69].

5.2.4 Biosynthesis of Nanoparticles by Actinomycetes

Actinomycetes, especially *Streptomyces* species, are notable for their ability to biosynthesize a wide array of NPs with diverse properties. For example, *Streptomyces griseus* has been used for the synthesis of copper oxide nanoparticles (CuO NPs), which demonstrate antimicrobial and antifungal activities, making them suitable for use in agricultural applications to control plant diseases [70]. Similarly, *Streptomyces albus* has been explored for producing selenium nanoparticles (Se NPs), known for their antioxidant properties, which can be utilized in therapeutic and cosmetic formulations [70].

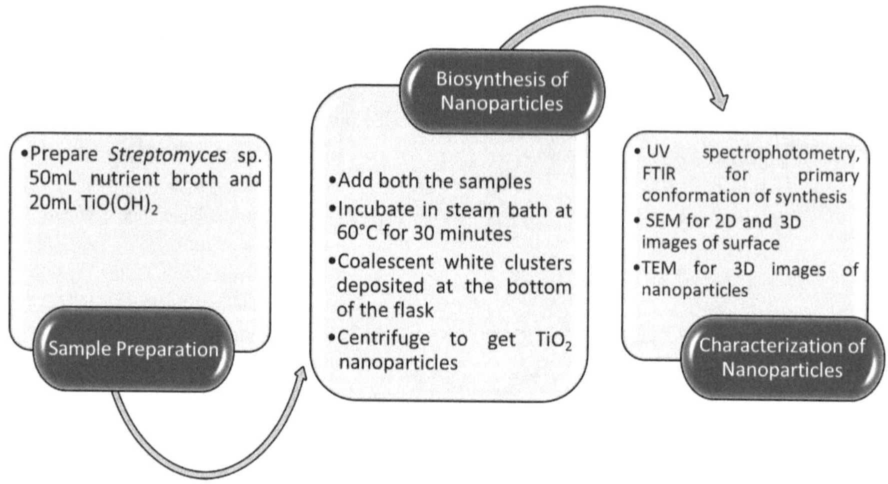

FIGURE 5.8 Biosynthesis of TiO_2 nanoparticles by *Streptomyces* sp.

In addition, *Streptomyces aureofaciens* has been shown to synthesize silver nanoparticles (Ag NPs) that exhibit potent antibacterial and anticancer activities, ideal for biomedical applications [71]. *Streptomyces lavendulae* produces iron oxide nanoparticles (Fe_3O_4 NPs), which are highly magnetic and have applications in drug delivery and diagnostics [71]. Furthermore, *Streptomyces* sp. has been involved in the biosynthesis of titanium dioxide nanoparticles (TiO_2 NPs), which are widely used in antimicrobial, antibiofilm, UV protection, and environmental remediation due to their photocatalytic properties as shown in Figure 5.8 [72].

These microorganisms are increasingly recognized for their potential in NP production, providing an eco-friendly, cost-effective alternative to chemical synthesis methods. The diverse applications of these NPs span fields like medicine, agriculture, and environmental management, underlining the importance of actinomycetes in nanotechnology.

5.3 CHARACTERIZATION OF NANOPARTICLES

Characterizing NPs is a crucial aspect of understanding their properties and ensuring their suitability for specific applications. This section explores various techniques employed in the characterization of antimicrobial NPs, shedding light on their size, shape, composition, and surface characteristics [1,73,74]. Almost all metal NPs will have a positive charge which will interact with the negative charge of microbes. Because of electrostatic interaction, metal NPs break the cell wall of microbes and kill them. Among all silver NPs are widely used for their excellent antimicrobial properties. Thus, in this chapter, we will explain silver NPs as an example.

5.3.1 UV–Vis Spectrophotometer

The UV–Vis spectrophotometer is a vital tool for the primary confirmation of NPs due to its simplicity, speed, and cost-effectiveness. This technique leverages the unique surface plasmon resonance (SPR) properties of NPs, which vary based on their size, shape, composition, and aggregation state, allowing researchers to confirm their synthesis without extensive sample preparation. By analyzing the UV–Vis absorption spectra, critical insights can be obtained about the presence and characteristics of NPs. The principle of a UV–Vis spectrophotometer is depicted in Figure 5.9.

One significant advantage of UV–Vis spectrophotometry is its ability to provide rapid and non-destructive analysis. It enables real-time monitoring of NP synthesis without altering the

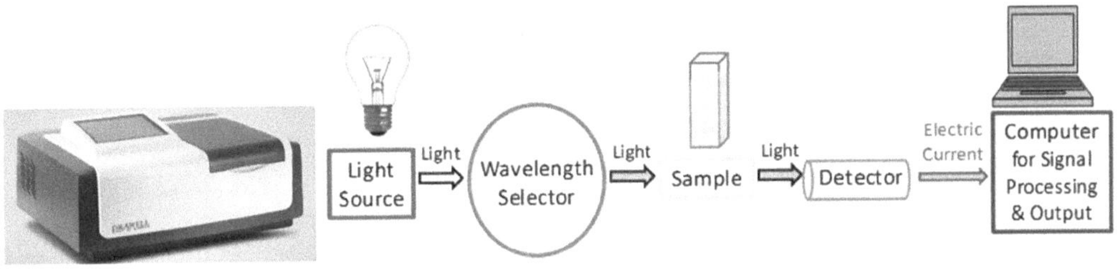

FIGURE 5.9 UV–Vis spectrophotometer and its components [75].

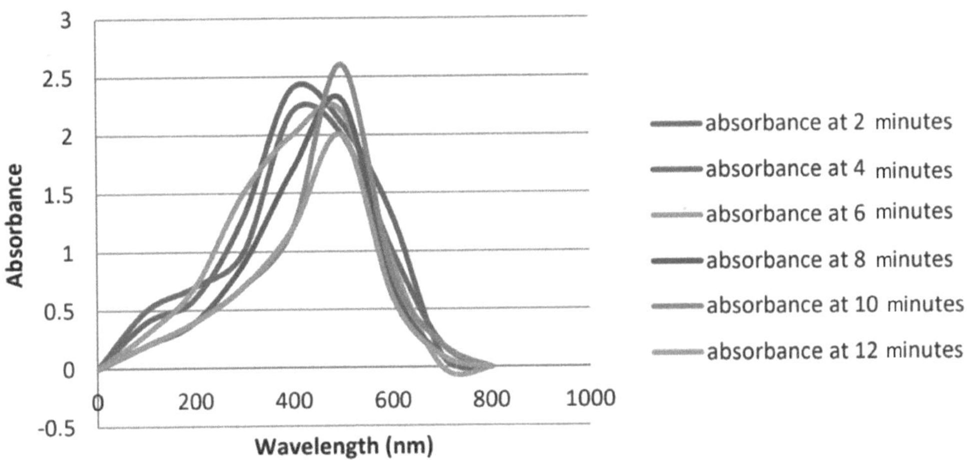

FIGURE 5.10 UV–Vis spectrophotometer readings for the green synthesis of silver nanoparticles. (Reprinted with permission from article [76].)

sample, making it ideal for dynamic studies. The technique also offers specificity, as metallic NPs such as silver, gold, and zinc oxide exhibit distinct SPR bands that serve as fingerprints for their identification. Furthermore, the spectrophotometer provides both qualitative and quantitative data in a single measurement, as the intensity of the absorption peak correlates with NP concentration.

The method's versatility is another key benefit, as it is applicable to various types of NPs, including metals, metal oxides, and sulfides. Additionally, UV–Vis spectrophotometry helps monitor NP stability and aggregation by detecting shifts or broadening of SPR peaks, which are critical parameters for ensuring consistent performance in applications. Compared to more advanced techniques like transmission electron microscopy (TEM), scanning electron microscopy (SEM), or XRD, UV–Vis spectrophotometry is more affordable and widely accessible, making it an excellent choice for initial NP characterization.

The UV–Vis spectrophotometry is a widely used technique to measure the absorbance of light by NPs. This method aids in determining the concentration of NPs in a solution and provides insights into their optical properties. By analyzing the absorption spectra, researchers can gather information about the size and stability of NPs [69]. Silver NPs, being one of the most widely used antimicrobial NPs, give the maximum absorbance at a wavelength range of 400–480 nm as shown in Figure 5.10 [76]. So, anyone who wants to confirm the presence of silver NPs can use a UV–Vis spectrophotometer as a primary conformation of NPs synthesis. If the UV absorbance is found in the range between wavelength 400 and 480 nm, then the silver NPs may be present in the given sample.

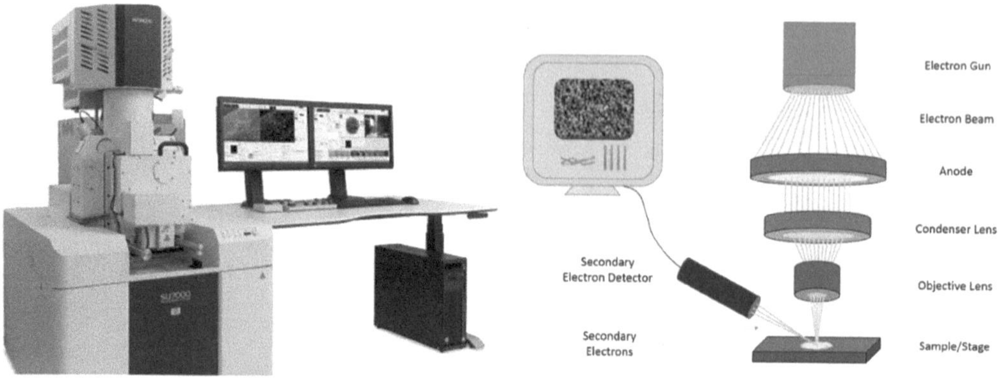

FIGURE 5.11 Scanning electron microscope and its various parts [78].

5.3.2 Scanning Electron Microscope

SEM is an essential technique for the characterization of NPs, offering high-resolution imaging and detailed morphological analysis. SEM allows researchers to observe the size, shape, and surface topography of NPs, which are critical parameters for their functional applications. The components of an SEM are depicted in Figure 5.11. Unlike UV–Vis spectrophotometry, SEM provides direct visualization, enabling precise measurement of particle dimensions and confirmation of uniformity and dispersion within the sample. This is particularly useful for identifying aggregation or irregularities in synthesis. Additionally, SEM can be paired with energy-dispersive X-ray spectroscopy (EDS) to determine the elemental composition of NPs, ensuring their chemical identity. Although SEM requires specialized equipment and preparation, its ability to deliver nanoscale insights with unparalleled clarity makes it indispensable for validating the structural properties of NPs [1,77].

The SEM analysis result in Figure 5.12 shows that the average particle sizes of prepared NPs are 25–35 nm. On the other hand, Shareef et al., Ukkund et al., and Dhabalia et al. got the particle size of 75, 30–45, 55–65, and 30–70 nm, respectively [1,6,69,76], with the similar method by fungi, wherein by using Panchakavya, smaller silver NPs were obtained. Thus, Panchakavya acted as more efficient reducing agent than the materials used in previous studies [79].

5.3.3 Transmission Electron Microscope

TEM is a powerful tool for the detailed characterization of NPs, providing ultra-high-resolution imaging that reveals their size, shape, and crystalline structure at the atomic level. The components of a TEM are depicted in Figure 5.13. TEM is particularly effective for confirming the nanoscale dimensions of particles and analyzing their morphological uniformity. Beyond imaging, TEM can be combined with techniques like selected area electron diffraction (SAED) to determine the crystal structure and phase of NPs, offering comprehensive structural and material insights. Additionally, high-resolution TEM (HR-TEM) allows researchers to visualize lattice fringes, confirming crystallinity and providing information on interplanar spacing. Although TEM requires meticulous sample preparation and access to advanced instrumentation, its ability to deliver unparalleled structural details makes it a cornerstone technique for NP characterization and validation.

By transmitting electrons through the sample, TEM generates detailed images of the internal structure of NPs with 3D images. This method allows researchers to observe the crystallinity, size, and shape of individual NPs with exceptional resolution [81]. One of the studies including TEM image and EDX pattern of silver NPs is depicted in Figure 5.14 [82].

FIGURE 5.12 SEM image of silver nanoparticles showing 25–35 nm sized spherical nanoparticles. (Reprinted with permission from article [79].)

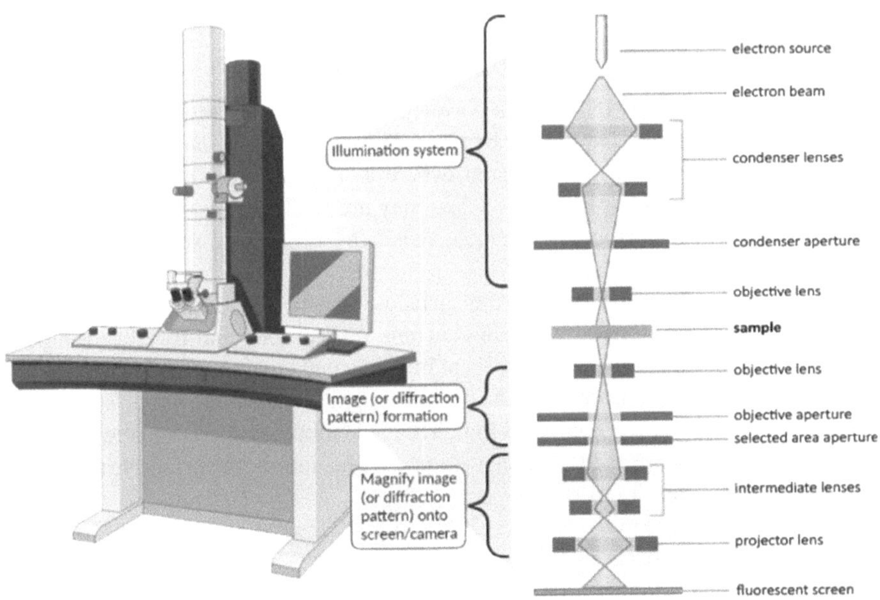

FIGURE 5.13 Simplified TEM diagram highlighting the major components throughout the column [80].

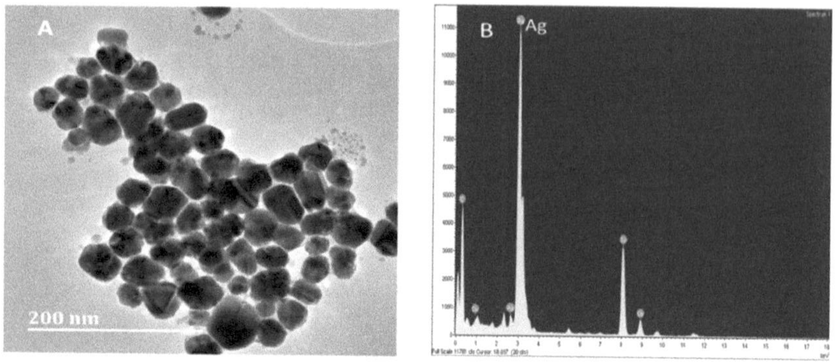

FIGURE 5.14 TEM picture of (A) silver nanoparticles and (B) EDX pattern. (Reprinted with permission from article [82].)

The morphology of NPs typically exhibits a quasi-spherical shape. In the image at higher magnification, silver NPs ranging from 5 to 33 nm in size were also identified (Figure 5.14). Nevertheless, the average NP size was determined to be 13 nm. EDX analysis corroborated the elemental composition of the particles as silver (Figure 5.12).

Comparing the strengths and limitations of these characterization techniques is essential for a comprehensive understanding of antimicrobial NPs. UV–Vis spectrophotometry provides a quick assessment of concentration, while SEM and TEM offer detailed structural information. By combining multiple techniques, researchers can gain a holistic view of NP properties, facilitating informed decisions about their suitability for specific applications.

Discussion and comparison also extend to the interpretation of results obtained through these techniques. Factors such as sample preparation, instrument calibration, and data analysis methods contribute to the reliability of characterization outcomes. Researchers must carefully consider these aspects to draw accurate conclusions about the properties of antimicrobial NPs.

5.3.4 Scanning Tunneling Microscope

Scanning tunneling microscopy (STM) is a powerful technique that allows imaging and analysis of surfaces at the atomic scale. Developed by Gerd Binnig and Heinrich Rohrer in 1981, STM is based on the quantum tunneling phenomenon, where a small electrical current flows between a sharp conductive tip and the sample surface when they are in close proximity, even without direct contact [83]. The tunneling current is extremely sensitive to the tip-to-surface distance, enabling atomic-scale resolution.

The STM utilizes a sharp tip that scans the surface with sub-angstrom precision, controlled by piezoelectric elements. Variations in tunneling current are recorded and translated into a detailed topographic map of the sample. STM can operate in a constant current mode, where the tip adjusts to maintain a stable tunneling current, or in a constant height mode, where the tip's position remains fixed, and current fluctuations are measured [84]. This dual mode of operation allows STM to study conductive and semiconductive materials, providing insights into both surface structures and electronic properties, such as density of states.

STM is integral to NP research due to its ability to provide atomic-level details of surface morphology, structure, and electronic properties. For metallic NPs, STM enables the identification of atomic arrangements, grain boundaries, and surface defects, which are crucial for understanding their catalytic and electronic characteristics. For example, STM has been employed to study molecule adsorption on gold NPs, providing insights into their catalytic mechanisms in chemical reaction [85].

Beyond structural analysis, STM is highly effective for probing the electronic properties of nanomaterials. It has been used to study quantum dots and semiconductor NPs, revealing details about their electronic band structures and charge distribution. These findings are essential for optimizing NPs for applications in electronics, sensors, and photonics [83]. Moreover, STM's capability to manipulate individual atoms and molecules on surfaces makes it invaluable in nanofabrication, particularly for designing nanoscale devices involving NPs.

The study by Quate [83] provides high-resolution STM images of mixed monolayer-protected gold NPs, revealing crucial insights into the arrangement of ligands on the NP surface. In this study, large-scale STM imaging displays nanoscale domain structures, characterized by a stripe-like morphology, with molecular thicknesses and spacings of approximately 1 nm. The study examined three distinct types of gold nanoparticles (Au NPs) coated with varying ligand mixtures to analyze their structural arrangements via STM imaging. The first type, labeled NP1, was coated with a blend of 11-mercapto-1-undecanol (C11ol) and 4-mercapto-1-butanol (C4ol). The second type, NP2, utilized a mixture of 1-octanethiol (C8) and mercaptopropionic acid (C2COOH), while the third type, NP3, was coated exclusively with C8.

Figure 5.15 highlights a representative STM image of the NP1 sample, showcasing NPs with identifiable stripe-like patterns. A digitally zoomed image of a single-striped particle (Inset 5.15c) provides a closer look at this morphology, which is visually represented in another illustration (Inset 5.15b). The interstrip distance is measured at approximately 1.0 nm, a characteristic consistent

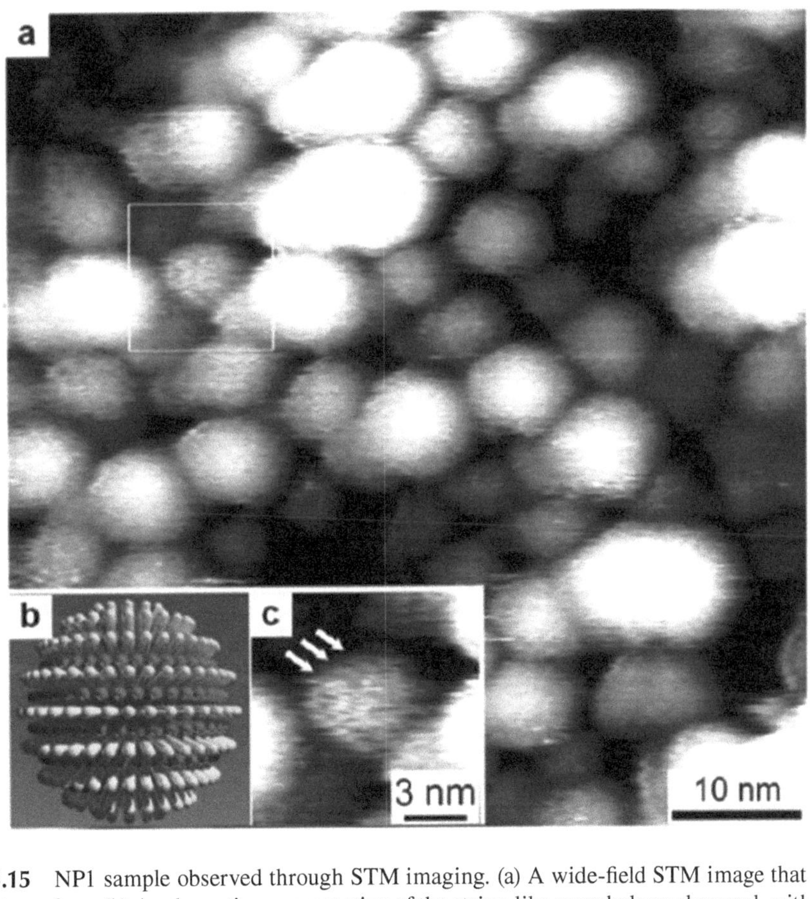

FIGURE 5.15 NP1 sample observed through STM imaging. (a) A wide-field STM image that captures the nanoparticle surface. (b) A schematic representation of the stripe-like morphology observed, with both molecular thickness and interstrip spacing measured at approximately 1 nm. (c) A digitally enhanced close-up view of a section from (a), highlighting the structural details of the nanoparticle's surface arrangement [83].

across varying scanning speeds, as confirmed by power spectral density (PSD) analysis. This observation aligns with previously documented stripe-like structures, shedding light on the nanoscale organization within the ligand shells of these NPs, offering potential insights into their chemical and physical properties.

The combination of UV–Vis spectrophotometry, SEM, TEM, and STM in the characterization process enhances the precision and reliability of data, contributing to a more thorough understanding of antimicrobial NPs and their potential applications.

5.4 ANTIMICROBIAL ACTIVITY

Antimicrobial activity assessment is a critical aspect of understanding the effectiveness of NPs in inhibiting microbial growth. This section delves into the methodologies used to evaluate the antimicrobial activity of NPs and provides a comparative analysis of their efficacy [86]. Almost all metal NPs carry a positive charge, which interacts with the negatively charged surfaces of microbes. This electrostatic interaction causes the metal NPs to disrupt the cell walls of the microbes, ultimately leading to their destruction. Recently, silver nanoparticles (Ag NPs) have been recognized as promising candidates for the development of novel antimicrobial agents [53]. The creation of nanomaterials-based drug delivery carriers with antimicrobial properties is of great importance, as these carriers could provide new platforms for the treatment of various infectious diseases. The effectiveness of Ag NPs in antimicrobial applications is largely dependent on their morphology, which directly influences the total surface area and the amount of silver ions released.

The antimicrobial mechanism of Ag NPs can be explained primarily by the penetration of silver ions into bacterial cells, leading to efficient bacterial killing. However, it is important to note that the antimicrobial actions of Ag NPs are generally considered to involve multiple pathways. The precise morphology of NPs plays a crucial role in determining the extent of their antimicrobial activity, making it a key factor in their design and application.

5.4.1 Zone of Inhibition Studies

Zone of inhibition studies are a common method employed to assess the antimicrobial activity of NPs. In this approach, a sample of NPs is applied to a medium inoculated with microorganisms. The formation of clear zones around the NP-treated area indicates the inhibition of microbial growth. The size of these zones correlates with the effectiveness of the NPs against specific microorganisms [1,86].

The choice of microorganisms is crucial, considering the diverse range of microbes with varying susceptibilities. Standardized tests against common pathogens provide a baseline for comparison, and researchers often expand their studies to include a spectrum of microorganisms to assess broad-spectrum antimicrobial activity.

5.4.2 Silver Nanoparticles–based Studies

Comparative studies of antimicrobial activity involve evaluating the performance of different types of NPs under similar conditions. Factors such as concentration, size, and surface characteristics influence the antimicrobial efficacy of NPs. This section discusses the outcomes of such comparative studies, providing insights into the strengths and limitations of various antimicrobial NPs.

The discussion encompasses the interaction of NPs with microbial cells, exploring mechanisms such as cell membrane disruption, ROS generation, and interference with cellular processes. Understanding these mechanisms contributes to the design and optimization of NPs for enhanced antimicrobial activity.

One of the studies [1] explaining zone of inhibition is explained here; the effectiveness of Ag NPs against both gram-positive and gram-negative bacteria was evaluated to gauge their antibacterial action. Inhibitory zone measurements and fold region increment rates were recorded. Positive

TABLE 5.4
Zone of Inhibition for Different Antibiotics against Test Organisms [1]

	Methicillin (10 μg/disc)			Erythromycin (10 μg/disc)		
	Met (Zone in mm)	Met+Ag NP (Zone in mm)	Fold Increase (in %)	Ery (Zone in mm)	Ery+Ag NP (Zone in mm)	Fold Increase (in %)
S. aureas	0	0	0	0	0	0
P. mirabilis	0	0	0	2	4	100
	Vancomycin (10 μg/disc)			**Bacitracin (10 μg/disc)**		
	Van (Zone in mm)	**Van+Ag NP (Zone in mm)**	**Fold Increase (in %)**	**Bac (Zone in mm)**	**Bac+Ag NP (Zone in mm)**	**Fold Increase (in %)**
S. aureas	13	16	63.07	18	20	11.11
P. mirabilis	0	0	100	0	0	0

controls using antibiotics were employed for comparison with Ag NPs. Ag NPs, when combined with erythromycin, vancomycin, and bacitracin, exhibited enhanced antibacterial activity, whereas methicillin showed no inhibitory zone against any test organism (Table 5.4).

This enhancement was also assessed by measuring the increase in fold zone. Erythromycin displayed the highest fold region increase. A similar investigation was conducted to assess the antifungal activity of silver NPs when combined with drugs and the impact of silver NPs on gram-negative bacteria was found to be superior to that on gram-positive bacteria [87].

As researchers continue to refine and expand upon methods for assessing antimicrobial activity, ongoing comparative studies play a crucial role in advancing our understanding of how NPs can effectively combat microbial threats. This knowledge is instrumental in guiding the development of antimicrobial NPs with improved performance and broader applications [88].

5.4.3 Gold Nanoparticles–based Studies

The zone of inhibition (ZOI) assay is a reliable and widely used method to assess the antimicrobial potential of gold nanoparticles (Au NPs). This method involves applying a solution of Au NPs to an agar plate inoculated with the target microorganism. The NPs diffuse through the agar, inhibiting microbial growth, and forming a clear area, or ZOI, around the site of application. The size of this zone reflects the antimicrobial effectiveness of Au NPs. Au NPs exhibit antimicrobial properties due to mechanisms such as disrupting microbial cell walls, generating ROS, and interfering with intracellular processes like DNA replication and protein synthesis. Additionally, the synthesis parameters of Au NPs such as size, shape, and surface functionalization play critical roles in their antimicrobial efficacy. Functionalization of Au NPs with biomolecules or chemical capping agents, such as antibiotics or plant extracts, enhances their antimicrobial activity, broadening their application against resistant microbial strains [89,90].

Studies have demonstrated the effectiveness of Au NPs against a wide range of microorganisms, including gram-positive bacteria like *S. aureus* and gram-negative bacteria like *E. coli*, as well as fungal strains such as *C. albicans*. Functionalized Au NPs often show larger zones of inhibition compared to unfunctionalized particles, emphasizing the importance of surface modification in enhancing antimicrobial properties. The ZOI assay has also been used to compare Au NP efficacy across different synthesis methods and microbial strains, offering a versatile approach to evaluate their therapeutic potential.

Results from ZOI studies highlight that Au NPs, particularly those below 20 nm, demonstrate enhanced antimicrobial activity due to their larger surface area-to-volume ratio, which facilitates stronger interactions with microbial cells. In one study, plant extract-functionalized Au NPs exhibited ZOI diameters of up to 15 mm against *S. aureus*, compared to smaller zones observed with unfunctionalized particles [91]. The effectiveness was also found to vary with microbial type, as gram-positive bacteria showed greater susceptibility due to their simpler cell wall structure. Furthermore, combining Au NPs with antibiotics enhanced the antimicrobial effect, with the ZOI increasing by 30%–50%, highlighting a synergistic interaction. These findings underscore the potential of Au NPs as versatile agents in combating microbial infections, especially those involving resistant strains.

5.4.4 Antimicrobial Studies of Other Nanoparticles

Beyond gold and silver NPs, various other NPs have demonstrated significant antimicrobial efficacy in ZOI studies. Zinc oxide (ZnO) NPs are well documented for their antimicrobial properties. In a study involving *E. coli* and *S. aureus*, ZnO NPs with an average size of 20 nm showed ZOI measurements of 18 and 14 mm, respectively, at a concentration of 100 µg/mL. The antimicrobial action of ZnO NPs is primarily attributed to their generation of ROS and the release of Zn^{2+} ions, which compromise bacterial membrane integrity and disrupt vital intracellular processes [92].

Titanium dioxide (TiO_2) NPs, known for their photocatalytic activity, have also been tested in ZOI assays. When exposed to UV light, TiO_2 NPs induce oxidative stress, damaging bacterial cells. For example, TiO_2 NPs with a size range of 15–30 nm produced ZOI diameters of 12 mm against *P. aeruginosa* and 10 mm against *Salmonella typhimurium* at 200 µg/mL. This enhanced activity under UV light makes TiO_2 NPs particularly useful for applications in environments with controlled lighting conditions [93].

Another promising candidate is copper oxide (CuO) NPs. In a study on *Klebsiella pneumoniae* and *C. albicans*, CuO NPs with an average size of 40 nm demonstrated ZOI measurements of 15 and 13 mm, respectively, at 150 µg/mL. The antimicrobial mechanism of CuO NPs is believed to involve membrane disruption and oxidative damage caused by the generation of copper ions (Cu^{2+}) [94]. These findings highlight the potential of non-noble metal and metal oxide NPs as robust antimicrobial agents, offering a cost-effective alternative to precious metal NPs like silver and gold [1–3].

5.4.5 Antimicrobial Activities of Ag and Au Nanoparticles for Clinical and Food Safety

5.4.5.1 Clinical Applications

Silver nanoparticles (Ag NPs) have revolutionized the treatment of microbial infections due to their broad-spectrum antimicrobial properties. One of their significant real-time applications is in wound dressings. Ag NPs are embedded into bandages and creams to prevent infections in burn wounds and surgical sites. Products such as Acticoat™ incorporate Ag NPs, providing sustained antimicrobial activity against pathogens like *S. aureus* and *P. aeruginosa*. These NPs disrupt bacterial cell membranes, leading to cytoplasmic leakage and cell death, making them invaluable for infection control [91]. Additionally, Ag NPs are being explored in catheters and implantable devices to prevent biofilm formation, a common cause of device-associated infections. Gold nanoparticles (Au NPs), though less commonly used for direct antimicrobial action, are utilized in diagnostic applications. Functionalized Au NPs enhance sensitivity in pathogen detection systems, such as lateral flow assays for *E. coli* or *Salmonella*, providing quick and accurate results for clinical diagnostics [95].

5.4.5.2 Applications in Food Safety

In the food industry, Ag NPs and Au NPs are deployed to extend shelf life and ensure microbial safety. Ag NPs are incorporated into food-packaging materials to prevent microbial contamination. For instance, silver-coated films are used to wrap perishable items such as meat and dairy products, inhibiting the growth of bacteria like *Listeria monocytogenes* and *S. typhimurium*. These

NPs interact with microbial DNA and proteins, disrupting their replication and metabolic functions, which reduces spoilage and foodborne illnesses [96,97]. Similarly, Au NPs are applied in food safety monitoring systems. Biosensors using Au NPs can detect low concentrations of pathogens or toxins in food samples, ensuring timely intervention to prevent outbreaks. Their high surface area and excellent conductivity enhance the sensitivity and efficiency of detection processes [89].

5.4.5.3 Dual Clinical and Food Applications

The versatility of Ag NPs and Au NPs extends to dual applications in both clinical and food-related contexts. For example, Ag NPs synthesized using green methods have demonstrated antimicrobial effects against both clinical pathogens like *C. albicans* and foodborne microbes such as *B. cereus*. A comparative study highlighted that biosynthesized Ag NPs reduced microbial colony counts by over 95% in both scenarios [98]. Likewise, Au NPs functionalized with curcumin not only inhibited biofilm formation of *P. aeruginosa* in clinical samples but also reduced bacterial contamination in packaged food products by over 80% [91]. These studies underscore the potential of NPs as versatile antimicrobial agents capable of addressing global challenges in healthcare and food safety.

5.5 NANOPARTICLES USED IN BIOFILMS

Biofilms, which are structured microbial communities encased in an extracellular polymeric substance (EPS) matrix, pose significant challenges in healthcare and industrial settings due to their resistance to conventional antibiotics and disinfectants. NPs, owing to their unique physicochemical properties, have emerged as potent agents for biofilm disruption and prevention.

5.5.1 Silver Nanoparticles (Ag NPs)

Ag NPs are well known for their potent antibiofilm activities. These NPs penetrate biofilm matrices and interfere with bacterial quorum sensing, a key mechanism in biofilm formation. Ag NPs disrupt the structural integrity of the EPS matrix, thereby exposing bacteria to antimicrobial agents. In a study by Kalishwaralal et al. [99], Ag NPs significantly inhibited biofilm formation of *P. aeruginosa* and *Staphylococcus epidermidis* at concentrations as low as 10 µg/mL, with a biofilm reduction of over 80% observed within 24 hours. Additionally, Ag NPs also prevent bacterial adhesion on medical devices, reducing the risk of infections associated with catheters and implants [99].

5.5.2 Gold Nanoparticles (Au NPs)

Although less studied compared to Ag NPs, Au NPs have shown promising antibiofilm activity, especially when functionalized with biomolecules. Functionalized Au NPs disrupt biofilm stability by interfering with bacterial communication pathways or directly binding to components of the EPS matrix. For example, Au NPs conjugated with antibiotics such as vancomycin exhibited enhanced penetration into *Enterococcus faecalis* biofilms, reducing biomass by approximately 70% [95]. This synergistic effect highlights the potential of Au NPs in addressing biofilm-associated infections in healthcare.

5.5.3 Zinc Oxide Nanoparticles (ZnO NPs)

ZnO NPs exhibit antibiofilm properties by generating ROS that damage microbial cells within the biofilm. Studies have shown that ZnO NPs disrupt biofilms of *E. coli* and *C. albicans* at concentrations of 50–100 µg/mL. ZnO NPs were also effective in preventing biofilm formation on surfaces, making them suitable for coatings in medical and industrial applications. A study by Azam et al. [100] reported a 60%–80% reduction in biofilm biomass for *E. coli* and *S. aureus*, emphasizing their broad-spectrum activity.

5.5.4 Titanium Dioxide Nanoparticles (TiO_2 NPs)

TiO_2 NPs, particularly under UV light, demonstrate antibiofilm properties through photocatalytic activity that generates ROS. This mechanism effectively disrupts bacterial membranes and the biofilm matrix. Research indicates that TiO_2 NPs reduce biofilm formation of *P. aeruginosa* on industrial surfaces by up to 90%, making them valuable in water treatment and food processing industries. Furthermore, the photodynamic properties of TiO_2 NPs allow for non-toxic, environmentally friendly antibiofilm strategies [101].

5.5.5 Copper Oxide Nanoparticles (CuO NPs)

CuO NPs are known for their dual antibacterial and antibiofilm activities. These NPs exhibit efficacy against biofilms formed by *K. pneumoniae* and *Acinetobacter baumannii*, both of which are notorious for causing hospital-acquired infections. A study by Khairy et al. [102] demonstrated a 70% reduction in biofilm mass when CuO NPs were applied to contaminated surfaces. Their oxidative stress-inducing properties and ability to inhibit bacterial adhesion are key factors in their antibiofilm activity.

In conclusion, NPs such as silver, gold, zinc oxide, titanium dioxide, and copper oxide have shown strong potential in disrupting biofilm formation across various microorganisms, presenting valuable applications in healthcare and industry. Silver NPs, in particular, are effective in preventing biofilm development on medical devices and wound care products. Gold NPs, often enhanced with functional molecules, can increase the effectiveness of antibiotics against biofilm-producing bacteria. Zinc oxide and titanium dioxide NPs work by generating ROS or utilizing photocatalytic activity to break down biofilms, while copper oxide NPs help reduce biofilm mass and bacterial adhesion. These mechanisms highlight the versatile role of NPs in addressing biofilm-related challenges and improving antimicrobial strategies in clinical and food safety environments.

5.6 INDUSTRIAL APPLICATIONS OF MICROBIAL-DERIVED ANTIMICROBIAL NANOPARTICLES

Antimicrobial NPs find diverse applications across various domains due to their unique properties and broad-spectrum efficacy (Figure 5.16). This section explores the practical implementations of antimicrobial NPs in different industries and highlights specific examples and case studies (Table 5.5).

Microbial-derived antimicrobial NPs are gaining significant attention for their diverse industrial applications due to their eco-friendly synthesis process and potent antimicrobial properties. In the food industry, these NPs are used as effective agents for packaging materials to enhance shelf life and reduce microbial contamination. Silver NPs synthesized using microbial processes, for instance, are incorporated into food storage films to inhibit bacterial growth, ensuring food safety and quality [103,104]. Additionally, these NPs are applied in water treatment systems for their ability to remove or neutralize harmful microorganisms. Their biocompatibility and high antimicrobial activity make them suitable for treating water in industrial and municipal settings.

Beyond food and water industries, microbial-derived NPs find applications in healthcare and cosmetics. For example, NPs like zinc oxide and gold, synthesized through bacterial or fungal methods, are used in topical creams, wound dressings, and antimicrobial coatings on medical devices [96]. In cosmetics, these NPs are utilized for their dual roles as antimicrobial agents and UV protectors, enhancing product effectiveness and safety. Moreover, the environmentally sustainable synthesis process involving microbes aligns with industrial goals of reducing carbon footprints, positioning these NPs as a green alternative for antimicrobial applications.

Microbial-derived NPs offer a sustainable and efficient approach to combating microbial contamination across various industries. In the **food industry**, silver nanoparticles (Ag NPs) synthesized

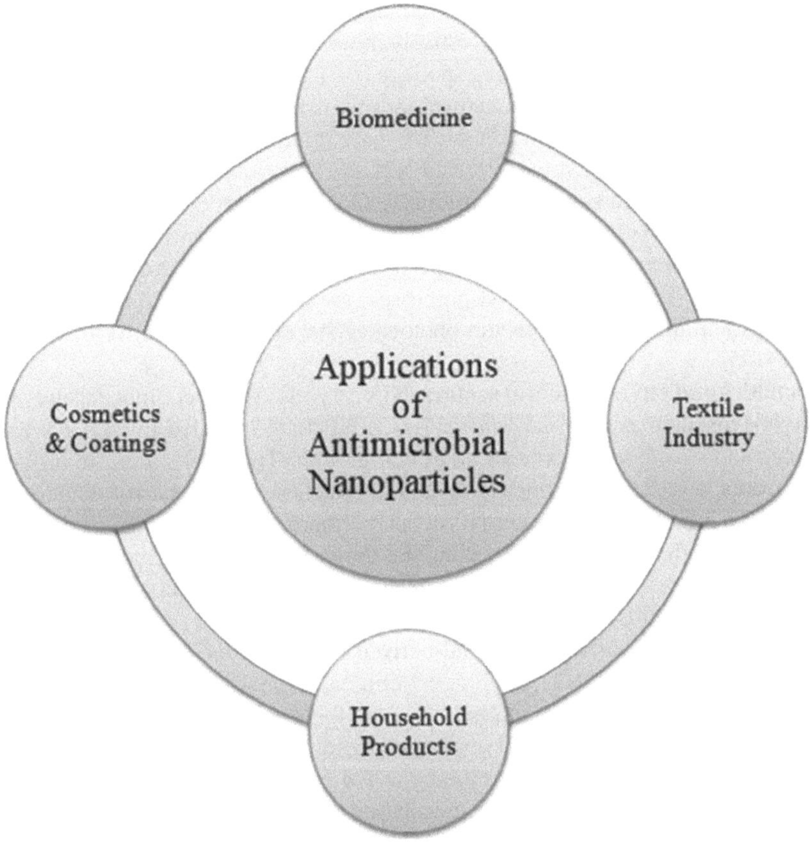

FIGURE 5.16 Industrial applications of microbial-derived antimicrobial nanoparticles.

TABLE 5.5
Microbial-Derived Antimicrobial Nanoparticles across Various Industrial Domains: Biomedicine, Textiles, Household Products, Cosmetics, and Coatings

Area	Applications	Reference
Biomedicine	Antimicrobial nanoparticles play a crucial role in biomedicine, where controlling microbial contamination is paramount. Applications include the development of antimicrobial coatings for medical devices, drug delivery systems, and wound dressings.	[108]
Textiles	The incorporation of antimicrobial nanoparticles in textiles has revolutionized the textile industry. Nanoparticles such as silver have been integrated into fabrics to impart antimicrobial properties, reducing the growth of bacteria and fungi.	[109]
Household products	Antimicrobial nanoparticles find their way into various household products, including cleaning agents, kitchenware, and personal care items. Their incorporation enhances the antimicrobial properties of these products, contributing to a healthier living environment.	[110]
Cosmetics	The cosmetic industry benefits from the antimicrobial properties of nanoparticles, particularly in the formulation of skincare and personal care products. Nanoparticles contribute to the preservation of cosmetics by inhibiting the growth of harmful microorganisms.	[111]
Coatings	Antimicrobial coatings have gained prominence in various industries, including healthcare, food processing, and transportation. Surfaces coated with antimicrobial nanoparticles resist microbial colonization, reducing the risk of contamination and infection.	[112]

through microbial processes are integrated into food-packaging materials to prevent bacterial growth and extend shelf life. These NPs effectively inhibit pathogens like *E. coli* and *Salmonella* on food surfaces, ensuring safety and reducing spoilage [103]. Similarly, zinc oxide nanoparticles (ZnO NPs) are used as food preservatives due to their antimicrobial properties and ability to degrade into biocompatible products, making them suitable for direct food contact [96].

In the **healthcare sector**, microbial-derived NPs are incorporated into wound dressings, coatings for medical devices, and surgical instruments. Gold nanoparticles (Au NPs), produced using bacteria like *P. aeruginosa*, are employed in wound-healing and antimicrobial coatings due to their biocompatibility and antibacterial properties [70]. Titanium dioxide nanoparticles (TiO_2 NPs), synthesized via fungal or bacterial processes, are used in hospital sanitation products for their dual role as antimicrobial agents and photocatalysts, enabling effective disinfection of hospital surfaces [105].

In the **cosmetics and environmental sectors**, NPs serve as critical components for antimicrobial applications. For instance, copper nanoparticles (Cu NPs) are included in personal care products like deodorants and shampoos to prevent microbial growth. These NPs are also utilized in **water treatment systems**, where their potent antimicrobial effects neutralize harmful microorganisms in drinking water and industrial wastewater [106]. Furthermore, iron oxide nanoparticles (Fe_3O_4 NPs) are applied in bioremediation processes, aiding the removal of contaminants and microbial pathogens from polluted water sources [107]. The versatility of microbial-derived NPs highlights their potential for enhancing industrial efficiency and sustainability.

A recent account of the various microbial-derived antimicrobial NPs under clinical trials has been compiled by Mondal et al. [113]. In Phase 1 clinical trials, silver NPs are being tested to prevent COVID-19, fight against *nosocomial* bacteria, and for topical applications. PLGA and chitosan NPs are also observed as smart gels and respirators, respectively. In Phase 4 clinical trials, PLGA NPs that have entrapped ciprofloxacin are being tested for the treatment of *E. faecalis* infections in endodontics. Among other metallic NPs, iron-based ferumoxytol NPs are being tested as antibiofilm treatment. The antimicrobial effect of titanium dioxide-based NPs is observed in complete dentures made for edentulous patients. Lastly, zinc oxide–based NPs are being capped for industrial usage in the treatment of fungal feet infection (WhitfieldZinc) [113].

5.7 FUTURE DIRECTIONS AND POTENTIAL APPLICATIONS

Anticipating future directions and potential applications is crucial for staying at the forefront of research and innovation in the field of antimicrobial NPs.

5.7.1 Technological Advancements

5.7.1.1 Nanoscale Imaging Techniques

Technological advancements in nanoscale imaging techniques, such as super-resolution microscopy and *in situ* imaging, offer unprecedented insights into the behavior of antimicrobial NPs at the molecular level.

Nanoscale imaging techniques have revolutionized our understanding of the behavior of antimicrobial NPs by providing detailed insights at the molecular level. Technological advancements in methods such as super-resolution microscopy and *in situ* imaging allow scientists to observe and analyze NPs with remarkable precision. Super-resolution microscopy, for instance, breaks the diffraction limit of conventional light microscopy, enabling visualization of structures and interactions at a resolution previously thought unattainable. This level of detail is crucial for studying the intricate mechanisms by which antimicrobial NPs interact with microbial cells. By observing these interactions in real time, researchers can gain a deeper understanding of how NPs penetrate microbial membranes, disrupt cellular processes, and ultimately lead to cell death. This knowledge is invaluable for the development and optimization of NP-based antimicrobial agents [53,114].

One of the prime examples of the application of these imaging techniques is in the study of silver nanoparticles (Ag NPs), which are renowned for their potent antimicrobial properties. Using super-resolution microscopy, researchers can observe how Ag NPs attach to bacterial cell membranes, penetrate the cells, and release silver ions that cause cellular damage. *In situ* imaging further enhances this understanding by allowing scientists to monitor these processes within the natural environments where NPs operate, such as in biological tissues or biofilms. Other NPs, such as zinc oxide (ZnO) and titanium dioxide (TiO_2), have also been studied using these techniques. ZnO NPs are widely used for their antibacterial and antifungal properties, particularly in medical and cosmetic applications. TiO_2 NPs, on the other hand, are known for their photocatalytic properties, which are harnessed in applications ranging from self-cleaning surfaces to antimicrobial coatings in medical devices. The ability to image these NPs at the nanoscale has not only improved our understanding of their antimicrobial mechanisms but also guided the design of more effective and targeted NP-based therapies [115].

5.7.1.2 High-Throughput Screening

The development of high-throughput screening methods for evaluating antimicrobial NPs accelerates the discovery process. High-throughput screening (HTS) methods have significantly advanced the evaluation of antimicrobial NPs, enabling rapid and efficient assessment of their efficacy against a wide range of pathogens. By allowing the simultaneous testing of thousands of NP formulations, HTS accelerates the discovery process, making it possible to identify the most potent antimicrobial candidates in a fraction of the time required by traditional methods [116,117]. For example, HTS has been used to screen various silver NP formulations to determine their effectiveness against drug-resistant bacteria, such as *S. aureus* and *E. coli*. This approach not only speeds up the identification of promising antimicrobial agents but also aids in optimizing their properties, such as particle size, shape, and surface charge, to enhance their therapeutic potential. The application of HTS in NP research is crucial for developing next-generation antimicrobial therapies that can address the growing challenge of antibiotic resistance [118].

5.7.2 Societal and Environmental Considerations

5.7.2.1 Safety and Toxicity Assessment

The increasing application of antimicrobial NPs in medicine, food preservation, and environmental disinfection necessitates an in-depth understanding of their safety profiles. The unique physicochemical properties of NPs, such as their high surface-to-volume ratio and ability to cross biological barriers, can lead to unexpected toxicological effects. These include oxidative stress, inflammation, and potential genotoxicity [119]. Moreover, the size, shape, coating, and surface charge of NPs can significantly influence their interaction with cellular components and their toxicity levels [120]. As NPs are integrated into products intended for human use, rigorous assessment frameworks are essential to evaluate acute and chronic exposure risks.

Future research aims to address these safety concerns by developing standardized toxicity assessment protocols tailored to NPs. Such protocols would include *in vitro* and *in vivo* models that reflect realistic exposure scenarios, ensuring the reliability of toxicity data. Additionally, advanced computational models and HTS methods are expected to complement experimental studies, providing a more comprehensive understanding of NP safety [120]. By prioritizing safety evaluations, researchers can minimize risks and enhance the acceptability of antimicrobial NPs across diverse applications.

5.7.2.2 Environmental Impact

As the utilization of antimicrobial NPs expands across industries, their environmental impact becomes a critical area of concern. NPs released into the environment through industrial effluents, agricultural applications, or wastewater discharge can accumulate in soil and water. Their persistence and unique properties, such as reactivity and bioavailability, may disrupt ecological balance by affecting microbial communities essential for nutrient cycling [121]. For instance, NPs like silver and zinc oxide are known to exhibit antimicrobial effects, potentially altering soil and aquatic ecosystems [122]. Understanding the fate, transport, and transformation of NPs in various environmental matrices is key to assessing their long-term impact.

Addressing these concerns requires interdisciplinary efforts to develop eco-friendly NPs and establish environmental safety guidelines. Current research focuses on biocompatible coatings, degradable nanomaterials, and green synthesis methods to minimize harm [121]. Furthermore, predictive models and environmental monitoring systems are being developed to evaluate the interactions between NPs and biological systems in real-world conditions [122]. Such initiatives will pave the way for sustainable NP applications while protecting environmental health.

5.7.3 Patents

NPs have found a huge success in various fields, particularly for biomedical applications due to their unique properties. This is the reason why NPs have been the focus of several patents in the past decade. These patents aim at enhancing the stability and antimicrobial efficacy of the NPs, while ensuring easy and favorable synthesis processes. They have contributed to tackling the problems associated with varied clinical outcomes and antibiotic resistance. Though there are a number of patents related to NPs, we have focused on the bio-based and microbial synthesis of NPs targeting antimicrobial applications.

The patent US 9567610 B2 describes the biological synthesis of gold NPs from the strains of *Botrytis cinerea* and the molecules produced by the strains of *B. cinerea* to target a number of applications such as therapies against some types of cancer, medical diagnostics, and biomedicine in general, to name a few [123].

Biogenic silver NPs prepared using extracts of *Alternanthera bettzickiana* (US 10828328 B1) have shown strong antimicrobial activity against various strains including drug-resistant microorganisms. Majorly, they have also shown anticancer activity against certain human cancer cell lines [124].

As per the patent US 10590438 B1, metal NPs, specifically, copper sulfate NPs, have been synthesized from fungal extract for successful applications as antimicrobial agents [125]. Also, selenium nanostructures synthesized from bacterial strains (WO 2020206459 A1) have proved to be successful in studies for inhibiting cancer cells and bacterial cells which are resistant to drugs. They eliminate the use of toxic components for chemical synthesis and do not produce by-products that are toxic in nature [126].

These patents clearly symbolize the advancements in the field of antimicrobial NPs, enabling its successful application in medical fields by enhancing the stability of the particles and refining the synthesis methods of the particles.

5.8 CHALLENGES AND CONSIDERATIONS

There are many challenges and considerations to be made regarding antimicrobial NPs as shown in Table 5.6.

Addressing these challenges and considerations is integral to advancing the field of antimicrobial NPs responsibly. By acknowledging potential obstacles and providing strategies for overcoming

TABLE 5.6
Challenges and Considerations of Antimicrobial Nanoparticles

Challenges	Considerations
Nanoparticle stability and aggregation	One of the challenges in the production of antimicrobial nanoparticles lies in maintaining their stability and preventing aggregation. Factors such as pH, temperature, and interaction with surrounding environments can influence nanoparticle stability.
Biocompatibility and cytotoxicity	The biocompatibility of antimicrobial nanoparticles is a critical consideration, especially in biomedical applications. Assessing their impact on living cells and tissues is essential to avoid adverse effects.
Resistance and adaptation	Microorganisms have the potential to develop resistance to antimicrobial agents, including nanoparticles.
Scalability and production costs	Scaling up the production of antimicrobial nanoparticles for practical applications can pose challenges.
Regulatory and ethical considerations	The application of antimicrobial nanoparticles in diverse industries raises regulatory and ethical considerations.
Environmental impact and sustainability	Understanding the environmental impact of antimicrobial nanoparticles is crucial for responsible technological development.
Public perception and acceptance	Public perception and acceptance of products containing antimicrobial nanoparticles play a vital role in their successful integration into society.

them, this section contributes to the development of a comprehensive and ethical framework for the use of antimicrobial NPs.

5.9 CONCLUSION

In conclusion, the production and utilization of antimicrobial NPs represent a significant advancement in the field of nanotechnology, offering unprecedented opportunities for addressing microbial challenges across diverse industries. Throughout this chapter, we have explored the foundational principles of nanotechnology, the synthesis methods of antimicrobial NPs, their characterization techniques, antimicrobial activity assessment methodologies, and their wide-ranging applications.

Nanotechnology, with its ability to manipulate materials at the nanoscale, has enabled the engineering of NPs with unique physicochemical properties, particularly suited for antimicrobial applications. Through various synthesis methods including physical, chemical, and biological approaches, NPs such as silver, gold, copper, and titanium dioxide have been tailored to exhibit potent antimicrobial activity.

Characterization techniques such as UV–Vis spectrophotometry, SEM, and TEM provide valuable insights into the size, shape, and composition of NPs, essential for understanding their antimicrobial efficacy. Antimicrobial activity assessment methodologies, including ZOI studies, offer a quantitative measure of the effectiveness of NPs against a spectrum of microorganisms.

The applications of antimicrobial NPs span across biomedicine, textiles, household products, cosmetics, and coatings, where they contribute to infection control, product preservation, and environmental hygiene. Future directions in the field involve the development of advanced synthesis techniques, targeted delivery systems, and smart nanomaterials, alongside technological advancements in nanoscale imaging and HTS.

As we move forward, it is imperative to address societal and environmental considerations, including safety and toxicity assessment, as well as the environmental impact of antimicrobial NPs. By embracing interdisciplinary collaboration and responsible innovation, we can harness the full potential of antimicrobial NPs to create a safer, healthier, and more sustainable future.

REFERENCES

1. Shareef, J., et al., Synthesis and characterization of silver nanoparticles from *Penicillium sps. Materials Today: Proceedings*, 2017. **4**(11): pp. 11923–11932.
2. Silva, G.A., Introduction to nanotechnology and its applications to medicine. *Surgical Neurology*, 2004. **61**(3): pp. 216–220.
3. Ray, M.K., et al., Nanotechnology as a promising tool against phytopathogens: A futuristic approach to agriculture. *Agriculture*, 2023. **13**: p. 1856. https://doi.org/10.3390/agriculture13091856.
4. Nayak, D., et al., Eco-friendly mycogenic synthesis of iron oxide nanoparticles using volvariella volvacea and unveiling its potential as antibiofilm and anticancer agents. *ECS Journal of Solid State Science and Technology*, 2024. **13**(8): p. 087003.
5. Sriram, P. and A. Suttee, Nanotechnology advances, benefits, and applications in daily life, in Cherry Bhargava, Amit Sachdeva (eds.) *Nanotechnology*. 2020, Boca Raton: CRC Press. pp. 23–44.
6. Dhabalia, D., et al., Antifungal activity of biosynthesized silver nanoparticles from Candida albicans on the strain lacking the CNP41 gene. *Materials Research Express*, 2020. **7**(12): p. 125401.
7. Ukkund, S.J., et al., Adsorption method for the remediation of brilliant green dye using halloysite nanotube: Isotherm, kinetic and modeling studies. *Applied Sciences*, 2021. **11**(17): p. 8088.
8. Afzal, A. and S.J. Ukkund. Experimental investigation on physical and thermal properties of graphite nanofluids. in *AIP Conference Proceedings*. 2018. Bangalore, India: AIP Publishing.
9. Mohanta, Y.K., et al., Promising applications of phyto-fabricated silver nanoparticles: Recent trends in biomedicine. *Biochemical and Biophysical Research Communications*, 2023. **688**: p. 149126.
10. Kumar, R., M. Kumar, and G. Luthra, Fundamental approaches and applications of nanotechnology: A mini review. *Materials Today: Proceedings*, 2023. https://doi.org/10.1016/j.matpr.2022.12.172.
11. Rai, M., A. Yadav, and A. Gade, Silver nanoparticles as a new generation of antimicrobials. *Biotechnology Advances*, 2009. **27**(1): pp. 76–83.
12. Ukkund, S.J., et al., Role of microbial nanotechnology in energy devices, in *Handbook of Microbial Nanotechnology*. 2022, Elsevier. pp. 517–547. https://doi.org/10.1016/B978-0-12-823426-6.00017-6.
13. Khan, I., K. Saeed, and I. Khan, Nanoparticles: Properties, applications and toxicities. *Arabian Journal of Chemistry*, 2019. **12**(7): pp. 908–931.
14. Rizwan, M., et al., Types and classification of nanomaterials, in *Nanomaterials: Synthesis, Characterization, Hazards and Safety*. 2021, Elsevier. pp. 31–54. https://doi.org/10.1016/B978-0-12-823823-3.00001-X.
15. Mohanta, Y.K., et al., Amelioration of gold nanoparticles mediated through Ocimum oil extracts induces reactive oxygen species and mitochondrial instability against MCF-7 breast carcinoma. *RSC Advances*, 2024. **14**(38): pp. 27816–27830.
16. Bhat, N., et al., GO/CuO nanohybrid-based carbon dioxide gas sensors with an arduino detection unit. *ACS Omega*, 2023. **8**(36): pp. 32512–32519.
17. Contera, S., J.B. de la Serna, and T.D. Tetley, Biotechnology, nanotechnology and medicine. *Emerging Topics in Life Sciences*, 2020, **4**(6): pp. 551–554.
18. Kumar, P., et al., Nanotechnology and its challenges in the food sector: A review. *Materials Today Chemistry*, 2020. **17**: p. 100332.
19. Prasad, P., et al., Azobenzene dye-sensitised solar cells using TiO_2 nanoparticles and carbon nanotubes. *International Journal of Advanced Research Science and Engineering*, 2017. **6**: pp. 743–749.
20. Seil, J.T. and T.J. Webster, Antimicrobial applications of nanotechnology: Methods and literature. *International Journal of Nanomedicine*, 2012: pp. 2767–2781.
21. Bhushan, B., Introduction to nanotechnology, in *Springer Handbook of Nanotechnology*. 2017: pp. 1–19. https://doi.org/10.1007/978-3-642-02525-9_1.
22. Omran, B.A., *Nanobiotechnology: A Multidisciplinary Field of Science*. 2020: Springer, Cham.
23. Garg, P., et al., Advances and perspective on antimicrobial nanomaterials for biomedical applications. *Frontiers in Nanotechnology*, 2022. **4**: p. 898411.
24. Kimta, N., et al., Investigation on the influence of silver doping on structural and biomedical properties of zinc oxide nanoparticles synthesized using crude ethanolic extract of *Adiantum venustum* D. Don. *Emergent Materials*, 2024. **7**: pp. 1–27.
25. Ghotekar, S., et al., A bio-waste-mediated novel fabrication of $ZnCr_2O_4$ nanoparticles: A heterogeneous catalyst in Biginelli reaction. *Research on Chemical Intermediates*, 2024. **50**(5): pp. 2199–2219.
26. Wiley, B., Y. Sun, and Y. Xia, Synthesis of silver nanostructures with controlled shapes and properties. *Accounts of Chemical Research*, 2007. **40**(10): pp. 1067–1076.

27. Lakshmayya, N., et al., Cultivation and conservation of crops: Perspectives for agriculture through nanotechnology, in *Sustainable Green Nanotechnology*. CRC Press. pp. 152–163. https://doi.org/10.1201/9781003389408-9.
28. Ahmad, A., et al., Extracellular biosynthesis of silver nanoparticles using the fungus Fusarium oxysporum. *Colloids and surfaces B: Biointerfaces*, 2003. **28**(4): pp. 313–318.
29. Divya, K., et al., Green synthesis of silver nanoparticles by Escherichia coli: Analysis of antibacterial activity. *Journal of Water and Environmental Nanotechnology*, 2016. **1**(1): pp. 63–74.
30. Rónavári, A., et al., Green silver and gold nanoparticles: Biological synthesis approaches and potentials for biomedical applications. *Molecules*, 2021. **26**(4): p. 844.
31. Sharifabady, S.M., et al., Biosynthesis of zinc oxide nanoparticles by lactobacillus spp. and investigation of their antimicrobial effect. *Current Drug Discovery Technologies*, 2023. **20**(5): pp. 10–18.
32. MubarakAli, D., et al., Plant extract mediated synthesis of silver and gold nanoparticles and its antibacterial activity against clinically isolated pathogens. *Colloids and Surfaces B: Biointerfaces*, 2011. **85**(2): pp. 360–365.
33. Rajakumar, G., et al., Fungus-mediated biosynthesis and characterization of TiO_2 nanoparticles and their activity against pathogenic bacteria. *Spectrochimica Acta Part A: Molecular and Biomolecular Spectroscopy*, 2012. **91**: pp. 23–29.
34. Kharchenko, Y., et al., Antibacterial activity of green synthesised silver nanoparticles on Saccharomyces cerevisiae. *Applied Sciences*, 2022. **12**(7): p. 3466.
35. Manikprabhu, D. and K. Lingappa, Synthesis of silver nanoparticles using the Streptomyces coelicolor klmp33 pigment: An antimicrobial agent against extended-spectrum beta-lactamase (ESBL) producing Escherichia coli. *Materials Science and Engineering: C*, 2014. **45**: pp. 434–437.
36. Manivasagan, P., et al., Actinobacteria mediated synthesis of nanoparticles and their biological properties: A review. *Critical Reviews in Microbiology*, 2016. **42**(2): pp. 209–221.
37. Pradhan, N., Explorations towards the production of novel biogenic nanomaterial using Rhodococcus aetherivorans BCP1. 2023. https://doi.org/10.11575/PRISM/40758.
38. Jacob, J.M., et al., Microalgae: A prospective low cost green alternative for nanoparticle synthesis. *Current Opinion in Environmental Science & Health*, 2021. **20**: p. 100163.
39. Yue, Y., et al., Virus-templated synthesis of metal nanomaterials and their application. *Progress in Chemistry*, 2019. **31**(7): p. 1007.
40. Wadhwani, S.A., et al., Biogenic selenium nanoparticles: Current status and future prospects. *Applied Microbiology and Biotechnology*, 2016. **100**: pp. 2555–2566.
41. Musini, A., et al., Microbiologically synthesized nanoparticles and their role in biofilm inhibition, in *Microbial Processes for Synthesizing Nanomaterials*. 2023, Springer. pp. 285–315. https://doi.org/10.1007/978-981-99-2808-8_13.
42. Brierley, C.L., Microbiological mining. *Scientific American*, 1982. **247**(2): pp. 44–53.
43. Klaus, T., et al., Silver-based crystalline nanoparticles, microbially fabricated. *Proceedings of the National Academy of Sciences*, 1999. **96**(24): pp. 13611–13614.
44. Mukherjee, P., et al., Bioreduction of $AuCl_4$-ions by the fungus, Verticillium sp. and surface trapping of the gold nanoparticles formed. *Angewandte Chemie-International Edition*, 2001. **40**(19): pp. 3585–3588.
45. Riddin, T., M. Gericke, and C. Whiteley, Analysis of the inter-and extracellular formation of platinum nanoparticles by Fusarium oxysporum f. sp. lycopersici using response surface methodology. *Nanotechnology*, 2006. **17**(14): p. 3482.
46. Bhattacharya, D. and R.K. Gupta, Nanotechnology and potential of microorganisms. *Critical Reviews in Biotechnology*, 2005. **25**(4): pp. 199–204.
47. Mohanpuria, P., N.K. Rana, and S.K. Yadav, Biosynthesis of nanoparticles: Technological concepts and future applications. *Journal of Nanoparticle Research*, 2008. **10**: pp. 507–517.
48. Ray, M.K., et al., Mycoremediation of heavy metals and/or metalloids in soil, in *Land Remediation and Management: Bioengineering Strategies*, H. Sarma and S. Joshi, Editors. 2023, Springer Nature, Singapore. pp. 161–190.
49. Narayanan, K.B. and N. Sakthivel, Biological synthesis of metal nanoparticles by microbes. *Advances in Colloid and Interface Science,* 2010. **156**(1–2): pp. 1–13.
50. Panda, J., et al., Exploring biopolymer for food and pharmaceuticals application in the circular bioeconomy: An agro-food waste-to-wealth approach. *Waste and Biomass Valorization*, 2024. **15**: pp. 1–31.
51. Nisar, P., et al., Antimicrobial activities of biologically synthesized metal nanoparticles: An insight into the mechanism of action. *JBIC Journal of Biological Inorganic Chemistry*, 2019. **24**: pp. 929–941.
52. Kaur, M., K. Singh, and V. Kumar, Green synthesis of silver nanoparticles using *penicillium camemberti* and its biological applications. *BioNanoScience*, 2024. **14**: pp. 1–15.

53. Prakruthi, R., et al., Silver vanadate nanoparticles: Green synthesis, enhanced photocatalytic and antibacterial activity. *AIP Advances*, 2024. **14**(8): p. 085320.
54. Siddiqi, K.S., A. Husen, and R.A. Rao, A review on biosynthesis of silver nanoparticles and their biocidal properties. *Journal of Nanobiotechnology*, 2018. **16**: pp. 1–28.
55. Patil, M.P., et al., Extracellular synthesis of gold nanoparticles using the marine bacterium Paracoccus haeundaensis BC74171T and evaluation of their antioxidant activity and antiproliferative effect on normal and cancer cell lines. *Colloids and Surfaces B: Biointerfaces*, 2019. **183**: p. 110455.
56. Kandikonda, R., et al., Biosynthesis of Ag and Cu nanoparticles and their interactions with cyanobacteria, microalgae, and macroalgae. *Importance & Applications of Nanotechnology*. 2021. **3**: pp. 15–25.
57. Singh, P., et al., Biological synthesis of nanoparticles from plants and microorganisms. *Trends in Biotechnology*, 2016. **34**(7): pp. 588–599.
58. Wen, A.M. and N.F. Steinmetz, Design of virus-based nanomaterials for medicine, biotechnology, and energy. *Chemical Society Reviews*, 2016. **45**(15): pp. 4074–4126.
59. Rasal, R.K., et al., Synthesis and characterization of nanoparticles for antimicrobial applications-a review: Special column on microbial nanotechnology. *Applied Microbiology: Theory & Technology*, 2023. **4**: pp. 73–98.
60. Vigneshwaran, N., et al., Biomimetics of silver nanoparticles by white rot fungus, Phaenerochaete chrysosporium. *Colloids and Surfaces B: Biointerfaces*, 2006. **53**(1): pp. 55–59.
61. Sajjad, S., et al., Green synthesis of metal-based nanoparticles and their applications, in *Green Metal Nanoparticles: Synthesis, Characterization and Their Applications*, 2018. pp. 23–77. https://doi.org/10.1002/9781119418900.ch2.
62. Sweeney, R.Y., et al., Bacterial biosynthesis of cadmium sulfide nanocrystals. *Chemistry & Biology*, 2004. **11**(11): pp. 1553–1559.
63. Prasad, K. and A.K. Jha, Biosynthesis of CdS nanoparticles: An improved green and rapid procedure. *Journal of Colloid and Interface Science*, 2010. **342**(1): pp. 68–72.
64. Bai, H.-J., Z.-M. Zhang, and J. Gong, Biological synthesis of semiconductor zinc sulfide nanoparticles by immobilized Rhodobacter sphaeroides. *Biotechnology Letters*, 2006. **28**: pp. 1135–1139.
65. Ahmad, A., et al., Enzyme mediated extracellular synthesis of CdS nanoparticles by the fungus, Fusarium oxysporum. *Journal of the American Chemical Society*, 2002. **124**(41): pp. 12108–12109.
66. Dameron, C., et al., Biosynthesis of cadmium sulphide quantum semiconductor crystallites. *Nature*, 1989. **338**(6216): pp. 596–597.
67. Saravanakumar, K., et al., Fungal enzyme-mediated synthesis of chitosan nanoparticles and its biocompatibility, antioxidant and bactericidal properties. *International Journal of Biological Macromolecules*, 2018. **118**: pp. 1542–1549.
68. Shamim, A., T. Mahmood, and M.B. Abid, Biogenic synthesis of zinc oxide (ZnO) nanoparticles using a fungus (Aspergillus niger) and their characterization. *International Journal of Chemistry*, 2019. **11**(2): p. 119.
69. Ukkund, S.J., et al., Synthesis and characterization of silver nanoparticles from Fuzarium Oxysporum and investigation of their antibacterial activity. *Materials Today: Proceedings*, 2019. **9**: pp. 506–514.
70. Prashanth, K., *Biosynthesis of nanoparticles from actinomycetes and their biomedical applications*. 2023, Doctoral thesis, Kuvempu University.
71. Alavi, M., Bacteria and fungi as major bio-sources to fabricate silver nanoparticles with antibacterial activities. *Expert Review of Anti-infective Therapy*, 2022. **20**(6): pp. 897–906.
72. Mechouche, M.S., et al., Biosynthesis, characterization, and evaluation of antibacterial and photocatalytic methylene blue dye degradation activities of silver nanoparticles from Streptomyces tuirus strain. *Environmental Research*, 2022. **204**: p. 112360.
73. Jain, P.K., et al., Calculated absorption and scattering properties of gold nanoparticles of different size, shape, and composition: Applications in biological imaging and biomedicine. *The Journal of Physical Chemistry B*, 2006. **110**(14): pp. 7238–7248.
74. Nel, A., et al., Toxic potential of materials at the nanolevel. *Science*, 2006. **311**(5761): pp. 622–627.
75. Technology Networks, https://www.technologynetworks.com/analysis/articles/uv-vis-spectroscopy-principle-strengths-and-limitations-and-applications-349865. UV-vis spectroscopy characterization tool 2024 28/11/2024].
76. Ukkund, S.J., et al., Microwave assisted green synthesis and characterization of silver nanoparticles from Hibiscus leaf extract and investigation of their antimicrobial activities. in *AIP Conference Proceedings*. 2019. Mangalore, India: AIP Publishing.
77. Kirdat, P., et al., Synthesis and characterization of ginger (Z. officinale) extract mediated iron oxide nanoparticles and its antibacterial activity. *Materials Today: Proceedings*, 2021. **43**: pp. 2826–2831.

78. Tribonet. https://www.tribonet.org/wiki/scanning-electron-microscope/. SEM characterization tool 2024 28/11/2024].
79. Ukkund, S.J., et al., Antimicrobial coating of fabric by biosynthesized silver nanoparticles from Panchakavya. *Nano Express*, 2021. **2**(1): p. 010033.
80. Nanoscience. https://www.nanoscience.com/techniques/transmission-electron-microscopy/. TEM characterization tool 2024 28/11/2024].
81. Wilson, B.K. and R.K. Prud'homme, Nanoparticle size distribution quantification from transmission electron microscopy (TEM) of ruthenium tetroxide stained polymeric nanoparticles. *Journal of Colloid and Interface Science*, 2021. **604**: pp. 208–220.
82. Murei, A., et al., Synthesis, characterization and in vitro antibacterial evaluation of Pyrenacantha grandiflora conjugated silver nanoparticles. *Nanomaterials*, 2021. **11**(6): p. 1568.
83. Quate, C.F., *Scanning Tunneling Microscopy*. N for N, 1990: p. 22. https://apps.dtic.mil/sti/tr/pdf/ADA227576.pdf#page=24
84. Chen, C., *Introduction to Scanning Tunneling Microscopy*. 2008: Oxford University Press, Oxford.
85. Chen, C.J., *Introduction to Scanning Tunneling Microscopy*, Third Edition. Vol. 69. 2021: Oxford University Press, Oxford.
86. Feroze, N., et al., Fungal mediated synthesis of silver nanoparticles and evaluation of antibacterial activity. *Microscopy Research and Technique*, 2020. **83**(1): pp. 72–80.
87. Singh, M., et al., Nanotechnology in medicine and antibacterial effect of silver nanoparticles. *Digest Journal of Nanomaterials and Biostructures*, 2008. **3**(3): pp. 115–122.
88. Alsamhary, K.I., Eco-friendly synthesis of silver nanoparticles by Bacillus subtilis and their antibacterial activity. *Saudi Journal of Biological Sciences*, 2020. **27**(8): pp. 2185–2191.
89. Graily-Moradi, F., A. Maadani Mallak, and M. Ghorbanpour, Biogenic synthesis of gold nanoparticles and their potential application in agriculture, in *Biogenic Nano-particles and Their Use in Agro-Ecosystems*, 2020: pp. 187–204. https://doi.org/10.1007/978-981-15-2985-6_11.
90. Zhang, X., Gold nanoparticles: Recent advances in the biomedical applications. *Cell Biochemistry and Biophysics*, 2015. **72**: pp. 771–775.
91. Sánchez-López, E., et al., Metal-based nanoparticles as antimicrobial agents: An overview. *Nanomaterials*, 2020. **10**(2): p. 292.
92. Padmavathy, N. and R. Vijayaraghavan, Enhanced bioactivity of ZnO nanoparticles—An antimicrobial study. *Science and Technology of Advanced Materials*, 2008. **9**: p. 035004.
93. Anucha, C.B., et al., Titanium dioxide (TiO_2)-based photocatalyst materials activity enhancement for contaminants of emerging concern (CECs) degradation: In the light of modification strategies. *Chemical Engineering Journal Advances*, 2022. **10**: p. 100262.
94. Ren, G., et al., Characterisation of copper oxide nanoparticles for antimicrobial applications. *International Journal of Antimicrobial Agents*, 2009. **33**(6): pp. 587–590.
95. Dykman, L. and N. Khlebtsov, Gold nanoparticles in biomedical applications: Recent advances and perspectives. *Chemical Society Reviews*, 2012. **41**(6): pp. 2256–2282.
96. Sharma, V.K., R.A. Yngard, and Y. Lin, Silver nanoparticles: Green synthesis and their antimicrobial activities. *Advances in Colloid and Interface Science*, 2009. **145**(1–2): pp. 83–96.
97. Mishra, B., et al., Recent advances in sustainable biopolymer-based nanocomposites for smart food packaging: A review. *International Journal of Biological Macromolecules*, 2024. **279**: p. 135583.
98. Swidan, N.S., et al., Antibiofilm activity of green synthesized silver nanoparticles against biofilm associated enterococcal urinary pathogens. *Scientific Reports*, 2022. **12**(1): p. 3869.
99. Kalishwaralal, K., et al., Silver nanoparticles impede the biofilm formation by Pseudomonas aeruginosa and Staphylococcus epidermidis. *Colloids and Surfaces B: Biointerfaces*, 2010. **79**(2): pp. 340–344.
100. Azam, A., et al., Antimicrobial activity of metal oxide nanoparticles against Gram-positive and Gram-negative bacteria: A comparative study. *International Journal of Nanomedicine*, 2012: pp. 6003–6009.
101. Chand, K., et al., Photocatalytic and antimicrobial activity of biosynthesized silver and titanium dioxide nanoparticles: A comparative study. *Journal of Molecular Liquids*, 2020. **316**: p. 113821.
102. Khairy, T., et al., Antibacterial activity of green synthesized copper oxide nanoparticles against multidrug-resistant bacteria. *Scientific Reports*, 2024. **14**(1): p. 25020.
103. Ali, S., et al., Advancements and challenges in phytochemical-mediated silver nanoparticles for food packaging: Recent review (2021–2023). *Trends in Food Science & Technology*, 2023. **141**: p. 104197.
104. Kungwani, N.A., et al., Combating bacterial biofilms and related drug resistance: Role of phyto-derived adjuvant and nanomaterials. *Microbial Pathogenesis*, 2024. **195**: p. 106874.

105. Sunny, N.E., et al., Green synthesis of titanium dioxide nanoparticles using plant biomass and their applications: A review. *Chemosphere*, 2022. **300**: p. 134612.
106. Joseph, T.M., et al., Nanoparticles: Taking a unique position in medicine. *Nanomaterials*, 2023. **13**(3): p. 574.
107. Quici, N., J. Crespi, and V.N. Montesinos, LCA applied to nanomaterials for environmental remediation: Advances and challenges. 2023. https://doi.org/10.1039/BK9781837670215-00264.
108. Nikzamir, M., A. Akbarzadeh, and Y. Panahi, An overview on nanoparticles used in biomedicine and their cytotoxicity. *Journal of Drug Delivery Science and Technology*, 2021. **61**: p. 102316.
109. Shah, M.A., et al., Applications of nanotechnology in smart textile industry: A critical review. *Journal of Advanced Research*, 2022. **38**: pp. 55–75.
110. Huang, W., et al., Antibacterial nanomaterials for environmental and consumer product applications. *NanoImpact*, 2020. **20**: p. 100268.
111. Carrouel, F., et al., Nanoparticles as anti-microbial, anti-inflammatory, and remineralizing agents in oral care cosmetics: A review of the current situation. *Nanomaterials*, 2020. **10**(1): p. 140.
112. Rodrigues, M.C., et al., Biogenic synthesis and antimicrobial activity of silica-coated silver nanoparticles for esthetic dental applications. *Journal of Dentistry*, 2020. **96**: p. 103327.
113. Mondal, S.K., et al., Antimicrobial nanoparticles: Current landscape and future challenges. *RSC Pharmaceutics*, 2024. **1**: p. 388.
114. Bruna, T., et al., Silver nanoparticles and their antibacterial applications. *International Journal of Molecular Sciences*, 2021. **22**(13): p. 7202.
115. Marinescu, L., et al., Optimized synthesis approaches of metal nanoparticles with antimicrobial applications. *Journal of Nanomaterials*, 2020. **2020**(1): p. 6651207.
116. Ibraheem, D.R., et al., Inhibitory effect of silver nanoparticles–conjugated PEG-nystatin against some resistance pathogenic bacteria. *Inorganic Chemistry Communications*, 2024. **168**: p. 112952.
117. Torres-Ramos, M.I., et al., Cu2O nanoparticles as nanocarriers and its antibacterial efficacy. *Pharmaceuticals*, 2024. **17**(9): p. 1124.
118. Fang, Z., et al., High-throughput screening and rational design of biofunctionalized surfaces with optimized biocompatibility and antimicrobial activity. *Nature Communications*, 2021. **12**(1): p. 3757.
119. Zhang, N., G. Xiong, and Z. Liu, Toxicity of metal-based nanoparticles: Challenges in the nano era. *Frontiers in Bioengineering and Biotechnology*, 2022. **10**: p. 1001572.
120. Noga, M., et al., Toxicological aspects, safety assessment, and green toxicology of silver nanoparticles (AgNPs)—Critical review: State of the art. *International Journal of Molecular Sciences*, 2023. **24**(6): p. 5133.
121. Umapathi, A., M. Kumawat, and H.K. Daima, Engineered nanomaterials for biomedical applications and their toxicity: A review. *Environmental Chemistry Letters*, 2022. **20**(1): pp. 445–468.
122. Pucciarelli, M., Life Cycle Thinking for the sustainability assessment of nanoparticles' manufacturing and applications. 2023, UCL (University College London).
123. Retamal, M.C. and A.C. Nara, Use of Botrytis cinerea for obtaining gold nanoparticles. 2017, Google Patents.
124. Ramalingam, R.J. and H. Al Lohedan, Method of preparing biogenic silver nanoparticles. 2020, Google Patents.
125. Alharbi, N.S.H.A., et al., Biosynthesis of metal nanoparticles. 2020, Google Patents.
126. Cruz, D.M. and T.J. Webster, Biosynthesis of selenium nanoparticles having antimicrobial activity. 2022, Google Patents.

6 Potential Applications of Microbial Nano Biosensors

Minisrang Daimary, Yugal Kishore Mohanta, Abdullah G. Al-Sehemi, and Hemen Sarma

6.1 INTRODUCTION

Nanotechnology is a multidisciplinary field that has developed over the past few decades and plays a significant role in the environment, industry, agriculture, and pharmacology. Currently, nanomaterials are being extensively investigated for use in various applications [1,2]. A sensor is a device created to directly measure a specific compound, known as an analyte, within a sample. Ideally, this device should offer continuous and reversible responses without causing any harm to the sample. A nano sensor, on the other hand, is a system that integrates at least one nanostructure to detect various factors like gases, chemicals, biological agents, electric fields, light, and heat. The inclusion of nanomaterials significantly enhances the sensitivity of the system. In biosensors, the element responsible for binding to the analyte and enabling precise detection is a biological component, such as a DNA strand, antibody, enzyme, or entire cell [3]. Nanomaterials are vital in developing highly selective and sensitive biosensors for detecting bioreceptors such as cells, enzymes, and antibodies. Their unique properties distinguish them from fine particles. The integration of nano-scaled materials into biosensors, known as nano biosensor technology, is transforming healthcare by enhancing the identification, observation, and diagnosis of pathogens, viruses, and bacteria [4].

Nano biosensors are devices that utilize a biological component as the detection component while an electrode functions as the transducer. Nanostructures are typically incorporated into these systems to bridge the gap between the bioreceptor, which operates at the nanoscale, and the converter. Based on the nature of the biomaterial detection process, electrochemical biosensors are categorized into catalytic and propulsion types [3]. Nanomaterials have gained prominence in life sciences, healthcare, electronics, and farming due to their wide-ranging applications. Biological synthesis, particularly involving microorganisms, offers a reliable, sustainable, and environmentally friendly alternative to chemical and physical methods. Bacteria, algae, fungi, and yeast absorb metals and transform them into nanoparticles, which are either accumulated or discharged. Additionally, viruses have been utilized to develop robust nanocarriers. During biosynthesis, capping agents are often secreted to prevent nanoparticle aggregation and enhance stabilization [5]. Incorporating nanomaterials into biosensors has significantly increased their sensitivity and boosted their overall functionality, enabling new signal transduction technologies. Advances in the fabrication, measurement, and imaging of nanoscale objects have led to sensors that interact with tiny molecules. Different nano biosensor designs, including mechanical systems, optical cavities, modified nanoparticles, nanowires, nanotubes, and nanofibers, are currently utilized. Research is actively focused on gold nanoparticles, carbon nanotubes, magnetic nanoparticles, and quantum dots for their potential in biosensor applications [6].

Microbial nano biosensors are emerging as powerful tools in various fields due to their unique combination of biological recognition elements and nanomaterials. They utilize the specificity of microbial cells and the enhanced sensitivity of nanomaterials to identify and measure a broad spectrum of analytes [7,8]. A microbial nano biosensor is a diagnostic device that integrates microorganisms

DOI: 10.1201/9781003487142-6

with a transducer to detect specific substances. The development of nanotechnology has facilitated the use of nanomaterials to enhance immobilization, making microbial biosensors more reliable and selective. Additionally, advancements in transducer technology have significantly improved sensitivity. Microbial nano biosensors have become highly effective for monitoring environmental, food, and clinical samples [9]. Microorganisms have been engineered as biosensors, offering several benefits including the capability to detect various substances, lower costs, potential for scalable production, and simpler genetic modification compared to systems that use enzymes and mammalian cells [10]. A microbial nano biosensor typically features a transducer paired with either viable or dead microbial cells. Dead microbial cells, which can be obtained through permeabilization or those containing periplasmic enzymes, are often used as cost-effective alternatives to enzymes. By contrast, living cells use their respiratory and metabolic activities, where the analyte acts as either a reactant or an inhibitor of these functions. Microbial biosensors utilizing bioluminescence have been created employing genetically modified organisms that produce the luciferase gene combined with a regulated promoter, enabling the evaluation of toxicity and bio accessibility [11]. A range of bioreceptors, such as cofactors, enzymes, antibodies, microbes, cellular organelles, tissue samples, and cells from more complex organisms, have been utilized in the development of biosensors. Of these elements, enzymes are the most commonly employed materials due to their exceptional specificity and sensitivity [11,12]. Although purified enzymes demonstrate high specificity for their substrates or inhibitors, their application in biosensor development can be restricted by the intricate, lengthy, and costly enzyme purification process, the need for multiple enzymes to generate a detectable signal, or the dependence on cofactors or coenzymes. Microorganisms offer a promising alternative to overcome these limitations [13]. The variety of enzymes and cofactors present in cells enables them to detect and process a broad range of chemicals, though this can sometimes reduce their selectivity. Nonetheless, cells can be readily modified and adjusted to handle and break down new substrates under specific cultivation conditions [12]. Furthermore, advancements in molecular biology and recombinant DNA technologies have provided countless opportunities to customize microorganisms, either by enhancing the function of existing enzymes or by introducing external enzymes or proteins into recipient cells. These developments make microbes highly effective as biosensing elements [14].

A key application of microbial nano biosensors is in environmental monitoring, where they are utilized to identify pollutants, harmful substances, and microorganisms in air, water, and soil. For instance, these biosensors have been employed to monitor heavy metals, pesticides, and organic pollutants, offering real-time and on-site detection capabilities [15]. In the field of medical diagnostics, microbial nano biosensors have shown promise in early disease detection, monitoring of disease progression, and personalized medicine. They can detect biomarkers, pathogens, and other disease-related molecules with high accuracy, providing valuable information for timely medical intervention [16]. Food safety is another critical area where microbial nano biosensors are making a significant impact. They are used to detect contaminants such as pathogens, toxins, and food allergens, maintaining safety and quality throughout the food distribution [17]. Additionally, in industrial processes, these biosensors can monitor the presence of microorganisms that could affect product quality and safety, thereby improving the efficiency and reliability of production processes [18].

Although microbial biosensors offer significant potential for a range of detection applications, they continue to encounter challenges like limited specificity, limited sensitivity, and difficulties with portability. To address these issues, microbial biosensors have been combined with modern micro- and nanoscale technologies, broadening their range of detection purposes [19]. Microbial nano biosensors represent a versatile and powerful tool in modern detection technologies, offering unparalleled sensitivity, specificity, and rapid response times across various applications. The ongoing progress in nanotechnology and microbiology continues to drive the advancement of more sophisticated and efficient biosensors, opening new avenues for their application and enhancing their impact on society (Figure 6.1).

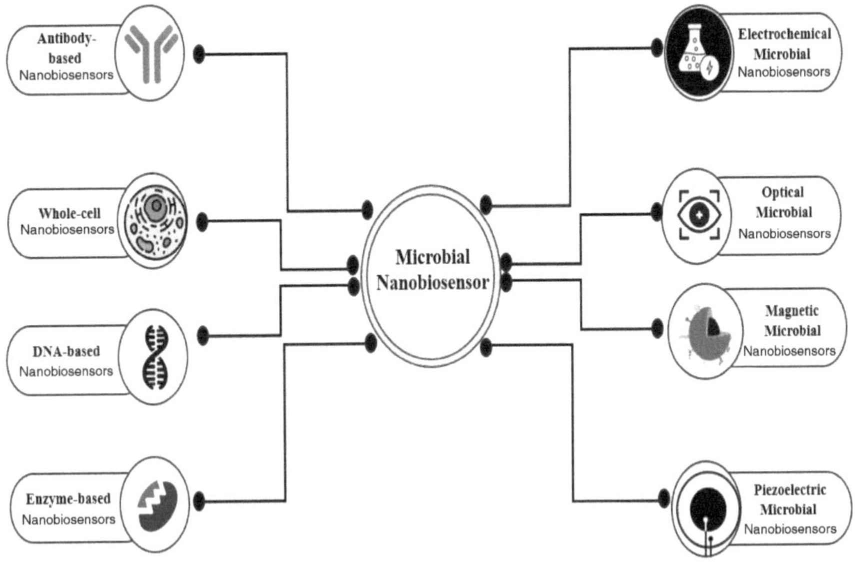

FIGURE 6.1 Schematic diagram of microbial nano biosensor.

6.2 PRINCIPLES AND MECHANISMS

Microbial nano biosensors utilize microbial cells or their components as recognition elements, providing high specificity for target analytes. These interactions are converted into measurable signals through optical, electrochemical, piezoelectric, or thermal methods. Nanomaterials enhance sensitivity and response time because of their large surface area and distinctive electronic characteristics [7,8,11]. The biological component of a biosensor is tasked with the specific interaction with the target analyte, while the transducer transforms this interaction into a quantifiable signal. For example, mechanical transducers convert the interaction between the analyte and bioreceptor into changes in deformation or resonant frequency. Optical transducers usually convert this interaction into variations in light wavelength or intensity, while electrochemical transducers convert it into variations in current or voltage. The measurement system then assesses the extent of these changes [3].

6.2.1 Types of Microbial Nano Biosensors

Microbial nano biosensors can be categorized based on their sensing mechanisms and the types of interactions they detect.

6.2.1.1 Electrochemical Nano Biosensors

Electrochemical microbial nano biosensors are highly popular for microbial detection owing to their exceptional sensing accuracy and versatility, making them ideal for applications like point-of-care testing devices [10]. Electrochemical biosensors have been employed in the analysis of food and beverages, detecting genetically modified organisms (GMO) content in food, and assessing food freshness [19]. An electrochemical nano biosensor is a self-contained, integrated device designed to deliver specific quantitative or semi-quantitative analytical information. It achieves this by utilizing a biological recognition element in direct spatial contact with an electrochemical transduction element [20]. Because of their low cost and ease of miniaturization, electrochemical nano biosensors are highly promising for applications where reducing size and price is essential, such as online monitoring of allergen contamination and point-of-care allergy diagnosis [21].

6.2.1.2 Optical Nano Biosensors

An optical biosensor is a compact small analytical device that integrates a biorecognition element with an optical transducer system [22]. Its primary objective is to generate a signal proportional to the concentration of the analyte being measured. Various biological materials, such as enzymes, antibodies, antigens, receptors, nucleic acids, whole cells, and tissues, can be used as biorecognition elements in an optical biosensor. Optical nano biosensors are highly effective detection tools with applications spanning food safety, healthcare, and environmental monitoring [23]. By integrating the selectivity of biological elements with the advanced processing capabilities of modern micro-electronics and optoelectronics, optical biosensors present significant analytical potential, particularly in food safety. Their ability to detect analytes in complex matrices with minimal sample preparation is a key advantage. Most optical biosensors operate by measuring changes in the surface properties of a sensor chip when the analyte binds to a sensing layer through sorption or complex formation with the receptor-target pair [23].

6.2.1.3 Magnetic Nano Biosensor

Magnetic nano sensing, an emerging field at the forefront of biosensing and magnetics, has rapidly advanced over the past decade, advancing biological and biomedical applications. Innovations in magnetic nanomaterials and nanoscale technologies have revolutionized diagnostics, enabling early-stage disease detection. Techniques such as magnetoresistance, magnetic particle analysis, and nuclear magnetic imaging are central to these advancements [24]. Magnetic biosensors based on surface and volume measurements have been designed for detecting viruses, disease-causing agents, cancer indicators, metal ions, and various other substances [24–26].

6.2.1.4 Piezoelectric Nano Biosensors

Piezoelectric nano biosensors are constructed with a slender quartz crystal sheet, which has metallic electrodes on both sides. The sensing region is defined by the overlap of these electrodes, which are usually made from gold. The thickness of the quartz plate influences the resonant frequency of the device. The operational principle of piezoelectric biosensors is based on the oscillation of the crystal, which resonates at its inherent frequency [27]. Piezoelectric microbial nano biosensors fundamentally rely on detecting shifts in the resonant frequency of a piezoelectric crystal, which result from alterations in mass on the crystal surface. These mass changes are usually caused by biochemical interactions, such as the interaction between an antibody and antigen or the association of a DNA fragment with its complementary sequence [28].

6.2.1.5 Enzyme-Based Nano Biosensors

Enzymes are frequently preferred among biomolecules in biosensor design because of their inherent capability to catalyse reactions with remarkable selectivity, precision, and effectiveness—key factors essential for enzyme-based nano biosensors [29]. Enzyme-based nano biosensors offer potential applications in a range of fields, such as environmental monitoring, pharmaceuticals, food safety, and biorefineries [30]. Enzymatic biosensor nanoparticles (EBNPs), containing enzymes within a 10–100 nm range, can be employed in diverse applications. These include amperometric enzyme sensors for blood sugar monitoring, surface plasmon resonance analysis in pharmaceuticals, and DNA microarrays for enzymatic assays [31].

6.2.1.6 DNA-Based Nano Biosensors

DNA nano biosensors employ DNA nanotechnology to detect target DNA, transforming its presence into measurable electrical, optical, or acoustic signals. These sensors consist of two main components: a molecular recognition component and a transducer. The recognition element detects the target DNA within the sample, while the transducer converts this detection into observable and recordable signals [32]. DNA nano biosensors are ideal for developing highly sensitive, specific, and fast techniques into compact and affordable devices [33].

6.2.1.7 Whole-Cell Nano Biosensors

Microbial whole-cell nano biosensors use microbial species to regulate a signalling protein in response to a target substance. The resulting signal quantifies the analyte's bioavailable levels at a lower cost than traditional analytical methods. They are developed for biomedical, environmental, and agricultural monitoring [34]. Whole cell–based biosensors can measure the status of living animal or bacterial cells by transforming homeostatic signals into detectable electrical and optical outputs. Unlike traditional chemical biosensors, these whole-cell biosensors deliver real-time, rapid, and distinctive data streams about the cell's homeostatic condition and its surrounding microenvironment with exceptional sensitivity and selectivity [35]. Whole cell–based nano biosensors can offer comprehensive testing and evaluation for disease diagnosis [36].

6.2.1.8 Antibody-Based Nano Biosensors

Antibody-based nano biosensors, known as immunosensors, have transformed diagnostics by allowing the identification of various analytes, including disease markers, food and environmental pollutants, biological warfare pathogens, and illegal drugs [37]. Double antibody immune sensing devices were developed and utilized for medical diagnosis. The sensor's active surface, composed of gold nanoparticles/tungsten oxide/carbon nanotubes (AuNPs/WO_3/CNTs), facilitated the chemical stabilization of a combination of SARS-CoV-2-specific antibodies, including RBD-S-specific monoclonal antibodies and anti-RBD-S-anti-Llama antibodies [38]. Antibody-driven sensors enable the quick and responsive detection of various disease-causing agents and their associated toxins [39] (Table 6.1).

TABLE 6.1
Classifications of Microbial Nano Biosensors

Classification	Description	Examples	Applications	References
Electrochemical nano biosensors	These biosensors detect changes in electrical signals from microbial interactions	Glucose metres, pregnancy strips	Microbial detection, cancer diagnosis, health monitoring	[40]
Optical nano biosensors	Optical biosensors detect light changes from analyte interactions	Quantum dot–based microbial biosensors	Pathogen detection, environmental monitoring	[41]
Magnetic nano biosensor	Magnetic nano biosensors detect analytes via magnetic changes	Planar Hall effect–based magnetic sensors	Disease diagnosis, cancer screening, drug monitoring	[42]
Piezoelectric nano biosensors	Piezoelectric sensors detect mass changes via shifts in surface oscillations	Quartz crystal microbalance (QCM)	Identification of pesticides in water and milk	[43,44]
Enzyme-based nano biosensors	These biosensors use microbial enzymes to generate detectable signals.	Horseradish peroxidase biosensors	Food, health, environmental, and clinical applications.	[45]
DNA-based nano biosensors	Uses microbial DNA to detect specific genetic material.	Gold and quantum dot DNA biosensors	Genetic testing, cancer diagnosis, and resistance monitoring.	[46]
Whole-cell nano biosensors	Uses whole microbial cells to detect specific analytes.	*Escherichia coli* and carbon nanotube whole-cell biosensor.	Heavy metal water contamination monitoring.	[35]
Antibody-based nano biosensors	Detects target molecules using antibody-coated nanomaterials.	Bacteriophage-based nano biosensors	Fast impedance-based pathogen detection in food.	[38]

6.2.2 Mechanisms

Microbial nano biosensors are analytical tools that use microorganisms to identify a specific target substance and translate this detection into a measurable response, which can be physiological, electrical, or biochemical. Sensory-regulative biosensors are capable of detecting different cell signals and then translating these signals into optical or electrochemical outputs [10]. Microbial biosensors utilize different sensing and recognition mechanisms, including traditional optical methods such as fluorescence or biological luminescence [11,12], electrochemical techniques [47], and sensory-based devices [48].

6.3 APPLICATIONS

6.3.1 Environmental Applications

There is a significant demand for fast, reliable, and cost-effective systems to detect, monitor, and diagnose environmental and agricultural pollutants. Traditional methods like chromatography and spectroscopy, while accurate, require costly equipment, expert personnel, and complex procedures. To address these limitations, new biosensors utilizing nanotechnology are being developed for real-time, on-site, and frequent contaminant monitoring [49,50]. Nano biosensors offer high sensitivity and specificity in detecting environmental pollutants, toxins, and pathogens. Their application is vital for ensuring the safety of water supplies, food products, and air quality. Moreover, they are essential for monitoring environmental changes and tackling ecological issues [51]. Microbial nano biosensors offer key advantages for environmental applications compared to conventional analytical techniques, including portability, miniaturization, and efficiency. They can be employed as tools for monitoring environmental quality, assessing biological and ecological health, and detecting both inorganic and organic priority pollutants [52]. Nano sensors detect allergens, pathogens, and contaminants, monitoring both atmospheric conditions and internal factors like chemical changes and microbial growth in packaged food products [53]. Nano biosensors can be applied across the agri-food spectrum, from soil assessment and natural resource evaluation to monitoring soil moisture and pH, managing diseases, detecting pathogens, harmful chemicals, adulterants, and throughout the commercialization process [54]. Microbial and cellular nano biosensors are crucial for environmental monitoring. In ecotoxicity testing, microbial biosensors offer advantages such as lower cultivation costs and large-scale production under strict quality control. They quickly respond to toxic compounds, indicating bioavailability more effectively than chemical analysis. Engineered microorganisms, like algae and bacteria, and macromolecules, such as enzymes and proteins, are key in multi-response biosensor design [55]. Microbe-based nano sensors have been utilized to diagnose heavy metals and toxicity in soil and water samples [56]. Algae are frequently used in biochemistry for detecting trace environmental toxins due to their sensitivity to toxicants, revealing pollutants' effects on living organisms. In the biosensor industry, microalgae and photosynthetic cyanobacteria provide pollution data through changes in photosynthetic or cellular metabolism. Monitoring species such as reactive oxygen species, H_2O_2, and pH-sensitive ions—key components in algae's metabolism and photosynthesis—can reveal the presence of herbicides, heavy metals, pathogens, and contaminants [57]. Crop productivity is constantly threatened by pests, weeds, and pathogens, impacting farm economies, and necessitating effective protection measures. Nanostructured biosensors can enhance smart farming by tracking soil conditions and plant growth over large areas and detecting crop diseases before visible symptoms appear [58]. However, nanotechnology-driven biosensors are still in the early stages of development (Figure 6.2).

6.3.2 Agricultural Applications

Amid climate change, global agriculture faces significant challenges in boosting production while minimizing impact on the environment. Nano-enhanced technologies can transform farming by enhancing input efficiency, reducing losses, and supporting sustainable practices. Microbial nano

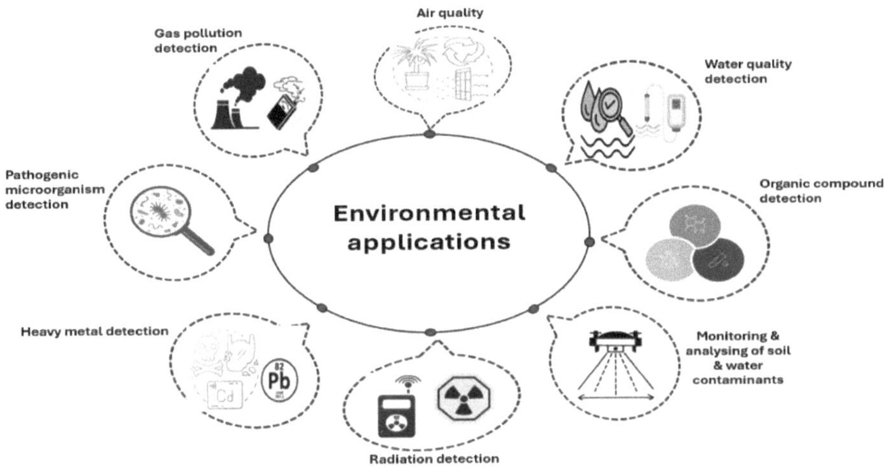

FIGURE 6.2 Environmental applications of microbial nano biosensor.

biosensors can identify both biotic and abiotic stresses in plants before they affect crop yields, providing real-time information on nutrient concentrations, metabolites, pesticides, pathogens, soil moisture, and temperature. This helps in precision agriculture and enhances resource management [59]. A major challenge in crop production is overseeing plant health. Timely interventions for diseases or deficiencies can save costs and reduce environmental impact. Smart farming uses advanced technologies, including nano biosensors, to detect specific stresses and issues at the individual plant level. Nano biosensors can be applied to develop innovative methods for sustainable agriculture. They can provide broad responses across entire fields through smart sensing or offer precise data using nanobionics. Nanobionics involves using mathematical models to detect changes in phytohormones and signalling molecules [60]. Early pathogen detection using novel biosensing methods helps manage diseases and reduce productivity losses. Precise pathogen identification aids in developing effective treatments. Microbial nano biosensors offer faster detection compared to traditional methods; a technique known as nano diagnostics. Gold nanoparticles–based nano biosensors have been used to identify agroterrorism agents and plant pathogens. Typically, these sensors use DNA as the biological component and can monitor plant–pathogen interactions, pathogen genetics, and gene transfer between pathogens and hosts [54,61]. Farmers can use nano biosensors to carefully monitor environmental conditions for ideal plant growth and protection [56]. Low-cost, portable microbial nano biosensors are highly effective for quickly assessing plant and soil health on-site. They measure plant stress levels, nutritional status, and toxic chemicals in the soil, which helps maintain optimal farming and crop productivity. Nano biosensors detect physiological signals and convert them into standardized outputs. Nanoparticles enhance signal amplification and sensor sensitivity, improving detection limits. Their high specificity and responsiveness allow for early detection and management of specific issues [62]. Soil health is dynamic, requiring on-site monitoring of parameters like pH, nutrients, and moisture for effective agriculture. Nanoparticle-based sensors are vital for precision farming. For instance, lipid-based nano biosensors can detect nitrites in soil at 2.1 μg/L, mimicking natural environments for stable detection [63–65]. Nano sensors diagnose soil diseases by measuring oxygen consumption of "good" and "bad" microbes in soil samples. Sensors with each microbe type compare their oxygen consumption rates to identify microbial favourability and predict potential disease outbreaks, offering an innovative, semi-quantitative soil health assessment [66]. The integration of zeolite (a microporous element) with nano biosensors has revolutionized agriculture, significantly advancing the field. Nano biosensors detect nutrients or water deficiencies in soil or plants and regulate their release from zeolite. Additionally, pesticides encapsulated within nanoparticles are designed to release their contents based on time or

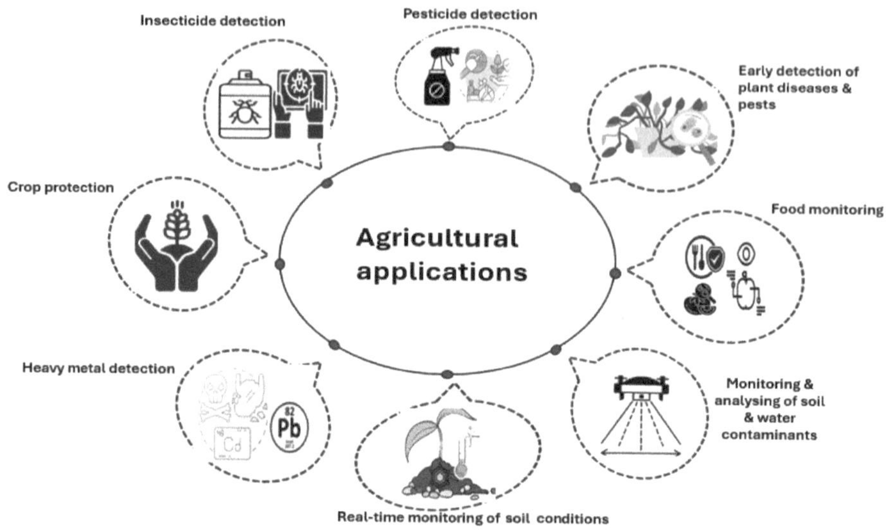

FIGURE 6.3 Various applications of microbial nano biosensor in agricultural development and sustainable development.

environmental changes. Similarly, herbicides combined with nano biosensors are applied only when needed, enhancing crop yields [67] (Figure 6.3).

6.3.3 Medical and Healthcare Applications

Effective healthcare management requires quick decision-making supported by rapid diagnostic tools, advanced data analysis, and informatics. The swift point-of-care analyte detection is made possible by combining microfluidics with nanotechnology and biotechnology. Microbial nano biosensors utilize nanotechnology for diagnostics, swift disease detection, monitoring, and oversight. They sense biomolecules using bioreceptors, which translate the interaction of a biomolecule into an optical or electrical signal using a transducer. These nano biosensors are highly sensitive and selective, making them essential for the early detection of diseases [68,69]. Nano biosensors possess significant potential for detecting a wide range of microorganisms, disease indicators, environmental contaminants, and enhancing drug delivery (such as identifying pathogens in food or blood pathogen biomarkers). Additionally, they have the capability to detect early-stage cancers in body fluids, such as breast cancer [68,70,71]. Microfluidic-integrated nano biosensor technology has provided significant advancements in personalized healthcare. This emerging technology is extensively utilized for early detection and detection of infectious agents, including viral entities, as well as for the analysis and treatment of persistent diseases. It is also applied in overall health management, retinal implants, cardiovascular conditions, diabetes management, parental testing, MRI contrast imaging, cardiac diagnostics, medical fungal studies, and extensive health monitoring, and among other fields [72,73]. Nano biosensors are extensively utilized for the early identification of assorted biological disorders in patients, contributing to reduced healthcare costs and resolving many healthcare challenges [74,75]. With the advent of nanotechnology, the use and advancement of nano biosensors have emerged as widespread in numerous scientific fields such as environmental monitoring, food security, biomedicine, clinical diagnostics, and sustainable healthcare. Various sensing techniques incorporating nanotechnology in biosensors and nano biosensors have enabled the recognition and analysis of biological markers related to both communicable and non-communicable diseases [76]. Microbial nano biosensor technology is capable of providing fast and accurate detection, along with clear and reliable images of cancer progression. Many nano biosensors have also been researched for their effectiveness in identifying disease-causing microorganisms in both biological and non-biological samples [77]. Accurate monitoring of biological analytes, including biomolecules (proteins, lipids, DNA, and RNA) and biological cells (blood cells, viruses, and bacteria), is crucial for evaluating and mitigating health risks. Nano biosensors, which are analytical devices that integrate a biologically sensitive element with a nanostructured transducer, are extensively utilized for the molecular identification of biomarkers related to disease diagnosis and the identification of infectious organisms [78]. Nano biosensors, small and intelligent technology, are extensively used across various biomedical fields, including tissue engineering, 3D printing, biotechnology, biomaterials, sports, and medical technology. They serve multiple purposes such as diagnosing, preventing, clinical testing, treating, and monitoring human diseases. A typical example is a transistor-based biosensor with nanowire characteristics and protein–nanoparticle biosensors designed for ultra-sensitive detection of DNA and proteins without needing PCR (polymerase chain reaction) or bisulphite treatment for DNA. These sensors can also detect plant diseases and address unique issues related to biomarkers and mineral deficiencies [79]. Nano biosensors exhibit significant enhancements in sensitivity, specificity, and parallelism to address the future requirements of various fields, including in vitro medical diagnostics, pharmaceutical research, and pathogen detection. Their application in diagnosing diseases such as tuberculosis, malaria, and AIDS promises to revolutionize biomedical research by enabling rapid, precise, affordable, and non-invasive diagnosis. However, nano biosensors must undergo evaluation and obtain approval as medical devices before they can be marketed [80] (Figure 6.4).

6.3.4 Industrial Applications

With the requirement for high-quality products from both customers and regulatory bodies, there is a need for quick and cost-effective methods to guarantee product quality and process control. Fermentation, a widely used process for producing food and beverages, necessitates careful management of the fermentation system [81,82]. Microbial biosensors are utilized to monitor substances for controlling the fermentation process. Due to the critical role of ethanol in various fermentation processes, these biosensors are used to sensitively detect ethanol, ensuring proper monitoring of the fermentation [83]. Microbial nano biosensors are extensively used in environmental, food, and diagnostic industries due to their low cost, stability, and quick response. Unlike enzymes,

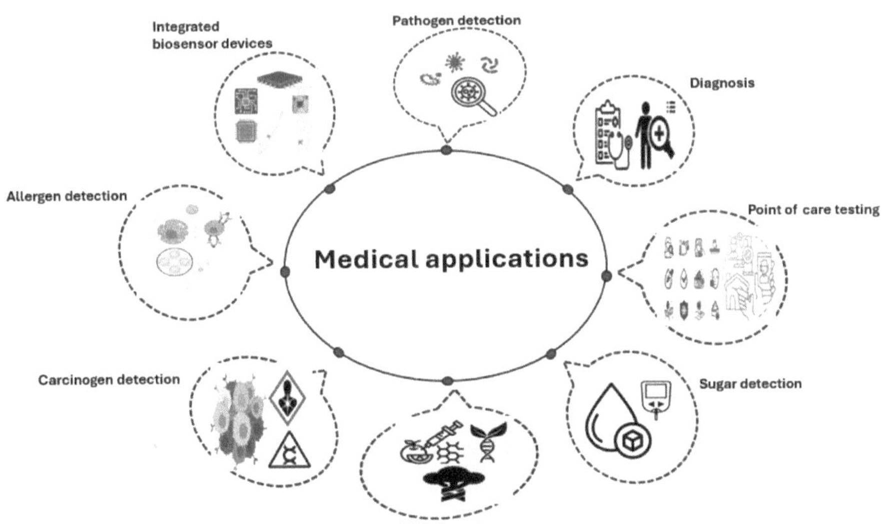

FIGURE 6.4 Medical and health care applications of microbial nano biosensor.

microorganisms employed as bioelements can specifically react to analytes without requiring time-consuming and costly purification. These advantageous properties have led to several promising developments in the field of microbial nano biosensors [9]. Agricultural waste can serve as an innovative material for nano biosensors due to its environmental abundance and cost-effectiveness. Nanocellulose fibres have been produced and can be utilized for various applications. Rice husk, a byproduct of rice milling, can also be used to develop nano biosensors. When combined with silica, rice husks offer enhanced applications. These materials are often employed in agricultural practices. Despite their efficiency for various uses, a significant challenge remains the high cost of their design and processing [84,85]. Thus, a variety of nanostructures, from expensive options to affordable agricultural by-products, have been incorporated into the design of robust nano biosensors [54]. Microbial nano biosensors have the capability to enhance detection processes across the chemical and food industries. In food production, they can monitor the mixing of substrates in biological reactors, optimizing nutrient feeding for various uses. On an industrial scale, these sensors can improve the separation and preparation of different commercial ingredients. Furthermore, nano biosensors can detect specific pollutants in the food industry during harvesting, storage, and transportation [4].

6.3.5 Food Safety and Quality Control

Investing in the development of food quality biosensors is crucial, as they can enhance monitoring and control processes. However, only a limited number of nano biosensors currently play a significant role in the food industry. Therefore, substantial efforts are needed to develop reliable, cost-effective nano biosensors capable of operating effectively in real-world conditions [86,87]. Biosensors in the food sector have crucial applications, ensuring safety by detecting xenobiotics, bacterial toxins, and pathogens. The food sector relies on testing methods to ensure safety, quality, and efficiency, detecting contaminants and essential components such as alcohols, sugars, and amino acids. Microbial nano biosensors are increasingly used for these purposes, offering a viable alternative to traditional methods like. They monitor food quality by analysing composition, shelf life, and safety of fresh products, and measure pesticide levels in dairy. Technologically, they track amino acids and sugars during fermentation and pasteurization. Primarily, enzyme-based biosensors in the beverage industry detect carbohydrates, alcohol, and other compounds, enhancing food safety and quality [88]. Nano biosensors, which merge nanotechnology with biotechnology, are used to oversee and control food quality, safety, and shelf life. Embedded in packaging materials or devices, they continuously track factors such as temperature, humidity, gas levels, and microbial activity, allowing for prompt actions to maintain food freshness and safety. Using optical, electrochemical, or nanomechanical principles, these sensors identify and examine specific substances. Additionally, they enhance intelligent packaging by embedding sensors to monitor conditions, detect issues such as temperature increases or bacterial growth, and maintain optimal conditions for food integrity and quality [17]. Nano biosensors enable continuous, real-time tracking of temperature, humidity, gas levels, and microbial activity within packaging, facilitating immediate identification of deviations and prompt corrective measures. They enhance quality control through the detection of degradation indicators like VOCs and microbial activity, enabling swift decisions to prevent substandard products. For shelf-life optimization, they deliver exact real-time insights into packaging status, addressing spoilage factors quickly. Additionally, they detect harmful contaminants and allergens, enhancing food safety and authenticity control [89–92] (Figure 6.5).

6.3.6 Biodefence and Security Applications

The development of nanomaterial-based (bio)sensing systems is driving significant research and advancements in safety and security applications. Notably, nanosized and nanomaterial-based biosensors, referred to as nano biosensors, are a modern and highly efficient class of detection systems.

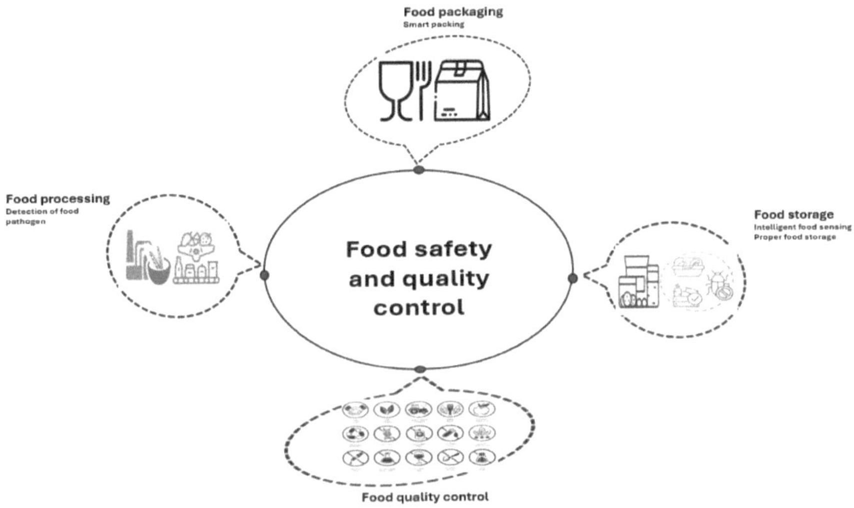

FIGURE 6.5 Microbial nano biosensor applications in food safety and quality control.

Their application in the food industry, environmental monitoring, and clinical diagnostics has the potential to greatly enhance safety and protection measures against bioterrorism agents. Presently, laboratories and institutions in the field of nanobiotechnology are working together to improve their capabilities in detecting and responding to biological and chemical weapon threats [93]. The emergence of biological threats and their increasing severity has created an urgent need for the development of rapid diagnostic tools. Early detection of these threats is essential for the prevention, protection, and mitigation of pathogens. Consequently, the demand for biosensors to identify biothreats is surging, driving the advancement of detection strategies and the creation of portable, point-of-use devices [94]. A biosensor is designed to specifically target bacteria even when other non-target organisms are present. By modifying the specificity of the antibody on the capture pad, the biosensor can be adapted to detect other analytes relevant to biosecurity. Additionally, its applications extend to medical diagnostics and biodefence measures [95]. Nano biosensors have potential applications in security and defence, such as detecting biological and chemical threats, monitoring battlefield conditions, and identifying hazardous materials [51] (Figure 6.6).

6.3.7 Innovative and Emerging Applications

Nano sensors have become crucial in the food industry because of their high specificity and sensitivity. These biosensors integrate a transducer with a biorecognition element to convert biological signals into digital ones. Due to their precise sensitivity and affordability, nano biosensors are widely used to detect toxic chemicals, such as pesticide residues and pathogenic microbes, offering significant advantages to the food industry. They use nanocarriers to bind to impurities, pollutants, foodborne microorganisms, and their toxins, like mycotoxins. This innovative technology ensures food safety in food processing industries [51]. Lab-on-a-chip technology is a key area of research interest, particularly for its potential in biotic-micro electrochemical systems. By integrating microbes onto microfluidic chips, these systems present the benefits of immediate reaction times and compact size. Lab-on-a-chip technology is extensively used in creating microfluidic devices that can detect analytes at extremely low concentrations by efficiently delivering target analytes to the surface of microbial biosensors. This technology holds significant promise for advancing the development of microbial biosensors [9]. Forensic identification is an emerging area where microbial biosensors show significant potential. This process involves distinguishing the unique physical characteristics of an object from others of its kind. Conventional forensic identification methods

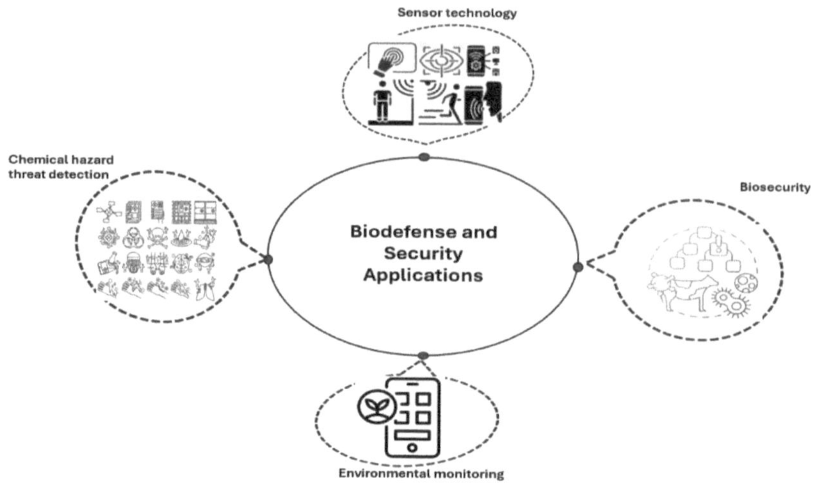

FIGURE 6.6 Applications of microbial nano biosensor in biodefence and security.

TABLE 6.2
Comparison of Traditional versus Nano Biosensor

Parameter	Traditional Techniques	Microbial Nano Biosensor Techniques	References
Principle	Based on established methods like spectrophotometry, chromatography, and immunoassays	Utilizes microbial cells functionalized with nanomaterials to detect specific analytes	[98,99]
Sensitivity	Moderate sensitivity, often limited by the detection method's resolution	High sensitivity due to enhanced microbial response and nanoscale interactions	[99,100]
Response time	Slower due to multiple processing steps and longer reaction times	Faster response due to the direct interaction between microbes and nanomaterials	[10,101]
Specificity	High specificity, but potential cross-reactivity with similar compounds	Enhanced specificity by leveraging microbial metabolic pathways combined with targeted nanomaterials	[10,101]
Environmental impact	Can generate significant chemical waste, especially in chromatographic and spectrometric techniques	Lower environmental impact due to minimal reagent use and the potential for green synthesis of nanomaterials	[102,103]
Applications	Widely used in clinical diagnostics, environmental monitoring, and food safety	Emerging applications in environmental monitoring, bioremediation, and biosensing of pollutants	[9,54]
Detection method	Relies on optical, electrochemical, or mass-based detection techniques	Combines microbial metabolic responses with nanoscale electrochemical or optical detection methods	[10,104]

are often intricate and time-consuming, complicating criminal investigations. However, microbial biosensors, with their DNA analysis capabilities, could offer a more efficient solution for forensic identification [9,96,97] (Table 6.2).

6.4 CHALLENGES AND FUTURE PERSPECTIVES

While microbial biosensors provide numerous benefits compared to conventional detection tools, their broad adoption is limited by certain intrinsic drawbacks of microorganisms. These include relatively low responsiveness, which is closely linked to cell population size and optical signal, limited selectivity in detecting multiple targets simultaneously, intrinsic cellular heterogeneity at both the genotypic and phenotypic levels, and the stochastic nature of protein expression [10]. While nano biosensors hold great promise for applications in agriculture and food sectors, helping to ensure the resilience and long-term sustainability of these industries, they also face several challenges. One major concern is the potential toxicity and ecotoxicity of nanomaterials, stemming from their distinctive features such as size, structure, high surface-to-volume ratio, and composition. Therefore, it is crucial to accurately assess their environmental impact, considering factors like size, dosage, accumulation characteristics, immune reaction, and retention duration [105]. Another challenge is developing miniaturized versions of nano biosensors to address portability issues. If nano biosensors can be integrated with information technology, they could benefit remote areas in the agricultural and food industries by reducing costs, enhancing productivity, improving early detection of disease epidemics, and optimizing the use of natural resources like soil, water, and climate conditions. However, the cost of fabricating nano biosensors remains high. To address this, alternative and cost-effective biological elements like enzymes or cells, along with novel matrices like chitosan for stabilizing and immobilizing nanomaterials, can be employed to enhance the overall durability and reusability of nano biosensors [106].

Enhanced identification of foodborne pathogens, pesticides, and antibiotics, and other contaminants will be crucial for ensuring food safety, as their presence will present a risk to human health. Nano biosensors in agriculture will not only identify biological analytes in food products but also generate high-quality food to satisfy both local and global demands [107]. Biosensors and nanoscopic analysis/visualization tools represent novel and promising technologies that have the potential to be life-saving treatment decisions and a deeper understanding of pandemic variants. As nanotechnology continues to advance rapidly, sensing mechanisms will adopt a multidisciplinary approach, enabling rapid, on-site virus detection for point-of-care diagnostics and early epidemic prevention [108]. Microbial nanotechnology is an exciting and rapidly growing field poised to drive future breakthroughs in nanomaterial synthesis. By employing a "green" and sustainable approach, this field has the potential to significantly impact various areas, including sensor technology and biomedicine, fostering innovation in nanomanufacturing [109].

6.5 CONCLUSION

Microbial nano biosensors represent a transformative advancement in the field of biosensing, offering significant potential across multiple applications, such as ecological surveillance, food quality assurance, and healthcare diagnostics. Their ability to identify specific biological analytes with high sensitivity and precision, combined with the advantages of nanotechnology, positions them as powerful tools for addressing contemporary challenges. The integration of micro-/nanotechnologies has markedly advanced the capabilities of microbial nano biosensors, overcoming many limitations of traditional detection methods. These technological advancements have notably improved the performance of both optical and electrochemical detection systems, enhancing sensitivity and selectivity. Successful integration of these technologies has met diverse industrial and environmental demands through miniaturization and high-throughput methods. Furthermore, the computerization and miniaturization of tools and bioreactors have allowed microbial nano biosensors to be employed more effectively and practically. The advancement of portable detectors and auxiliary tools has also expanded the scope of applications, making previously impractical uses feasible. Overall, these innovations position microbial nano biosensors as a powerful tool for advancing detection capabilities and addressing contemporary challenges in various fields.

However, the widespread adoption of microbial nano biosensors is contingent upon overcoming several key challenges, including issues related to sensitivity, selectivity, and the environmental impact of nanomaterials. Continued research and innovation are essential to enhance their performance, reduce costs, and ensure their safe and effective integration into real-world applications. By addressing these challenges, microbial nano biosensors can make a profound impact on improving health outcomes, enhancing food safety, and advancing environmental sustainability.

Future research should focus on improving the stability, reusability, and affordability of these sensors, as well as enhancing their integration with existing technologies. Addressing these challenges will be crucial for fully realizing the transformative impact of microbial nano biosensors on public health, environmental conservation, and food security.

REFERENCES

1. Fariq, A.; Khan, T.; Yasmin, A. Microbial synthesis of nanoparticles and their potential applications in biomedicine. *J Appl Biomed* 2017, *15*, 241–248.
2. Ray, M.K.; Mishra, A.K.; Mohanta, Y.K.; Mahanta, S.; Chakrabartty, I.; Kungwani, N.A.; Avula, S.K.; Panda, J.; Pudake, R.N. Nanotechnology as a promising tool against phytopathogens: A futuristic approach to agriculture. *Agriculture* 2023, *13*, 1856.
3. Huang, X.; Zhu, Y.; Kianfar, E. Nano biosensors: Properties, applications and electrochemical techniques. *J Mater Res Technol* 2021, *12*, 1649–1672.
4. Kulkarni, M.B.; Ayachit, N.H.; Aminabhavi, T.M. Recent advancements in nanobiosensors: Current trends, challenges, applications, and future scope. *Biosensors (Basel)* 2022, *12*, 892.
5. Ghosh, S.; Ahmad, R.; Zeyaullah, M.; Khare, S.K. Microbial nano-factories: Synthesis and biomedical applications. *Front Chem* 2021, *9*, 626834.
6. Dewan, N.; Ahmed, P.; Chowdhury, G.; Pandit, S.; Dasgupta, D. Nanotechnology based biosensors and its application. *Pharma Innov J* 2016, *5*, 18.
7. Qian, L.; Durairaj, S.; Prins, S.; Chen, A. Nanomaterial-based electrochemical sensors and biosensors for the detection of pharmaceutical compounds. *Biosens Bioelectron* 2021, *175*, doi:10.1016/j.bios.2020.112836.
8. Gavrilaş, S.; Ursachi, C.Ş.; Perţa-Crişan, S.; Munteanu, F.D. Recent trends in biosensors for environmental quality monitoring. *Sensors* 2022, *22*, 1513.
9. Dai, C.; Choi, S. Technology and applications of microbial biosensor. *Open J Appl Biosens* 2013, *02*, 83–93, doi:10.4236/ojab.2013.23011.
10. Lim, J.W.; Ha, D.; Lee, J.; Lee, S.K.; Kim, T. Review of micro/nanotechnologies for microbial biosensors. *Front Bioeng Biotechnol* 2015, *3*, 61.
11. D'souza, S.F. Microbial Biosensors. *Biosensors and Bioelectronics* 2001, *16*(6), 337–353, https://doi.org/10.1016/S0956-5663(01)00125-7.
12. Lei, Y.; Chen, W.; Mulchandani, A. Microbial biosensors. *Anal Chim Acta* 2006, *568*, 200–210.
13. Ka, A.; Ia, K. Biosensors for Monitoring of the Environmental Heavy Metals. *J Sci Eng* 2016, *3*(3), 31–38.
14. Rensing, C.; Maier, R.M. Issues underlying use of biosensors to measure metal bioavailability. *Ecotoxicol Environ Saf* 2003, *56*, 140–147, doi:10.1016/S0147-6513(03)00057-5.
15. Hernandez-Vargas, G.; Sosa-Hernández, J.E.; Saldarriaga-Hernandez, S.; Villalba-Rodríguez, A.M.; Parra-Saldivar, R.; Iqbal, H.M.N. Electrochemical biosensors: A solution to pollution detection with reference to environmental contaminants. *Biosensors (Basel)* 2018, *8*, 29.
16. Shikha, D.; Monga, V.; Bhatia, R. Nano-biosensors: A new future for diagnosis and modern analysis. *Curr Pharm Anal* 2022, *18*, 739–744, doi:10.2174/1573412918666220602122727.
17. Kakimova, Z.; Orynbekov, D.; Zharykbasova, K.; Kakimov, A.; Zharykbasov, Y.; Mirasheva, G.; Toleubekova, S.; Muratbayev, A.; Ntsefong, G.N. Advancements in nano bio sensors for food quality and safety assurance – A review. *Potr S J Food Sci* 2023, *17*, 728–747, doi:10.5219/1903.
18. Prasad, K.; Ranjan, R.K.; Lutfi, Z.; Pandey, H. Biosensors: Applications and overview in industrial automation. *Int J Appl Bio-Eng* 2009, *3*, 66–70, doi:10.18000/ijabeg.10041.
19. Devi Lamabam, S.; Thangjam, R. Emerging trends in the application of nanobiosensors in the food industry. In *Novel Approaches of Nanotechnology in Food*; Elsevier, 2016; pp. 663–696. ISBN 9780128043080.
20. Ronkainen, N.J.; Halsall, H.B.; Heineman, W.R. Electrochemical biosensors. *Chem Soc Rev* 2010, *39*, 1747–1763, doi:10.1039/b714449k.
21. Pilolli, R.; Monaci, L.; Visconti, A. Advances in biosensor development based on integrating nanotechnology and applied to food-allergen management. *TrAC – Trends Anal Chem* 2013, *47*, 12–26.
22. Damborský, P.; Švitel, J.; Katrlík, J. Optical biosensors. *Essays Biochem* 2016, *60*, 91–100, doi:10.1042/EBC20150010.
23. Homola, J. Surface plasmon resonance sensors for detection of chemical and biological species. *Chem Rev* 2008, *108*, 462–493.

24. Wu, K.; Cheeran, M.C.J.; Wang, J.P.; Saha, R.; Su, D.; Krishna, V.D.; Liu, J. Magnetic-nanosensor-based virus and pathogen detection strategies before and during Covid-19. *ACS Appl Nano Mater* 2020, *3*, 9560–9580.
25. Shan, B.; Broza, Y.Y.; Li, W.; Wang, Y.; Wu, S.; Liu, Z.; Wang, J.; Gui, S.; Wang, L.; Zhang, Z.; et al. Multiplexed nanomaterial-based sensor array for detection of COVID-19 in exhaled breath. *ACS Nano* 2020, *14*, 12125–12132, doi:10.1021/acsnano.0c05657.
26. Aytur, T.; Foley, J.; Anwar, M.; Boser, B.; Harris, E.; Beatty, P.R. A novel magnetic bead bioassay platform using a microchip-based sensor for infectious disease diagnosis. *J Immunol Methods* 2006, *314*, 21–29, doi:10.1016/j.jim.2006.05.006.
27. Zuber, A.A.; Klantsataya, E.; Bachhuka, A. Biosensing. In *Comprehensive Nanoscience and Nanotechnology*; Elsevier, 2019; Vol. 1–5, pp. 105–126. ISBN 9780128122952.
28. Apetrei, C.; Ghasemi-Varnamkhasti, M. Biosensors in food PDO authentication. In *Comprehensive Analytical Chemistry*; Elsevier B.V., 2013; Vol. 60, pp. 279–297. doi:10.1016/B978-0-444-59562-1.00011-6.
29. Staiano, M.; Pennacchio, A.; Varriale, A.; Capo, A.; Majoli, A.; Capacchione, C.; D'Auria, S. Enzymes as sensors. In *Methods in Enzymology*; Academic Press Inc., 2017; Vol. 589, pp. 115–131. doi:10.1016/bs.mie.2017.01.015.
30. Soy, S.; Sharma, S.R.; Nigam, V.K. Bio-fabrication of thermozyme-based nano-biosensors: Their components and present scenario. *J Mater Sci Mater Electron* 2022, *33*, 5523–5533, doi:10.1007/s10854-022-07741-9.
31. Esmaeili, A. Enzyme-based biosensors with nanoparticles as analytical tools. In *Nano-Enzyme Incorporated Particles*; Elsevier, 2024; pp. 253–304. doi:10.1016/B978-0-443-18810-7.00010-7.
32. Yuwen, L.; Zhang, S.; Chao, J. Recent advances in DNA nanotechnology-enabled biosensors for virus detection. *Biosensors (Basel)* 2023, *13*, 822.
33. Soukarié, D.; Ecochard, V.; Salomé, L. DNA-based nanobiosensors for monitoring of water quality. *Int J Hyg Environ Health* 2020, *226*, 113485.
34. Moraskie, M.; Roshid, M.H.O.; O'Connor, G.; Dikici, E.; Zingg, J.M.; Deo, S.; Daunert, S. Microbial whole-cell biosensors: Current applications, challenges, and future perspectives. *Biosens Bioelectron* 2021, *191*, 113359.
35. Gui, Q.; Lawson, T.; Shan, S.; Yan, L.; Liu, Y. The application of whole cell-based biosensors for use in environmental analysis and in medical diagnostics. *Sensors (Switzerland)* 2017, *17*, 1623.
36. Shyam, S.; Nava, A.; Sarma, H. Potential microbes for environment and agriculture: Bioengineering strategies for a sustainable future. In *Biotechnology of Emerging Microbes*; Elsevier, 2024; pp. 1–29, doi:10.1016/B978-0-443-15397-6.00001-2.
37. Sharma, S.; Kennedy, R.J. Antibodies and antibody-derived analytical biosensors. *Essays Biochem* 2016, *60*, 9–18, doi:10.1042/EBC20150002.
38. Hussein, H.A.; Kandeil, A.; Gomaa, M.; Hassan, R.Y.A. Double-antibody-based nano-biosensing system for the onsite monitoring of SARS-CoV-2 variants. *Microsyst Nanoeng* 2023, *9*, doi:10.1038/s41378-023-00578-0.
39. Byrne, B.; Stack, E.; Gilmartin, N.; O'Kennedy, R. Antibody-based sensors: Principles, problems and potential for detection of pathogens and associated toxins. *Sensors (Switzerland)* 2009, *9*, 4407–4445.
40. Hassan, R.Y.A. Advances in electrochemical nano-biosensors for biomedical and environmental applications: From current work to future perspectives. *Sensors* 2022, *22*, 7539.
41. Song, M.; Yang, M.; Hao, J. Pathogenic virus detection by optical nanobiosensors. *Cell Rep Phys Sci* 2021, *2*, 100288.
42. Wang, M.; Jin, L.; Hang-Mei Leung, P.; Wang-Ngai Chow, F.; Zhao, X.; Chen, H.; Pan, W.; Liu, H.; Li, S. Advancements in magnetic nanoparticle-based biosensors for point-of-care testing. *Front Bioeng Biotechnol* 2024, *12*, 1393789.
43. Pohanka, M. Overview of piezoelectric biosensors, immunosensors and DNA sensors and their applications. *Materials* 2018, *11*, 448, doi:10.3390/ma11030448.
44. Marrazza, G. Piezoelectric biosensors for organophosphate and carbamate pesticides: A review. *Biosensors (Basel)* 2014, *4*, 301–317, doi:10.3390/bios4030301.
45. Alvarado-Ramírez, L.; Rostro-Alanis, M.; Rodríguez-Rodríguez, J.; Sosa-Hernández, J.E.; Melchor-Martínez, E.M.; Iqbal, H.M.N.; Parra-Saldívar, R. Enzyme (single and multiple) and nanozyme biosensors: Recent developments and their novel applications in the water-food-health nexus. *Biosensors (Basel)* 2021, *11*, 410.
46. Abu-Salah, K.M.; Zourob, M.M.; Mouffouk, F.; Alrokayan, S.A.; Alaamery, M.A.; Ansari, A.A. DNA-based nanobiosensors as an emerging platform for detection of disease. *Sensors (Switzerland)* 2015, *15*, 14539–14568.

47. Su, L.; Jia, W.; Hou, C.; Lei, Y. Microbial biosensors: A review. *Biosens Bioelectron* 2011, *26*, 1788–1799.
48. Zhang, F.; Keasling, J. Biosensors and their applications in microbial metabolic engineering. *Trends Microbiol* 2011, *19*, 323–329.
49. Salouti, M.; Derakhshan, F.K. Biosensors and nanobiosensors in environmental applications. In *Biogenic Nano-Particles and Their Use in Agro-Ecosystems*; Springer Singapore, 2020; pp. 515–591. ISBN 9789811529856.
50. Gogoi, B.; Chowdhury, N.K.; Shyam, S.; Choudhury, R.; Sarma, H. Application of nanoparticles in environmental monitoring. In *Technology for a Sustainable Environment*; Bentham Science Publishers, 2023; pp. 108–128. https://doi.org/10.2174/9789815124033123010010.
51. Guruprasath, N.; Sankarganesh, P.; Adeyeye, S.A.O.; Babu, A.S.; Parthasarathy, V. Review on emerging applications of nanobiosensor in food safety. *J Food Sci* 2024, *89*, 3950–3972.
52. Maria da Costa Silva, L.; Farias Melo, A.; Salgado, A.M. *Biosensors for Environmental Applications*; InTech Open, 2010, doi:10.5772/20154.
53. Fuertes, G.; Soto, I.; Carrasco, R.; Vargas, M.; Sabattin, J.; Lagos, C. Intelligent packaging systems: Sensors and nanosensors to monitor food quality and safety. *J Sens* 2016, *2016*, 1–8.
54. Thakur, M.; Wang, B.; Verma, M.L. Development and applications of nanobiosensors for sustainable agricultural and food industries: Recent developments, challenges and perspectives. *Environ Technol Innov* 2022, *26*, doi:10.1016/j.eti.2022.102371.
55. Scognamiglio, V.; Pezzotti, I.; Pezzotti, G.; Cano, J.; Manfredonia, I.; Buonasera, K.; Rodio, G.; Giardi, M.T. A new embedded biosensor platform based on micro-electrodes array (MEA) technology. *Sens Actuators B Chem* 2013, *176*, 275–283, doi:10.1016/j.snb.2012.09.101.
56. Kanjana, D. Advancement of nanotechnology applications on plant nutrients management and soil improvement. In *Nanotechnology: Food and Environmental Paradigm*; Springer Singapore, 2017; pp. 209–234. ISBN 9789811046780.
57. Tsopela, A.; Lale, A.; Vanhove, E.; Reynes, O.; Séguy, I.; Temple-Boyer, P.; Juneau, P.; Izquierdo, R.; Launay, J. Integrated electrochemical biosensor based on algal metabolism for water toxicity analysis. *Biosens Bioelectron* 2014, 290–297, doi:10.1016/j.bios.2014.05.004ï.
58. Antonacci, A.; Arduini, F.; Moscone, D.; Palleschi, G.; Scognamiglio, V. Nanostructured (bio)sensors for smart agriculture. *TrAC – Trends Anal Chem* 2018, *98*, 95–103.
59. Miguel-Rojas, C.; Pérez-De-Luque, A. Nanobiosensors and nanoformulations in agriculture: New advances and challenges for sustainable agriculture. *Emerg Top Life Sci* 2023, *7*, 229–238.
60. Butnariu, M.; Butu, A. Plant nanobionics: Application of nanobiosensors in plant biology. In *Nanotechnology in the Life Sciences*; Springer Science and Business Media B.V., 2019; pp. 337–376.
61. Ali, Q.; Ahmar, S.; Aamir Sohail, M.; Kamran, M.; Ali, M.; Hamzah Saleem, M.; Rizwan, M.; Mushtaque Ahmed, A.; Mora-Poblete, F.; Teixeira do Amaral Júnior, A.; et al. Research advances and applications of biosensing technology for the diagnosis of pathogens in sustainable agriculture, doi:10.1007/s11356-021-12419-6/Published.
62. Mondal, R.; Dam, P.; Chakraborty, J.; Paret, M.L.; Katı, A.; Altuntas, S.; Sarkar, R.; Ghorai, S.; Gangopadhyay, D.; Mandal, A.K.; et al. Potential of nanobiosensor in sustainable agriculture: The state-of-art. *Heliyon* 2022, *8*, e12207.
63. Siontorou, C.G.; Georgopoulos, K.N. A biosensor platform for soil management: The case of nitrites. *J Clean Prod* 2016, *111*, 133–142, doi:10.1016/j.jclepro.2015.07.038.
64. Nettleton, W.D.; Peterson, F.F. Landform, soil, and plant relationships to nitrate accumulation, central nevada. *Geoderma* 2011, *160*, 265–270, doi:10.1016/j.geoderma.2010.08.005.
65. Ji, R.; Brune, A. Nitrogen mineralization, ammonia accumulation, and emission of gaseous NH_3 by soil-feeding termites. *Biogeochemistry* 2006, *78*, 267–283, doi:10.1007/s10533-005-4279-z.
66. Rai, V.; Acharya, S.; Dey, N. Implications of nanobiosensors in agriculture. *J Biomater Nanobiotechnol* 2012, *3*, 315–324, doi:10.4236/jbnb.2012.322039.
67. Javad, S. *Nanoagronomy*; Springer International Publishing, 2020; ISBN 9783030412753.
68. Kumari, M.; Gupta, V.; Kumar, N.; Arun, R.K. Microfluidics-based nanobiosensors for healthcare monitoring. *Mol Biotechnol* 2024, *66*, 378–401.
69. Chowdhury, N.K.; Choudhury, R.; Gogoi, B.; Chang, C.-M.; Pandey, R.P. Microbial synthesis of gold nanoparticles and their application. *Curr Drug Targets* 2022, *23*, 752–760, doi:10.2174/1389450123666220128152408.
70. Alba-Patiño, A.; Vaquer, A.; Barón, E.; Russell, S.M.; Borges, M.; de la Rica, R. Micro- and nanosensors for detecting blood pathogens and biomarkers at different points of sepsis care. *Microchimica Acta* 2022, *189*, 74.

71. Gopinath, S.C.B.; Lakshmipriya, T. An update on nanomaterial-based biosensing systems for discriminative diagnosis of breast cancer. *Cancer Plus* 2023, *5*, 2589, doi:10.36922/cp.2589.
72. Haleem, A.; Javaid, M.; Singh, R.P.; Suman, R.; Rab, S. Biosensors applications in medical field: A brief review. *Sens Int* 2021, *2*, 100100.
73. Aqra, M.W.; Ramanathan, A.A. Review of the recent advances in nano-biosensors and technologies for healthcare applications. *Chem Proc* 2022, *5*, 76.
74. Ahmed, M.U.; Saaem, I.; Wu, P.C.; Brown, A.S. Personalized diagnostics and biosensors: A review of the biology and technology needed for personalized medicine. *Crit Rev Biotechnol* 2014, *34*, 180–196.
75. Upadhyay, L.S.B.; Verma, N. Alkaline phosphatase inhibition based conductometric biosensor for phosphate estimation in biological fluids. *Biosens Bioelectron* 2015, *68*, 611–616, doi:10.1016/j.bios.2015.01.064.
76. Kumar, S.; Singh, H.; Feder-Kubis, J.; Nguyen, D.D. Recent advances in nanobiosensors for sustainable healthcare applications: A systematic literature review. *Environ Res* 2023, *238*, 117177, doi:10.1016/j.envres.2023.117177.
77. Preetam, S.; Dash, L.; Sarangi, S.S.; Sahoo, M.M.; Pradhan, A.K. Application of nanobiosensor in health care sector. In *Bio-Nano Interface: Applications in Food, Healthcare and Sustainability*; Springer, Singapore, 2021; pp. 251–270. ISBN 9789811625169.
78. Gholve, S.; Ghodake, K.; Gore, S.; Bhusnure, O. Development & applications of nanobiosensors for biomedical diagnosis. *Int J ChemTech Res* 2017, *10*, 683–695.
79. Banigo, A.T.; Azeez, T.O.; Ejeta, K.O.; Lateef, A.; Ajuogu, E. Nanobiosensors: Applications in biomedical technology. In *Proceedings of the IOP Conference Series: Materials Science and Engineering*; 2020, *805*(1). doi:10.1088/1757-899X/805/1/012028.
80. Rai, M.; Gade, A.; Gaikwad, S.; Marcato, P.D.; Durán, N. Biomedical applications of nanobiosensors: The state-of-the-art. *J. Braz. Chem. Soc.* 2012, *23*. doi:10.1590/S0103-50532012000100004.
81. Mello, L.D.; Kubota, L.T. Review of the Use of Biosensors as Analytical Tools in the Food and Drink Industries. *Food Chemistry*, *77*(2), 237–256. doi: 10.1016/S0308-8146(02)00104-8.
82. Kim, Y.H.; Park, J.S.; Jung, H.I. An impedimetric biosensor for real-time monitoring of bacterial growth in a microbial fermentor. *Sens Actuators B Chem* 2009, *138*, 270–277, doi:10.1016/j.snb.2009.01.034.
83. Akyilmaz, E.; Dinçkaya, E. An amperometric microbial biosensor development based on Candida tropicalis yeast cells for sensitive determination of ethanol. *Biosens Bioelectron* 2005, *20*, 1263–1269, doi:10.1016/j.bios.2004.04.010.
84. Shen, Y. Rice husk silica derived nanomaterials for sustainable applications. *Renew Sustain Energy Rev* 2017, *80*, 453–466.
85. Marchiol, L. Nanotechnology in agriculture: New opportunities and perspectives. In *New Visions in Plant Science*; InTech, 2018, doi:10.5772/intechopen.74425.
86. Eggins, B.R. *Chemical Sensors and Biosensors*; John Wiley & Sons, Hoboken, NJ, 2012.
87. Korotkaya, E. V. Biosensors: Design, classification, and applications in the food industry. *Foods Raw Mater* 2014, *2*, 161–171, doi:10.12737/5476.
88. Singh, R.; Ashraf, J.; Manzoor, S.; Dar, M.H. Application of biosensors in food quality control. *J Post Harvest Technol* 2020, *8*(1), 53–74.
89. Prakash, A.; Isa, G. The emerging usage and applications of nanotechnology in food processing industries: The new age of nanofood. *Int J Pharm Sci Rev Res* 2013, *22*, 107–111.
90. Berekaa, M.M. Nanotechnology in food industry: Advances in food processing, packaging and food safety. *Int.J.Curr.Microbiol.App.Sci* 2015, *4*(5), 345–357.
91. Prakash, H.G.; Devi, S.; Sonkar, S.; Mishra, P. Potential application of nanotechnology in food processing and packaging: A review. *Int J Curr Microbiol Appl Sci* 2021, *10*, 601–613, doi:10.20546/ijcmas.2021.1010.070.
92. Wasilewska, A.; Bielicka, M.; Klekotka, U.; Kalska-Szostko, B. Nanoparticle applications in food – a review. *Food Funct* 2023, *14*, 2544–2567, doi:10.1039/D2FO02180C.
93. Pérez-López, B.; Merkoçi, A. Nanomaterials-based (bio)sensing systems for safety and security applications. In *Proceedings of the NATO Science for Peace and Security Series A: Chemistry and Biology*, 2012; pp. 43–61. doi: 10.1007/978-94-007-2872-1_3.
94. Ahamed Fazil, A.H.; Reddy, U.S.; Bhargavi Gumpu, M. Biosensor for biothreat detection and defense application. In *Health and Environmental Applications of Biosensing Technologies*; Elsevier, 2024; pp. 267–291. doi:10.1016/B978-0-443-19039-1.00013-4.
95. Muhammad-Tahir, Z.; Alocilja, E.C. A conductometric biosensor for biosecurity. In *Proceedings of the Biosensors and Bioelectronics*; Elsevier Ltd, May 1 2003; Vol. 18, pp. 813–819. doi:10.1016/s0956-5663(03)00020-4.

96. Biedermann, A.; Bozza, S.; Taroni, F. Decision theoretic properties of forensic identification: Underlying logic and argumentative implications. *Forensic Sci Int* 2008, *177*, 120–132, doi:10.1016/j.forsciint.2007.11.008.
97. Saks, M.J. Forensic identification: From a faith-based "Science" to a scientific science. *Forensic Sci Int* 2010, *201*, 14–17, doi:10.1016/j.forsciint.2010.03.014.
98. Samal, S.; Mohanty, R.P.; Mohanty, P.S.; Giri, M.K.; Pati, S.; Das, B. Implications of biosensors and nanobiosensors for the eco-friendly detection of public health and agro-based insecticides: A comprehensive review. *Heliyon* 2023, *9*, e15848.
99. Bhardwaj, N.; Gianni, E.; Bhardwaj, S.; Gulati, A.; Gulati, S. Development of microbial-based nanosensors and their applications: The state of the art. In *Microbial Nanotechnology*; De Gruyter, 2022; pp. 211–242. doi:10.1515/9783110754476-010.
100. Rassaei, L.; Marken, F.; Sillanpää, M.; Amiri, M.; Cirtiu, C.M.; Sillanpää, M. Nanoparticles in electrochemical sensors for environmental monitoring. *TrAC Trends Anal Chem* 2011, *30*, 1704–1715, doi:10.1016/j.trac.2011.05.009.
101. Ramesh, M.; Janani, R.; Deepa, C.; Rajeshkumar, L. Nanotechnology-enabled biosensors: A review of fundamentals, design principles, materials, and applications. *Biosensors (Basel)* 2023, *13*, 40.
102. Pechyen, C.; Tangnorawich, B.; Toommee, S.; Marks, R.; Parcharoen, Y. Green synthesis of metal nanoparticles, characterization, and biosensing applications. *Sens Int* 2024, *5*, 100287.
103. Bilal, M.; Iqbal, H.M.N. Microbial-derived biosensors for monitoring environmental contaminants: Recent advances and future outlook. *Process Saf Environ Prot* 2019, *124*, 8–17, doi:10.1016/j.psep.2019.01.032.
104. Barreiros dos Santos, M.; Rodriguez-Lorenzo, L.; Queirós, R.; Espiña, B. Fundamentals of biosensors and detection methods. *Adv Exp Med Biol* 2022, *1379*, 3–29.
105. Prasad, R.; Gupta, N.; Banka, A. Rapid diagnosis and shorter regimen for multidrug-resistant tuberculosis: A priority to improve treatment outcome. *Lung India* 2017, *34*, 1–2.
106. Alvarado, K.; Bolaños, M.; Camacho, C.; Quesada, E.; Vega-Baudrit, J. Nanobiotechnology in agricultural sector: Overview and novel applications. *J Biomater Nanobiotechnol* 2019, *10*, 120–141, doi:10.4236/jbnb.2019.102007.
107. Singh, R.P. Recent trends, prospects, and challenges of nanobiosensors in agriculture. Biosensors in Agriculture: Recent Trends and Future Perspectives. *Concepts and Strategies in Plant Sciences.* Cham: Springer. 2021; pp. 3–13. doi:10.1007/978-3-030-66165-6_1.
108. Bhalla, N.; Pan, Y.; Yang, Z.; Payam, A.F. Opportunities and challenges for biosensors and nanoscale analytical tools for pandemics: COVID-19. *ACS Nano* 2020, *14*, 7783–7807.
109. Grasso, G.; Zane, D.; Dragone, R. Microbial nanotechnology: Challenges and prospects for green biocatalytic synthesis of nanoscale materials for sensoristic and biomedical applications. *Nanomaterials* 2020, *10*, 11.

7 Nanobiotechnology-Based E-Waste Management for Environmental Sustainability

Sanjeeb Kumar Mandal, Alwina G., Bodika Shynisha, B. Sneha, Vennela Lakavath, B. Sumithra, Mahaboob Basha D., Pooja Aich, and Anuranjeeta

7.1 INTRODUCTION

The worldwide use of electrical and electronic equipment (EEE) is closely associated with the expansion of the world economy. EEE has become indispensable in modern societies and is enhancing living standards [1]. Because of the complexity and rapid advancement of technology, it is difficult to group them into sensible and practically useful categories [2]. When EEE is disposed of after usage, valuable and dangerous items are included in the waste stream. This waste stream is referred to as e-waste, or waste electrical and electronic equipment (WEEE), a term used mainly in Europe [1]. Electronic waste (e-waste) is a general term that includes various forms of old EEE, i.e., electronic equipment that is no longer durable and no longer valuable to its owner [3]. Technologies that became obsolete because the technology is old and replaced by a new technology drive the production of e-waste [2]. Quick advancement and cost reduction on electronic gadget made a drastic transformation among regular users to access the new electronic products and digital technology [4]. The high rate of electronic waste generation causes the accumulation process of pollution to be fast and has a very dangerous impact on health and the environment [3].

E-waste is categorized as hazardous waste because it contains components that have dangerous and toxic properties such as mercury, lead, cadmium, chromium, arsenic, and polychlorinated biphenyls, which are harmful to human health and the environment [3]. E-waste is globally recognized as a resource because of the potential for recovering valuable materials including iron, aluminium, copper, gold, silver, and rare earth metals [5]. Ferrous metals are the most common valuable materials in WEEE, next to plastics. In addition to its economic value, the recovery of plastics from WEEE is encouraged by new policies and regulations [6]. Consumer behaviour is a key factor affecting the effectiveness of e-waste recycling [7]. E-waste is a global, interregional, and domestic problem [8]. A number of toxins and hazardous substances in e-waste are linked to serious environmental pollution [9].

Getting rid of electronic waste is a complicated process that calls for more advanced technology and expertise. A major factor that leads to environmental deterioration and a potential health risk is the waste's hazardous and poisonous content. Management of electronic waste that ignores environmental management will cause uncontrolled contamination of heavy metals and toxic compounds that are cross-border [3]. Traditional e-waste processing through inappropriate networks has resulted in an immense quantity of heavy metals and other contaminants into the natural environment, which negatively affects natural environments such as soil, water, dust, and plants [10]. Documentation is increasing about health effects related to contamination in air, soil, and water for people working and living at or near informal e-waste processing sites. Evidence of adverse health effects and the increasing number of e-waste sites make protecting human health and the environment from e-waste contamination an expanding challenge [5]. It is imperative to develop unique technology that is both

DOI: 10.1201/9781003487142-7

ecologically benign and suitable for treating electronic wastes. The technologies most frequently used for WEEE secondary treatment are hydro- and pyrometallurgy. Electrochemical processes may also be an alternative to reduce the consumption of chemicals and wastewater generation [11]. Many of the methods have not been put into practice on a commercial scale due to a variety of reasons, including the necessity of handling the massive amount of waste acid liquid that was produced throughout the operations with caution. Thus, it is crucial to create new, environmentally friendly methods for recycling copper that use less energy and reagent and don't emit any pollutants [12]. Instead of incremental improvements to existing technologies, nanobiotechnology offers disruptive, game-changing breakthroughs and innovations that can provide immediate answers and solutions to help our society, environment, and the planet [13]. Nanomaterials remove hazardous metals from waste or contaminated water [14]. Nanotechnology proved to be one of the best technologies in management of waste because of its unique characteristics [15]. Engineered nanomaterials can be synthesized either top-down or bottom-up. Using cutting, grinding, and scattering processes, the bulk material is reduced to nanosized in a top-down manner. It consists of mechanical milling, vacuum vapour deposition, pulsed wire discharge, and laser (pulse) ablation. Atoms, molecules, or clusters create nanomaterials via the bottom-up method. Micro-emulsion techniques, chemical reduction, microwave-assisted techniques, and hydrothermal, electrochemical, and biosynthetic processes are all included [16]. A more effective and resource-rich recycling process can result from the deliberate separation of important metals and rare earth elements from e-waste using nanotechnology. A more sustainable and circular economy for electronic devices can be facilitated by nanotechnology, which makes it possible to recover and reuse precious materials at the nanoscale.

7.2 TYPES AND SOURCES OF E-WASTE

E-waste can also be produced throughout the electronics production process in the form of scraps, broken goods, etc. Three main categories are typically used to separate the e-waste produced from discarded EEE: consumer equipment (TVs, DVD players, mobile phones, mp3 players, and leisure and sporting equipment) (Figure 7.1), information technology (IT) and telecom (personal computers, monitors, and laptops), and large household appliances (washing machines and refrigerators). Equipment components including batteries, circuit boards, plastic casings, cathode-ray tubes, activated glass, and lead capacitors are also considered to be e-waste [8]. It covers a broad range of electronic devices, ranging from large household appliances, IT and telecommunications equipment, lighting equipment, medical devices, monitoring and control instruments, automatic dispensers, and consumer electronics, such as electrical and electronic tools, toys, leisure and sports equipment, and mobile phones to computers [17].

E-waste items from household uses are microwaves, housing entertainment products, electric cookers, heaters and fans, etc. [18]. Electronics such as ovens, grills, refrigerators, fax machines, and sewing machines, and the like washing machines, fans, air conditioners, grinders, irons, heaters,

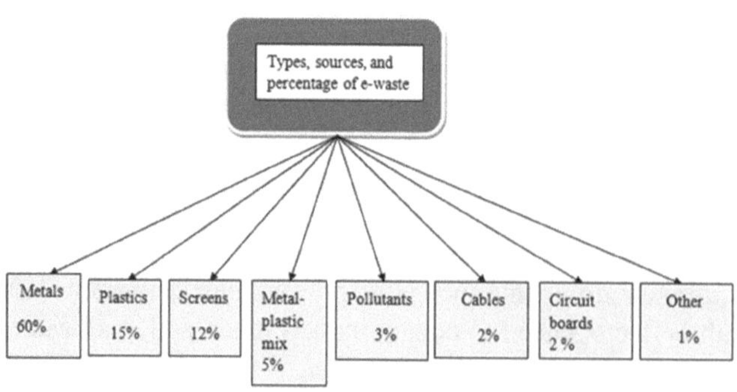

FIGURE 7.1 Types and sources of e-waste.

military, aviation, automobile, and laboratory electronics are examples of both industrial and consumer electronics [19]. E-waste from the medical business includes electronic medical equipment such as electrocardiogram (ECG) machines, patient monitors, computed tomography scan, magnetic resonance imaging, X-ray, ultrasound, and dialysis devices.

7.3 HAZARDOUS IMPLICATIONS OF E-WASTE

The negligent outdoor disposal of electronic waste, which includes burning, is undoubtedly a factor in the rapidly increasing threat of pollution that harms both the environment and human health. The various components of a computer include highly hazardous substances that cause metal leaching, which pollutes soil and groundwater on a huge scale. For areas that have been subjected to long-term dumping, the issue only gets worse with time. The recovery of lead from circuit boards also emits dioxin and other chlorine compounds into the air [20]. The e-waste stream is categorized as hazardous due to the presence of several toxic chemicals and metal components and halogens in component materials [17]. E-waste is a hazardous industrial waste because it contains several types of persistent substances, such as tetrabromobisphenol A and decabromodiphenyl ether [21]. Sophisticated electronics can contain up to 60 components. E-waste contains toxic substances like lead, mercury, cadmium, and polycyclic aromatic hydrocarbons (PAHs) that have adverse impacts on human health and the environment [22].

7.3.1 Toxic Effects on Environment

An enormous number of dangerous metals (Cd, Hg, Pb, and Cr) from e-waste may contribute to increase the toxicity levels of ecosystem. The open disposal of the e-waste-containing toxic substances in water bodies and landfills pollutes the ground water [23]. Restoration techniques that are too archaic have the potential to discharge harmful substances into the soil, water, and air. Environmental contamination as a result of improper e-waste recycling and illegal disposal activities can lead to both direct and indirect exposures via soil, air, dust, and water around e-waste-recycling sites, afterwards affecting the biota zone [24]. Computing activities contribute to global warming through carbon emissions released during the manufacture, use, and recycling of the resultant e-waste through the release of harmful chemicals into the environment [25].

7.3.2 Toxic Effects on Human and Animal Health

People living near e-waste-processing areas are stated to get easily exposed to the hazardous substances released and transported through all kinds of natural pathways [24]. Long-term recycling activities in the informal sector, compounded with long-term persistence of e-waste pollutants in the environment, increase the exposure risk to the waste-recycling workers and the population inhabiting nearby localities and thus are a health hazard [26]. Heavy metal debris, including lead, arsenic, mercury, cadmium, and nickel, will damage the body's primary metabolic system. Carcinogenic metal ions like arsenic and nickel trigger redox reactions in biological systems. The free radicals produced by these reactions cause oxidative damage to proteins and DNA [33]. The heavy metals make their way into the human body and are known to have serious detrimental effects on human health at elevated levels [27]. These metals can harm the liver, kidneys, and neurological system.

Hexabromocyclododecane (HBCD) and tetrabromobisphenol A (TBBP-A) are brominated flame retardants (BFRs) widely used in electronics to reduce flammability [28]. It is generally recognized that the crude electronic waste (e-waste) recycling/dismantling is a main source of BFR pollution [29]. Bromophenolic flame retardants have been scientifically defined through toxicological research to have negative impacts on living things. For instance, TBBP-A given to rats caused a decrease in serum thyroxine level, induced CYP2B1 activity, and increased liver weight and constitutive expression of androstane receptors in the liver and has also been shown to induce development of uterine cancer in rats and adenocarcinoma, adenoma, and Muller's mixed tumour in mice and rats [30]. Several studies have also confirmed that BFRs are indeed endocrine disruptors, with the potential to disrupt male and female reproductive development and adult reproductive function by having antiandrogenic actions (males) and by altering steroidogenic activities [31]. BFRs negatively affect the environment and human health, especially in the sensitive (developing) nervous system [32]. The organs of kids remain developing, rendering them especially vulnerable to the adverse effects of e-waste. Children who are exposed to e-waste chemicals may experience a variety of issues, such as birth deformities, behavioural issues, learning impairments, and growth issues.

7.3.3 Toxic Effects on Aquatic Life

The accumulation of heavy metals such as lead, arsenic, mercury, cadmium, and nickel will destroy the main metabolic process [33]. Even in low concentrations, these metals are highly toxic to aquatic organisms. Apart from hindering growth and reproduction, they have the potential to kill fish. Heavy metals are easily absorbed by aquatic life from tainted water. As they go up the food chain, these metals then collect in their bodies and grow more concentrated. Electronic parts and plastics can accumulate near water bodies as a result of improper e-waste disposal. Fish and other aquatic life's ability to survive and propagate is impacted by the physical pollution that deteriorates aquatic ecosystems.

7.4 ELEMENTAL COMPOSITION OF E-WASTE

Quantification of the elemental composition of versatile e-waste streams to build the related knowledge database is the starting point for developing any approaches for value recovery [34]. E-waste contains a range of materials, such as plastics, metals, and glass [17,38]. E-waste contains many valuable, recoverable materials such as aluminium, ferrous metals, copper, gold, and silver [20]. The components of electronic waste are diverse and fall into the "dangerous" and "non-dangerous" categories [35]. A typical composition pattern for e-waste is 30% organics (e.g., polymers, flame retardants, and glass fibre), 30% ceramics (e.g., silica, mica, and alumina), and 40% inorganics (e.g., ferrous and non-ferrous metals) [36].

7.5 EXISTING METHODS FOR THE DISPOSAL OF E-WASTES

With the implementation of the WEEE Directive, which attempts to eradicate this waste and enhance environmental quality, the disposal of electronic trash in developed nations has advanced (Figure 7.2). In waste management, numerous techniques were developed and utilized such as life cycle assessment, Material Flow Analysis (MFA), Multiple Criteria Analysis (MCA), and Extended Producer Responsibility (EPR) [35]. It is handled through various approaches like conventional disposal in landfills, incineration, and recycling [20].

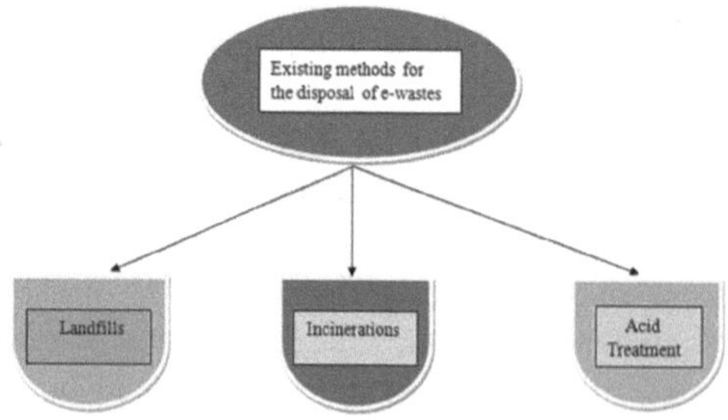

FIGURE 7.2 Existing methods for the disposal of e-wastes.

7.5.1 Landfills

The product is dumped in landfill sites, where it may remain indefinitely [20]. A landfill is an engineered ditch in which layers of solid waste are packed, compressed, and enclosed for complete disposal. To prevent contaminating groundwater, it has insulation at the base. Landfills require skilled design as well as expert operators and appropriate organization to assurance their functionality [18]. E-waste disposed of in landfills is often described as a toxic time bomb because it releases toxic metals and gases through natural processes into the environment after several months of disposal [9].

7.5.2 Incineration or Open-Air Burning

One method of treating waste is burning, which involves setting the components in the waste products on fire. Such procedure is beneficial in the low-level handling of the electronic waste with the more content of valuable metals for additional processing by hydrometallurgical routes [18].

7.5.3 Acid Treatment

It was crushed under 0.2 mm and was mixed with H_2SO_4 acid in varying concentrations from 30% to 95%. It is likely to obtain almost complete leaching of Cu and precious metals at a 95% concentration of H_2SO_4 acid. [18].

Valuable metals can be extracted from electronic waste (e-waste) using an acid treatment technique. Strong acids are used in this procedure to dissolve the metals from the waste material so that they can be extracted and processed further. This technique can be useful, but because of the possibility of hazardous waste production, poisonous emissions, and acid spills, it also poses serious environmental risks [37,38].

Unintentional spills have the potential to seriously harm the ecosystem by contaminating water and soil resources. Hazardous gases including sulphur dioxide and hydrogen chloride may be released into the atmosphere as a result of the process. The resulting acidic liquids and sludge can be difficult to dispose of safely and can constitute a harm to human health and the environment. To reduce environmental concerns, the use of acids in e-waste processing is governed by stringent legislation in many countries. Limiting the amount of acid used, using particular acids or acid mixtures to target particular metals, and reducing environmental impact. Combining acid treatment with other techniques, such solvent extraction and precipitation, can improve metal recovery efficiency and reduce waste output. Using microorganisms to extract metals from electronic trash offers a less dangerous and more environmentally friendly approach [39,40].

7.6 ROLE OF NANOTECHNOLOGY TO REMOVE OR DEGRADE OR DISPOSE E-WASTES

The environmental pollutants and environmental impacts that may be generated in different stages of e-waste recycling are depicted to remind possible ecological and environmental risks [21]. It is necessary to dispose of hazardous material in an environmentally responsible and safe manner, adhering to all trash elimination norms. However, these elements can be successfully removed or degraded by nanoparticles (NPs), which can help to curb the menace of e-waste metal toxicity [41]. Nanotechnology is an area of science, engineering, and technology that deals with nanoscale dimensions and tolerances, specifically the manipulation of single atoms and molecules with diameters ranging from 1 to 100 nm [42]. The nanotechnology in waste management is gaining momentum globally owing to the unique properties of nanomaterials and their high compatibility with the environment compared to the existing conventional treatment technologies [43]. Nanotechnology proved to be one of the best technologies in management of waste because of its unique characteristics like high surface area to volume ratio, freedom to control and manipulate the material or synthesizing fewer toxic substances, time efficient, and labour efficient [15]. Instead of incremental improvements to existing technologies, nanotechnology offers disruptive, game-changing breakthroughs and innovations that can provide immediate answers and solutions to help our society, environment, and the planet [13]. The amazing physiochemical features of nanomaterials have garnered significant attention in the degradation and mineralization of organic contaminants that are hazardous. In the process of degradation, the organic pollutants are splitting or decomposed into several products, while in the case of mineralization, the complete destruction of organic pollutant takes place into water, carbon dioxide, and some inorganic ions [44].

7.6.1 Recovering Valuable Materials

Electrical waste contains hazardous but also valuable and scarce materials [22]. The extraction of noble metals (Au, Pt, Pd, Ta, Te, Ge, and Se) from e-waste should be given major priorities (Table 7.1). However, the realization of recycling should be the basis of maximum recovery and minimum negative impact to the environment [23]. Scrap from IT and telecommunication systems

TABLE 7.1
Elemental Composition of E-waste

Category	Elements	Significance	References
Rare earth elements (REE)	Lanthanides (La), scandium (Sc), neodymium (Nd), praseodymium (Pr), dysprosium (Dy), and yttrium (Y)	Electrical and electronic equipment such as cathode-ray tubes, printed circuit boards, permanent magnets, and batteries contain the highest fraction of REEs.	[50,51]
Brominated flame retardants (BFRs)	Hexabromocyclododecane (HBCD), polybrominated diphenyl ethers (PBDEs), decabromodiphenyl ether (deca-BDE), and tetrabromobisphenol A (TBBP-A)	BFRs are added to textiles, and electronic circuit boards to reduce their flammability and prevent the rapid spread of fire.	[28,52]
Heavy metals	Copper (Cu), iron (Fe), zinc (Zn), nickel (Ni), aluminium (Al), manganese (Mn), cobalt (Co), chromium (Cr), molybdenum (Mo), and cadmium (Cd)	Heavy metals are used for their conductivity, durability, etc.	[50]
Precious metals	Palladium (Pd), platinum (Pt), silver (Ag), and gold (Au)	Precious metals are used for their superior electrical conductivity, ductility, corrosion resistance, etc.	[50]
Glass	SiO_2 and Al_2O_3	Glass foams are widely used in structural and thermal insulation applications due to their low density, high thermal stability, low thermal conductivity, and high chemical resistance.	[53]
Polymer	Acrylonitrile butadiene styrene (ABS), high-impact polystyrene (HIPS), polycarbonate (PC), polyamide (PA), polypropylene (PP) and epoxies, polycarbonate (PC), polypropylene (PP), polyethylene (PE), polyvinyl chloride (PVC), polybutylene terephthalate (PBT), polyamide (PA), poly(methyl methacrylate) (PMMA), and polystyrene (PS)	Polymers may serve as supporting substrates for conductive components, protective encapsulants, adhesives, matrix, or active materials, thus have attracted great interest by virtue of their tunable electronic properties, and are used in electronic devices due to their remarkable flexibility.	[52,54]

contains a higher number of precious metals [19]. Precious metals like gold, copper, and lithium can be extracted from e-waste using nanoparticles more effectively than with conventional techniques. Their high surface area makes them very reactive, allowing them to bind to specific metals. This can reduce the environmental impact of mining new resources. E-waste contains rare earth elements essential for modern electronics but difficult to extract. Nanoparticles show promise in separating these elements from complex e-waste mixtures [45].

It is possible to create nanoparticles to have particular affinities for particular elements or molecules. Selective adsorption of desired elements from a mixture can be achieved by modifying the surface characteristics of the nanoparticles, thereby leaving behind unwanted elements. Electrophoresis and electrodialysis are two electrokinetic techniques that can be employed with nanoparticles to separate components according to their size and charge. The components can be moved via a nanoporous channel or membrane by providing an electric field [46,47].

7.6.2 Greener Processing

Nanotechnology can contribute to developing cleaner e-waste-recycling processes. This approach has the potential to reduce energy consumption and hazardous waste generation compared to traditional methods [48].

Because electronics are updated so frequently, a large amount of e-waste accumulates. This debris can be recycled to extract key metals using conventional methods, but doing so can be problematic because it releases toxic compounds into the environment, especially in developing nations. In a similar vein, physical and chemical methods of producing metal nanoparticles are too harsh and not environmentally benign. These nanometals also have a limited shelf life. The recovery of valuable metals like copper, gold, and silver is essential to the e-waste-recycling industry's financial stability. As a result, it is possible to recycle e-waste and create metal nanoparticles in an environmentally responsible way. For the past few decades, it has been known for years that bacteria, actinomycetes, and fungus have the ability to extract metals from their surroundings. Lately, weeds—sometimes referred to as "enemies of the farmer"—have also been effectively employed in the production of gold, copper, and silver nanoparticles. The process of producing nanoparticles can be achieved economically and practically by using natural, environmentally benign sources like weed extracts and microbes. There are several uses for the metal nanoparticles that are produced from bioleaching e-waste, particularly in the medical industry. Tissue engineering, gene therapy, antimicrobials, drug delivery, and medical prostheses are a few of these uses. As a result, it is conceivable to "marry" the two disparate domains as "e-waste to nano" for the biological sciences, creating a dynamic arena for nanotechnology research. The conventional approaches to managing e-waste have shown to be harmful to the environment. The utilization of microorganisms and weed extracts in the bioremediation of hazardous e-waste to recover valuable metals is a cost-effective and environmentally sustainable method. Compared to those produced by artificial capping, these biosynthesized nanoparticles with natural protein caps are renowned for having a remarkably longer shelf life and stability. Enhancing the quality of current medical practices can be greatly aided by obtaining these metals in pure nano form from e-waste and investigating their various medicinal applications [49].

7.6.3 Detoxification

Metals can be hazardous because of their non-degradative nature, which causes them to linger in the environment for longer. The detoxification contaminant is considered to be a prime concern in ecological remediation [55]. Metal pollution has been treated using a variety of physical, chemical, and biological detoxification techniques in recognition of the potential hazard. Due to their high cost, high level of technical complexity, tendency to produce secondary pollutants, and inconsistent outcomes, these procedures have not proven entirely successful. Nanotechnology, a rapidly evolving field, has recently been used to remediate hazardous metals [56]. Products obtained through photocatalysis-assisted engineered nanotechnology have successfully proven upgraded performance characteristics like increased surface area, easy recovery, reuse capability, better volumetric potential, reduced mechanical stress, enhanced shelf life, excellent sorption behaviour, stability under operational conditions, easy separation, no secondary pollution, and many more [55]. By preventing the release of dangerous compounds into the air, water, and soil—substances that can have a major negative impact on health—detoxification helps protect the environment.

Detoxification procedures can be used to recover a variety of valuable elements from e-waste, including metals and polymers. This promotes the preservation of natural resources. Numerous health problems might arise from exposure to harmful materials included in e-waste. These hazards are reduced for both employees and the general public by detoxification. Detoxification can support the circular economy and provide jobs in the recycling sector. Procedures for detoxification can be costly, particularly for small-scale businesses. Because of this, implementing it in developing nations may be difficult. Employees engaged in the detoxification of e-waste could be exposed to dangerous materials if appropriate safety precautions are not taken. Due to their intricate composition, some e-waste components, such printed circuit boards, might be challenging to fully detoxify. The management of e-waste is being regulated, yet illicit dumping is still a major issue in many areas. This may make detoxification less effective.

7.7 CONCLUSION

The rapid growth of the global economy cannot be achieved without the contribution of electronic and electrical products. Keeping in mind the rapidly escalating scenario and change in lifestyle pattern, future safety with respect to environment and life, evolving sustainable and scientific e-waste management in a focused manner with sufficient infrastructure and financial resources, is imperative. The increasing levels of e-waste, low collection rates, and non-environmentally sound disposal and treatment of this waste stream pose significant risks to the environment and human health. The removal of contaminant from polluted resources is a significant focus for environmental stability and sustainability. Most of the e-waste has been disposed to landfills and incinerators, and it has also been exported from the developed countries to the developing countries. Most of these developments do not promote environmental sustainability by lowering environmental degradation and reducing global warming. Most consumers do not know about proper disposal of the older versions of the product once they go for the latest updated versions. There is not much information on medical personnel awareness of efficient e-waste management practices following the establishment of WEEE policy. Effective management of e-waste is not just a legal necessity but also a major social responsibility. There should be a proper enlightenment programme for both manufacturers and end users on the inherent danger of the used electronic equipment. We have achieved ground-breaking discoveries in every field of study thanks to nanotechnology. By selectively extracting rich metals like copper, silver, and gold from e-waste components, nanotechnology-based methods can increase recovery rates and decrease the need for mining. It is possible to create nanoparticles from e-waste materials, which opens up new uses and increases their longevity. For example, antimicrobial coatings can be made using silver nanoparticles that were removed from printed circuit boards.

An effective arsenal for addressing the escalating problem of e-waste management is provided by nanotechnology. We can transform the way that e-waste is processed and advance a circular economy and environmental sustainability by utilizing its capacity to modify matter at the nanoscale. Furthermore, it's important to keep developing waste treatment technologies based on nanotechnology since they may offer reliable answers to the challenges associated with treating e-waste. Nanomaterials are predicted to present significant opportunities in the treatment of e-waste and environmental remediation in the future.

REFERENCES

1. Forti, V., Baldé, C. P., Kuehr, R., & Bel, G. (2020). The global e-waste monitor 2020. United Nations University (UNU), International Telecommunication Union (ITU) & International Solid Waste Association (ISWA), Bonn/Geneva/Rotterdam, 120.
2. Ramanayaka, S., Keerthanan, S., & Vithanage, M. (2020). Urban mining of E-waste: Treasure hunting for precious nanometals. In *Handbook of Electronic Waste Management* (pp. 19–54). Butterworth-Heinemann. https://doi.org/10.1016/B978-0-12-817030-4.00023-1.
3. Raharjo, S., & Utomo, A. H. (2021). Comparative study of electronic waste management in developed countries and Indonesia. *Andalasian International Journal of Applied Science, Engineering and Technology*, *1*(1), 21–32.
4. Kemp, S. (2020). Digital use around the world in July 2020. *We are social [Электронный ресурс]*, 21.
5. Heacock, M., Kelly, C. B., Asante, K. A., Birnbaum, L. S., Bergman, Å. L., Bruné, M. N., ... & Suk, W. A. (2016). E-waste and harm to vulnerable populations: A growing global problem. *Environmental Health Perspectives*, *124*(5), 550–555.
6. Baldé, C. P., Forti, V., Gray, V., Kuehr, R., & Stegmann, P. (2017). *The global e-waste monitor 2017: Quantities, flows and resources*. United Nations University, International Telecommunication Union, and International Solid Waste Association.
7. Wang, F., Liu, Q., & Yin, X. (2023). Exploring influencing factors of consumers' wasted household appliances disposal behavior considering product heterogeneity in China. *Journal of Cleaner Production*, *418*, 138065.
8. Perkins, D. N., Drisse, M. N. B., Nxele, T., & Sly, P. D. (2014). E-waste: A global hazard. *Annals of Global Health*, *80*(4), 286–295.
9. Andeobu, L., Wibowo, S., & Grandhi, S. (2023). Environmental and health consequences of E-waste dumping and recycling carried out by selected countries in Asia and Latin America. *Sustainability*, *15*(13), 10405.
10. Rajput, R., & Nigam, N. A. (2021). An overview of E-waste, its management practices and legislations in present Indian context. *Journal of Applied and Natural Science*, *13*(1), 34–41.
11. Ambaye, T. G., Vaccari, M., Castro, F. D., Prasad, S., & Rtimi, S. (2020). Emerging technologies for the recovery of rare earth elements (REEs) from the end-of-life electronic wastes: A review on progress, challenges, and perspectives. *Environmental Science and Pollution Research*, *27*, 36052–36074.
12. Huang, J., Chen, M., Chen, H., Chen, S., & Sun, Q. (2014). Leaching behavior of copper from waste printed circuit boards with Brønsted acidic ionic liquid. *Waste Management*, *34*(2), 483–488.
13. Pokrajac, L., Abbas, A., Chrzanowski, W., Dias, G. M., Eggleton, B. J., Maguire, S., ... & Mitra, S. (2021). Nanotechnology for a sustainable future: Addressing global challenges with the international network4sustainable nanotechnology. *ACS Nano*, *15*, 18608.
14. Bhat, S. A., Sher, F., Hameed, M., Bashir, O., Kumar, R., Vo, D. V. N., ... & Lima, E. C. (2022). Sustainable nanotechnology based wastewater treatment strategies: Achievements, challenges and future perspectives. *Chemosphere*, *288*, 132606.
15. Vyas, S., Dhiman, N., Negi, S., & Kumar, R. (2021). Management of waste using nanotechnology. In *New Frontiers of Nanomaterials in Environmental Science* (pp. 253–279). https://doi.org/10.1007/978-981-15-9239-3_12.
16. Ealia, S. A. M., & Saravanakumar, M. P. (2017, November). A review on the classification, characterisation, synthesis of nanoparticles and their application. *IOP Conference Series: Materials Science and Engineering*, *263*(3), 032019.
17. Ilankoon, I. M. S. K., Ghorbani, Y., Chong, M. N., Herath, G., Moyo, T., & Petersen, J. (2018). E-waste in the international context–A review of trade flows, regulations, hazards, waste management strategies and technologies for value recovery. *Waste Management*, *82*, 258–275.
18. Dey, S., Veerendra, G. T. N., Padavala, S. S. A. B., & Manoj, A. P. (2023). Recycling of e-waste materials for controlling the environmental and human heath degradation in India. *Green Analytical Chemistry*, *7*, 100085.
19. Mmereki, D., Li, B., Baldwin, A., & Hong, L. (2016). The generation, composition, collection, treatment and disposal system, and impact of E-waste. *E-waste in Transition-from Pollution to Resource*, *10*, 61332.
20. Ari, V. (2016). A review of technology of metal recovery from electronic waste. *E-Waste in Transition—From Pollution to Resource*, 122–158. https://doi.org/10.5772/61569.
21. Liu, K., Tan, Q., Yu, J., & Wang, M. (2023). A global perspective on e-waste recycling. *Circular Economy*, *2*(1), 100028.

22. Subhaprada, C. S., & Kalyani, P. (2017). Study on awareness of e-waste management among medical students. *International Journal of Community Medicine and Public Health*, *4*(2), 506–510.
23. Islam, A., Ahmed, T., Awual, M. R., Rahman, A., Sultana, M., Abd Aziz, A., ... & Hasan, M. (2020). Advances in sustainable approaches to recover metals from e-waste – A review. *Journal of Cleaner Production*, *244*, 118815.
24. Li, W., & Achal, V. (2020). Environmental and health impacts due to e-waste disposal in China – A review. *Science of the Total Environment*, *737*, 139745.
25. Maphosa, M., & Maphosa, V. (2022). A bibliometric analysis of the effects of electronic waste on the environment. *Global Journal of Environmental Science and Management*, *8*(4), 589–606.
26. Ahirwar, R., & Tripathi, A. K. (2021). E-waste management: A review of recycling process, environmental and occupational health hazards, and potential solutions. *Environmental Nanotechnology, Monitoring & Management*, *15*, 100409.
27. Zahra, N., & Kalim, I. (2017). Perilous effects of heavy metals contamination on human health. *Pakistan Journal of Analytical & Environmental Chemistry*, *18*(1), 1–17.
28. Feiteiro, J., Mariana, M., & Cairrão, E. (2021). Health toxicity effects of brominated flame retardants: From environmental to human exposure. *Environmental Pollution*, *285*, 117475.
29. Zhao, X., Chen, T., Yang, B., Wang, D., Sun, W., Wang, Y., ... & Shi, Z. (2021). Serum levels of novel brominated flame retardants (NBFRs) in residents of a major BFR-producing region: Occurrence, impact factors and the relationship to thyroid and liver function. *Ecotoxicology and Environmental Safety*, *208*, 111467.
30. Włuka, A., Woźniak, A., Woźniak, E., & Michałowicz, J. (2020). Tetrabromobisphenol A, terabromobisphenol S and other bromophenolic flame retardants cause cytotoxic effects and induce oxidative stress in human peripheral blood mononuclear cells (in vitro study). *Chemosphere*, *261*, 127705.
31. Kodavanti, P. R. S., Stoker, T. E., Fenton, S. E., & Curras-Collazo, M. (2022). Brominated flame retardants. In *Reproductive and Developmental Toxicology* (pp. 691–726). Academic Press. https://doi.org/10.1016/B978-0-323-89773-0.00036-9.
32. Dong, L., Wang, S., Qu, J., You, H., & Liu, D. (2021). New understanding of novel brominated flame retardants (NBFRs): Neuro (endocrine) toxicity. *Ecotoxicology and Environmental Safety*, *208*, 111570.
33. Fu, Z., & Xi, S. (2020). The effects of heavy metals on human metabolism. *Toxicology Mechanisms and Methods*, *30*(3), 167–176.
34. Buechler, D. T., Zyaykina, N. N., Spencer, C. A., Lawson, E., Ploss, N. M., & Hua, I. (2020). Comprehensive elemental analysis of consumer electronic devices: Rare earth, precious, and critical elements. *Waste Management*, *103*, 67–75.
35. Ajekwene, K. K., Aigbokhan, E. E., Akindele, O. E., Yibowei, M. E., Momoh, F. P., & Ugonna, U. K. (2022). Electronic waste (E-Waste): Sources, proliferation, effects & management in developing nations. *IOSR Journal of Engineering*, *12*, 12–27.
36. Rautela, R., Arya, S., Vishwakarma, S., Lee, J., Kim, K. H., & Kumar, S. (2021). E-waste management and its effects on the environment and human health. *Science of the Total Environment*, *773*, 145623.
37. Liu, R., Li, J., & Ge, Z. (2016). Review on Chromobacterium violaceum for gold bioleaching from e-waste. *Procedia Environmental Sciences*, *31*, 947–953.
38. Li, J., Wen, J., Guo, Y., An, N., Liang, C., & Ge, Z. (2020). Bioleaching of gold from waste printed circuit boards by alkali-tolerant Pseudomonas fluorescens. *Hydrometallurgy*, *194*, 105260.
39. Işıldar, A., van de Vossenberg, J., Rene, E. R., van Hullebusch, E. D., & Lens, P. N. (2016). Two-step bioleaching of copper and gold from discarded printed circuit boards (PCB). *Waste Management*, *57*, 149–157.
40. Adetunji, A. I., Oberholster, P. J., & Erasmus, M. (2023). Bioleaching of metals from e-waste using microorganisms: A review. *Minerals*, *13*(6), 828.
41. Dodamani, S., Kurangi, B., Teradal, N., & Kurjogi, M. (2021). Nanobiotechnology for e-waste management. In *Nanobiotechnology* (pp. 271–281). Elsevier. https://doi.org/10.1016/B978-0-12-822878-4.00017-1.
42. Mariappan, B., Prakash, S., & Binesh, A. (2023). Probiotic nanoparticles for food. In *Recent Advances in Aquaculture Microbial Technology* (pp. 307–338). Academic Press. https://doi.org/10.1016/B978-0-323-90261-8.00008-0.
43. Dermatas, D., Mpouras, T., & Panagiotakis, I. (2018). Application of nanotechnology for waste management: Challenges and limitations. *Waste Management & Research*, *36*(3), 197–199.
44. Yaqoob, A. A., Parveen, T., Umar, K., & Mohamad Ibrahim, M. N. (2020). Role of nanomaterials in the treatment of wastewater: A review. *Water*, *12*(2), 495.

45. Pineda-Vásquez, T., Rendón-Castrillón, L., Ramírez-Carmona, M., & Ocampo-López, C. (2023). From E-waste to high-value materials: Sustainable synthesis of metal, metal oxide, and MOF nanoparticles from waste printed circuit boards. *Nanomaterials*, *14*(1), 69.
46. Wang, Y., Zhou, Y., & Liu, J. (2016). Selective adsorption of mercury(II) from aqueous solutions using gold nanoparticles functionalized with thiol groups. *Journal of Hazardous Materials*, *318*, 342–350.
47. Melchor-Durán, A. P., Moreno-Virgen, M. R., Bonilla-Petriciolet, A., Reynel-Ávila, H. E., Lucio Ortiz, E., & González-Vázquez, O. F. (2024). Heavy metal removal from water using graphene oxide in magnetic-assisted Adsorption Systems: Characterization, adsorption properties, and modelling. *Separations*, *11*(10), 294.
48. Mukherjee, T., Mitra, S., Herat, S., & Kaparaju, P. (2023). Nanotechnology as a cleaner and greener approach toward e-waste recycling and valorization. In *Development in E-waste Management* (pp. 87–103). CRC Press. https://doi.org/10.1201/9781003301899-7.
49. Majumder, D. R. (2013). Waste to health: Bioleaching of nanoparticles from e-waste and their medical applications. *Indian Journal of Applied Research*, *3*(2): 277–288.
50. Tunali, M., Tunali, M. M., & Yenigun, O. (2021). Characterization of different types of electronic waste: Heavy metal, precious metal and rare earth element content by comparing different digestion methods. *Journal of Material Cycles and Waste Management*, 23, 149–157.
51. Ramprasad, C., Gwenzi, W., Chaukura, N., Azelee, N. I. W., Rajapaksha, A. U., Naushad, M., & Rangabhashiyam, S. (2022). Strategies and options for the sustainable recovery of rare earth elements from electrical and electronic waste. *Chemical Engineering Journal*, *442*, 135992.
52. Das, P., Zeng, Q., Leybros, A., Gabriel, J. C. P., Tay, C. Y., & Lee, J. M. (2023). Enhanced extraction of brominated flame retardants from e-waste plastics. *Chemical Engineering Journal*, *469*, 144126.
53. Assefi, M., Maroufi, S., Mansuri, I., & Sahajwalla, V. (2021). High strength glass foams recycled from LCD waste screens for insulation application. *Journal of Cleaner Production*, *280*, 124311.
54. Gong, M., Zhang, L., & Wan, P. (2020). Polymer nanocomposite meshes for flexible electronic devices. *Progress in Polymer Science*, *107*, 101279.
55. Mazumder, A., Bhattacharya, S., & Bhattacharjee, C. (2020). Role of nano-photocatalysis in heavy metal detoxification. In *Nanophotocatalysis and Environmental Applications: Detoxification and Disinfection* (pp. 1–33). https://doi.org/10.1007/978-3-030-12619-3_1.
56. Saleem, S., Rizvi, A., & Khan, M. S. (2023). Microbiome-mediated nano-bioremediation of heavy metals: A prospective approach of soil metal detoxification. *International Journal of Environmental Science and Technology*, *20*(11), 12823–12846.

8 Genetically or Metabolically Modified Microbes for Nanoparticle Production

Anitha T. Simon

8.1 INTRODUCTION

Nanotechnology facilitates the fabrication and modification of organic and inorganic matters to nanoscale range. When the material size is reduced to nano range, the surface to volume ratio increases allowing more number of atoms or molecules per volume, thereby enhancing the characteristics. Therefore, as compared to bulk materials, nanomaterials show remarkable thermal, chemical, optical and catalytic properties, desirable for various applications [1]. Engineered nanomaterials exhibit significant promises in agriculture, electronics, environment and medicine, by improving various limitations of their conventional approaches. In the field of electronics, nanomaterials have been given profound interest in the application of conversion and storage of energy through semiconductors [2]. Conductive nanomaterials have been developed for inkjet technology, and metal nanoparticles are used to create electronic circuits [3]. Agricultural applications include the use of nanoparticles to enhance mechanical properties of plant cell walls, prevent parasitic attack and increase water retention of soil [4]. Furthermore, various nanoparticles are extensively studied and employed for environmental applications such as disinfection, toxic dye degradation, heavy metal removal and treatment of pharmaceutical waste [5]. The success stories of nanoparticles in medical applications have elevated the interest of using them for future clinical therapeutics. Therapeutic efficiency of different nanomaterials has been documented in various branches of medicine such as immunology, dentistry, oncology, and ophthalmology [6]. Nanomaterials have revolutionised the medical field through their applications in the form of therapeutic formulations, drug and gene delivery vehicles, scaffolds, bioimaging agents and biosensors [7].

Nanoparticles can be fabricated using top-down or bottom-up approach [8] (Figure 8.1a). Top-down approach is a physical process of miniaturising bulk components to nanosized materials. The top-down approach includes high-energy mechanical milling, laser ablation, lithography, sputtering and electro-spraying. The process of mechanical milling mainly involves the use of heavy balls for size reduction of bulk materials by providing a high amount of energy. Laser ablation uses laser beam as the source for synthesising nanosized materials, whereas lithography employs electrons or light beams to develop nanoparticles. Sputtering is a process based on physical vapour deposition and is widely used for depositing nanostructured particles and films. Electro-spraying is used for the formation of nanoparticles from mostly polymers with the usage of high voltage [9]. The bottom-up approach, which includes chemical and biological methods, develops nanostructures by building them up from basic units such as atoms, molecules or clusters. Chemical methods in the bottom-up approach involve chemical vapour deposition, polyol synthesis, sol-gel technique, hydrothermal synthesis and microemulsion method. The process of chemical vapour deposition deals with various chemical reactions, which drive the formation of thin films. Polyol synthesis mainly uses poly(ethylene glycol)s, which not only serve as solvents but also act as reducing agents for the formation of metallic nanoparticles. Sol-gel is a wet chemical synthesis route that produces a network of polymers or particles from colloidal particles suspension and polymers.

 DOI: 10.1201/9781003487142-8

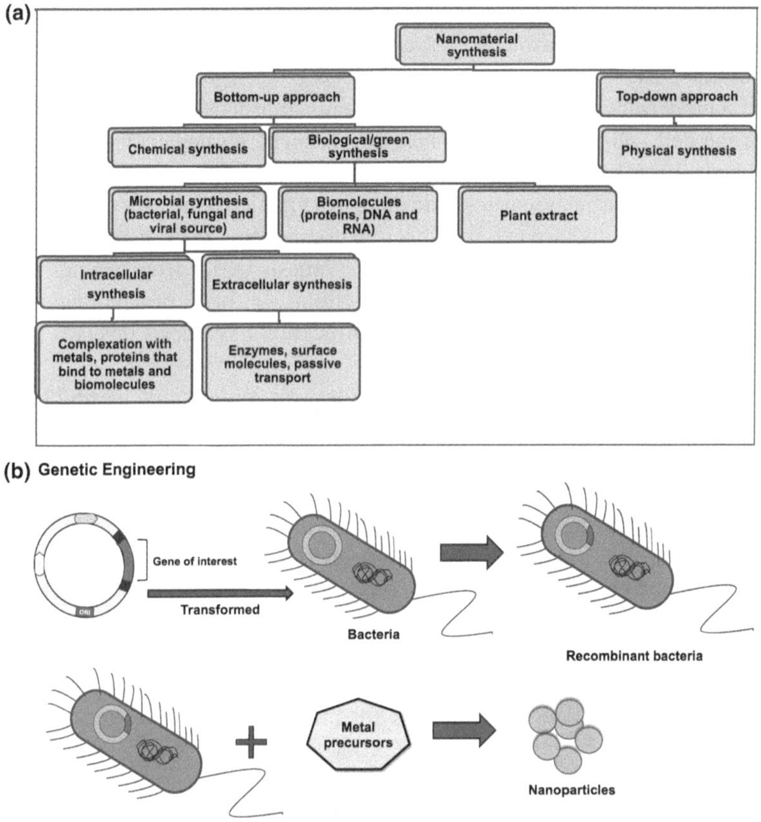

FIGURE 8.1 Diagrammatic representation of (a) various approaches to nanoparticle synthesis and (b) genetic engineering method for the formation of nanoparticles.

Hydrothermal process focuses on altering parameters such as pressure and temperature of supercritical water to fabricate nanoparticles with tailored characteristics. Furthermore, microemulsion is applied for the formation of composite, magnetic and metal nanoparticles by employing the aqueous phase, oil phase and surfactant [10].

Although some of the physical and chemical methods offer advantages such as large-scale production, durability and options to alter reaction parameters, most of them essentially need the usage of reagents which are corrosive, explosive and toxic [11]. Some of the methods demand high energy, longer reaction time and resources that might escalate the expense. Moreover, the physical techniques tend to incorporate contaminants within nanostructures, diminish their purity and cause lack of reproducibility [12].

As compared to physical and chemical methods, the biological route of synthesis is eco-friendly and reduces the formation of toxic by-products, which is an essential strategy for various applications, especially in the medical field. The biological method is also called green synthesis approach, where nanoparticles are fabricated using biologically active agents such as biomolecules, plants and microbes. Among biologically active agents, biomolecules such as proteins, DNA and RNA are used as templates for the synthesis of nanoparticles [13]. Green synthesis through plants is carried out by using their extract rich in phytochemicals, metabolites and biomolecules, which can actively reduce metal ions to form biocompatible nanoparticles [14].

Among the microbial sources, fungi, actinomycetes, viruses and bacteria are widely explored for the synthesis of nanoparticles. Microbes, which are designated as nano-factories, can potentially control the shape and size of biologically synthesised nanoparticles. The biomolecules secreted by microbes act as capping agents or reducing agents to give rise to nanoparticles. They can synthesise metallic and non-metallic nanoparticles extracellularly and intracellularly, through the process of enzymatic reduction [15].

Synthesis of nanoparticles by microbes is a biochemical process where positively charged metal ions present in the cell wall or in the cytoplasm are reduced by negatively charged biomolecules. Most of these biochemical processes are meant to enhance microbial resistance, where toxic ions are converted to non-toxic metal nanoparticles through enzymatic reduction, resulting in intracellular detoxification [16]. Oxidoreductase and transporting components majorly take part in the microbial synthesis of nanoparticles. The site of nanoparticle synthesis varies from species to species, which might be either extracellular or intracellular. Extracellular synthesis of nanomaterials is facilitated by surface proteins or secreted enzymes, whereas intracellular synthesis involves biomolecules and metal-binding proteins [17].

The abundance in nature, ability to multiply relatively faster, and feasibility to culture and maintain *in vitro* make microbes the alternative and sustainable source for nanoparticle synthesis. Moreover, microbes reduce bio-waste, enable larger nanoparticle yield and do not require highly toxic chemicals [18]. Some prokaryotes can grow in oxygen-deprived environments with minimal stress and produce metallic nanoparticles. Microbes, which multiply in a faster rate, can speed up the nanoparticle production, and those that survive extreme conditions can produce highly stable nanoparticles [19]. Since microbes harbour an active metabolic system that functions under ambient growth conditions, the biological route of synthesis is considered to be a less energy-intensive method.

Nanomaterials generated through the microbial source have defined morphology, size and composition, and they serve as excellent candidates for biosensing, drug delivery, therapeutic studies and bioimaging. However, achieving large-scale production of microbially synthesised nanomaterials is challenging as there are limitations in maintaining the uniformity of chemical composition, size, shape and functionality [20]. Nanotechnology, combined with biotechnology and microbiology, has evolved to form "nanobiotechnology" which aims to integrate various ideas to overcome the aforementioned limitations. Through genetic modification, the field has helped in manipulating the molecular machineries of microbes, which result in production of nanomaterials with desired properties (Figure 8.1b). Various microbes that are explored for the synthesis of nanomaterials through genetic engineering are discussed further in detail.

8.2 GENETICALLY MODIFIED MICROORGANISMS FOR THE SYNTHESIS OF NANOMATERIALS

8.2.1 Bacteria

Prokaryotes and actinomycetes are widely investigated for the generation of metallic nanoparticles [21] (Figure 8.2). Bacteria are unicellular, prokaryotic microbes, which do not possess nuclear membrane and are not structurally elaborate as eukaryotes. Though they appear as simple living organisms, they are highly sophisticated, metabolically active and adaptable. Most of the bacterial species can live as parasites or free-forms and act as pathogenic or non-pathogenic agents. Compared to other living forms, bacteria can divide rapidly by binary fission, have the potential to maintain high culture density and can be genetically manipulated. As mentioned earlier, for more than a decade, bacteria are used for the synthesis of various nanoparticles which are applied for environmental remediation, biosensing, diagnostic purpose, and antioxidant, antibacterial and anticancer applications [22] (Table 8.1). Some of the interesting reports on the bacteria mediated nanoparticle synthesis are discussed further.

Wang et al. synthesised monoclinic selenium nanoparticles from *Bacillus subtilis*. The spherically shaped nanoparticles had diameter ranging from 50 to 400 nm. Here, selenium nanoparticles were formed after 48 hours in the presence of activated bacteria and sodium selenite at 35°C.

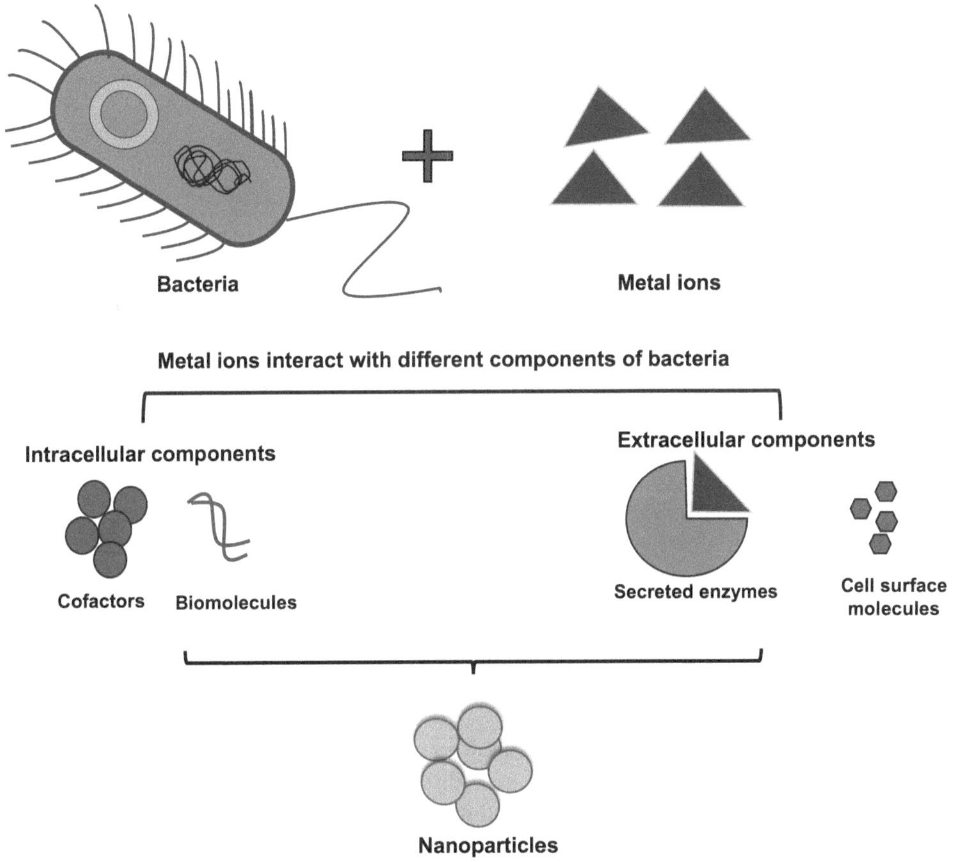

FIGURE 8.2 A general schematic representation of the overall mechanism of nanoparticle synthesis from microbes (e.g. bacteria).

TABLE 8.1
List of Bacterial Species Reported for the Synthesis of Various Nanoparticles

Bacteria	Nanoparticles	Size Distribution of Nanoparticles	Applications/Properties
Bacillus subtilis [23]	Selenium nanoparticles	50–400 nm	Catalytic property
Bacillus brevis [24]	Silver nanoparticles	41–68 nm	Antimicrobial application
Escherichia sp. SINT7 [25]	Copper nanoparticles	22.33–39.0 nm	Biodegradation
Proteus vulgaris [26]	Iron oxide nanoparticles	19.23–30.51 nm	Antimicrobial, antioxidant and anticancer applications
Vibrio alginolyticus [27]	Gold nanoparticles	100–150 nm	Antioxidant property and anticancer applications
	Genetically Engineered Bacteria		
Escherichia coli (strain JM109 and R189) [32]	Cadmium sulphide nanocrystals	2–6 nm (strain JM109) 3–4 nm (strain R189)	Fluorescence property
Escherichia coli (DH5α) [33]	Silver nanoparticles	20 nm	Could be explored for potential biological application

(*Continued*)

TABLE 8.1 (*Continued*)
List of Bacterial Species Reported for the Synthesis of Various Nanoparticles

Bacteria	Nanoparticles	Size Distribution of Nanoparticles	Applications/Properties
Escherichia coli [34]	CdZn nanoparticles CdSe nanoparticles CdTe nanoparticles SeZn nanoparticles	10.01±0.97 nm 5.01±0.89 nm 7.01±0.66 nm 3.92±0.35 nm	Fluorescence property, magnetic property, bioimaging function
Escherichia coli [35]	Arsenic sulphide nanoparticles	381 ± 246 nm	Semiconducting property, which could be explored for applications such as electronics and optoelectronics
Pseudomonas savastanoi *Pseudomonas syringae* *Ralostonia solanacerum* [36]	Gold nanoparticles	40 nm 30 nm 15 nm	Potential biological applications
Geobacter sulfurreducens [37]	Protein nanowires	1.5 nm	Electrical conductivity
Rhodospirillum rubrum [38]	Magnetosome particles	<100 nm	Magnetogenetics, bioimaging

It is assumed that the biochemical agents released by bacteria as a result of precursor stress probably acted as reducing agents to form selenium nanoparticles. These nanoparticles showed biocompatibility, adhesive capability and catalytic properties for the detection of H_2O_2 [23]. Another bacterial species, namely, *Bacillus brevis*, produced silver nanoparticles with a diameter of 41–68 nm. Silver nanoparticles were formed from cell-free filtrate that contained $AgNO_3$. These nanoparticles possessed antimicrobial property against multidrug resistance strains such as *Staphylococcus aureus* and *Salmonella typhi* [24]. Copper-resistant strain, namely *Escherichia* strain SINT7, was shown to synthesise spherical copper nanoparticles with a size distribution of 22.33–39 nm. These nanoparticles could be of considerable importance for efficient degradation of azo dye and treatment of textile effluents [25]. Furthermore, spherical iron oxide nanoparticles could be synthesised from *Proteus vulgaris*. The reported study suggests the size of as-synthesised nanoparticles to be between 19.23 and 30.51 nm. Besides antibacterial property, these bacteria-derived nanoparticles display good antioxidant property. Additionally, these nanoparticles exhibit anticancer properties against glioblastoma cells with appreciable cytocompatibility towards normal cell lines [26]. Gold nanoparticles have been synthesised from a marine microbe, namely, *Vibrio alginolyticus*, with an estimated size of 100–150 nm. They are suggested to have a biomedical property with appreciable antioxidant characteristics and anticancer property against colon cancer cells [27]. Numerous studies have reported the usage of a large number of bacterial species belonging to *Pseudomonas*, *Rhodococcus*, *Lactobacillus*, *Streptomyces*, etc. for the formation of nanoparticles (palladium, zinc, titanium, selenium, etc.) [28]. However, the remarkable feasibility of genome or metabolic modification makes bacteria the suitable candidates for genetic or metabolic engineering which leverages the yield of nanoparticles with multifaceted importance.

Genetically engineered bacteria are manipulated to express proteins or molecular substances for a desired function. Genetic engineering can be done by phage infection, plasmid-based transformation, recombination through transposition, conjugation and protoplast fusion. These methods, which facilitate genetic modification, transform the bacterial characteristics. Although *Escherichia coli*, *Salmonella* and *Lactobacillus* are popular in genetic engineering studies, species like *Listeria monocytogenes*, *B. subtilis*, *Lactobacillus brevis* and *Bifidobacterium* are also engineered for different medical applications and scientific studies [29,30].

Having investigated profoundly of the cellular and molecular mechanism, *E. coli* is commonly used as a biofactory for the synthesis of a variety of nanoparticles [31]. A study reports that the synthesis of cadmium sulphide (CDS) nanocrystals is possible from genetically engineered *E. coli* (strain JM109 and R189) which produce phytochelatins. The gene responsible for the synthesis of phytochelatins is called phytochelatins synthase. Now, phytochelatin synthase is seen in the yeast *Schizosaccharomyces pombe* which produces phytochelatins to mitigate cadmium toxicity. Based on the report, it is understood that in addition to transforming *E. coli* with phytochelatins synthase gene, the yield of phytochelatins could be elevated by the introduction of an additional plasmid containing g-glutamylcysteine synthetase. Furthermore, upon incubating the modified bacterial strains with cadmium chloride and sodium sulphide, cadmium sulphide nanocrystals could be formed. These cadmium nanocrystals have been reported to show fluorescence properties (strain JM109–384 nm and strain R189–330 nm) and have an average size of 2–6 nm (strain JM109) and 3–4 nm (strain R189). Thus, the study demonstrates a tunable method for the fabrication of cadmium sulphide nanocrystals with uniform size distribution [32].

Genetically modified *E. coli* (DH5α) could be transformed to biosynthesise silver nanoparticles by the expression of metallothionein, a metal-binding protein rich in cysteine. It is known that metallothionein can chelate and reduce metal ions. Due to their significant affinity towards metal, metallothionein are keenly considered for the formation of metallic nanoparticles. As reported specifically, metallothionein gene of *Candida albicans* was inserted into plasmid pUC19 in the region between HindIII and BamHI. The expression of metallothionein in the recombinant bacteria increased their tolerance towards metal ions and led to a higher yield of silver nanoparticles. The thiol group of metallothionein protein probably played a key role in the reduction of Ag^+ ions to Ag^0. The silver nanoparticles synthesised were spherically shaped with an average diameter of 20 nm. This study therefore suggests that in addition to the metal-binding proteins and enzymes, functional groups of organic compounds present in the recombinant bacteria could also drive the synthesis of metallic nanoparticles [33].

Another successful manipulation of *E. coli* strains has been done to express phytochelatin synthase from *Arabidopsis thaliana* and metallothionein from *Pseudomonas putida*. In their study, plasmids were constructed for the expression of these genes individually or together. Once transformed, the recombinant strains were found to synthesise varieties of nanoparticles within their cells, in the presence of various metal ions. Nanoparticles made of SeZn, CdZn, CdTe and CdSe with prominent crystalline properties were formed. Upon analysing the fluorescence property of nanomaterials, the emission intensity was significantly higher in CdSe nanoparticles synthesised by bacterial cells which co-expressed both the genes than those which were synthesised by recombinant cells expressing individual genes. In addition, iron-based nanoparticles, silver nanocrystallites and rare earth–based nanoparticles were also obtained from these recombinant *E. coli* cells. The study suggests that the co-expression of phytochelatin synthase and metallothionein genes facilitated the fabrication of highly ordered nanoparticles, whose characteristics could be controlled by varying metal ion concentrations. This study highlights the importance of exploring recombinant *E. coli* as a biofactory for the *in vivo* synthesis of nanoparticles with desired characteristics [34].

Genetic modification of *E. coli* strain can also result in the production of arsenic sulphide nanomaterials. The property of microorganism to perform redox reaction in the presence of enzymes as a part of their respiratory or detoxification mechanism was explored in one of the studies. Here, genes expressing arsenate reductase and thiosulfate reductase were taken from two different bacterial strains such as *Shewanella* species (ANA-3) and *Salmonella enterica* serovar *typhimurium*, respectively. Once the *E. coli* strain was engineered with the reductase genes, the genes were induced to act upon arsenate and thiosulfate present in the bacterial culture. Bacterial cells that expressed the reductase genes formed arsenic sulphide nanomaterials. The study also showed the importance of cells and cellular components in the formation of nanomaterials. Interestingly, the influence of biological components and the redox mechanism played a major role in the determination of the size and shape of arsenic sulphide nanomaterials. The resultant nanostructures were formed with an average size of 381 ± 246 nm [35].

Apart from *E. coli*, research studies have been conducted to formulate and characterise nanoparticles generated from other bacterial strains. A group of research team has demonstrated the synthesis of gold nanoparticles using genetically modified bacteria such as *Pseudomonas savastanoi*, *Pseudomonas syringae* and *Ralostonia solanacerum* [36]. These bacterial strains were genetically modified by incorporation of plasmid that carries luxAB genes. Plasmid transformation was carried out by electroporation method, and the bacteria were marked with lux genes through miniTn-5 luxAB transposon. These genetically engineered bacteria were further used to form spherically shaped gold nanoparticles at room temperature in the presence of gold chloride solution. The bacterial culture supernatant could reduce Au^{3+} ions to produce nanoparticles which showed surface plasmon resonance corresponding to gold nanoparticles. These genetically modified bacteria facilitated the formation of gold nanoparticles with a size distribution of 15–40 nm. The study suggests that the presence of NADH or NADH-dependent enzymes would probably have catalysed the reduction of Au (3+) to Au (0) to favour the biosynthesis of stable gold nanoparticles.

Genetic manipulation of *Geobacter sulfurreducens* led to the formation of protein nanowires. *G. sulfurreducens* has pili which are electrically conductive and are called microbial nanowires. The peptide unit of pili is PilA which is made up of amino acids such as tyrosine and phenylalanine. These aromatic amino acids promote transportation of electrons in pili. However, as compared to phenylalanine and tyrosine, tryptophan helps in electron transport at a much faster rate. Based on this idea, *G. sulfurreducens* was reported to be modified by replacing its chromosomal allele with a mutant allele that caused substitution of phenylalanine and tyrosine with tryptophan in the PilA protein. The transformed bacterial cells were harvested by growing them as biofilms on a graphite electrode. The grown cells were gently removed from the electrode surface and washed. The pili were collected by subjecting the cells to blending process. The characterisation studies showed a mass production of thin pili fibres having a diameter of 1.5 nm. These protein nanowires from the genetically engineered *G. sulfurreducens* are suggested to be non-toxic with consistent morphology, size distribution and composition. Of utmost significance, it has been demonstrated that the nanowires show excellent electrical conductivity, up to 2000-folds [37].

Interestingly, one of the investigations suggests that magnetic nanoparticles can be synthesised from a non-magnetic bacterial strain, *Rhodospirillum rubrum*. *R. rubrum* is a photosynthetic bacterium into which a group of genes from *Magnetospirillum gryphiswaldense* could be introduced. Once transformed, the bacterial strain shows intracellular formation of magnetosomes and exhibits magnetotactic properties. As compared to the wild-type strain, the recombinant *R. rubrum* is suggested to be metabolically active, with a stable phenotype observed up to 40 generations. Furthermore, the formation of magnetosome in the recombinants follows a similar pattern to that of donor organism. In the future, these resultant biogenic magnetic nanoparticles can hold greater potential in the field of magnetogenetics and medical bioimaging [38].

8.2.2 Fungi

Among the lower categories of eukaryote, fungi have been used for the biofabrication of nanoparticles [28]. Fungi can be cultured in simple media and maintained efficiently, with the possibility of achieving relatively faster growth. Most importantly, the fugal cell wall exhibits high interaction and tolerance towards high metal concentrations [39]. The characteristics of accumulating more metals by fungi and their larger biomass suggest the use of fungi as a promising biofactory for nanoparticle synthesis (Table 8.2). Fungi are also known to secrete enzymes that can reduce metal ions to metal nanoparticles and can produce nanoparticles in larger amounts as compared to bacteria [40]. Considering its tolerance to metals, the extracellular formation of gold and silver nanoparticles has been conducted in a rhizospheric fungus. For the study, various rhizospheric fungal strains were collected from a metal-enriched region. The fungal strains were subjected to screening for their tolerance to silver and gold metal ions. Among various strains, *Penicillium janthinellum* DJP06 was

TABLE 8.2
List of Fungal Species Reported for the Synthesis of Various Nanoparticles

Fungus	Nanoparticles	Size Distribution of Nanoparticles	Applications/Properties
Penicillium janthinellum DJP06 [41]	Gold nanoparticles Silver nanoparticles	19±3.2 nm 13±2.7 nm	Could be explored for potential biological application
Cochliobolus geniculatus [42]	Zinc oxide nanoparticles	2–6 nm	Photocatalytic application
Penicillium pinophilum [43]	Platinum nanoparticles	2–30 nm	Antimicrobial and anticancer application
	Genetically Engineered Fungi		
Saccharomyces cerevisiae [44]	Cadmium–selenium quantum dots	3.0±0.3 nm	Fluorescent property
Pichia pastoris [46]	Gold nanoparticles and palladium nanoparticles	100 nm	Biocompatible property
Pichia pastoris [47]	Silver and selenium nanoparticles	70–180 nm	Could be explored for different applications

found to be tolerant towards high metal ion concentrations, and thus was further selected for the synthesis of nanoparticles. For the extracellular synthesis of gold and silver nanoparticles, the fungus was grown, and the biomass was processed to obtain the secretome. The collected fungal secretome was incubated in the presence of metal precursors to obtain uniformly dispersed quasi-spherical gold and silver nanoparticles. The resultant gold and silver nanoparticles were reported to have an average size of 19±3.2 and 13±2.7 nm, respectively. Through various characterisation techniques, the study suggested the influence of fungal secretome proteins in the formation of stable metallic nanoparticles [41]. Having the inherent capability of producing biological components such as proteins and enzymes, the endophytic fungus *Cochliobolus geniculatus* is proven to be effective for the synthesis of zinc oxide nanoparticles. In addition to showing antioxidant, antimicrobial and anti-tumour properties, the fungal-derived biological components act as capping agents, thus stabilising nanoparticles. A study with *C. geniculatus* reports that the substrate free of fungal mycelia upon exposure to precursor zinc acetate salt, formed protein-coated zinc oxide nanoparticles with an average size of 2–6 nm. The characteristic quasi-shape of zinc oxide nanoparticles might be attributed to the cysteine groups of protein present in nanoparticles. Such protein-capped zinc oxide nanoparticles can replace nanoparticles with artificial capping and could be the better options for various applications (e.g. photocatalytic studies) [42]. Platinum nanoparticles could be synthesised from cell-free filtrate of *Penicillium pinophilum.* In addition to metal precursors, the presence of secondary metabolites, proteins and enzyme could drive the myco-synthesis of platinum nanoparticles. These nanoparticles are reported to be crystalline in nature and thermally stable. Moreover, these biogenic platinum nanoparticles exhibit good antimicrobial property towards bacterial and fungal pathogens, as well as anticancer properties towards mammalian cancer cell line [43].

Similarly, numerous studies have reported fungi as a suitable source for green synthesis of a variety of nanoparticles. Over time, researchers have explored the possibilities of using genetically engineered fungi for obtaining functional nanoparticles. *Saccharomyces cerevisiae* was engineered to biosynthesise higher amounts of cadmium–selenium quantum dots (in the size of 3.0±0.3 nm) by modifying its biochemical metabolism. A study [44] reported selenocysteine as the essential precursor for the formation of cadmium–selenium quantum dots intracellularly. With the enhanced expression of MET6 gene in the engineered organism, the yield was improved. This suggests the importance of understanding metabolic pathway in obtaining cost-effective, functional nanomaterials from modified microbes.

Like *E. coli* among prokaryotic microbes, *Pichia pastoris* is an easily modifiable microbe among eukaryotes. Owing to its high expression level, *P. pastoris* is used as a suitable tool for the production of various important proteins [45]. Because *P. pastoris* can be manipulated using biochemistry, genetic and molecular biology techniques, it is used to express intracellular proteins or secreted proteins. Another study [46] has reported the use of genetically modified *P. pastoris* for the bioleaching and biosynthesis of gold and palladium nanoparticles. *P. pastoris* was engineered to express cytochrome b5 reductase and was further cultured in the presence of metal precursors ($HAuCl_4$ and $PdCl_2$). The recombinant *P. pastoris* was found to be metal resistant showing appreciable growth at higher gold and palladium concentrations as compared to the parental strain. The recombinant microbe was further evaluated for the biosorption capacity and nanoparticle production. The biosynthesised gold and palladium nanoparticles were characterised to be spherical in shape with an average diameter of 100 nm. When applied towards mammalian cells, palladium nanoparticles were found to be least toxic. Based on the results obtained, the recombinant *P. pastoris* is understood to be a suitable and robust system for biosorption of heavy metals.

Genetically modified *P. pastoris* [47] has also been reported for the biogenic synthesis of silver and selenium nanoparticles. *P. pastoris* was modified by transferring cytochrome b5 reductase gene from *Mucor racemosus*. The recombinant strain exhibited tolerance to silver and selenium, and showed biosorption of these metals. The recombinant *P. pastoris* could efficiently synthesise silver and selenium nanoparticles intracellularly. The reduction of metal ions in the presence of cytochrome b5 reductase enzyme resulted in the formation of stable and spherical nanoparticles with an average size of 70–180 nm.

8.2.3 Viruses

Viruses are obligate parasites that reside intracellularly and possess RNA or DNA as their genetic material. The genetic material is protected within capsid proteins which may or may not have a membrane envelope. Although viruses are usually identified as disease-causing agents, scientists have discovered various ways for exploring them as valuable tools for multidisciplinary applications. Genetically modified viruses are also known as viral nanoparticles, and those without their genetic material are defined as viral-like particles [48]. Various studies have reported the use of viruses and their components as scaffolds for biogenic synthesis of multifunctional nanoparticles [49]. Bacteriophages, various plant viruses such as chlorotic mottle virus [50], squash leaf curl China virus [51] and tobacco mosaic virus [52] are reported for the fabrication of semiconductor nanoparticles such as cadmium and zinc sulphide and gold–silver composite nanoparticles. Additionally, the non-pathogenic plant viral components are applied for the synthesis of nanoparticles for drug delivery applications. Utilising the aspects of genetic engineering, viral capsid proteins can be independently manipulated to create unique nanoarchitectures (Table 8.3).

Huang et al. have presented the assembly of gold and cadmium selenide nanostructures using engineered M13 bacteriophage. The filamentous M13 bacteriophage has its single-stranded DNA encased within capsid which is made up of functional proteins. M13 bacteriophage has pIII, pVI, pVIII and pIX proteins in several copies. In this study, pIII and pVIII proteins were modified by insertion of corresponding genes into the viral genome. The genetically modified bacteriophage acted as a template for fabrication of different hetero nanoparticle structure and arrays. Moreover, the group has employed the resultant nanoparticle arrays as template for further nucleation of gold nanowires of conductive nature [53]. Another report demonstrated the synthesis of gold nanoparticles using genetically engineered tobacco mosaic virus (TMV). The rod-shaped TMV has single-stranded RNA as its genetic material enclosed within a capsid made of identical protein subunits. TMV was modified to express a peptide that can interact with metal ions and reduce them to form gold nanoparticles. Without the addition of external reducing agent, the modified viral construct in the presence of potassium tetrachloroaurate formed gold nanoparticles with an average size of 10–40 nm [54]. Another group has genetically modified filamentous viruses (fd viruses) which

TABLE 8.3
List of Genetically Modified Viruses Explored for the Synthesis of Various Nanoparticles

Genetically Engineered Virus	Nanoparticles	Size Distribution of Nanoparticles	Applications/Properties
M13 Bacteriophage [53]	Gold nanowires	~40 nm	Could be applied for various biotechnology and nanoscale applications
Tobacco mosaic virus [54]	Gold nanoparticles	10–40 nm	Could be applied for various biotechnology and nanoscale applications
M13 phage [56]	Gold nanowires	10–50 nm	Catalytic function
T7 phage [57]	Clusters of gold nanoparticles		Photothermal therapy for cancer cells

expressed specific binding peptides for gold and platinum. The coat protein present in the head area such as P9 was manipulated to display binding peptides, namely, V5 and p7A, specific to gold and platinum, respectively. Experimental and characterisation studies suggested significant interaction of genetically engineered viruses with gold particles as compared to wild type. Such site-specific binding peptides expressed via genetic engineering can play excellent roles in applications such as biosensing, targeted drug delivery and nanowire formation [55]. Nanowire synthesis of genetically modified M13 phage was reported by Lee et al. The M13 phage was modified to express binding peptides for gold particles. The study showed the site-specific binding of modified M13 phage with Au ions, which resulted in the formation of uniformly sized gold nanowires of 10–50 nm. The generated gold nanowires exhibited significant catalytic function against the oxidation of carbon monoxide molecules. The gold nanowires generated in higher yield were further used for the fabrication of gold–platinum core shell nanowires, which were explored for oxidation of ethanol [56]. Gold nanoparticles of therapeutic application were developed by genetically engineered T7 phage. The engineered T7 phage displayed peptides which bind to gold ions. T7 phage, which has modified capsid protein, acted as a template for the assembly of gold nanoparticles. The clustered gold nanoparticles were found to be functionally specific and exhibited photothermal therapy towards prostate cancer cells [57], thereby holding advantages for future medical applications.

8.3 KEY CHALLENGES

Genetically engineered microbes facilitate the exclusion of external reducing agents, because their own intracellular machineries or molecules reduce the precursor metal ions. Though genetically modified microbes undoubtedly hold significant importance in the field of nanobiotechnology, achieving reproducibility, purity of the extracted nanoparticles, and long-term stability are key challenges. For clinical applications, detailed investigations of nanoparticles from genetically engineered microbes are very essential for understanding the immune tolerance, effects on vital organs, retention time, degradation, and renal clearance.

8.4 FUTURE PROSPECTS AND CONCLUSION

A great deal of attention has been given to the novel and innovative strategies for nanomaterial production. Although various physical/chemical methods yield functional nanoparticles, green synthesis using microbes showcases the possibility of opting affordable, less toxic, readily available resources for nanomaterial fabrication. Nanoparticles synthesised via non-biological routes require additional chemical modifications for specific physiochemical attributes and targeted studies. In contrast, biogenic nanoparticles do not demand much further modifications as they tend to have functional groups derived from microbial biomolecules. It should be noted that growth medium for microbes, temperature, pH and concentration of metal precursors could also be controlled in obtaining desired nanoparticles. The inherent biological, molecular and metabolic characteristics of microbes direct the scientific world to discover more about the pathways and utilise them for research, industrial and medicinal purposes. This profound understanding about the molecular mechanisms helps in the genetic engineering of microbes to produce desired nanomaterials with enhanced yield and essential properties. By carefully manipulating their functional genes or pathways, nanostructures with unique electronic, optical, and catalytic nature can be generated.

REFERENCES

1. Singh, N.A., et al., Nanoparticles synthesis via microorganisms and their prospective applications in agriculture. *Plant Nano Biol*, 2023. 5: p. 100047.
2. Hossain, N., et al., Advances and significances of nanoparticles in semiconductor applications – A review. *Results Eng*, 2023. 19: p. 101347.
3. Kamyshny, A. and S. Magdassi, Conductive nanomaterials for printed electronics. *Small*, 2014. 10(17): pp. 3515–3535.
4. Rastogi, A., et al., Application of silicon nanoparticles in agriculture. *3 Biotech*, 2019. 9(3): p. 90.
5. Khan, F., et al., Prospects of algae-based green synthesis of nanoparticles for environmental applications. *Chemosphere*, 2022. 293: p. 133571.
6. Zhang, L., et al., Nanoparticles in medicine: Therapeutic applications and developments. *Clin Pharmacol Ther*, 2008. 83(5): pp. 761–769.
7. Rudramurthy, G.R. and M.K. Swamy, Potential applications of engineered nanoparticles in medicine and biology: An update. *J Biol Inorg Chem*, 2018. 23(8): pp. 1185–1204.
8. Altammar, K.A., A review on nanoparticles: Characteristics, synthesis, applications, and challenges. *Front Microbiol*, 2023. 14: p. 1155622.
9. Jamkhande, P.G., et al., Metal nanoparticles synthesis: An overview on methods of preparation, advantages and disadvantages, and applications. *J Drug Deliv Sci Technol*, 2019. 53: p. 101174.
10. Dhand, C., et al., Methods and strategies for the synthesis of diverse nanoparticles and their applications: A comprehensive overview. *RSC Adv*, 2015. 5(127): pp. 105003–105037.
11. Kumari, S., et al., A comprehensive review on various techniques used for synthesizing nanoparticles. *J Mater Res Technol*, 2023. 27: pp. 1739–1763.
12. Abid, N., et al., Synthesis of nanomaterials using various top-down and bottom-up approaches, influencing factors, advantages, and disadvantages: A review. *Adv Colloid Interface Sci*, 2022. 300: p. 102597.
13. Kapoor, R.T., et al., Exploration of microbial factories for synthesis of nanoparticles - A sustainable approach for bioremediation of environmental contaminants. *Front Microbiol*, 2021. 12: p. 658294.
14. Ahmad, F., et al., Biological synthesis of metallic nanoparticles (MNPs) by plants and microbes: Their cellular uptake, biocompatibility, and biomedical applications. *Appl Microbiol Biotechnol*, 2019. 103(7): pp. 2913–2935.
15. Dhanker, R., et al., The emerging trend of bio-engineering approaches for microbial nanomaterial synthesis and its applications. *Front Microbiol*, 2021. 12: p. 638003.
16. Mughal, B., et al., Biogenic nanoparticles: Synthesis, characterisation and applications. *Appl Sci*, 2021. 11(6): p. 2598.
17. Ghosh, S., et al., Microbial nano-factories: Synthesis and biomedical applications. *Front Chem*, 2021. 9: p. 626834.
18. Bhardwaj, B., et al., Eco-friendly greener synthesis of nanoparticles. *Adv Pharm Bull*, 2020. 10(4): pp. 566–576.
19. Carmona, M., et al., Opportunities and obstacles in microbial synthesis of metal nanoparticles. *Microb Biotechnol*, 2023. 16(5): pp. 871–876.
20. Manivasagan, P., S.Y. Nam, and J. Oh, Marine microorganisms as potential biofactories for synthesis of metallic nanoparticles. *Crit Rev Microbiol*, 2016. 42(6): pp. 1007–1019.
21. Pandit, C., et al., Biological agents for synthesis of nanoparticles and their applications. *J King Saud Univ Sci*, 2022. 34(3): p. 101869.
22. Grasso, G., D. Zane, and R. Dragone, Microbial nanotechnology: Challenges and prospects for green biocatalytic synthesis of nanoscale materials for sensoristic and biomedical applications. *Nanomaterials (Basel)*, 2019. 10(1): p. 11.

23. Wang, T., et al., Extracellular biosynthesis and transformation of selenium nanoparticles and application in H_2O_2 biosensor. *Colloids Surf B Biointerfaces*, 2010. 80(1): pp. 94–102.
24. Saravanan, M., et al., Synthesis of silver nanoparticles from Bacillus brevis (NCIM 2533) and their antibacterial activity against pathogenic bacteria. *Microb Pathog*, 2018. 116: pp. 221–226.
25. Noman, M., et al., Use of biogenic copper nanoparticles synthesized from a native Escherichia sp. as photocatalysts for azo dye degradation and treatment of textile effluents. *Environ Pollut*, 2020. 257: p. 113514.
26. Majeed, S., et al., Bacteria mediated synthesis of iron oxide nanoparticles and their antibacterial, antioxidant, cytocompatibility properties. *J Clust Sci*, 2021. 32(4): pp. 1083–1094.
27. Shunmugam, R., et al., Biosynthesis of gold nanoparticles using marine microbe (Vibrio alginolyticus) and its anticancer and antioxidant analysis. *J King Saud Univ Sci*, 2021. 33(1): p. 101260.
28. Koul, B., et al., Microbe-mediated biosynthesis of nanoparticles: Applications and future prospects. *Biomolecules*, 2021. 11(6): p. 886.
29. Liu, Y., et al., Genetically engineered bacterium: Principles, practices, and prospects. *Front Microbiol*, 2022. 13: p. 997587.
30. Iravani, S. and R.S. Varma, Genetically engineered organisms: Possibilities and challenges of heavy metal removal and nanoparticle synthesis. *Clean Technol*, 2022. 4(2): pp. 502–511.
31. Iravani, S. and R.S. Varma, Biofactories: engineered nanoparticles via genetically engineered organisms. *Green Chem*, 2019. 21(17): pp. 4583–4603.
32. Kang, S.H., et al., Microbial synthesis of CdS nanocrystals in genetically engineered E. coli. *Angew Chem Int Ed Engl*, 2008. 47(28): pp. 5186–5189.
33. Yuan, Q., M. Bomma, and Z. Xiao, Enhanced silver nanoparticle synthesis by Escherichia Coli transformed with Candida Albicans metallothionein gene. *Materials (Basel)*, 2019. 12(24): p. 4180.
34. Park, T.J., et al., In vivo synthesis of diverse metal nanoparticles by recombinant Escherichia coli. *Angew Chem Int Ed Engl*, 2010. 49(39): pp. 7019–7024.
35. Chellamuthu, P., et al., Engineering bacteria for biogenic synthesis of chalcogenide nanomaterials. *Microb Biotechnol*, 2019. 12(1): pp. 161–172.
36. Attaran, N., et al., Genetically modified luminescent bacteria Ralostonia solanacerum, Pseudomonas syringae, Pseudomonas savastanoi, and wild type bacterium Vibrio fischeri in biosynthesis of gold nanoparticles from gold chloride trihydrate. *Artif Cells Nanomed Biotechnol*, 2016. 44(1): pp. 263–269.
37. Tan, Y., et al., Synthetic biological protein nanowires with high conductivity. *Small*, 2016. 12(33): pp. 4481–4485.
38. Kolinko, I., et al., Biosynthesis of magnetic nanostructures in a foreign organism by transfer of bacterial magnetosome gene clusters. *Nat Nanotechnol*, 2014. 9(3): pp. 193–197.
39. Subramaniyan, S.A., et al., One-pot facile synthesis of Pt nanoparticles using cultural filtrate of microgravity simulated grown P. chrysogenum and their activity on bacteria and cancer cells. *J Nanosci Nanotechnol*, 2018. 18(5): pp. 3110–3125.
40. Saravanan, A., et al., A review on biosynthesis of metal nanoparticles and its environmental applications. *Chemosphere*, 2021. 264(Pt 2): p. 128580.
41. Pareek, V., A. Bhargava, and J. Panwar, Biomimetic approach for multifarious synthesis of nanoparticles using metal tolerant fungi: A mechanistic perspective. *Mater. Sci. Eng. B*, 2020. 262: p. 114771.
42. Kadam, V.V., J.P. Ettiyappan, and R. Mohan Balakrishnan, Mechanistic insight into the endophytic fungus mediated synthesis of protein capped ZnO nanoparticles. *Mater Sci Eng B*, 2019. 243: pp. 214–221.
43. Gholami-Shabani, M., et al., Platinum nanoparticles as potent anticancer and antimicrobial agent: Green synthesis, physical characterization, and in-vitro biological activity. *J Clust Sci*, 2023. 34(1): pp. 501–516.
44. Shao, M., et al., Living cell synthesis of CdSe quantum dots: Manipulation based on the transformation mechanism of intracellular Se-precursors. *Nano Res*, 2018. 11(5): pp. 2498–2511.
45. Higgins, D.R., Overview of protein expression in Pichia pastoris. *Curr Protoc Protein Sci*, 2001. doi:10.1002/0471140864.ps0507s02.
46. Elahian, F., et al., Genetically modified Pichia pastoris, a powerful resistant factory for gold and palladium bioleaching and nanostructure heavy metal biosynthesis. *Artif Cells Nanomed Biotechnol*, 2020. 48(1): pp. 259–265.
47. Elahian, F., et al., High-throughput bioaccumulation, biotransformation, and production of silver and selenium nanoparticles using genetically engineered Pichia pastoris. *Nanomedicine*, 2017. 13(3): pp. 853–861.
48. Chen, M.Y., et al., Physical, chemical, and synthetic virology: Reprogramming viruses as controllable nanodevices. *Wiley Interdiscip Rev Nanomed Nanobiotechnol*, 2019. 11(3): p. e1545.

49. Smith, M.T., A.K. Hawes, and B.C. Bundy, Reengineering viruses and virus-like particles through chemical functionalization strategies. *Curr Opin Biotechnol*, 2013. 24(4): pp. 620–626.
50. Slocik, J.M., et al., Viral templates for gold nanoparticle synthesis. *J Mater Chem*, 2005. 15(7): pp. 749–753.
51. Thangavelu, R.M., et al., Fabrication of virus metal hybrid nanomaterials: An ideal reference for bio semiconductor. *Arab J Chem*, 2020. 13(1): pp. 2750–2765.
52. Kobayashi, M., et al., Chiral meta-molecules consisting of gold nanoparticles and genetically engineered tobacco mosaic virus. *Opt Express*, 2012. 20(22): pp. 24856–24863.
53. Huang, Y., et al., Programmable assembly of nanoarchitectures using genetically engineered viruses. *Nano Lett*, 2005. 5(7): pp. 1429–1434.
54. Love, A.J., et al., A genetically modified tobacco mosaic virus that can produce gold nanoparticles from a metal salt precursor. *Front Plant Sci*, 2015. 6: p. 984.
55. Korkmaz, N. and T. Arslan, Genetically engineered Fd viruses for site specific material binding. *BioNanoScience*, 2019. 9(4): pp. 862–872.
56. Lee, Y., et al., Virus-templated Au and Au/Pt Core/shell nanowires and their electrocatalytic activitives for fuel cell applications. *Energy Environ Sci*, 2012. 5(8): pp. 8328–8334.
57. Oh, M.H., et al., Genetically programmed clusters of gold nanoparticles for cancer cell-targeted photothermal therapy. *ACS Appl Mater Interfaces*, 2015. 7(40): pp. 22578–22586.

9 Significance of Microbial Nanotechnology for Environmental Sustainability

Srinivasan Kameswaran, Bellamkonda Ramesh, A. Sivashankar Reddy, N.O. Gopal, and Manjunatha Bangeppagari

9.1 INTRODUCTION

The current way of life and the environment are undoubtedly impacted by modern civilisation, its technologies, and the industries that support them. The loss and dissolution of resources from the earth, which compromises "long-term environmental quality" or "environmental sustainability," constitutes a few of those modifications and their effects, either directly or indirectly [1]. A great deal of work, research, planning, and initiative have been shown to preserve our ecosystem and its resources for future generations. As per Darwesh et al. [2,3], the most promising approaches to protect the environment involve utilising biotechnologies, namely microbial nanotechnology applications, to enhance and expedite their environmental sustainability across several domains. Microbiological biotechnology is known as a science of using microbes to produce goods and services. According to Abou-Shanab et al. [4] and Darwesh et al. [5–7], it involves the green synthesis of nanoparticles and the management of microorganisms and their enzymes for the breakdown of farm debris, biological remediation of xenobiotics, and production of biofuels, medicines, and nutritional additives.

The creation and use of materials that are in the size of nanoscale (ranging from 1 to 100 nm) is known as nanotechnology. These materials can be nanoparticles, nanocomposites, or nanotubes, all of which have a wide range of technical and biological uses. A relatively recent development, the integration of nanotechnology into microbial and ecological biotechnology has demonstrated promising outcomes and applications in numerous fields [8,9]. Utilising nanomaterials improves the effectiveness and calibre of various production processes because of their particular "size, shape, surface area, and structure of particles," which provide them unique features and properties as compared to their bulk counterparts. By facilitating the reusability and recycling of certain resources or by lowering the amount expected from others, the application of nanotechnologies to routine tasks also significantly improves resource sustainability. Additionally, several nanomaterials have been found to have lesser environmental toxicity than non-nanoscale substances, including silver nanoparticles [10,11].

One potential solution to technical and environmental concerns, while promoting economic development and sustainability in many countries, is the careful and efficient use of nanotechnology [12–14]. The primary use of microbial nanoproducts has been as durable carriers for the intact transport and utilisation of bioactive components for customised applications in the medicine and agri-food industries [15]. Microbial nanomaterials have been utilised to biocatalytically break down hazardous contaminants released from industrial effluents into harmless by-products, helping to clean up environmentally hazardous materials [16,17]. Consequently, microbial nanobiotechnology has a wide range of applications and is a cost-effective approach to microbial nanomanufacturing, holding the potential for significant benefits for humanity in the near future.

DOI: 10.1201/9781003487142-9

9.2 THE NECESSITY OF SUSTAINABILITY IN THE ENVIRONMENT

Neglecting the environment could endanger human existence in addition to "compromising the capacity of the next generation to meet their needs." Certain global hazards, such as pollution, global warming, and ozone layer degradation, are seen as serious and persistent issues that affect both the present and future generations [18]. Numerous resources have been depleted throughout time because of careless consumption patterns of humans, continuously increasing world population, and governments' competition over scarce resources. Humanity has also developed and used advanced technologies that, in general, destroy and damage our surroundings to increase the productivity and efficiency of different production units such as factories and farms, thus disregarding the privileges of their posterity to healthy environment. Additionally, humans have created a plethora of strategies and tactics to boost the output and effectiveness of diverse manufacturing facilities. Unfortunately, a great deal of these technologies disregards the privileges of the next generation to natural resources by contributing to environmental pollution.

Major drawbacks of relying solely on traditional sources of energy, such as fossil fuels, are the potential for these resources to run out and the denial of the opportunity to benefit from them for future generations. Moreover, the ongoing burning of these fuels will lead to global warming by increasing the proportion of greenhouse gases in the atmosphere. According to El-Baz et al. [19], Matter et al. [20], Sohail et al. [21], and Darwesh et al. [22], depending more on renewable energy sources like solar and wind power in addition to biofuels has become imperative for energy and environmental sustainability. An additional instance involves the infiltration of various xenobiotics into soils, including pesticides and aliphatic and aromatic hydrocarbons, as a result of frequent and heavy pesticide applications and the widespread use of low-quality irrigation waters [4,23,24]. There is least attention to monitor the levels of pollution and determine the best course of action for detoxifying, eliminating, or minimising its harmful effects to the environment. Prevention of pollution at the outset will benefit both human sustainability and the surroundings they inhabit. It is also urgently necessary to preserve and protect the finite resources for future generations, given the depletion and restrictions of several resources. To avoid environmental degradation and maintain its sustainability, modern technology should be eco-friendly in this aspect. To comply with the ethics for a clean and sustainable environment, traditional technologies need also be updated and modified.

To enhance crop nutrition, nanoparticles are utilised in various agricultural applications such as food packaging, nanostructured ingredients, scratch-resistant nano-coatings, nanofiltration, and nano-delivery systems [24–29]. Food microbiologists and scientists rely heavily on nanoparticles (NPs) to help identify foodborne infections and extend the shelf life of food. NPs also play a significant role in root elongation and seed germination. They are utilised in cosmetics, such as skin creams, to combat skin ageing by using proteins produced from stem cells.

Nanoparticles are utilised in wastewater treatment, a widely used method worldwide. They help reduce pesticides, fertilisers, and heavy metals to provide clean water for irrigation and drinking [30]. This method is both cost-effective and sustainable. Nanoparticles (NPs) can inhibit specific pathogenic microbes and eliminate dyes such as methylene blue. NPs enhance various processes by serving as catalysts.

In agriculture, crop growth cannot be sustained without the use of pesticides, herbicides, and fertilisers for crop protection. It is crucial to develop sustainable nanofertilisers, nano-insecticides, and nano-herbicides due to the significant losses in food and money caused by plant diseases and insect pests [31]. Nano-herbicides and nanopesticides help prevent plants from degrading too quickly, control pests for longer periods, and improve solubility. On the other hand, nanofertilisers assist in maintaining soil fertility. Pesticides can also be directly sprayed on grains and seeds to prevent plant diseases [32].

Scientists study silver nanoparticles (AgNPs) due to their distinctive properties and wide range of applications. They are commonly used in the textile industry, in wastewater treatment, for increasing crop yields, for controlling diseases, and for influencing microorganisms to promote plant growth and nutrient cycling in the soil.

9.3 MICROBIOLOGICAL SYNTHESIS PROCESSES AND NANOMATERIALS

To improve the biological applications of nanomaterials, it is thought to be important to develop environmentally safe and eco-friendly technologies for the synthesis and manufacture of superior chemicals. Biofuel production, treatment of wastewater, agricultural product improvement, and biofertiliser provision are just a few of the future applications for the microbially synthesised nanoparticles [33,34]. The bacteria that target ions from their surroundings and use an enzyme system to transform those ions into element metals are the ones that synthesise nanomaterials. Different microorganisms generate nanomaterials via different ways. However, the most common method for producing nanomaterials is to trap metal ions inside or on the surface of microbial cells. The enzyme system breaks down the trapped ions into nanosize particles. In general, there are two ways that microbes affect the production of minerals. Changing the formula to become supersaturated or more supersaturated than it was before about a certain phase is the first step. Microorganisms can affect mineral production by generating natural polymers that can either facilitate or obstruct the stabilisation of original mineral seeds and impact the nucleation process.

The microbiological production of metal nanomaterials is said to be significantly influenced by the "NADH" and "NADH-dependent" nitrate reductase enzymes. According to Darwesh et al. [6], the nitrate reductase enzyme may be in charge of the bio-reduction of Ag^+ to Ag^0 and the subsequent production of silver nanoparticles. Metallic nanoparticles have been created for many purposes using microbial enzymes such as xylanase, keratinase, and laccase, among others [35–38]. Microorganisms have various metal-resistance gene groups that help in cell detoxification through complexation, reductive precipitation, and efflux mechanisms. Rock piles from mine waste, naturally mineralised areas, and the efflux streams of metal-processing facilities are examples of habitats where metallophilic bacteria can survive and grow in massive amounts of heavy metals [39]. Numerous microorganisms, including the genera of *Bacillus*, *Enterococcus*, *Pseudomonas*, *Fusarium*, *Trichoderma*, and *Aspergillus*, have been reported for use in the manufacture of nanomaterials [40–45]. These microorganisms were used to create nanoparticles of copper, zinc, iron, silver, gold, and selenium (Table 9.1).

TABLE 9.1
Nanoparticle Synthesis and Applications Categorised by Microbes

Microorganism	Nanoparticles	Size (nm)	Application	References
Aeromonas hydrophila	ZnO	57.7	Antimicrobial activity against *Pseudomonas aeruginosa* and *Aspergillus flavus*	[46]
Acinetobacter sp. SW30	Au	15–40	–	[47]
Actinomycetes sp.	Ag	10–20	Antibacterial activity	[48]
Actinomycetes	Ag	5–50	Antibacterial activity	[49]
Brevundimonas diminuta	Ag	5–50	Antimicrobial activity	[50]
Bacillus licheniformis	Cadmium sulphide	20–40	Antibacterial activity	[51]
Bacillus sp.	Selenium nanoparticles	80–220	Antioxidant and cytotoxic effects	[52]
Bacillus subtilis	TiO_2	10–30	Photocatalytic effect on aquatic biofilm	[53]
Escherichia coli	Ag	5–50	Antimicrobial activity	[50]
Exiguobacterium aurantiacumm	Ag	5–50	Antimicrobial activity	[50]

(Continued)

TABLE 9.1 (*Continued*)
Nanoparticle Synthesis and Applications Categorised by Microbes

Microorganism	Nanoparticles	Size (nm)	Application	References
Gordonia amicalis	Ag	5–25	Antioxidant scavenging activity	[54]
Gordonia amarae	Au	15–40	Application in rapid sensing of copper ions	[55]
Idiomarina sp. strain PR58-8	Lead(IV) sulphide	6–10	Bioimaging	[56]
Lactobacillus sporogenes	ZnO	145.70	Antimicrobial activity	[57]
Lactobacillus kimchicus DCY51	Au	5–30	Antioxidant activity	[58]
Lactobacillus acidophilus	Ag	45–60	Genomic toxicity	[59]
Lactobacillus plantarum	ZnO	7–19	–	[60]
Lactobacillus sp.	CdS	2.5–5.5	–	[61]
Lactobacillus sp.	TiO_2	50–100	Immobilisation and refolding of enzyme	[62]
Lactobacillus sp.	TiO_2	50–100	Antibacterial activity	[63]
Lactobacillus fermentum	Iron oxide	10–15	–	[64]
Mycobacterium sp.	Au	5–55	Anticancer	[65]
Micrococcus yunnanensis	Au	53.8	Antibacterial, Anticancer	[66]
Pantoea agglomerans	Selenium nanoparticles	90–110	Antioxidant activity	[67]
Paracoccus haeundaensis BC74171	Au	20.93±3.46	Antioxidant activity and antiproliferative effect	[68]
Serratia nematodiphila	Zinc sulphide	80±10	Antibacterial activity	[69]
Shewanella loihica	Cu	10–16	Antibacterial activity	[70]
Streptomycetes viridogens HM10	Au	18–20	Antibacterial activity	[71]
Streptomyces sp. LK3	Ag	5	Acaricidal activity against *Rhipicephalus microplus* and *Haemaphysalis bispinosa*	[72]
Streptomyces sp. VITPK1	Ag	20–45	Anticandidal activity	[73]
Streptomyces sp. JAR1	Ag	60–70	Antimicrobial activity	[74]

9.4 APPLICATIONS OF MICROBIALLY SYNTHESISED NPS IN THE ENVIRONMENT

The efficacy of the entire procedure is anticipated to be impacted by the NPs' particular features, which make them suitable for a variety of applications. There has been much research and evaluation of numerous NP-assisted environmental sustainability-related operations. As previously said, many researchers concentrated on employing biosynthesised NPs in various environmental studies since they are more environmentally friendly and sustainable than chemically synthesised ones [6,75,76]. Biological sensors, "nano-pesticides," "nanofertilisers," and biological remediation are the most common uses of NPs in environmental sustainability (Figure 9.1).

9.4.1 Bioremediation Using Microbiological Nanoparticles

Given the global water shortage, recycling and purification of water have emerged as critical environmental sustainability pillars [77,78]. Wastewater from industries and cities is distributed and grows every day. It is comprised of many toxins, including organic pollutants (chemicals and dyes), heavy metals, and different harmful microorganisms. Many methods for detoxifying wastewater and eliminating pollutants from it were identified, including the creation of non-conventional materials. An effective and affordable remediation method for heavy metals, radioactive materials,

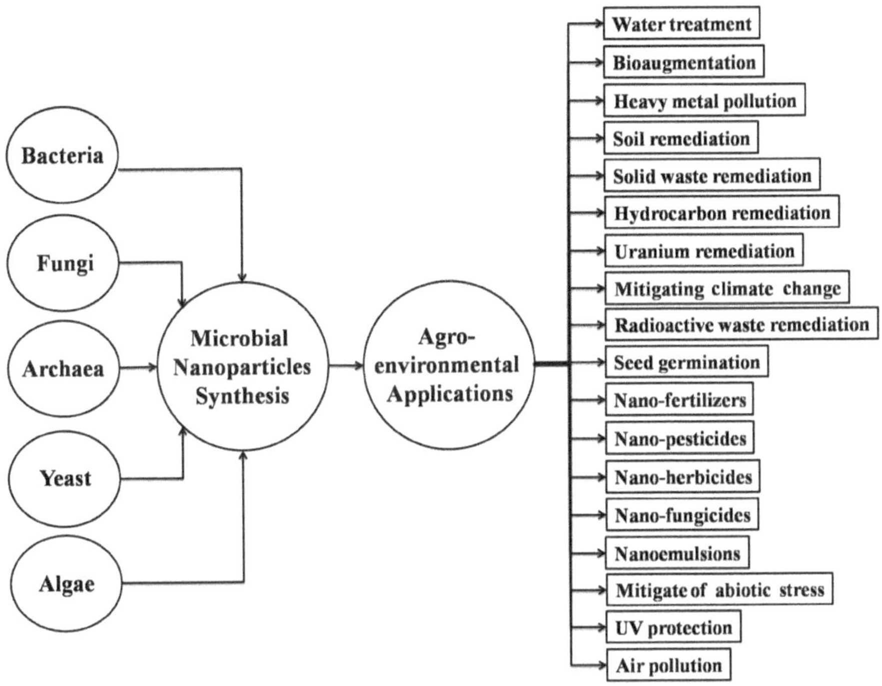

FIGURE 9.1 Environmental use of microbial nanoparticles.

and organic pollutants has been introduced by nanotechnology, which is one of several wastewater treatment technologies that has garnered significant attention worldwide in recent years [79–81]. Nanomaterials have garnered significant attention from the scientific community in various environmental sciences fields due to their distinctive chemical and physical characteristics and their potential for environmentally clean applications such as bioremediation [82,83].

Salvadori et al. [84] demonstrated how promising the use of nanotechnology in bioremediation methods can be by removing wastes and synthesising NPs through the application of fungal dead mycelium. Furthermore, *Rhizopus oryzae*-produced AuNPs showed amazing efficacy in reducing the negative environmental effects of organophosphorus pesticide contamination in both terrestrial and aquatic habitats [85]. NPs produced by microorganisms as a bioremediation technique offer a fantastic, environmentally beneficial, and commercially viable way to clean up ecosystems damaged by metal pollution. This technique turns harmful metallic pollutants into useful nanostructures, which raises the pollutants' economic value. In order to recover and use metals from contaminated water, some microbes are used to synthesise useful nanomaterials. This subject is thought to be a newly identified area with significant economic and environmental implications [86]. Specifically, Antarctic *Pseudomonas* strains were used to produce quantum dots (QDs), which are nano-sized structures that are used to remove contaminants like arsenic and cadmium from wastewater. Due to their possible semiconductor and spectroscopic qualities, the generated Cd and As nanoparticles are significant [87].

Research suggests that microbes that produce nanoparticles can absorb and accumulate "heavy metals" including As, Co, Cu, Fe, Ni, and Mg from aquatic environments. This is particularly true for "magnetotactic microbes," a type of bacteria that produce NPs [88,89]. Magnetosomes and magnetotactic bacteria are known for their enormous surface areas and magnetic qualities, which facilitate the adsorption of contaminants and their subsequent elimination from sewage [81,90]. Even while an external magnetic field can merely impact these bacteria, it can be used to remove toxic metals from sewage [91,92]. The "magnetosome" efficiently removes "up to 38 g/kg" of "fluoride" from water.

It also adsorbs Cu^{2+} from wastewater by forming a magnetic polymer with polyvinyl acetate. When compared to other methods of extracting metals, magnetosome's artificial magnetic particles yield more pure extracted metals, which is why it is superior as a noble metal adsorbent [91,93]. Thus, according to Liu and Zhou [94], Wang et al. [95], and Yan et al. [91], microbial magnetosomes play a crucial role in the treatment of wastewater, metal elimination, and recycling of resources.

Another hazardous source of polluting the environment that jeopardises the sustainability of natural resources, alongside heavy metals, is dye effluents released into the ecosystem [96]. It is possible to use NPs safely to produce dye breakdown. One powerful technique for this process can be the use of biogenic nanoparticles, which are produced by microorganisms [97]. Sharma and Shirkot [98] verified that biogenic iron NP may be used for the biological remediation of "azo dye" from the environment. Additionally, azo dye bioremediation from textile effluents was characterised as a bioprocess using CuNPs produced by a native *Escherichia* sp. After 5 hours of exposure to direct sunlight, most "biogenic CuNPs" were able to decolorise dyes "Congo red," "Malachite green," "Direct blue-1," and "Reactive black-5" with a degradation percentage of "97.07%", "90.55%", "88.42%", and "83.61%", respectively. As the dye concentration was increased to 100 mg/L, the removal efficacy reduced, but these results were still obtained at a lower dye concentration (25 mg/L) [99]. Analogously, it has been reported that the biogenic-synthesised AuNPs utilising select non-pathogenic *Enterococcus* species degrade methylene blue and malachite green dyes by 24 hours with a degradation percentage of 24.3%–57.6% and 88.85%–97.36%, respectively [42]. According to multiple studies [40,43,44], microbe-mediated NPs can efficiently degrade dyes.

9.4.2 Microbially Synthesised Nanoparticles: Their Use in Biosensing

Pollution detection and monitoring is a critical first step in addressing, protecting, and maintaining the environment for present and future generations. Understanding and evaluating the issue is a crucial first step in finding a solution. Conventional techniques such as spectroscopy and chromatography are typically employed in the analysis of pollution detection and quantification procedures. Those traditional methods are thought to be costly and time-consuming, and require expensive equipment and highly skilled individuals. Additionally, samples from the environment need to be brought into the lab for examination. Since the aforementioned standard approaches have drawbacks, the researchers concentrated on creating novel analysis techniques, the most significant of which are biosensors and nanobiosensors [100]. A transducer transmits optical, electrical, and mechanical signals generated by a particular receptor into a detector or read-out unit. Any device used to detect or measure an element or parameter is referred to as a sensor [100–102]. "Biosensors" are devices equipped with "bio-receptors" such as enzymes, antibodies, nucleic acids, or living/dead microbiological, plant, or animal cells. Biomimetic receptors, as defined by Ejeian et al. [103], Zare and Shekari [104], and Sinha et al. [102], are synthetic molecules that include molecularly imprinted polymers, aptamers, and synthetic antibodies that can be employed as bio-receptor elements.

Nanobiosensors, which incorporate certain nanomaterials (nanoparticles, nanocomposites, and nanotubes) with the transducer and/or bio-receptors, are the result of combining nanotechnology and biosensors. By enhancing signal transformation, the special qualities of nanomaterials are utilised to decrease the total size of the sensor and increase sensitivity and efficiency [105,106]. The receptors used, the signal transduction pathways, and the applications could all be used to classify nanobiosensors. However, according to Saini et al. [107], nanobiosensors can be divided into three categories based on the transducer's analysis mechanism: optical, electrochemical, and piezoelectric (mass-sensitive). Pesticides, herbicides, heavy metals, and other environmental contaminants that are present in the soil, water, and air can be quickly and accurately detected and monitored using biosensors and nanobiosensors. Due to their special qualities, which include a wide surface area and extremely luminous, electrical, magnetic, and plasmon resonance capabilities, many potential nanomaterials were reported to have developed effectively as probes with conjugation to proteins for biological tests. According to Chen et al. [108] and Matter et al. [100], nanomaterials such as magnetic nanoparticles, gold nanoparticles, "nano-sized porous alumina," hybrid "gold–platinum nanocomposites," "multiwall carbon nanotubes," "carbon nanoparticles," and "grapheme" are frequently utilised in the production of nanobiosensors.

Since the application of nanobiosensors contributes to environmental sustainability by identifying potentially harmful substances, using green-synthesised nanoparticles produced by microorganisms to create these biosensors is thought to be more environmentally friendly. Using the activity of microorganisms, numerous attempts have been undertaken to synthesise the nanomaterial that makes up nanobiosensors. To our current knowledge, however, there is no proof that the chemical, physical, or biological synthesis of nanoparticles can have an impact on the effectiveness of nanosensors. In addition to being able to be produced chemically, magnetic nanoparticles (NPs) are also widely used in a variety of applications, including nanosensors. Moreover, a large number of microorganisms rapidly biosynthesise NPs. It is utilised in the production of various antibiotics, such as "kanamycin" [109], "heavy metals" such as mercury ("II") [110], and magnetic nanoparticle–urease nanosensors, which are capable of detecting harmful bacteria [111]. Fe_3O_4 magnetic nanoparticles have been integrated with alkaline phosphatase into a "sol gel/chitosan biosensor membrane" in order to detect and quantify herbicides, such as 2,4-dichlorophenoxyacetic acid (2,4-D) [112].

It has been claimed that gold nanoparticles can be used as a component in nanosensors that identify the herbicide glyphosate. This nano-biosensor was created by depositing "poly(2,5-dimethoxyaniline)" doped with "poly(4-styrenesulfonic acid)" onto a gold electrode. The enzyme horseradish peroxidase horseradish peroxidase (HRP) was then attached to the composite film [100,113]. Silver nanoparticles are being used to create the detecting layer of an optical fibre sensor. The sensor is coated with tantalum(V) oxide particles that are sequestered in nano-scaled matrices of reduced graphene oxide. This sensor is capable of identifying and detecting the organophosphate pesticide fenitrothion, as stated in Kant's 2020 research. To create nanosensors, a variety of nanocomposites have been used. Using chitosan as a cross-linker, Yang et al. [114] developed an "amperometric response nano-biosensor" to identify a few pesticides based on AChE/SiO_2 nanosheet–Nafio nanocomposites. Using the principle of AChE inhibition, this nano-biosensor identified low doses of pesticides (1.0×10^{-12} to 1×10^{-8} M) such as carbofuran, methyl parathion, and chlorpyrifos. Improved efficacy, speed, and accuracy of these nano-based sensors/biosensors are the result of advancements in the biogenic production of nanomaterials, which have occurred concurrently with advancements in sensing and biosensing techniques. This means that the advancement of sensing techniques is a step towards risk assessment and appropriate handling of environmentally hazardous compounds, which is a crucial step beyond the sustainability of the environment.

9.4.3 Recycling of Wastes with the Use of Microbially Synthesised NPs

The use of nanotechnology can help in the recycling of hazardous wastes in addition to decreasing the amount of trash produced by different industrial and agricultural processes. One of the key objectives of environmental sustainability is the introduction of useful goods (from low-value wastes), which lessens reliance on finite resources. As previously stated, this sustainability might be enhanced if the nanoparticles used to recycle trash were produced sustainably, for example, by employing microbial activity. For example, lignocellulosic materials found mostly in the agricultural wastes are abundant and can be used as raw materials to make a variety of valuable items, including paper and several types of furniture. It is also possible to employ these cellulose-rich materials as feedstock to produce a variety of advantageous products, including chemicals, biofuel, single-cell protein, organic fertilisers, and animal feed [115]. Fermentable sugars produced by the efficient breakdown of cellulose-rich wastes by cellulases serve as the building block for subsequent conversion into advantageous products [22]. When enzymes cannot be recycled for more cycles of degradation, cellulose synthesis may prove to be a financially advantageous procedure. Since the encapsulation of enzymes, especially "celluloses," permits utilisation of numerous catalysts in various phases, improvements to the encapsulation process will benefit the process as a whole.

As a result of their capacity to increase enzyme stability, storage efficiency, and recycling, nanomaterials—particularly magnetic NPs—are seen to be the best possible materials for the immobilisation of enzymes. With the use of magnetic nanoparticles, the sustainability of breaking down and recycling enzymes might be enhanced in the next biofuel generation process from lignocellulosic biomass. Moreover, iron oxide nanoparticles and an acidic pretreatment for maize stover, followed by its degradation into fermentable sugars, are two examples of how nanoparticles could be used as a catalyst for pretreatment for an enzyme reaction linked to waste recycling [7,116].

9.4.4 Using "Biogenic" Nanoparticles as "Nanocatalysts"

Through their ability to create bonds with reactant molecules to produce the "product(s)" and break free from the outcome to make room for more reaction cycles, catalysts speed up chemical reactions. Chemicals, solid surfaces, atoms, and biological enzymes can all act as catalysts. Since catalytic reactions generate 85%–90% of the products in the chemical industry, catalysts are crucial for industrial chemical transformations, which are widely regarded as the workhorses of the industry. Many pharmaceutical and necessary chemical products would not be possible without catalysts, and other products would require additional time, reactants, reaction temperature, and waste generation [117,118]. Put another way, this may imply that the costs to mankind and the environment would increase in the absence of catalysts (Figure 9.2).

When it comes to their internal or external architecture, catalytic materials with at least one nanoscale dimension are called nanocatalysts [119]. Consequently, the use of nanotechnology in catalysts can be achieved by the production of nanoparticles for heterogeneous catalysis or the insertion of artificial nanoparticles (such as magnetic nanoparticles) to retain or improve the activity of enzymes or facilitate their reutilisation. For example, a homogeneous catalyst (either acidic, alkaline, or enzyme) might be present during the esterification reaction to convert fatty acids utilising methanol into methyl ester of fatty acids (biodiesel), or a heterogeneous solid catalyst

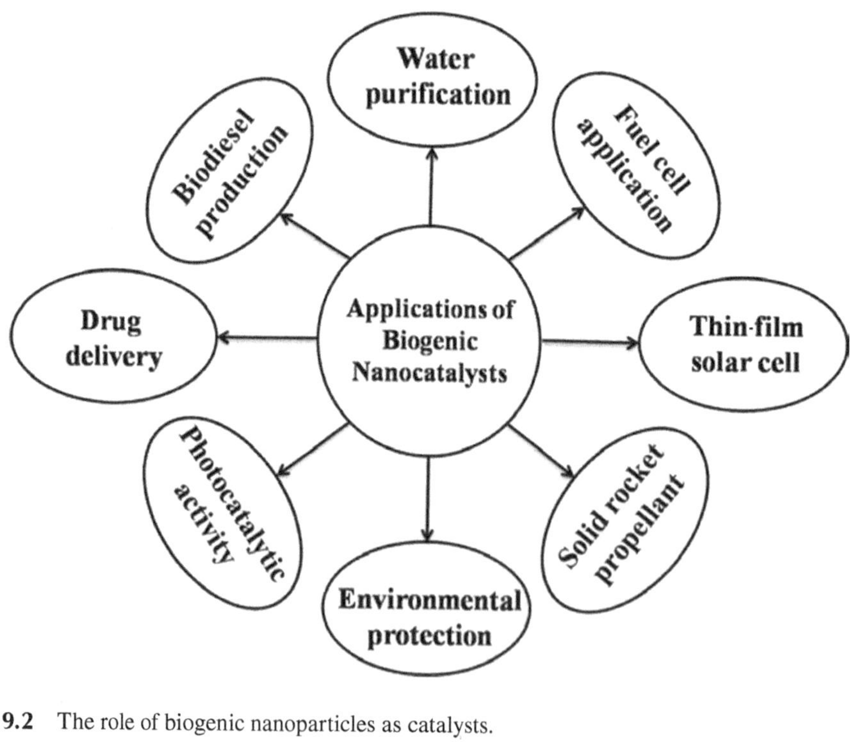

FIGURE 9.2 The role of biogenic nanoparticles as catalysts.

could be used [19,119]. According to Farrokheh et al. [120], several magnetic nanocatalysts, like CaO/KOHFe_3O_4 and KF/KOH-Fe_3O_4, are potential heterogeneous nanocatalysts for electrolysing *Chlorella vulgaris* and *Spirulina platensis* microalgae to create biodiesel. High-grade biodiesel product was produced with a high-yielding method of 96.8% at 25°C without the use of dangerous substances like sulphuric acid or sodium hydroxide and any side reactions related to saponification. On the other hand, an acid and a base are needed as catalysts in traditional esterification and transesterification, respectively, in addition to the high heat that is required; these catalysts are thought to be less sustainable overall.

Additionally, regenerating, recovering, and reusing the catalyst for other production cycles are simple using magnetic nanocatalysts (with the aid of an external magnetic field). Darwesh et al. [7], on the other hand, employed "peroxidase enzymes" immobilised on "magnetic" nanoparticles as a "nano-biocatalyst" for the "bioremediation of textile" effluent "dye in a reactor." Because the biocatalyst was immobilised with magnetic nanoparticles, they reported increased stability, bioremediation efficiency, and reusability of the catalyst. In summary, the use of nanocatalysts can lead to increased environmental sustainability, reduced environmental waste discharge, and catalyst sustainability.

9.4.5 Using Biogenic Nanoparticles to Generate Renewable Energy

Energy is essential for all living activities, and its presence is a sign of life on earth. It is also closely linked to industrial progress and population growth. Unfortunately, most of the energy we use comes from finite carbon-based fossil fuels like coal, oil, and natural gas, which contribute to greenhouse gas emissions and global warming. Apart from the disruption of numerous manufacturing and energy facilities worldwide, the decreasing supply of fossil fuels will also pose a threat to the current way of life by triggering conflicts over scarce resources. In order to ensure a sustainable future for our planet, it is crucial that we prioritise the use of clean and renewable energy sources. Among the most sustainable and clean energy sources are solar and wind power. Other significant clean, environmentally friendly, and renewable energy sources include hydrothermal, marine, biomass, and solar energy. By utilising these sources, we can achieve a sustainable environment for generations to come. Enhancing the production, transmission, and consumption efficiency of new and renewable energy was directly impacted by the application of nanotechnology. The use of carbon nanotubes, graphene, silicon dioxide (SiO_2), titanium-based nanoparticles (NPs), and silicon dioxide (NPs) in solar and bacterial fuel cells allows for the effective production of energy due to the special features of NPs [21]. In addition to the primary advantages of using renewable and sustainable energy sources, it should be highlighted that using microorganisms to produce nanomaterials linked to the generation of renewable energy is an environmentally benign method. One way to categorise biofuels would be as solid (like biochar), liquid (like bioethanol and biodiesel), or gaseous biofuels that are primarily made from bio-renewable feedstock. Since their primary sources, such as plants and algae, or indirectly through animal sources that consume plants, are products of photosynthesis, biofuels are seen as renewable and environmentally beneficial. Bioethanol and biodiesel are the two types of liquid biofuels most commonly found. It is possible to create bioethanol—a second-generation biofuel—by microbiologically breaking down lignocellulosic wastes into sugars that may be fermented. On the contrary, conversion of lipids from plants, animals, microbes, and algae into methyl ester of fatty acids is the process by which biodiesel is made. The application of nanotechnology in the biofuel production process presents numerous opportunities, including increasing biofuel yield [121], harvesting biomass through flocculation [122,123], and acting as a growth enhancer and stress inducer to optimise lipid contents (in microalgae, for example). Additionally, in order to achieve environmental sustainability, bioremediation of wastewater and municipal solid wastes could be combined with the generation of biofuels [8,116,124].

9.5 MICROBIAL NANOPARTICLES FOR AGRICULTURAL APPLICATIONS

Nanotechnology, a rapidly advancing scientific and technological sector, has the potential to significantly improve agricultural regulating processes. By reducing agricultural inputs, generating safer and better-quality food, and applying nanoscale nutrients, nanotechnology holds promise for enhancing sustainability in the agroecosystem. Agro-environmental sustainability can be greatly affected by the use of nanotechnology in managing pests and plant nutrition. Nanotechnology has advanced the management of agrochemicals, such as fertilisers and pesticides, in farming. This technology helps agri-inputs acquire intelligent properties, making the agro-processes more effective. By improving solubility, regulating release, targeting distribution, and increasing the shelf life of agrochemicals, nanotechnology not only increases efficacy but also reduces the risk of environmental contamination [125–127].

9.5.1 Ingenious Nanofertilisers

Nanofertilisers are a recent innovation in agriculture that utilises nanotechnology. Due to their small size and high surface area, they have been shown to be more efficient than traditional fertilisers in various experiments. Many studies have examined the use of different materials, including "silver," "iron," "zinc," "calcium," "titanium," "carbon nanotubes," "molybdenum," and "silica," as nanofertilisers for various farming crop systems. Based on various studies, it has been observed that nanomaterials have the potential to increase the length of roots and shoots, vegetative biomass, chlorophyll content, phytochemicals, and seed germination to a great extent, provided they are present at certain concentrations. Some of the authors who support this claim include Salama [128], Sharma et al. [129], Taha et al. [130], Azeez et al. [131–133], Chhipa [126], and Hussein et al. [134]. Additionally, research has shown that the application of carbon nanotubes and CuO as nanofertilisers can increase plant growth by more than 50% in crops such as tobacco and maize. For instance, Adhikari et al. [135] and Khodakovskaya et al. [136] have reported such improvements. It's important to not only increase the effectiveness and availability of nanofertilisers in agriculture but also decrease their harmful impacts on the environment. This is why nanotechnology plays a crucial role in the production of fertilisers [137].

Tarafdar et al. [138] found that *Aspergillus tubingensis* is capable of producing nano-phosphorus from tricalcium phosphate, which is a vital macronutrient. Many microorganisms are also involved in the synthesis of various nano-micronutrients, including Zn, Fe, Mn, Mg, and Co. These elements find extensive applications in farming for the production of both plants and animals, as reported by Sohair et al. [139] and Yusof et al. [140]. After being cultivated for 14 days, *Aspergillus flavus* TFR 7–mediated TiO_2 NPs were sprayed on leaves. The study conducted by Raliya et al. [141] found that mung beans exhibited significant increases in various "vegetative" characteristics such as "shoot length," "root length," "root area," "root nodule," "chlorophyll content," and "total protein content." Compared to the control group, the application of TiO_2 nanoparticles increased soil "microbiota" (21.4–48.1%), "acid and alkaline phosphatase" (72% and 67.3%, respectively), "phytase" (64%), and "dehydrogenase" (108.7%).

9.5.2 Microbiological Nano-insecticides

In recent years, there has been a growing interest in finding safer substitutes for chemical pesticides. This is due to the need to reduce the use of agrochemicals in agriculture and to prevent production loss caused by plant diseases. Nanoparticles have evolved as an eco-friendly substitute for toxic pesticides to address this issue. Studies conducted by Masum et al. [142], Suman et al. [143], and Elshahawy et al. [144] have demonstrated the potential of nanoparticles in the field of pest control. According to Ali et al. [145], metallic nanoparticles (NPs) like Ag, Cu, Au, and Zn that are biologically mediated have been found to have a preventative effect against bacteria that cause diseases like

Bacillus subtilis, *Escherichia coli*, and *Staphylococcus aureus*, and few disease-causing fungi like *Aspergillus niger*, *Fusarium oxysporum*, and *Aspergillus fumigatus*. The use of silver nanoparticles (NPs) has gained interest due to their eco-friendly manufacturing process using plants, bacteria, fungi, or yeasts [146]. As per the research conducted by Fouda et al. [147], *Streptomyces*-mediated AgNPs have shown significant potential in managing various plant pathogenic fungi including *Alternaria alternata*, *A. niger*, *F. oxysporum*, and *Pythium ultimum*. Additionally, a study by Ojo et al. [40] demonstrated that the growth of *A. fumigatus* and *A. niger* strains was inhibited by 66.67%–90.78% at a concentration of 200 μg/mL of AuNPs and Ag-AuNPs that were biosynthesised using *Bacillus safensis* LAU 13 cell-free extract. Lateef et al. [148] discovered that the cell-free extract of a bacterial strain can produce AgNPs that can suppress the growth of *S. aureus*, *E. coli*, *Klebsiella granulomatis*, and *Pseudomonas aeruginosa* at a concentration of 100 μg/mL. Nanoparticles can be used as a pesticide by spraying them on leaves or coating them on seeds. To minimise their adverse effects on the environment, nanoparticles can also be used as a carrier for regulating the release of chemical pesticides from their formulation, while still acting as a pesticide [145,149] (Table 9.2).

TABLE 9.2
A Summary of Research Using Nanoparticles to Deliver Insecticides or Active Ingredients Intended to Control Insect Pests

Nanoparticle	Pesticide	Target Pest	Crop	Reference
Polyvinylchloride (PVC)	Chlorpyrifos	Birch and southern yellow pine	–	[150]
Solid lipid nanoparticle (SLN)	*Artemisia arborescens* L. essential oil[a]	–	–	[151]
Porous hollow silica nanoparticles (PHSN)	Avermectin (6)	–	–	[152]
PHSN	Avermectin (6)	–	–	[153]
PHSN	Avermectin (6)	–	–	[154]
Silica	Chlorfenapyr (13)	Cotton bollworm (*Helicoverpa armigera*), *Plutella xylostella*	*Brassica* chinese	[155]
Chitosan-coated beeswax SLN	Deltamethrin (3A)	–	–	[156]
Corn oil (liquid lipid) and beeswax SLN	Deltamethrin (3A)	–	–	[157]
Polyethylene glycol (PEG)	Garlic essential oil[a]	Red flour beetle (*Tribolium castaneum*)	Rice (harvested)	[158]
Chitosan/polylactide (PLA)	Chlorpyrifos (1B)	–	–	[159]
Chitosan	Azadirachtin (UN)	–	–	[160]
Silica	α-Pinene[a] and linalool[a]	Tobacco cutworm (*Spodoptera litura*), castor semi-looper (*Achaea janata*)	Castor leaf discs	[161]
Chitosan	Azadirachtin (UN)	Tobacco cutworm (*Spodoptera litura*) culture ovarian cell lines Sl-1	–	[162]
Porous silica	Abamectin (6)	–	–	[163]
Sodium alginate	Imidacloprid (4A)	Leafhopper (Jassids)	Plants (unspecified)	[164]
Layered double hydroxide (LDH)	Anacardic acid[a] Abamectin (6)	Tobacco cutworm (*Spodoptera litura*)	Mustard leaves (*Brassica juncea*)	[165]
Polydopamine	Avermectin (6)	–	Cotton leaves and corn leaves	[166]

(*Continued*)

TABLE 9.2 (*Continued*)
A Summary of Research Using Nanoparticles to Deliver Insecticides or Active Ingredients Intended to Control Insect Pests

Nanoparticle	Pesticide	Target Pest	Crop	Reference
Silica shell (oil core)	Fipronil (2B)	Termites (*Coptotermes acinaciformis*)	–	[167]
Polydopamine	Avermectin (6)	–	–	[168]
Hollow polymeric shell	λ-Cyhalothrin (3A)	–	–	[169]
Dendrimers	Thiamethoxam (4A)	Cotton bollworm cells and larvae (*Helicoverpa armigera*)	–	[170]
Polymer-coated silver	Organochlorine pesticides (EPA 8080 pesticide mix) (2A)	–	–	[171]
Chitosan	Carvacrol[a] Linalool[a]	Mite (*Tetranychus urticae*)	–	[172]
Polydopamine	Avermectin (6)	Aphids (unspecified)	Cucumber Broccoli	[173]
Zinc oxide and chitosan	Azadirachtin (UN)	Groundnut bruchid (*Caryedon serratus* O.)	–	[174]
Castor oil–based polyurethane	Avermectin (6)	–	Corn leaves	[175]
Chitosan	PONNEEM[a]	Cotton bollworm larvae (*Helicoverpa armigera*)	–	[176]
Silver	*Suaeda maritima*–based herbal coils[a]	Tobacco cutworm (*Spodoptera litura*)	Cotton leaf discs	[177]
Zein	Geraniol and R-citronellal essential oils[a]	Mite (*Tetranychus urticae*)	–	[178]
Chitosan/gum arabic	Geraniol[a]	Whitefly (*Bemisia tabaei*)	–	[179]
Chitosan/TPP	Nicotine (4B)	House fly (*Musca domestica*)	–	[180]
Chitosan/TPP	*Satureja hortensis* L. essential oil[a]	Mite (*Tetranychus urticae*)	–	[181]
Chitosan	Carvacrol[a] Linalool[a]	Corn earworm (*Helicoverpa armigera*) and Mites (*Tetranychus urticae*)	Bean leaf discs	[182]
Porous hollow silica nanoparticles (PHSN)	Avermectin (6)	*Plutella xylostella* larvae	*Brassica oleracea*	[183]
Magnetic nanocomposites	Imidacloprid (4A) Cyfluthrin (3A)	–	–	[184]
Copper-doped ZIF-8 (Cu@ZIF-8)	Fludioxonil	*Fusarium graminearum*	Wheat (*Triticum aestivum*)	[185]

[a] Groups of pesticides do not yet contain insecticidal oils or extracts that are unknown to act at certain target areas.

9.6 FATE OF MICROBIAL NANOPARTICLES

The removal of heavy metals from the ground is achieved through phototherapeutic treatment and wastewater purification using magnetite nanoparticles (MNPs) that exhibit paramagnetic behaviour [186,187]. MNPs are non-cytotoxic micronuclear particles and provide a cost-effective and environmentally friendly solution. Various adsorbents including silica, carbon nanotubes, activated carbon, and rice husk are utilised to treat wastewater [188,189].

Many industries such as paints, vehicle coatings, food packaging, fabrics, medicine, cosmetics, and tanning lotions utilise nanotechnology. Nanosensors are capable of effectively monitoring pesticides and microorganisms in the environment [190–192]. Nanoparticles (NPS) find applications in solar, electrical, biological, and geothermal processes. They serve as catalysts to enhance reactions [193,194].

9.7 CONCLUSION AND FUTURE PERSPECTIVE

This chapter covers the advancements in microbial nanoparticle synthesis and its importance for environmental sustainability. The green approach towards nanotechnology has gained significant attention over the years due to its several advantages over chemical and physical processes. These benefits include environmental safety, cost-effectiveness, adaptability, and ease of scaling up. The discussions covered the benefits of microbial manufacturing of nanoparticles (NPs), the use of microbially mediated NPs in agriculture, and bioremediation. Due to the enhanced safety of biogenic nanoparticles compared to chemically and physically synthesised nanomaterials, the only direction for their utilisation is towards these purposes. The importance of biogenic nanoparticles in energy preservation and biofuel development cannot be overstated, especially when considering sustainability challenges related to energy. This chapter highlights how microbial NP production has helped improve environmental biosensing. NPs have shown great potential in the environment, particularly in the areas of biosensors and biomonitoring. Nevertheless, much more study and pilot testing were needed to increase the efficiency of these bioprocesses to close the gap between bench research and industrial application in the field of biogenic NPs for ecological uses. More research is needed on improving biometric extraction and utilisation methods, optimising parameters, and screening for more productive bionanofactories such as bacteria, actinomycetes, fungi, yeast, viruses, algae, and even plants to achieve these goals. The exact process by which nanoparticles (NPs) are formed biologically is still unknown. As a result, controlling the size, shape, and crystallinity of the generated NPs is a significant challenge in scaling up the environmental application of NPs. The development of biogenic NP synthesis has been a difficult problem that requires further investigation. Furthermore, additional studies are necessary to evaluate the toxicity of microbially mediated NPs, even if they seem to be less harmful.

REFERENCES

1. Davis AS, Prakash P, Thamaraiselvi K (2017) Nanobioremediation technologies for sustainable environment. In: *Bioremediation and Sustainable Technologies for Cleaner Environment*. Springer, Cham. pp. 13–33. https://doi.org/10.1007/978-3-319-48439-6_2.
2. Darwesh OM, Moawad H, Barakat OS, Abd El- Rahim WM (2015) Bioremediation of textile reactive blue azo dye residues using nanobiotechnology approaches. *Res J Pharm Biol Chem Sci* 6:1202–1211.
3. Darwesh OM, Ali SS, Matter IA, Elsamahy T, Mahmoud YA (2020) Enzymes immobilization onto magnetic nanoparticles to improve industrial and environmental applications. In: *Methods in Enzymology*, vol 630. Academic Press, Cambridge, pp. 481–502. https://doi.org/10.1016/bs.mie.2019.11.006.
4. Abou-Shanab RAI, Khalafallah MA, Emam NF, Aly MA, Abou-Sdera SA, Matter IA (2012) Characterisation and identification of carbofuran-utilising bacteria isolated from agricultural soil. *Chem Ecol* 28:193–203. https://doi.org/10.1080/02757540.2011.628317.
5. Darwesh OM, Eida MF, Matter IA (2018) Isolation, screening and optimization of L-asparaginase producing bacterial strains inhabiting agricultural soils. *Biosci Res* 15:2802–2812.
6. Darwesh OM, Matter IA, Eida MF, Moawad H, Oh YK (2019) Influence of nitrogen source and growth phase on extracellular biosynthesis of silver nanoparticles using cultural filtrates of Scenedesmus obliquus. *Appl Sci* 9:1465. https://doi.org/10.3390/app9071465.
7. Darwesh OM, Matter IA, Eida MF (2019) Development of peroxidase enzyme immobilized magnetic nanoparticles for bioremediation of textile wastewater dye. *J Environ Chem Eng* 7:102805. https://doi.org/10.1016/j.jece.2018.11.049.
8. Sangeetha J, Mundaragi A, Thangadurai D, Maxim SS, Pandhari RM, Alabhai JM (2019) Nanobiotechnology for agricultural productivity, food security and environmental sustainability. In:

Nanotechnology for Agriculture: Crop Production & Protection. Springer, Singapore, pp. 1–23. https://doi.org/10.1007/978-981-32-9374-8_1.
9. Besha AT, Liu Y, Bekele DN, Dong Z, Naidu R, Gebremariam GN (2020) Sustainability and environmental ethics for the application of engineered nanoparticles. *Environ Sci Policy* 103:85–98. https://doi.org/10.1016/j.envsci.2019.10.013.
10. Mourad R, Helaly F, Darwesh OM, Sawy SE (2019) Antimicrobial and physicomechanical natures of silver nanoparticles incorporated into silicone- hydrogel films. *Contact Lens Anter Eye* 42:325–333. https://doi.org/10.1016/j.clae.2019.02.007.
11. Abdel-Monem RA, Khalil AM, Darwesh OM, Hashim AI, Rabie ST (2020) Antibacterial properties of carboxymethyl chitosan Schiff-base nanocomposites loaded with silver nanoparticles. *J Macromol Sci* A 57:145–155. https://doi.org/10.1080/10601325.2019.1674666.
12. Simate GS, Ndlovu S, Iyuke SE, Walubita LF (2013) Biotechnology and nanotechnology: A means for sustainable development in Africa. In: *Chemistry for Sustainable Development in Africa*. Springer, Berlin, pp. 159–191. https://doi.org/10.1007/978-3-642-29642-0_9
13. Hadef F (2018) An introduction to nanomaterials. In: *Environmental Nanotechnology*. Environmental Chemistry for a Sustainable World, vol 14. Springer, Cham, pp. 1–58. https://doi.org/10.1007/978-3-319-76090-2_1.
14. Nasrollahzadeh M, Sajadi SM, Sajjadi M, Issaabadi Z (2019) Applications of nanotechnology in daily life. In: *Interface Science and Technology*, vol 28. Elsevier, Amsterdam, pp. 113–143. https://doi.org/10.1016/b978-0-12-813586-0.00004-3.
15. Chamundeeswari M, Jeslin J, Verma ML (2019) Nanocarriers for drug delivery applications. *Environ Chem Lett* 17:849–865. https://doi.org/10.1007/s10311-018-00841-1.
16. Verma ML (2017) Nanobiotechnology advances in enzymatic biosensors for the agri-food industry. *Environ Chem Lett* 15:555–560. https://doi.org/10.1007/s10311-017-0640-4.
17. Verma ML, Dhanya BS, Sukriti Thakur M, Jeslin J, Kushwaha R (2020) Carbohydrate and protein based biopolymeric nanoparticles: Current status and biotechnological applications. *Int J Biol Macromol* 154:390–412. https://doi.org/10.1016/j.ijbiomac.2020.03.105.
18. Durgalakshmi D, Rajendran S, Naushad M (2019) Current role of nanomaterials in environmental remediation. In: *Advanced Nanostructured Materials for Environmental Remediation*. Springer, Cham, pp. 1–20. https://doi.org/10.1007/978-3-030-04477-0_1.
19. El-Baz FK, Gad MS, Abdo SM, Abed KA, Matter IA (2016) Performance and exhaust emissions of a diesel engine burning algal biodiesel blends. *Int J Mechan Mechatr Eng* 16:150–157.
20. Matter IA, Darwesh OM, Eida MF (2018) Harvesting of microalgae Scenedesmus obliquus using chitosan-alginate dual flocculation system. *Biosci. Res.* 15:540–548.
21. Sohail MI, Waris AA, Ayub MA, Usman M, urRehman MZ, Sabir M, Faiz T (2019) Environmental application of nanomaterials: A promise to sustainable future. In: *Comprehensive Analytical Chemistry*, vol 87. Elsevier, Amsterdam, pp. 1–54. https://doi.org/10.1016/bs.coac.2019.10.002.
22. Darwesh OM, El-Maraghy SH, Abdel-Rahman HM, Zaghloul RA (2020) Improvement of paper wastes conversion to bioethanol using novel cellulose degrading fungal isolate. *Fuel* 262:116518. https://doi.org/10.1016/j.fuel.2019.116518.
23. Hoballah E, Saber M, Matter I, Zaghloul A (2014) Bioremediation of polychlorinated biphenyl (PCBs) in a sewaged soil by certain remediative amendments followed by phytoremediation. *Res J Pharm Biol Chem Sci* 5:91–103.
24. Saber M, Hoballah E, Matter I, Zaghloul A (2014) Bioremediation of chlorinated hydrocarbons in a sewaged soil by certain remediative amendments followed by phytoremediation. *Res J Pharm Biol Chem Sci* 5:83–90. https://rjpbcs.com/pdf/2014_5(2)/[11].pdf.
25. Cox A, Venkatachalam P, Sahi S, Sharma N (2017) Reprint of Silver and titanium dioxide nanoparticle toxicity in plants: A review of current research. *Plant Physiol Biochem PPB* 110:33–49.
26. Dasgupta N, Ranjan S, Mundekkad D, Ramalingam C, Shanker R, Kumar A (2015) Nanotechnology in agro-food: From field to plate. *Food Res Int* 69:381–400.
27. Awad MA, Eisa NE, Virk P, Hendi AA, Ortashi KMOO, Mahgoub ASA, Elobeid MA, Eissa FZ (2019) Green synthesis of gold nanoparticles: Preparation, characterization, cytotoxicity, and anti-bacterial activities. *Mater Lett* 256:126608.
28. Takeuchi MT, Kojima M, Luetzow M (2014) State of the art on the initiatives and activities relevant to risk assessment and risk management of nanotechnologies in the food and agriculture sectors. *Food Res Int* 64:976–981.
29. Marchiol L, Mattiello A, Poščić F, Giordano C, Musetti R (2014) In vivo synthesis of nanomaterials in plants: Location of silver nanoparticles and plant metabolism. *Nanoscale Res Lett* 9:101.

30. López-Valdez F, Fernández-Luqueño F (Eds.) (2018) *Agricultural Nanobiotechnology: Modern Agriculture for a Sustainable Future.* Springer International Publishing, Cham. pp. 3–16. ISBN 978-3-319-96719-6.
31. Chaudhry N, Dwivedi S, Chaudhry V, Singh A, Saquib Q, Azam A, Musarrat J (2018) Bio-inspired nanomaterials in agriculture and food: Current status, foreseen applications and challenges. *Microb Pathog* 123:196–200.
32. Prabhakar PK, Lakhanpal J (2020) Recent advances in the nucleic acid-based diagnostic tool for coronavirus. *Mol Biol Rep* 47:9033–9041.
33. Li X, Xu H, Chen ZS, Chen G (2011) Biosynthesis of nanoparticles by microorganisms and their applications. *J Nanomater* 1:2011. https://doi.org/10.1155/2011/270974.
34. Darwesh OM, Sultan YY, Seif MM, Marrez DA (2018) Bio-evaluation of crustacean and fungal nano-chitosan for applying as food ingredient. *Toxicol Rep* 5:348–356. https://doi.org/10.1016/j.toxrep.2018.03.002.
35. Lateef A, Adelere IA, Gueguim-Kana EB, Asafa TB, Beukes LS (2015) Green synthesis of silver nanoparticles using keratinase obtained from a strain of Bacillus safensis LAU 13. *Int Nano Lett* 5(1):29–35. https://doi.org/10.1007/s40089-014-0133-4.
36. Adelere IA, Lateef A (2016) A novel approach to the green synthesis of metallic nanoparticles: The use of agro-wastes, enzymes, and pigments. *Nanotechnol Rev* 5(6):567–587. https://doi.org/10.1515/ntrev-2016-0024.
37. Lateef A, Adeeyo AO (2015) Green synthesis and antibacterial activities of silver nanoparticles using extracellular laccase of Lentinus edodes. *Not Sci Biol* 7(4):405–411. https://doi.org/10.15835/nsb749643.
38. Elegbede JA, Lateef A, Azeez MA, Asafa TB, Yekeen TA, Oladipo IC, Aina DA, Beukes LS, Gueguim-Kana EB (2020) Biofabrication of gold nanoparticles using xylanases through valorization of corncob by Aspergillus niger and Trichoderma longibrachiatum: Antimicrobial, antioxidant, anti-coagulant and thrombolytic activities. *Waste Biomass Valoriz* 11(3):781–791. https://doi.org/10.1007/s12649-018-0540-2.
39. Nies DH (1999) Microbial heavy-metal resistance. *Appl Microbiol Biotechnol* 51(6):730–750. https://doi.org/10.1007/s002530051457.
40. Ojo SA, Lateef A, Azeez MA, Oladejo SM, Akinwale AS, Asafa TB, Yekeen TA, Akinboro A, Oladipo IC, Gueguim-Kana EB, Beukes LS (2016) Biomedical and catalytic applications of gold and silver-gold alloy nanoparticles biosynthesized using cell-free extract of Bacillus safensis LAU 13: Antifungal, dye degradation, anti-coagulant and thrombolytic activities. *IEEE Trans Nanobiosci* 15:433–442. https://doi.org/10.1109/TNB.2016.2559161.
41. Oladipo IC, Lateef A, Azeez MA, Asafa TB, Yekeen TA, Akinboro A, Akinwale AS, Gueguim-Kana EB, Beukes LS (2017) Green synthesis and antimicrobial activities of silver nanoparticles using cell-free extracts of Enterococcus species. *Not Sci Biol* 9(2):196–203. https://doi.org/10.15835/nsb929938.
42. Oladipo IC, Lateef A, Elegbede JA, Azeez MA, Asafa TB, Yekeen TA, Akinboro A, GueguimKana EB, Beukes LS, Oluyide TO, Atanda OR (2017) Enterococcus species for the one-pot biofabrication of gold nanoparticles: Characterization and nanobiotechnological applications. *J Photochem Photobiol B Biol* 173:250–257. https://doi.org/10.1016/j.jphotobiol.2017.06.003.
43. Elegbede JA, Lateef A, Azeez MA, Asafa TB, Yekeen TA, Oladipo IC, Adebayo EA, Beukes LS (2018) Fungal xylanases-mediated synthesis of silver nanoparticles for catalytic and biomedical applications. *IET Nanobiotechnol* 12:857–863. https://doi.org/10.1049/iet-nbt.2017.0299.
44. Elegbede JA, Lateef A, Azeez MA, Asafa TB, Yekeen TA, Oladipo IC, Hakeem AS, Beukes LS, Gueguim-Kana EB (2019) Silver-gold alloy nanoparticles biofabricated by fungal xylanases exhibited potent biomedical and catalytic activities. *Biotechnol Prog* 35:2829. https://doi.org/10.1002/btpr.2829.
45. El-Shanshoury AE, Darwesh OM, Sabae SZ, Awadallah OA, Hassan SH (2020) Bio-manufacturing of selenium nanoparticles by Bacillus subtilis isolated from Qarun lake and evaluation their activity for water remediation. *Biointerf Res Appl Chem* 10:5834–5842. https://doi.org/10.33263/BRIAC104.834842.
46. Jayaseelan C, Rahuman AA, Kirthi AV, Marimuthu S, Santhoshkumar T, Bagavan A, et al. (2012) Novel microbial route to synthesize ZnO nanoparticles using Aeromonas hydrophila andtheir activity against pathogenic bacteria and fungi. *Spectrochim Acta Part A Mol Biomol Spectrosc* 90:78–84. https://doi.org/10.1016/j.saa.2012.01.006.
47. Wadhwani SA, Shedbalkar UU, Singh R, Vashisth P, Pruthi V, Chopade BA (2016) Kinetics of synthesis of gold nanoparticles by Acinetobacter sp. SW30 isolated from environment. *Indian J Microbiol* 56:439–444. https://doi.org/10.1007/s12088-016-0598-0.
48. Abdeen S, Geo S, Praseetha PK, Dhanya RP (2014) Biosynthesis of silver nanoparticles from actinomycetes for therapeutic applications. *Int J Nano Dimens* 5:155–162.

49. Narasimha G, Alzohairy M, Khadri H, Mallikarjuna K (2013) Extracellular synthesis, characterization and antibacterial activity of Silver nanoparticles by Actinomycetes isolative. *Int J Nano Dimens* 4:77–83. https://doi.org/10.7508/IJND.2013.01.010.
50. Saeed S, Iqbal A, Ashraf MA (2020) Bacterial-mediated synthesis of silver nanoparticles and their significant effect against pathogens. *Environ Sci Pollut Res* 27:37347–37356. https://doi.org/10.1007/s11356-020-07610-0.
51. Shivashankarappa A, Sanjay KR (2015) Study on biological synthesis of cadmium sulfide nanoparticles by *Bacillus licheniformis* and its antimicrobial properties against food borne pathogens. *Nanosci Nanotechnol Res 3*(1):6–15.
52. Forootanfar H, Adeli-Sardou M, Nikkhoo M, Mehrabani M, Amir-Heidari B, Shahverdi AR, Shakibaie M (2014) Antioxidant and cytotoxic effect of biologically synthesized selenium nanoparticles in comparison to selenium dioxide. *J Trace Elem Med Biol* 28(1):75–79.
53. Dhandapani P, Maruthamuthu S, Rajagopal G (2012) Bio-mediated synthesis of TiO_2 nanoparticles and its photocatalytic effect on aquatic biofilm. *J Photochem Photobiol B Biol* 110:43–49.
54. Sowani H, Mohite P, Munot H, Shouche Y, Bapat T, Kumar AR, et al. (2016) Green synthesis of gold and silver nanoparticles by an actinomycete Gordonia amicalis HS-11: Mechanistic aspects and biological application. *Proc Biochem* 51:374–383. https://doi.org/10.1016/j.procbio.2015.12.013.
55. Bennur T, Khan Z, Kshirsagar R, Javdekar V, Zinjarde S (2016) Biogenic gold nanoparticles from the actinomycete Gordonia amarae: Application in rapid sensing of copper ions. *Sens Actuators B Chem* 233:684–690. https://doi.org/10.1016/j.snb.2016.04.022.
56. Srivastava P, Kowshik M (2017) Fluorescent lead (IV) sulfide nanoparticles synthesized by Idiomarina sp. strain PR58-8 for bioimaging applications. *Appl Environ Microbiol* 83:1–15. https://doi.org/10.1128/AEM.03091-16.
57. Mishra M, Paliwal JS, Singh SK, Selvarajan E, Subathradevi C, Mohanasrinivasan V (2013) Studies on the inhibitory activity of biologically synthesized and characterized zinc oxide nanoparticles using lactobacillus sporogens against Staphylococcus aureus. *J Pure Appl Microbiol* 7:1–6.
58. Markus J, Mathiyalagan R, Kim Y-J, Abbai R, Singh P, Ahn S, et al. (2016) Intracellular synthesis of gold nanoparticles with antioxidant activity by probiotic Lactobacillus kimchicus DCY51T isolated from Korean kimchi. *Enzyme Microb Technol* 95:85–93. https://doi.org/10.1016/j.enzmictec.2016.08.018.
59. Namasivayam SKR, Gnanendra EK, Reepika R (2010) Synthesis of silver nanoparticles by Lactobacillus acidophilus 01 strain and evaluation of its in vitro genomic DNA toxicity. *Nano Micro Lett* 2:160–163. https://doi.org/10.1007/BF03353635.
60. Selvarajan E, Mohanasrinivasan V (2013) Biosynthesis and characterization of ZnO nanoparticles using Lactobacillus plantarum VITES07. *Mater Lett* 112:180–182. https://doi.org/10.1016/j.matlet.2013.09.020.
61. Prasad K, Jha AK (2010) Biosynthesis of CdS nanoparticles: An improved green and rapid procedure. *J Colloid Interface Sci* 342:68–72. https://doi.org/10.1016/j.jcis.2009.10.003.
62. Ahmad R, Khatoon N, Sardar M (2013) Biosynthesis, characterization and application of TiO_2 nanoparticles in biocatalysis and protein folding. *J Proteins Proteomics* 4:115–121.
63. Ahmad R, Khatoon N, Sardar M (2014) Antibacterial effect of green synthesized TiO_2 nanoparticles. *Adv Sci Lett* 20:1616–1620. https://doi.org/10.1166/asl.2014.5563.
64. Park K-Y, Jeong J-K, Lee Y-E, Daily III JW (2014) Health benefits of kimchi (Korean fermented vegetables) as a probiotic food. *J Med Food* 17:6–20. https://doi.org/10.1089/jmf.2013.3083.
65. Camas M, Camas AS, Kyeremeh K (2018) Extracellular synthesis and characterization of gold nanoparticles using Mycobacterium sp. BRS2A-AR2 isolated from the aerial roots of the Ghanaian mangrove plant, Rhizophora racemosa. *Indian J Microbiol* 58:214–221. https://doi.org/10.1007/s12088-018-0710-8.
66. Jafari M, Rokhbakhsh-Zamin F, Shakibaie M, Moshafi MH, Ameri A, Rahimi HR, et al. (2018) Cytotoxic and antibacterial activities of biologically synthesized gold nanoparticles assisted by Micrococcus yunnanensis strain J2. *Biocatal Agric Biotechnol* 15:245–253. https://doi.org/10.1016/j.bcab.2018.06.014.
67. Torres SK, Campos VL, Leon CG, Rodriguez-Llamazares SM, Rojas SM, Gonzalez M, et al. (2012) Biosynthesis of selenium nanoparticles by Pantoea agglomerans and their antioxidant activity. *J Nanopart Res* 14:1–9. https://doi.org/10.1007/s11051-012-1236-3.
68. Patil MP, Kang M-J, Niyonizigiye I, Singh A, Kim J-O, Seo YB, et al. (2019) Extracellular synthesis of gold nanoparticles using the marine bacterium Paracoccus haeundaensis BC74171T and evaluation of their antioxidant activity and antiproliferative effect on normal and cancer cell lines. *Colloids Surf B Biointerfaces* 183:110455. https://doi.org/10.1016/j.colsurfb.2019.110455.
69. Malarkodi C, Annadurai G (2013) A novel biological approach on extracellular synthesis and characterization of semiconductor zinc sulfide nanoparticles. *Appl Nanosci* 3:389–395. https://doi.org/10.1007/s13204-012-0138-0.

70. Lv Q, Zhang B, Xing X, Zhao Y, Cai R, Wang W, et al. (2018) Biosynthesis of copper nanoparticles using Shewanella loihica PV-4 with antibacterial activity: Novel approach and mechanisms investigation. *J Hazard Mater* 347:141–149. https://doi.org/10.1016/j.jhazmat.2017.12.070.
71. Balagurunathan R, Radhakrishnan M, Rajendran RB, Velmurugan D (2011). Biosynthesis of gold nanoparticles by actinomycete Streptomyces viridogens strain HM10. *J Biochem Biophys* 48: 331–335.
72. Karthik L, Kumar G, Kirthi AV, Rahuman AA, Rao KVB (2014) Streptomyces sp. LK3 mediated synthesis of silver nanoparticles and its biomedical application. *Bioproc Biosyst Eng* 37:261–267. https://doi.org/10.1007/s00449-013-0994-3.
73. Sanjenbam P, Gopal JV, Kannabiran K (2014) Anticandidal activity of silver nanoparticles synthesized using Streptomyces sp. VITPK1. *J Mycol Med* 24:211–219. https://doi.org/10.1016/j.mycmed.2014.03.004.
74. Chauhan R, Kumar A, Abraham J (2013) A biological approach to the synthesis of silver nanoparticles with Streptomyces sp JAR1 and its antimicrobial activity. *Sci Pharm* 81:607–624. https://doi.org/10.3797/scipharm.1302-02.
75. Hasanin MS, Mostafa AM, Mwafy EA, Darwesh OM (2018) Eco-friendly cellulose nano fibers via first reported Egyptian Humicola fuscoatra Egyptia X4: Isolation and characterization. *Environ Nanotechnol Monit Manag* 10:409–418. https://doi.org/10.1016/j.enmm.2018.10.004.
76. Hussein HA, Darwesh OM, Mekki BB (2019) Environmentally friendly nano-selenium to improve antioxidant system and growth of groundnut cultivars under sandy soil conditions. *Biocatal Agric Biotechnol* 18:101080. https://doi.org/10.1016/j.bcab.2019.101080.
77. Nguyen CC, Hugie CN, Kile ML, Navab-Daneshmand T (2019) Association between heavy metals and antibiotic-resistant human pathogens in environmental reservoirs: A review. *Front Environ Sci Eng* 13:46. https://doi.org/10.1007/s11783-019-1129-0.
78. Yang J, Lei M, Chen T, Gao D, Zheng G, Guo G, Lee D (2014) Current status and developing trends of the contents of heavy metals in sewage sludges in China. *Front Environ Sci Eng* 8:719–728. https://doi.org/10.1007/s11783-013-0600-6.
79. Singh P, Kim YJ, Zhang D, Yang DC (2016) Biological synthesis of nanoparticles from plants and microorganisms. *Trends Biotechnol* 34:588–599. https://doi.org/10.1016/j.tibtech.2016.02.006.
80. de Castro AL, Yáñez-Vilar S, Piñeiro-Redondo Y, Rivas J (2019) Novel magnetic nanostructured beads for cadmium (II) removal. *Nanomaterials* 9:356. https://doi.org/10.3390/nano9030356.
81. Ranjan B, Pillai S, Permaul K, Singh S (2019) Simultaneous removal of heavy metals and cyanate in a wastewater sample using immobilized cyanatehydratase on magnetic-multiwall carbon nanotubes. *J Hazard Mater* 363:73–80. https://doi.org/10.1016/j.jhazmat.2018.07.116.
82. Liu M, Wang Z, Zong S, Chen H, Zhu D, Wu L, Hu G, Cui Y (2014) SERS detection and removal of mercury (II)/silver (I) using oligonucleotide-functionalized core/shell magnetic silica sphere@ Au nanoparticles. *ACS Appl Mater Interf* 6:7371–7379. https://doi.org/10.1021/am5006282.
83. Rizwan M, Singh M, Mitra CK, Morve RK (2014) Ecofriendly application of nanomaterials: Nanobioremediation. *J Nanopart* 2014:431787. https://doi.org/10.1155/2014/431787.
84. Salvadori MR, Ando RA, Nascimento CA, Corrêa B (2017) Dead biomass of Amazon yeast: A new insight into bioremediation and recovery of silver by intracellular synthesis of nanoparticles. *J Environ Sci Health Part A* 52:1112–1120. https://doi.org/10.1080/10934529.2017.1340754.
85. Das SK, Das AR, Guha AK (2009) Gold nanoparticles: Microbial synthesis and application in water hygiene management. *Langmuir* 25:8192–8199. https://doi.org/10.1016/j.hydromet.2010.03.016.
86. Órdenes-Aenishanslins NA, Saona LA, Durán-Toro VM, Monrás JP, Bravo DM, Pérez-Donoso JM (2014) Use of titanium dioxide nanoparticles biosynthesized by Bacillus mycoides in quantum dot sensitized solar cells. *Microb Cell Fact* 13:90. https://doi.org/10.1186/s12934-014-0090-7.
87. Glatstein DA, Bruna N, Gallardo-Benavente C, Bravo D, Carro Perez ME, Francisca FM, PérezDonoso JM (2018) Arsenic and cadmium bioremediation by Antarctic bacteria capable of biosynthesizing CdS fluorescent nanoparticles. *J Environ Eng* 144:04017107. https://doi.org/10.1061/(ASCE)EE.1943-7870.0001293.
88. Ambashta RD, Sillanpää M (2010) Water purification using magnetic assistance: A review. *J Hazard Mater* 180:38–49. https://doi.org/10.1016/j.jhazmat.2010.04.105.
89. Kabary H, Eida MF, Attia MM, Awad N, Easa SM (2017) Magnetotactic characterization and environmental application P. aeruginosa kb1 isolate. *Annu Res Rev Biol* 20:1–10. https://doi.org/10.9734/arrb/2017/37737.
90. Ali I, Peng C, Khan ZM, Naz I, Sultan M (2018) An overview of heavy metal removal from wastewater using magnetotactic bacteria. *J Chem Technol Biotechnol* 93:2817–2832. https://doi.org/10.1002/jctb.5648.

91. Yan L, Da H, Zhang S, López VM, Wang W (2017) Bacterial magnetosome and its potential application. *Microbiol Res* 203:19–28. https://doi.org/10.1016/j.micres.2017.06.005.
92. Zhu J, Li Q, Jiao W, Jiang H, Sand W, Xia J, Liu X, Qin W, Qiu G, Hu Y, Chai L (2012) Adhesion forces between cells of Acidithiobacillus ferrooxidans, Acidithiobacillus thiooxidans or Leptospirillum ferrooxidans and chalcopyrite. *Colloids Surf B* 94:95–100. https://doi.org/10.1016/j.colsurfb.2012.01.022.
93. Das N (2010) Recovery of precious metals through biosorption—A review. *Hydrometallurgy* 103:180–189. https://doi.org/10.1016/j.hydromet.2010.03.016.
94. Liu J, Zhou PG (2008) Application and development of magnetotactic bacteria in heavy metal wastewater treatment. *J Environ Sci Technol* 21:60–66.
95. Wang X, Li Y, Zhao J, Yao H, Chu S, Song Z, He Z, Zhang W (2020) Magnetotactic bacteria: Characteristics and environmental applications. *Front Environ Sci Eng* 14:1–14. https://doi.org/10.1007/s11783-020-1235-z.
96. Darwesh OM, Moawad H, El-Rahim WM, Barakat OS, Sedik MZ (2014) Bioremediation of textile reactive blue (RB) azo dye residues in wastewater using experimental prototype bioreactor. *Res J Pharm Biol Chem Sci* 5:1203–1219.
97. Sadhasivam S, Vinayagam V, Balasubramaniyan M (2020) Recent advancement in biogenic synthesis of iron nanoparticles. *J Mol Struct* 1217:128372. https://doi.org/10.1016/j.molstruc.2020.128372.
98. Sharma H, Shirkot P (2019) Bioremediation of azo dyes using biogenic iron nanoparticles. *J Microbiol Exp* 7:12–15. https://doi.org/10.15406/jmen.2019.07.00232.
99. Noman M, Shahid M, Ahmed T, Niazi MB, Hussain S, Song F, Manzoor I (2020) Use of biogenic copper nanoparticles synthesized from a native Escherichia sp. as photocatalysts for azo dye degradation and treatment of textile effluents. *Environ Pollut* 257:113514. https://doi.org/10.1016/j.envpol.2019.113514.
100. Matter IA, Darwesh OM, Matter HA (2021) Nanosensors for herbicides monitoring in soil. In: *Nanomaterials for Soil Remediation*. Elsevier. pp. 221–237. https://doi.org/10.1016/B978-0-12-822891-3.00011-6.
101. Arora K (2018) Advances in nano based biosensors for food and agriculture. In: *Nanotechnology, food security and water treatment*. Springer, Cham. pp. 1–52. https://doi.org/10.1007/978-3-319-70166-0_1.
102. Sinha R, Dwivedi S, Kumar A, Srivastava P (2020) Materials in bio-sensing of water pollutants. In: *Sensors in Water Pollutants Monitoring: Role of Material*. Springer, Singapore. pp. 187–211. https://doi.org/10.1007/978-981-15-0671-0_11.
103. Ejeian F, Etedali P, Mansouri-Tehrani HA, Soozanipour A, Low ZX, Asadnia M, Taheri-Kafrani A, Razmjou A (2018) Biosensors for wastewater monitoring: A review. *Biosens Bioelectron* 118:66–79. https://doi.org/10.1016/j.bios.2018.07.019.
104. Zare HR, Shekari Z (2019) Types of monitoring biosensor signals. In: *Electrochemical Biosensors*. Elsevier, Amsterdam. pp. 135–166. https://doi.org/10.1016/b978-0-12-816491-4.00006-1.
105. Ghaffar N, Farrukh MA, Naz S. (2020) Applications of nanobiosensors in agriculture. In: *Nanoagronomy*. Springer, Cham. pp. 179–196. https://doi.org/10.1007/978-3-030-41275-3_10.
106. Salouti M, Derakhshan FK (2020) Biosensors and nanobiosensors in environmental applications. In: *Biogenic Nano-particles and Their Use in Agro-Ecosystems*. Springer, Singapore. pp. 515–591. https://doi.org/10.1007/978-981-15-2985-6_26.
107. Saini RK, Bagri LP, Bajpai AK (2017) Smart nanosensors for pesticide detection. In: *New Pesticides and Soil Sensors*. Academic Press, Cambridge. pp. 519–559. https://doi.org/10.1016/B978-0-12-804299-1.00015-1.
108. Chen Y, Xianyu Y, Wu J, Yin B, Jiang X (2016) Click chemistry-mediated nanosensors for biochemical assays. *Theranostics* 6:969. https://doi.org/10.7150/thno.14856.
109. Chen YP, qiang Zou M, Qi C, Xie MX, Wang DN, Wang YF, Xue Q, Li JF, Chen Y (2013) Immunosensor based on magnetic relaxation switch and biotin–streptavidin system for the detection of kanamycin in milk. *Biosens Bioelectron* 39:112–117. https://doi.org/10.1016/j.bios.2012.06.056.
110. Yang H, Tian Z, Wang J, Yang S (2012) A magnetic resonance imaging nanosensor for Hg (II) based on thymidine-functionalized supermagnetic iron oxide nanoparticles. *Sens Actuat B Chem* 161:429–433. https://doi.org/10.1016/j.snb.2011.10.055.
111. Sun Y, Fang L, Wan Y, Gu Z (2018) Pathogenic detection and phenotype using magnetic nanoparticle-urease nanosensor. *Sens Actuat B* 259:428–432. https://doi.org/10.1016/j.snb.2017.12.095.
112. Loh KS, Lee YH, Musa A, Salmah AA, Zamri I (2008) Use of Fe_3O_4 nanoparticles for enhancement of biosensor response to the herbicide 2, 4-dichlorophenoxyacetic acid. *Sensors* 8:5775–5791. https://doi.org/10.3390/s8095775.

113. Songa EA, Arotiba OA, Owino JH, Jahed N, Baker PG, Iwuoha EI (2009) Electrochemical detection of glyphosate herbicide using horseradish peroxidase immobilized on sulfonated polymer matrix. *Bioelectrochemistry* 75:117–123. https://doi.org/10.1016/j.bioelechem.2009.02.007.
114. Yang L, Wang GC, Liu YJ, An JJ, Wang M (2013) Development of a stable biosensor based on a SiO_2 nanosheet–Nafion–modified glassy carbon electrode for sensitive detection of pesticides. *Anal Bioanal Chem* 405:2545–2552. https://doi.org/10.1007/s00216-012-6634-y.
115. Hasanin MS, Darwesh OM, Matter IA, El-Saied H (2019) Isolation and characterization of noncellulolytic Aspergillus flavus EGYPTA5 exhibiting selective ligninolytic potential. *Biocatal Agric Biotechnol* 17:160–167. https://doi.org/10.1016/j.bcab.2018.11.012.
116. Srivastava N, Srivastava M, Manikanta A, Singh P, Ramteke PW, Mishra PK (2017) Nanomaterials for biofuel production using lignocellulosic waste. *Environ Chem Lett* 15:179–184. https://doi.org/10.1007/s10311-017-0622-6.
117. Singh SB, Tandon PK (2014) Catalysis: A brief review on nano-catalyst. *J Energ Chem Eng* 2: 106–115.
118. Chorkendorff I, Niemantsverdriet JW (2017) *Concepts of Modern Catalysis and Kinetics*. John Wiley & Sons, Hoboken, 524 p. https://doi.org/10.1002/3527602658.
119. Saoud K (2018) Nanocatalyst for biofuel production: A Review. In: *Green Nanotechnology for Biofuel Production*. Springer, Cham. pp. 39–62. https://doi.org/10.1007/978-3-319-75052-1_4.
120. Farrokheh A, Tahvildari K, Nozari M (2020) Biodiesel production from the Chlorella vulgaris and Spirulina platensis microalgae by electrolysis using CaO/KOH-Fe_3O_4 and KF/KOH-Fe_3O_4 as magnetic nanocatalysts. *Biomass Convers Biorefin* 12:403–417. https://doi.org/10.1007/s13399-020-00688-z.
121. Sanusi IA, Suinyuy TN, Lateef A, Kana GE (2020) Effect of nickel oxide nanoparticles on bioethanol production: Process optimization, kinetic and metabolic studies. *Process Biochem* 92:386–400. https://doi.org/10.1016/j.procbio.2020.01.029.
122. Matter IA, Darwesh OM, El-baz FK (2016) Using the natural polymer chitosan in harvesting Scenedesmus species under different concentrations and cultural pH values. *Int J Pharm Biol Sci* 7: 254–260. https://doi.org/10.22376/ijpbs.2016.7.4.b254-260.
123. Matter IA, Bui VK, Jung M, Seo JY, Kim YE, Lee YC, Oh YK (2019) Flocculation harvesting techniques for microalgae: A review. *Appl Sci* 9:3069. https://doi.org/10.3390/app9153069.
124. Eida MF, Darwesh OM, Matter IA (2018) Cultivation of oleaginous microalgae Scenedesmus obliquus on secondary treated municipal wastewater as growth medium for biodiesel production. *J Ecol Eng* 19:38–51. https://doi.org/10.12911/22998993/91274.
125. Prasad R, Bhattacharyya A, Nguyen QD (2017) Nanotechnology in sustainable agriculture: Recent developments, challenges, and perspectives. *Front Microbiol* 8:1014. https://doi.org/10.3389/fmicb.2017.01014.
126. Chhipa H (2019) Applications of nanotechnology in agriculture. In: *Methods in Microbiology*, vol 46. Elsevier, Amsterdam. pp. 115–142. https://doi.org/10.1016/bs.mim.2019.01.002.
127. Mishra S, Singh A, Keswani C, Singh HB (2014) Nanotechnology: Exploring potential application in agriculture and its opportunities and constraints. *Biotech Today* 4:9–14. https://doi.org/10.5958/2322-0996.2014.00011.8.
128. Salama HM (2012) Effects of silver nanoparticles in some crop plants, common bean (Phaseolus vulgaris L.) and corn (Zea mays L.). *Int Res J Biotechnol* 3:190–197.
129. Sharma P, Bhatt D, Zaidi MG, Saradhi PP, Khanna PK, Arora S (2012) Silver nanoparticle-mediated enhancement in growth and antioxidant status of Brassica juncea. *Appl Biochem Biotechnol* 167: 2225–2233. https://doi.org/10.1007/s12010-012-9759-8.
130. Taha RA, Hassan MM, Ibrahim EA, Abou Baker NH, Shaaban EA (2016) Carbon nanotubes impact on date palm in vitro cultures. *Plant Cell Tissue Organ Cult* 127:525–534. https://doi.org/10.1007/s11240-016-1058-6.
131. Azeez L, Adejumo AL, Lateef A, Adebisi SA, Adetoro RO, Adewuyi SO, Tijani KO, Olaoye S (2019) Zero-valent silver nanoparticles attenuate Cd and Pb toxicities on Moringa oleifera via immobilization and induction of phytochemicals. *Plant Physiol Biochem* 139:283–292. https://doi.org/10.1016/j.plaphy.2019.03.030.
132. Azeez L, Lateef A, Wahab AA, Rufai MA, Salau AK, Ajayi IO, Ajayi EM, Maryam AK, Adebisi B (2019) Phytomodulatory effects of silver nanoparticles on Corchorus olitorius: Its antiphytopathogenic and hepatoprotective potentials. *Plant Physiol Biochem* 136:109–117. https://doi.org/10.1016/j.plaphy.2018.12.006.

133. Azeez L, Adejumo AL, Simiat OM, Lateef A (2020) Influence of calcium nanoparticles (CaNPs) on nutritional qualities, radical scavenging attributes of Moringa oleifera and risk assessments on human health. *J Food Meas Charact* 14:2185–2195. https://doi.org/10.1007/s11694-020-00465-6.
134. Hussein HA, Darwesh OM, Mekki BB, El-Hallouty SM (2019) Evaluation of cytotoxicity, biochemical profile and yield components of groundnut plants treated with nano-selenium. *Biotechnol Rep* 24:00377. https://doi.org/10.1016/j.btre.2019.e00377.
135. Adhikari T, Sarkar D, Mashayekhi H, Xing B (2016) Growth and enzymatic activity of maize (Zea mays L.) plant: Solution culture test for copper dioxide nano particles. *J Plant Nutr* 39:99–115. https://doi.org/10.1080/01904167.2015.1044012.
136. Khodakovskaya MV, De Silva K, Biris AS, Dervishi E, Villagarcia H (2012) Carbon nanotubes induce growth enhancement of tobacco cells. *ACS Nano* 6:2128–2135. https://doi.org/10.1021/nn204643g.
137. Shebl A, Hassan AA, Salama DM, El-Aziz A, AbdElwahed MS (2019) Green synthesis of nanofertilizers and their application as a foliar for Cucurbita pepo L. *J Nanomat* 2019:3476347. https://doi.org/10.3762/bxiv.2019.45.v1.
138. Tarafdar JC, Raliya R, Rathore I (2012) Microbial synthesis of phosphorous nanoparticle from tri-calcium phosphate using Aspergillus tubingensis TFR-5. *J Bionanosci* 6:84–89. https://doi.org/10.1166/jbns.2012.1077 [Google Scholar]
139. Sohair EED, Abdall AA, Amany AM, Hossain MF, Houda RA (2018) Evaluation of nitrogen, phosphorus and potassium nano-fertilizers on yield, yield components and fiber properties of Egyptian cotton (Gossypium barbadense L.). *J Plant Sci Crop Prot* 1:208.
140. Yusof HM, Mohamad R, Zaidan UH (2019) Microbial synthesis of zinc oxide nanoparticles and their potential application as an antimicrobial agent and a feed supplement in animal industry: A review. *J Anim Sci Biotechnol* 10:57. https://doi.org/10.1186/s40104-019-0368-z.
141. Raliya R, Biswas P, Tarafdar JC (2015) TiO_2 nanoparticle biosynthesis and its physiological effect on mung bean (Vigna radiata L.). *Biotechnol Rep* 5:22–26. https://doi.org/10.1016/j.btre.2014.10.009.
142. Masum M, Islam M, Siddiqa M, Ali KA, Zhang Y, Abdallah Y, Ibrahim E, Qiu W, Yan C, Li B (2019) Biogenic synthesis of silver nanoparticles using Phyllanthus emblica fruit extract and its inhibitory action against the pathogen Acidovorax oryzae strain RS-2 of rice bacterial brown stripe. *Front Microbiol* 10:820. https://doi.org/10.3389/fmicb.2019.00820.
143. Suman TY, Rajasree SR, Kanchana A, Elizabeth SB (2013) Biosynthesis, characterization and cytotoxic effect of plant mediated silver nanoparticles using Morinda citrifolia root extract. *Colloids Surf B* 106:74–78. https://doi.org/10.1016/j.colsurfb.2013.01.037.
144. Elshahawy I, Abouelnasr HM, Lashin SM, Darwesh OM (2018) First report of Pythiuma phanidermatum infecting tomato in Egypt and its control using biogenic silver nanoparticles. *J Plant Prot Res* 58:137–151. https://doi.org/10.24425/122929.
145. Ali M, Ahmed T, Wu W, Hossain A, Hafeez R, Masum IM, Wang Y, An Q, Sun G, Li B (2020) Advancements in plant and microbe-based synthesis of metallic nanoparticles and their antimicrobial activity against plant pathogens. *Nanomater* 10:1146. https://doi.org/10.3390/nano10061146.
146. Rafique M, Sadaf I, Rafique MS, Tahir MB (2017) A review on green synthesis of silver nanoparticles and their applications. *Artif Cells Nanomed Biotechnol* 45:1272–1291. https://doi.org/10.1080/21691401.2016.1241792.
147. Fouda A, Hassan SE, Abdo AM, El-Gamal MS (2020) Antimicrobial, antioxidant and larvicidal activities of spherical silver nanoparticles synthesized by endophytic Streptomyces spp. *Biol Trace Elem Res* 195:707–724. https://doi.org/10.1007/s12011-019-01883-4.
148. Lateef A, Ojo SA, Akinwale AS, Azeez L, Gueguim-Kana EB, Beukes LS (2015) Biogenic synthesis of silver nanoparticles using cell-free extract of Bacillus safensis LAU 13: Antimicrobial, free radical scavenging and larvicidal activities. *Biologia* 70:1295–306. https://doi.org/10.1515/biolog-2015-0164.
149. Khan MR, Rizvi TF (2014) Nanotechnology: Scope and application in plant disease management. *Plant Pathol J* 13:214–231. https://doi.org/10.3923/ppj.2014.214.231.
150. Liu Y, Yan L, Heiden P, Laks P (2001) Use of nanoparticles for controlled release of biocides in solid wood. *J Appl Polym Sci* 79:458–465.
151. Lai F, Wissing SA, Müller RH, Fadda AM (2006) Artemisia arborescens L essential oil-loaded solid lipid nanoparticles for potential agricultural application: Preparation and characterization. *AAPS PharmSciTech* 7:E10.
152. Wen LX, Li ZZ, Zou HK, Liu AQ, Chen JF (2005) Controlled release of avermectin from porous hollow silica nanoparticles. *Pest Manag Sci* 61:583–590.

153. Li Z-Z, Xu S-A, Wen L-X, Liu F, Liu A-Q, Wang Q, Sun H-Y, Yu W, Chen J-F (2006) Controlled release of avermectin from porous hollow silica nanoparticles: Influence of shell thickness on loading efficiency, UV-shielding property and release. *J Control Release* 111:81–88.
154. Li ZZ, Chen JF, Liu F, Liu AQ, Wang Q, Sun HY, Wen LX (2007) Study of UV-shielding properties of novel porous hollow silica nanoparticle carriers for avermectin. *Pest Manag Sci* 63:241–246.
155. Song M-R, Cui S-M, Gao F, Liu Y-R, Fan C-L, Lei T-Q, Liu D-C (2012) Dispersible silica nanoparticles as carrier for enhanced bioactivity of chlorfenapyr. *J Pestic Sci* 37:258–260.
156. Nguyen HM, Hwang IC, Park JW, Park HJ (2012) Photoprotection for deltamethrin using chitosan-coated beeswax solid lipid nanoparticles. *Pest Manag Sci* 68:1062–1068.
157. Nguyen H, Hwang I, Park J, Park H (2012) Enhanced payload and photo-protection for pesticides using nanostructured lipid carriers with corn oil as liquid lipid. *J Microencapsul* 29:596–604.
158. Yang F-L, Li X-G, Zhu F, Lei C-L (2009) Structural characterization of nanoparticles loaded with garlic essential oil and their insecticidal activity against Tribolium castaneum (Herbst)(Coleoptera: Tenebrionidae). *J Agric Food Chem* 57:10156–10162.
159. Zhang J, Li M, Fan T, Xu Q, Wu Y, Chen C, Huang Q (2013) Construction of novel amphiphilic chitosan copolymer nanoparticles for chlorpyrifos delivery. *J Polym Res* 20:107.
160. Feng B-H, Peng L-F (2012) Synthesis and characterization of carboxymethyl chitosan carrying ricinoleic functions as an emulsifier for azadirachtin. *Carbohydr Polym* 88:576–582.
161. Rani PU, Madhusudhanamurthy J, Sreedhar B (2014) Dynamic adsorption of α-pinene and linalool on silica nanoparticles for enhanced antifeedant activity against agricultural pests. *J Pest Sci* 87:191–200.
162. Lu W, Lu ML, Zhang QP, Tian YQ, Zhang ZX, Xu HH (2013) Octahydrogenated retinoic acid-conjugated glycol chitosan nanoparticles as a novel carrier of azadirachtin: Synthesis, characterization, and in vitro evaluation. *J Polym Sci Part A Polym Chem* 51:3932–3940.
163. Wang Y, Cui H, Sun C, Zhao X, Cui B (2014) Construction and evaluation of controlled-release delivery system of Abamectin using porous silica nanoparticles as carriers. *Nanoscale Res Lett* 9:2490.
164. Kumar S, Bhanjana G, Sharma A, Sidhu M, Dilbaghi N (2014) Synthesis, characterization and on field evaluation of pesticide loaded sodium alginate nanoparticles. *Carbohydr Polym* 101:1061–1067.
165. Nguyen TNQ, Le VA, Hua QC, Nguyen TT (2015) Enhancing Insecticide Activity of Anacardic Acid by Intercalating it into MgAl Layered Double Hydroxides Nanoparticles; Institut für Abfallwirtschaft und Altlasten, Fakultät Umweltwissenschaften, Technische Universität Dresden: Dresden, Germany.
166. Jia X, Sheng W-B, Li W, Tong Y-B, Liu Z-Y, Zhou F (2014) Adhesive polydopamine coated avermectin microcapsules for prolonging foliar pesticide retention. *ACS Appl Mater Interfaces* 6:19552–19558.
167. Wibowo D, Zhao C-X, Peters BC, Middelberg AP (2014) Sustained release of fipronil insecticide in vitro and in vivo from biocompatible silica nanocapsules. *J Agric Food Chem* 62:12504–12511.
168. Sheng W-B, Li W, Zhang G-X, Tong Y-B, Liu Z-Y, Jia X (2015) Study on the UV-shielding and controlled-release properties of a polydopamine coating for avermectin. *New J Chem* 39:2752–2757.
169. Meredith AN, Harper B, Harper SL (2016) The influence of size on the toxicity of an encapsulated pesticide: A comparison of micron-and nano-sized capsules. *Environ Int* 86:68–74.
170. Liu X, He B, Xu Z, Yin M, Yang W, Zhang H, Cao J, Shen J (2015) A functionalized fluorescent dendrimer as a pesticide nanocarrier: Application in pest control. *Nanoscale* 7:445–449.
171. Glinski A, Liebel S, Pelletier È, Voigt CL, Randi MAF, Campos SX, Oliveira Ribeiro CA, Filipak Neto F (2016) Toxicological interactions of silver nanoparticles and organochlorine pesticides in mouse peritoneal macrophages. *Toxicol Mech Methods* 26:251–259.
172. Campos EV, Proença PL, Oliveira JL, Melville CC, Vechia JF, Andrade DJ, Fraceto LF (2018) Chitosan nanoparticles functionalized with β-cyclodextrin: A promising carrier for botanical pesticides. *Sci Rep* 8:2067.
173. Liang J, Yu M, Guo L, Cui B, Zhao X, Sun C, Wang Y, Liu G, Cui H, Zeng Z (2018) Bioinspired development of P (St–MAA)–avermectin nanoparticles with high affinity for foliage to enhance folia retention. *J Agric Food Chem* 66:6578–6584.
174. Jenne M, Kambham M, Tollamadugu NP, Karanam HP, Tirupati MK, Balam RR, Shameer S, Yagireddy M (2018) The use of slow releasing nanoparticle encapsulated Azadirachtin formulations for the management ofCaryedon serratusO.(groundnut bruchid). *IET Nanobiotechnol* 12:963–967.
175. Zhang H, Qin H, Li L, Zhou X, Wang W, Kan C (2018) Preparation and characterization of controlled-release avermectin/castor oil-based polyurethane nanoemulsions. *J Agric Food Chem* 66:6552–6560.
176. Paulraj MG, Ignacimuthu S, Gandhi MR, Shajahan A, Ganesan P, Packiam SM, Al-Dhabi NA (2017) Comparative studies of Tripolyphosphate and Glutaraldehyde cross-linked chitosan-botanical pesticide nanoparticles and their agricultural applications. *Int J Boil Macromol* 104(Pt B):1813–1819.

177. Suresh U, Murugan K, Panneerselvam C, Rajaganesh R, Roni M, Al-Aoh HAN, Trivedi S, Rehman H, Kumar S, Higuchi A (2018) Suaeda maritima-based herbal coils and green nanoparticles as potential biopesticides against the dengue vector Aedes aegypti and the tobacco cutworm Spodoptera litura. *Physiol Mol Plant Pathol* 101:225–235.
178. Oliveira JLD, Campos ENV, Pereira AE, Pasquoto T, Lima R, Grillo R, Andrade DJD, Santos FAD, Fraceto LF (2018) Zein nanoparticles as eco-friendly carrier systems for botanical repellents aiming sustainable agriculture. *J Agric Food Chem* 66:1330–1340.
179. De Oliveira JL, Campos EVR, Pereira AES, Nunes LE, da Silva CC, Pasquoto T, Lima R, Smaniotto G, Polanczyk RA, Fraceto LF (2018) Geraniol encapsulated in chitosan/gum arabic nanoparticles: A promising system for pest management in sustainable agriculture. *J Agric Food Chem* 66:5325–5334.
180. Yang Y, Cheng J, Garamus VM, Li N, Zou A (2018) Preparation of an environmentally friendly formulation of the insecticide nicotine hydrochloride through encapsulation in chitosan/tripolyphosphate nanoparticles. *J Agric Food Chem* 66:1067–1074.
181. Ahmadi Z, Saber M, Akbari A, Mahdavinia GR (2018) Encapsulation of Satureja hortensis L. (Lamiaceae) in chitosan/TPP nanoparticles with enhanced acaricide activity against Tetranychus urticae Koch (Acari: Tetranychidae). *Ecotoxicol Environ Saf* 161:111–119.
182. Campos EV, Proença PL, Oliveira JL, Pereira AE, Ribeiro LN, Fernandes FO, Gonçalves KC, Polanczyk RA, Pasquoto-Stigliani T, Lima R (2018) Carvacrol and linalool co-loaded in β-cyclodextrin-grafted chitosan nanoparticles as sustainable biopesticide aiming pest control. *Sci Rep* 8:7623.
183. Kaziem AE, Gao Y, Zhang Y, Qin X, Xiao Y, Zhang Y, You H, Li J, He S (2018) α-Amylase triggered carriers based on cyclodextrin anchored hollow mesoporous silica for enhancing insecticidal activity of avermectin against Plutella xylostella. *J Hazard Mater* 359:213–221.
184. Wang X, Ma X, Huang P, Wang J, Du T, Du X, Lu X (2018) Magnetic Cu-MOFs embedded within graphene oxide nanocomposites for enhanced preconcentration of benzenoid-containing insecticides. *Talanta* 181:112–117.
185. Xu C, Cao L, Liu T, Chen H, Li Y (2023) pH-responsive copper-doped ZIF8 MOF nanoparticles for enhancing the delivery and translocation of pesticides in wheat plants. *Environ Sci Nano* 10:2578–2590. https://doi.org/10.1039/d3en00300k.
186. Ansari S, Ficiarà E, Ruffinatti F, Stura I, Argenziano M, Abollino O, Cavalli R, Guiot C, D'Agata F (2019) Magnetic iron oxide nanoparticles: Synthesis, characterization, and functionalization for biomedical applications in the central nervous system. *Materials* 12:465.
187. Mohammed L, Gomaa HG, Ragab D, Zhu J (2017) Magnetic nanoparticles for environmental and biomedical applications: A review. *Particuology* 30:1–14.
188. Singh NB, Nagpal G, Agrawal S (2018) Rachna water purification by using adsorbents: A review. *Environ Technol Innov* 11:187–240.
189. Zhu C, Zhao Q, Meng G, Wang X, Hu X, Han F, Lei Y (2020) Silver nanoparticle-assembled micro-bowl arrays for sensitive SERS detection of pesticide residue. *Nanotechnology* 31:205303.
190. Yuvaraj N, Srihari K, Dhiman G, Somasundaram K, Sharma A, Rajeskannan SMGSMA, Soni M, Gaba GS, AlZain MA, Masud M (2021) Nature-inspired-based approach for automated cyberbullying classification on multimedia social networking. *Math Probl Eng* 2021:1–12.
191. Mahesh KV, Singh SK, Gulati M (2014) A comparative study of top-down and bottom-up approaches for the preparation of nanosuspensions of glipizide. *Powder Technol* 256:436–449.
192. Kour D, Kaur T, Devi R, Yadav A, Singh M, Joshi D, Singh J, Suyal DC, Kumar A, Rajput VD, Yadav AN (2021). Beneficial microbiomes for bioremediation of diverse contaminated environments for environmental sustainability: Present status and future challenges. *Environ Sci Pollut Res* 28:24917–24939.
193. Ren X, Li C, Ma X, Chen F, Wang H, Sharma A, Gaba GS, Masud M (2021) Design of multi-information fusion based intelligent electrical fire detection system for green buildings. *Sustainability* 13(6):3405.
194. Singh G, Pruncu CI, Gupta MK, Mia M, Khan AM, Jamil M, Pimenov DY, Sen B, Sharma VS (2019) Investigations of machining characteristics in the upgraded MQL-assisted turning of pure titanium alloys using evolutionary algorithms. *Materials* 12(6):999.

10 Technology Transfer and Commercialisation Aspects of Biogenic Nanoparticles

Tanisha Kishan, M. Pooja, S. Simran, Ramya Sree Vasamsetty, Abinaya Krishnan, Mwalimu Raphael Jilani, Maheswari Vinodkumar, and Azhagu Saravana Babu Packirisamy

10.1 INTRODUCTION

Biogenic nanoparticles (NPs) are nanoscale particles produced by living organisms through natural biochemical processes, typically ranging from 1 to 100 nm. These NPs exhibit unique physical and chemical properties based on their biological origin. They are produced by a wide range of microorganisms such as bacteria, fungi, plants, and animals. Most commonly found NPs are AgNPs, CuNPs, AuNPs, FeONPs, ZnNPs, and TiO_2NPs [1]. These NPs have unparalleled properties and features that render them valuable for a multitude of applications. Biogenic NPs are tiny with diverse shapes. Their size, form, and surface properties influence interactions and stability when coated with biomolecules like proteins or polysaccharides. Nanoscale dimensions grant them an outstandingly high surface region to volume proportion, enhancing their reactivity and making them perfect for applications requiring high surface activity. Moreover, they display a broad range of chemical compositions, including elements like gold, silver, platinum, and various metal oxides typically present within their matrices [2]. The incorporation of biomolecules onto their surfaces imparts unique chemical functionalities, influencing their reactivity and potential uses. Furthermore, their stability is frequently enhanced by the introduction of capping or stabilising agents onto their surface preventing clumping and maintaining their colloidal stability [3]. Additionally, these NPs can be engineered to carry specific surface charges, dictating their fundamental interaction with biological entities (Figure 10.1).

The commercialisation and technology transfer of biogenic NPs present exciting opportunities for industry-academia collaborations. As research continues to uncover novel synthesis methods and applications, the potential for scaling up production and integrating these NPs into existing technologies grows exponentially. However, challenges remain in standardising production processes and ensuring consistent quality for large-scale applications.

Future research should focus on optimising synthesis protocols, exploring new biological sources, and expanding the range of applications. As we continue to unlock the full potential of biogenic NPs, their impact on sustainable development and technological innovation is poised to be transformative.

Development in NPs research has led to their increasing applications [4] in various fields. For example, biogenetic NPs can be used to deliver targeted drugs, produce imaging, and provide therapy. Biogenetic NPs are biocompatible and have controllable surface chemicals, making them perfect for diagnostics. In diagnostics, biogenic NP could be used to detect biomarkers or pathogens with high optical and magnetic sensitivity. In catalysts, they could be used to improve yield of crops, control pests, and improve soil fertility. In the industrial sector, they are useful catalysts

DOI: 10.1201/9781003487142-10

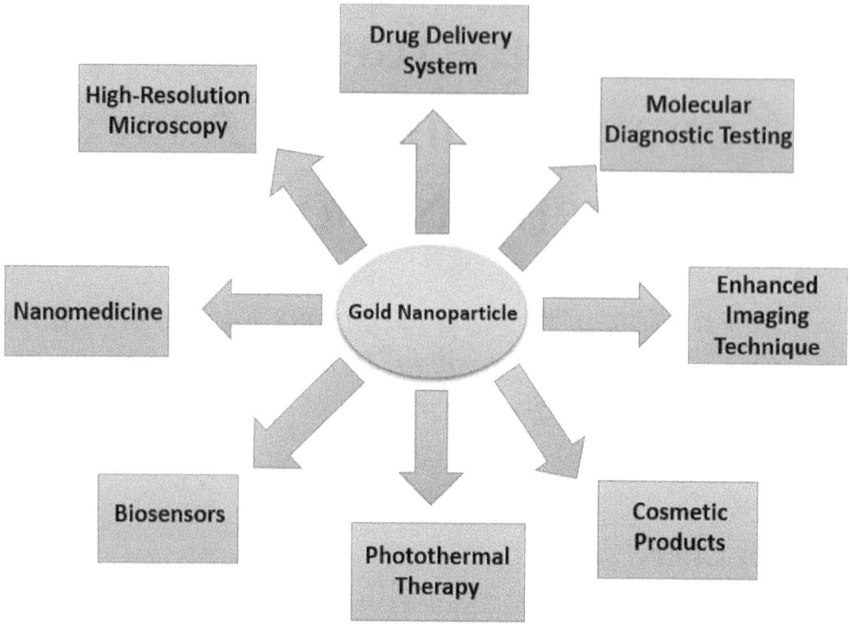

FIGURE 10.1 Commercialisation of gold nanoparticles.

for hydrogenation and the decomposition of pollutants. Their sustainable synthesis methods are in line with the increasing need for environmentally friendly manufacturing practices. In environmental remediation, biogenic NP are highly effective in removing pollutants from air and water, as well as pathogens, helping to create cleaner environments and improve public health. In agriculture, nano-phenolics formulated with biogenics can help to reduce environmental contamination, improve plant nutrient uptake, and foster innovation. Hence, biogenics offer a versatile, sustainable solution to address diverse challenges with a minimal environmental impact.

10.2 METHODS OF BIOGENIC NANOPARTICLE SYNTHESIS

10.2.1 Biological and Green Synthesis Methods

Biological synthesis of biogenic NPs utilises living organisms, such as plants, bacteria, and enzymes, for eco-friendly and sustainable NP production. This includes two main methods.

10.2.1.1 Bioreduction

The bioreduction approach of synthesis of biogenic NPs involves using biological agents, such as plant extracts or microorganisms, to reduce metal ions into NPs. This eco-friendly approach relies on the inherent reducing properties of biological entities to create NPs with controlled size and shape [5]. The bioreduction process is often considered greener compared to traditional chemical methods, as it eliminates the need for harsh chemicals and high temperatures. This method is widely explored for its simplicity, cost-effectiveness, and potential scalability in various applications, including medicine, catalysis, and electronics [6].

10.2.1.2 Biosorption

Biosorption is a technique within biogenic NP synthesis that involves using biological materials to adsorb metal ions from a solution, followed by their reduction to form NPs [7]. In this method, biological entities, such as bacteria, fungi, or plant extracts, serve as natural sorbents capable of binding metal ions onto their surfaces. For example, biomass of *Bacillus subtilis* is used to adsorb metal ions from a solution and subsequently reduce them, leading to the formation of NPs with specifically engineered properties. Once loaded with metal ions, these biological materials facilitate the reduction process, leading to the formation of biogenic NPs. The biosorption method is known for its simplicity, cost-effectiveness, and eco-friendly nature, as it utilises living organisms or their derivatives to both extract and reduce metal ions, eliminating the need for harsh chemicals and energy-intensive processes. This move is valuable for producing NPs with controlled properties for several applications, including medicine and environmental remediation.

10.2.2 Advantages of Biogenic Synthesis over Conventional Methods

Biogenic synthesis of NPs presents several clear-cut advantages over conventional methods, marking it a promising and environmentally conscious approach. One notable benefit is its inherent eco-friendliness, achieved by utilising natural materials like plant extracts, microorganisms, or enzymes, thereby reducing reliance on hazardous chemicals and minimising environmental impact. This method aligns with sustainable practices by tapping into renewable resources, contributing to a greener and more ethical production process. Biogenic synthesis operates under relatively low energy requirements compared to conventional techniques, making it a more energy-efficient alternative.

The controlled size and shape of NPs and improved biocompatibility achieved with biological entities make them particularly appropriate for medical applications. The simplicity and potential scalability of many biogenic methods make them cost-effective and versatile for many applications in medicine, catalysis, electronics, and environmental remediation [8]. Overall, biogenic synthesis stands out as an innovative and sustainable paradigm shift in NP production, offering many advantages that address both scientific and environmental aspects.

10.3 CHARACTERISATION

Biogenic NPs possess specific properties and features, which make them valued for a wide range of applications. This is accomplished by using various techniques and equipment. The most commonly used methods employed in structural analysis are **X-ray diffraction (XRD)**, which determines the phase composition and crystal structure by analysing the diffraction pattern of X-rays interacting with the material which are important for understanding the arrangement of atoms at nanoscale, and **transmission electron microscopy (TEM)**, which provides high-resolution images of internal structure of nanomaterials with exceptional clarity by using beam of electrons to pass through a thin sample and creating detailed images at atomic and molecular levels. This includes information about crystal defects, grain boundaries, and NP distribution. **Raman spectroscopy** is based on the inelastic scattering of monochromatic light interacting with the nanomaterial, providing information about vibrational modes and crystal structures. Nuclear magnetic resonance (NMR), a non-destructive analytical method, ascertains the atomic and molecular structure of nanomaterials and is sensitive to molecular bonds, particularly those based on carbon, such as graphene and carbon nanotubes [9]. This is achieved by subjecting the material to a strong magnetic field and radiofrequency radiation, providing insights about material's composition, chemical environment of nuclei, connectivity, and dynamics.

Biogenic nanomaterials typically have dimensions in the nanometre scale, with at least one dimension measuring 1–100 nm that can be determined by TEM which is not only employed for structural analysis but also for size and morphology characterisation, determining NP size, shape, and distribution. It is especially useful for visualising nanomaterials with intricate morphologies, such as nanorods, nanowires, and quantum dots. **Atomic force microscopy (AFM)** provides topographical information with high spatial resolution, allowing visualisation of surface features, measure particle heights, and obtain 3D images by systematically moving a fine-tip probe across the nanomaterial surface and gauging the interactions between the tip and the material [3]. This technique is suited for both liquid and dry samples. **Dynamic light scattering (DLS)** is a solution-based technique used for measuring the hydrodynamic size of NPs by analysing the fluctuations in light scattering caused by Brownian motion. This method is particularly used for characterising nanomaterials in colloidal systems, and **small-angle X-ray scattering (SAXS)** is used to determine size, shape, and arrangement of nanomaterials in the nanometre to micrometre range, providing information about the internal structure and organisation of nanomaterials.

On the other hand, surface functionalisation is analysed by employing **Fourier transform infrared spectroscopy (FTIR)**, which helps in the identification of functional groups and determining chemical bonds in nanomaterials by measuring the absorption of infrared light. **X-ray photoelectron spectroscopy (XPS)** analyses the elemental composition and chemical state of the outermost atomic layers of nanomaterials, providing insights into the chemical nature of attached groups. **Surface plasmon resonance (SPR)** provides real-time monitoring of fundamental interaction occurring at the surface of nanomaterials to study surface functionalisation, such as the binding of biomolecules or the attachment of specific ligands. **Chemical vapour deposition (CVD)** determines surface functionalisation by depositing thin films or functional groups onto nanomaterial surfaces [10]. It is commonly employed to modify the surface characteristics of nanomaterials, thereby improving their functionality for specific applications. **Zeta-potential measurement** determines the surface charge of NPs in a solution. It is crucial for understanding the stability and potential for surface functionalisation. Nanomaterials with appropriate zeta potentials are more amenable to modifications and functional coatings.

10.4 APPLICATIONS OF BIOGENIC NANOPARTICLES

Biogenic NPs are NPs that are integrated by biological bodies such as plants, animals, and microorganisms and have a mass significant attention due to their unique properties. Some of the applications of biogenic NPs are as follows:

1. **Biomedical Applications:** Increasing in the applications of biogenic NPs in biomedical science, these include drug delivery by liposomes and bio sensing. For example, magnetite with super magnetic iron oxide NPs is utilised in the research for improving drug delivery and immunoassays. For this, proper dimensions of not more than 100 nm with uniform size is essential. Semiconductors and metal NPs plays an important role in cancer detection with the help of advanced technology by using light scattering and adsorption technology [11].

   Gold NPs transforms the electromagnetic energy into heat by combining gold silica nano shells with MRI- Ultrasound imaging to remove the prostate tumorous cell. These NPs absorb the Near-Infrared Light and increase the temperature, which leads to the death of tumorous cells. In microorganism due to mutations antibiotic resistance occurs which can cause biofilms and efflux pump causes emergence to the multi-drug resistance pathogens. AgNPs are efficient against gram positive and gram-negative bacteria because they damage cell membranes and components and also produce Reactive Oxygen Species cells that effect DNA, RNA, and proteins [12].
2. **Environmental Remediation**: Environmental contamination is a very big problem that can affect land, water, and air and lead to contamination by lead, cadmium, and platinum.

   Since improper waste management, metallurgy, pesticides, fertilizers, and heavy metals like cobalt, nickel, and copper are on the rise, heavy metal pollution is, as we all know, the most prevalent issue in the globe. Heavy metals can cause grave environmental problems that can cause many health problems. To protect the environment and ensure human safety, it is important to remove heavy metals from the contaminated environments.

   Many treatment methods have been developed to clear the contaminants or heavy metals, including physical methods (such as adsorption, coagulation, evaporation, and filtration) and chemical methods (such as oxidation, ion exchange, and electrochemical processes). In these all known methods, adsorption is used because of its high removal efficiency, applicability, low cost, etc. Much research has been conducted for the removal of pollutants, heavy metals, and pathogens. Nanomaterials are good adsorbents and catalysts for bioremediation processes [11].

3. **Agriculture and Food Industry:** The production of food involves the use of nanomaterials, including inorganic (metal and metal oxide), composite (materials mixed with clay), and organic (materials obtained from natural products). Mostly, it is designed for delivery system and also used as ingredients or supplements in food industries.

   Food packaging, additives, preservatives, and other parts of customer prototypes or products are the most common areas they infiltrate. Following extensive testing and food safety confirmation, numerous traditional chemical or inorganic nanoparticles are used as additives or preservatives that are existing in the form of nanomaterials. For example: TiO_2 NPs are found as nanomaterials. Some nanoscale chemicals have also attracted people's attention regarding the potential risk.

Applications of nanoparticles in food include:

NPs are designed as flavour and colour additives and preservatives for food supplements. Materials used in food packaging come into direct contact with food items when they are being manufactured, stored, or transported. Nowadays, NPs are deliberately used and studied for food packaging. They possess many advantages compared to the conventional packaging materials. These are widely used because of their low cost, easy availability, and mechanical and thermal barrier properties, as listed in Table 10.1.

Many of the natural ingredients like lemongrass essential oil and chitosan have been developed as an edible coating to preserve food items or to increase their shelf life [19]. In agriculture, nanotechnology is widely used to increase the production of the best-quality crops with higher or equivalent nutritional values. Most common and effective way to improvise crop production is using plant growth regulators, herbicides, and pesticides. For example, atrazine-loaded poly (epsilon-caprolactone) can be used in the treatment of mustard plants (*Brassica juncea*). This improvised the herbicidal activity better than the commercial atrazine, which resulted in drastic decrease in net photosynthetic rate and stomatal conductance and thus led to weight loss and growth reduction of

TABLE 10.1
Various Applications of Nanomaterials

Sector	Application	Nanomaterials	Reference
Biomedical industries	Drug delivery, biosensing	Liposomes	[12]
Food industries	Additive or polymer production aid	• ZnO • Iron oxide • Aluminium oxide • Cobalt oxide • Manganese oxide	[13]
	Preservative	Silver silica	[13]
	Flavour carrier, marking fruits, and vegetables	Silicon dioxide Silicon dioxide	[14]
Agriculture sector	Pesticides detection	Zinc oxide QDs	[15]
Packaging material	Flame retardation additives	Montmorillonite	[16]
Environmental remediation	Water treatment	Titanium dioxide	[3]
	Soil remediation	Lead and cadmium	[17]
	Air purification	Platinum	[18]

tested plants. Nowadays, precision agriculture practices are used to improve crop yields without damaging soil and water in an effective manner. NPs-mediated gene or DNA transfer in crops or plants can be used to produce a variety of pest-resistant crops. ZnO, TiO_2, and CuO nanomaterials are studied as crop-protecting metal oxides [20].

Owing to many beneficial aspects of NPs, newly developed nanosensors are also introduced, which is recently very beneficial for crop monitoring, such as disease control, ensuring growth and nutrient availability, and checking the condition of soil and water content in the field. Gold and silver NPs are also under research for developing nanosensors for real-time monitoring growth and condition of crops.

10.5 CHALLENGES AND LIMITATIONS IN BIOGENIC NANOPARTICLE TECHNIQUES

Technology Transfer: Universities and businesses partner together to commercialise new knowledge by providing funding opportunities to both universities and businesses through patenting and licensing agreements or partnerships. For successful partnerships, universities need to provide faculty involvement and contribute courses and solutions alongside industry with funding by means of placements, internships, and workshops. These practices enable companies to protect their technologies, leading to reduced overheads, but experienced third-party involvement and may be exclusive or non-exclusive.

A joint venture is a partnership where two companies share their resources and risks for their mutual benefit. Although they propose long-term cooperation and have noteworthy technology transfer, their different goals and managerial issues are the main challenges they often encounter.

NPs with a specific shape, size, and distribution can be produced by changing the method of synthesis, reducing agent, and stabiliser. It is important to standardise variations in the plant extract used and the methodology adopted for optimising the synthesis protocol to get NPs with desired size, shape, and surface charges. Similarly, different synthesis methods, extract preparation techniques, different precursor and reducing agent concentrations, as well as changes in temperature and pH, result in different characteristics of the plant material employed in NP synthesis. These methods produce different features of NPs concerning size and shape. Among the plant materials, leaves can be extensively used for large-scale synthesis. The crop productivity should not be damaged, so plant materials like leaves are used, instead of other sources like flowers, fruits, seed, root, and latex. At some times, these other sources are also useful, but the volume of these materials should not affect crop productivity. In addition, plant-based NPs offer reproducibility, stability, and environmental friendliness. Various factors like pH, precursor and extract concentration, and time influence the shape and size of these NPs. Optimisation of these parameters can ensure the desired characteristics. Stability is crucial for long-term applications, with biological synthesis relying on capping agents for stability. pH plays a significant role, affecting both size and shape, with alkaline conditions favouring cluster distribution and preventing aggregation. Precursor and reducing agent concentration also impact NP size and shape, where the right balance is necessary for desired outcomes. Overall, careful parameter optimisation is key to achieving desired NP characteristics [3].

NPs are manufactured through the development of nanotechnology, which involves reducing metals to their atomic size. Several techniques are employed for NP synthesis, including physical, chemical, and biological methods. Physical methods offer the advantage of avoiding solvent contamination in NP synthesis, but they require a significant amount of energy for particle condensation and evaporation. Additionally, the manipulation of temperature and pressure adds to the overall cost of NP synthesis. Chemical techniques utilise reducing agents and protective agents to synthesise NPs and prevent agglomeration, ensuring high purity and stability. However, the extensive use of strong chemicals can lead to contamination of the synthesised NPs. By contrast, interest in biological NP synthesis has grown due to its eco-friendly and efficient nature. Biological synthesis, particularly through green synthesis, involves using microorganisms or plant extracts as reduction agents. This approach eliminates the need for powerful chemical agents and high-energy reactions, highlighting its significant advantages. Synthesising NPs with plant extracts using the biological method is particularly appealing for large-scale production, offering considerable potential in various medicinal applications [21].

This explores different pathways for the biosynthesis of NPs facilitated by various fungal organisms, delving into the molecular mechanisms involved. NPs find applications across diverse fields including medicine, agriculture, biosensors, and environmental treatment. While conventional methods like physical and chemical approaches can produce NPs, they come with drawbacks, prompting the search for more economical, non-hazardous, eco-friendly, and cost-effective synthesis processes. Mycological synthesis of NPs offers several advantages when compared to traditional methods, making it a promising alternative. This chapter reviews various fungi utilised in NP synthesis and discusses the process of synthesis along with their applications. Controlling critical parameters such as pH, temperature, and other medium conditions is crucial for successful NP synthesis. It also gives info about short comings that are observed in different fields, such as medicine, agriculture, drug delivery, and water treatment, when utilising NPs synthesised by fungi. By exploring these aspects, this chapter aims to provide insights into the potential of and challenges associated with fungal-mediated NP synthesis and its applications [22].

The usage of biogenic NPs across various sectors underscores the need for their industrial-scale production and formulation into stable compositions through eco-friendly processes. Consequently, evidential efforts are afoot to harness natural resources and implement biological synthesis methods due to their non-toxic, cost-effective, and easily scalable nature. The green synthesis of NPs using biological resources has proven highly advantageous. Green synthesis offers several benefits, including the production of NPs in stable forms with controlled sizes and shapes, absence of complex chemical synthesis, and avoidance of toxic reactants. Various biological resources such as medicinal plants, algae, microbes, and biosurfactants enable rapid NP synthesis. However, challenges remain, such as understanding the correlation between metal salt concentration and NP yield and optimising parameters to overcome polydispersity issues in different biological systems. Additionally, the lack of knowledge regarding the chemical components influencing synthesis, stabilisation, and mechanisms of action of biogenic NPs poses a challenge. Understanding how active groups of biogenic materials interact with NP surfaces and contribute to synthesis efficacy is crucial, particularly in terms of biocompatibility. While numerous microorganisms and plants have been successfully engaged for biogenic NP synthesis, further assessment of biological nanofactories is warranted to meet the demand for NP-based products. However, challenges persist in the biomedical applications of biogenic NPs, including pharmacokinetic profiling such as absorption, distribution, metabolism, excretion, and clearance from the body. Addressing these issues requires extensive research on the biocompatibility and bioavailability of biogenic NPs at early stages [23].

10.6 TECHNOLOGY TRANSFER STRATEGIES

Universities are increasingly involved in technology transfer in addition to teaching and research to support the growth of a nation. Technology transfer is a reflection of exchanges or ongoing partnerships and buyers. A technology is transferred in an academic setting between universities and businesses acting as suppliers and buyers of the technology. These partnerships are commonly known as academic–industry collaborations [24], as shown in Figure 10.2.

10.6.1 Academic–Industry Collaboration

Universities are a big source of innovation. This improves the economic development through knowledge generation and communication. Knowledge transfer between academics and industry is an important aspect that makes it easier for businesses to commercialise new scientific knowledge. Interaction between industries and researchers benefits them because they can generate new ideas for

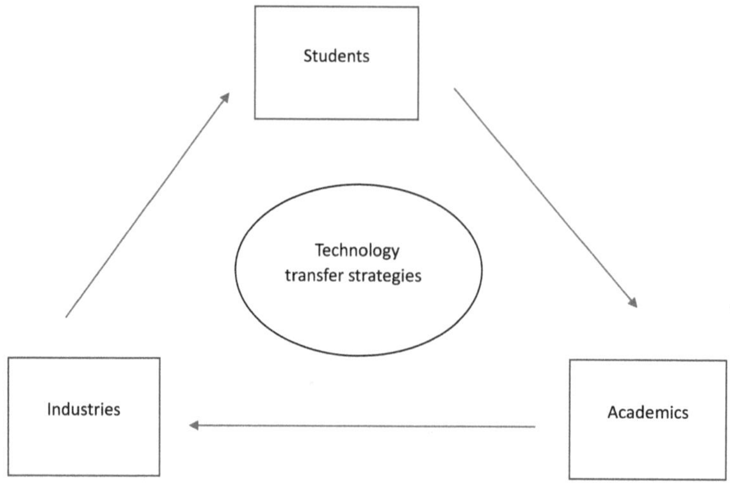

FIGURE 10.2 Technology transfer strategies.

study and lead to funding opportunities. This includes patenting, licensing, and joint ventures [24]. Weak connection with regional industrial clusters in terms of technology transfer can raise doubts among the taxpayer and government official about the role of university and the economic development. Programme strategy depends on some strategies that are followed by partnership projects between the institutes and the industries [25,26].

Academics

- Faculty representative
- Theory classes
- Industry-aligned electives
- Solution to concern

Industry

- Funding
- Placements
- Internship
- Workshops

10.6.2 Licensing Agreements

Licensing is an agreement in which a discovery or a technology is secured by intellectual property (IP) rights and by the owner of a patent, or any property can provide permission to another company to use the same technology or discovery for a certain period of time.

Licensing is divided into two types:

- An individual gives total rights to use the technology.
- Additionally, the owner has the ability to grant other businesses in the same region the same rights to utilize that technology.

Furthermore, this agreement can give a sublicensing term that gives or grants permission to someone else to use the technology. It will be less expensive than other technology transfer techniques and the technology will be safe (no one else can buy or use it without the owner's consent), but purchasing the patent requires knowledge, necessary expertise, and experience.

10.6.3 Joint Ventures and Partnerships

Both joint ventures and partnerships include an agreement or many arrangements that involve two or more companies working together. Joint venture involves more assets, management, risk, services, co-production, and more over a good marketing. We can get many benefits from joint venture like long-term partnership between companies, motivation for a successful transfer, and if companies work separately it will be in a reasonable rate or prize. There are some drawbacks. It often includes both the companies that would have different goals and vision and lack the ability to manage independently. Business relationship of today functions quite differently. Even though a partnership still requires two or more people, entrepreneurs, there's a whole new meaning to partnering that is a joint venture for the phrase partnership. Partnership is considered a long-term commitment between two or more organisations to achieve by making the most of each participant's resources in order to achieve a particular business objective [12].

10.7 COMMERCIALISATION PATHWAYS

Commercialisation pathways refer to the strategic routes and steps taken by businesses or organisations to bring a product or a technology from the research and development stage to the market. Theses pathways involve a series of activities aimed at turning innovative ideas or services as depicted in Figure 10.3. The field of nanotechnology exhibits great potential for transforming diverse facets of healthcare and product creation. Its applications include clinical diagnostics, drug delivery, nanomedicines, imaging, and the incorporation of nanomaterials into medical equipment. Nanotechnology has already advanced significantly, with many products available and more still in development. Drug delivery and *in vivo* imaging are two important areas of focus, with a focus on critical needs like drug reformulation for increased bioavailability and decreased toxicity. Medical nanotechnology is expected to advance through several stages, from simple improvements to current goods and procedures to the development of novel nanoproducts and nanodevices that are only constrained by imagination and technical limitations. A growing momentum towards technology commercialisation in the medical field is reflected in recent industry trends and commercial developments in medical nanotechnology. Advances in nanotechnology are transforming the manufacturing and production sectors by presenting new materials and methods for industrial use. These advancements are leading to the evolution of new products with unique characteristics, benefiting society at large. As a general-purpose technology, nanotechnology is poised to impact various sectors, including medicine, energy, environmental remediation, robotics, manufacturing, commerce, and space exploration, fostering rapid growth and transformation. It gives an explanation of how projected timescales for accomplishing each generation of advancement may have failed and suggests a revised timeline for expected advancements in nanotechnology.

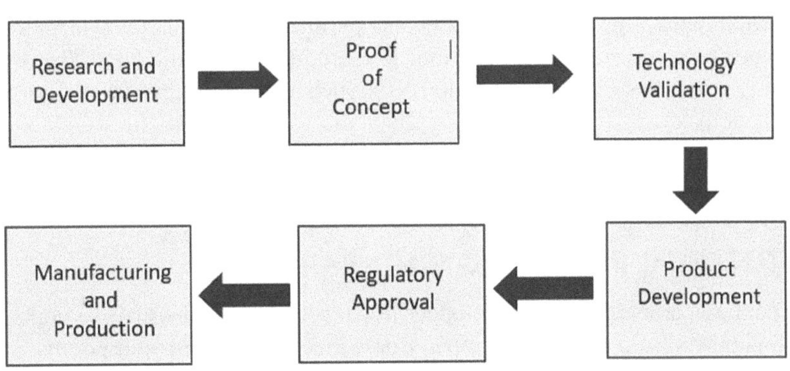

FIGURE 10.3 Pathways of commercialisation.

From the business perspective, a model for commercialising nanotechnology products and services is presented, exploring opportunities for market penetration. It gives information about the common challenges encountered by nanotechnology firms during commercialisation, such as time delays, funding gaps infrastructure limitations, absence of evaluation standards, bureaucratic hurdles, funding shortages, scarcity of skilled professionals, brand perception, and public support. Strategies for overcoming these challenges and achieving successful commercialisation are discussed, including the utilisation of various corporate business strategies [27].

10.7.1 Emerging Trends in Biogenic Nanoparticle Commercialisation

Nanotechnology has become pivotal across industries like electronics, catalysis, pharmaceuticals, textiles, and agriculture, relying heavily on precisely synthesised NPs with controlled sizes and shapes. While physical and chemical methods offer control over NP characteristics, they often involve hazardous chemicals and harsh conditions, posing environmental risks. In response, green synthesis methods utilising biological entities like bacteria, fungi, algae, and plants have emerged as eco-friendly alternatives. These methods minimise the use of toxic chemicals and energy, reducing environmental impact. Various organisms, including bacteria like *Lactobacillus* spp., fungi like *Trichoderma atroviride*, and plants like *Dioscorea bulbifera*, have demonstrated the ability to synthesise diverse NPs such as gold, silver, copper, and graphene oxide. Intellectual property rights (IPRs) and technology transfer mechanisms play crucial roles in protecting and disseminating innovations in NP synthesis, ensuring their commercialisation and widespread adoption. Optimisation of green synthesis methods, scale-up strategies, downstream processing, and functionalisation approaches are essential for bridging the gap between laboratory-scale synthesis and industrial production, enabling practical applications in medicine, agriculture, and environmental remediation. Incorporating drugs and targeting ligands onto NP surfaces enhances their therapeutic efficacy and targeting specificity, opening new avenues for biomedical applications [28].

The conventional methods for producing NPs in physical and chemical routes pose environmental concerns due to the need for specialised equipment, high temperatures, and pressure. By contrast, biosynthesis has gained significant attention for its eco-friendly approach and versatile applications in energy, nutrition, pharmaceuticals, and medicine. This review focuses on environmentally friendly methods for producing biogenic NPs using natural plant extracts and elucidates the mechanisms involved in their synthesis, growth, and stabilisation. It highlights recent advancements in utilising these NPs, particularly in combating diseases such as malaria, hepatitis, and cancer. Various characterisation techniques, including UV–visible spectrophotometry, Fourier transform infrared spectroscopy, XRD, scanning electron microscopy, TEM, and energy-dispersive X-ray spectroscopy, are discussed to analyse the properties of the synthesised NPs. Different parts of plants, including leaves, stems, barks, fruits, and flowers, are rich sources of bioactive compounds like flavonoids, phenols, steroids, terpenoids, enzymes, and alkaloids, which play crucial roles in reducing metal ions and producing metallic NPs. The review emphasises the unique properties of phytofabricated NPs, such as their antibacterial, antioxidant, cytotoxic, and drug delivery capabilities. Additionally, it discusses the challenges and future prospects in synthesising biogenic metal NPs for environmental and pharmaceutical advancements in the years to come [29].

10.8 INTELLECTUAL PROPERTY CONSIDERATIONS

Some of the common properties that come under IP banners are designs, artistic works, and brands. This will forbid others from using your property or technology without your permission. Once this work/design is created or made, then the property belongs to the creator who has all the rights over that property.

10.8.1 Patent and Trademarks

Patent is a legal protection given by the government to the people who invent or discover something innovative. This gives the patent holder all the rights to sell, use, or create and licence their discovery for a time period that is limited for 20 years. There are many types of patents like plant patent, design patent, and utility patent, but the most used patent is utility patent that covers more new machines and manufactures. Patents are very important to protect one's ideas and inventions that haven't been discovered before or publicly so that others should not use their idea without their permission or will.

A trademark is a type of the copyright/IP/patent that consists of an apparent size, design, and symbol that are used for identifying or differentiate between goods or services of a provider for those of others. This helps consumers or customers to identify their products easily that are associated to particular brands. One can get their trademark from the U.S. Patent and Trademark Office (USPTO). This is mostly applied to distinguishing the logo, phrases, and brands. Mostly, copyrights apply to original works.

10.8.2 IP Protection Strategy

IP is an essence of any organisation. IP can be anything from a particular manufacturing that is planned for a product launch. The World Intellectual Property Organization's (WIPO) meaning of IP is creativity of mind, literary invention, and invention of symbols, names, images, logo, and design. There are four types of IP [30].

10.8.2.1 Patent

This gives a legal right to eliminate anyone else from marketing or inventing one's own unique things.

10.8.2.2 Trademark

These are names, symbols, or designs used in an individual's product. This contributes to a brand, and the company's reputation will be enhanced by this degree of excellence.

10.8.2.3 Copyright

This protects written expressions fixed in a real-time medium like novels, poems, movies, or songs. This protects an expression of idea, but they can't protect the idea. The owner has the right to regenerate a work (like a movie based on a book) or to show it in an event as a public display.

10.8.2.4 Trade Secret

It can be anything like formula, device, or a pattern that refers to a confidential valuable information that provides a competitive advantage to its owner. Unlike patent, copyright, or trademark which needs registration for the protection, this does not require any registration as long as it remains confidential [31].

10.8.3 Patent Landscape in Biogenic Nanoparticles

Intellectual Property Issues in Nanotechnology delves into the crucial role of IP rights in driving innovation across various fields of nanotechnology. It explores the intersection of industry, law, and ethics, aiming to foster sustained innovation. By examining the implications of IP rights on omics-based research and discussing ethical and intellectual standards, the book offers a comprehensive perspective on navigating the complex landscape of nanotechnology innovation. With a unique integration of science and business, it provides an insightful analysis of the industry's foundations, making it valuable for both students and professionals seeking a deeper understanding of the field [32].

10.9 REGULATORY AND ETHICAL CONSIDERATIONS

10.9.1 Compliance with Regulatory Standards

Biogenic NPs are created using two approaches: top-down and bottom-up strategies. To synthesise these techniques, chemical, physical, and biological strategies are used. Physical strategies include discharge, milling, and ion implantation, whereas chemical strategies involve process such as reduction, solvothermal, coprecipitation, and sono chemical. Biological process includes the usage of seaweeds, plants and their extracts, microbes, and agricultural by-products. Scanning electron microscopy (SEM), TEM, and dynamic light scanning are few methods that are used for the characterisation of biogenic NPs. They exhibit properties like biocompatibility and chemical, electrical, and magnetic capabilities. They are used in pharmaceutical sectors in the form of anticancer agents, antifungal agents, biosensors, antibacterial agents, and antioxidants. Zinc oxide, gold, iron oxide, palladium, platinum, silver, and copper oxide are the most popular biogenic NPs used in the pharmaceutical sector. They are synthesised from fungi, bacteria, and plants as an alternative way. Numerous studies have shown that the size, stability, morphology, and chemical and physical properties of biogenic NPs are influenced by the stabilising agent's adsorption process, the experimental conditions, and the kinetics of the metal ion interaction with reducing agents. Due to their low melting point, high surface area, mechanical strengths, and magnetic and optical capabilities, they are different from other metals and are thus used in various industries [33].

10.9.2 Safety and Environmental Impact

Biogenic NPs are compatible for its various uses such as drug delivery, water treatment, diagnosis, and food preservation. Recently, uses of NPs have been significantly increased in the biotechnology field [34]. Development of NPs causes risks in the environment as they discharge harmful substances and raise negative consequences. Since they don't dissolve easily, they are stable. Biogenic NPs have different bioaccumulation potential, impact, and persistence on different organisms in the ecosystem. Hence, its environmental fate needs to be studied before its application. Though they are safer than synthetic NPs, they are less toxic due to their natural origin, but their toxicity level depends on their size, surface chemistry, and shape. Regulatory considerations should be adapted to ensure its proper risk and uses. They enhance the efficacy of drug delivery system and minimise the off-targeting effect by targeting the specific cells and tissues in the body. They are synthesised from renewable resources using environmentally friendly methods, thus reducing wastage production, pollution, and energy consumption and maintaining the sustainability. Here, the principles of green chemistry are applied where non-toxic reagents and solvents are used in order to minimise the environmental footprint associated with NP production and encourage the green manufacturing practices. Biogenic NPs also play an important role in environmental bioremediation such as removal of harmful pollutants, heavy metals, degrading organic pollutants, and the treatment of wastewater.

10.9.3 Ethical Implications in Commercialisation

Several ethical practices are developed in order to ensure the proper usage, development, and production of biogenic NP. Ethical considerations were started when the concern related to human health arises due to the usage of biogenic NPs in the product. These products are being used for medical purposes and commercial consuming. Thus, it is required to conduct various clinical trials and safety measurements to ensure that it doesn't harm patients and consumers. Their environmental impact is being studied in order to minimise the effect in the environment and ecosystem. It includes certain sustainable and environmentally friendly practices that are adopted to minimise environmental risks. Due to some socioeconomic factors and geographic location, the maximum usage of biogenic NP products in healthcare sector is restricted. Equal access and distribution of this biogenic NPs to all communities are required such that all can enjoy its benefits. The consent of patients should be received before performing clinical trials on them. Patients should be informed about clinical trials' advantages, disadvantages, and side effects such that their participation should be autonomous. It is also essential to protect patients' personal data used during clinical trials. Ethical conduct requires transparency and responsibility at each stage of the product lifecycle.

10.10 FUTURE TRENDS AND PERSPECTIVES

Biogenic NPs are biosynthesised, filled with plant extracts, and are therefore called green NPs. They are mostly used in drug delivery system, medical field, and research areas. In drug delivery system, these NPs are loaded with therapeutic compounds that are released in a controlled manner, which helps in targeting specific cells or tissues. They also play an important role in cancer detection and disease image processing. These characteristics of NPs allow drug action to last longer and reduce dosages. Biosynthesised NPs filled with plant extracts [35] are used in drug delivery system to target specific cells and tissues in the body. This method minimises side effects and enhances the therapeutic efficacy compared to other drug delivery methods. It provides drug resistance against many diseases. Since some drugs have low solubility and poor absorptivity, biogenic NPs help in improving the solubility and bioavailability of drug molecules. Green NPs have less side effects and are less toxic than other synthetic NPs. Biocompatible and biodegradable materials are used in order to minimise side effects and adverse reactions caused by long-term consumption of drugs. The combination of unique properties of metallic NPs and the medicinal properties of plant extracts gives rise to multifunctional NPs that possess therapeutic effects such as anti-inflammatory and antioxidant, and also help in the effective delivering of drugs [36].

10.10.1 Potential Disruptions and Transformations

Technology transfer and commercialisation of biogenic NPs have the ability to develop and transform in various sectors:

i. **Medicine and Healthcare**: They are used in the drug delivery system to target specific cells and tissues in the body while minimising side effects. They are also used in diagnoses as contrast agents that help detect diseases through, for example, CT scan, MRI, and ultrasound. They act as a regenerative medicine that repairs damaged tissues and facilitates the formation of new tissues.
ii. **Agriculture and Food Production**: They are used as food-packaging agents to extend the half-life of packaged foods and prevent them from spoilage. They are used for enhancing crop productivity and taking up of nutrients in the plants by developing biopesticides and herbicides, which kill pests and provide resistance to crops against diseases.
iii. **Environmental Remediation**: They enhance the removal of heavy metals and other organic waste from wastewater, and plays important roles in the wastewater treatment process. They help remove harmful pollutants from the environment and provide sustainability to the environment.
iv. **Energy and Electronics**: They are used in batteries, solar panels, and fuel cells to enhance their lifespan, efficacy, capacity, and performance.

10.10.2 Future Directions for Commercialisation

The biosynthesis of NPs using waste materials will help researchers not only to design safer nanomaterials but also to promote the understanding of health and safety considerations of NPs. Useful materials can be produced easily even at reasonable scale because the biomaterial-based routes eliminate the need to use toxic chemicals [36]. Considerable efforts are devoted on different capping agents such as biomolecules and polysaccharides, which can act as both chelating/reducing and capping agents for the synthesis of NPs [37]. Therefore, the resulting particles are protected from further reactions and aggregation, which increases their stability and longevity. Greener methods that have been used in the NP synthesis are generally single-pot reactions, without the use of additional surfactants, capping agents, and templates. Survey of literature reveals that these investigations have been carried out at laboratory scale, whereas there are no reports on the "pilot plant" synthesis of NPs using waste materials and algae. We believe that there are good opportunities for developing industrial-scale production, where NPs have important applications [38]. Overall, the use of algae and waste materials for the green synthesis of NPs/nanomaterials is an emerging and exciting area of nanotechnology and may have significant impacts on further advances in nanoscience. Moreover, with the availability of computational techniques, the area of green synthesis and development of nanomaterials will be broadened that could be used in the field of medicine as "nanodrugs" in the near future. Moreover, these NPs would provide a potential solution for the present energy crisis by finding their use as energy-driven devices [38] (Figure 10.4).

10.11 CONCLUSION

Technology transfer and commercialisation aspects of biogenic NPs include the development of their applications from laboratory to practical in agriculture, environmental remediation, and healthcare. Biogenic NPs that are derived from plants, bacteria, and fungi have diverse applications in the research field. They also offer promising commercialisation in medicines, agriculture, drug delivery system, and diagnostics. They help control air pollution, enhance crop productivity, treat wastewater, provide pest resistance, and minimise side effects. Regulatory and safety considerations are needed to ensure their application in the healthcare sector. Toxicity studies and risk assessments are necessary in order to approve them for human consumption. Sustainable production methods are important because they minimise environmental impacts. IP protection plays a major role in commercialisation of biogenic NPs. Investment and market dynamics are factors that influence the potential, growth, and commercialisation of biogenic NP products.

In summary, the commercialisation of biogenic NPs has both opportunities and challenges. Stakeholders play an important role in the commercialisation and development of biogenic NPs to ensure its safety and proper use. Researchers and academia conduct research on the properties

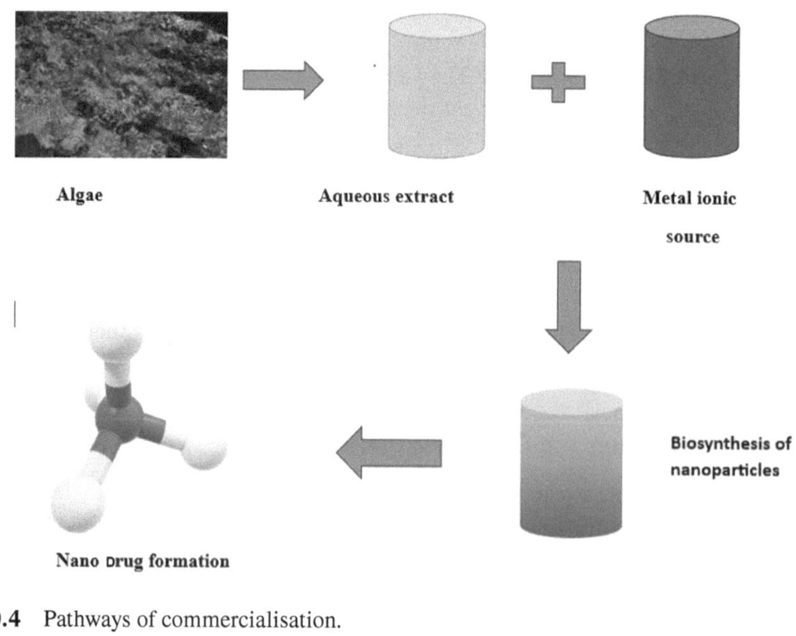

FIGURE 10.4 Pathways of commercialisation.

and applications of biogenic NPs and collaborate with industries and agencies to regulate their safety concern. Investment is made in research and development field in order to commercialise the products made from biogenic NPs. Moreover, collaboration with several institutions and research organisations, ensures its safety and regulatory standards. Regulatory agencies are involved in the development of guidelines and regulations which ensures the safety in production, commercial use, and disposing. Collaboration is made with stakeholders to evaluate the environmental impact and safety of using biogenic NPs. Healthcare professionals keep an update on the current breakthroughs achieved in the biogenic NPs utilized in diagnostic and therapeutics. They are involved in clinical trials and research studies to evaluate the side effects, safety, and efficacy. Patients should also be informed of the benefits and drawbacks of the biogenic NP-based products.

Environmental organisations are responsible for sustainable production and environmentally friendly practices in the use of biogenic NPs. They also monitor the impacts of biogenic NP products on the environment. Certain policies are introduced to create awareness to protect the environment.

LIST OF ABBREVIATION

AFM: atomic force microscopy
DLS: dynamic light scattering
NMR: nuclear magnetic resonance
SAXS: small-angle X-ray scattering
SEM: scanning electron microscopy
TEM: transmission electron microscopy
XRD: X-ray diffraction

REFERENCES

1. Tripathi, R. M., and Sang J. Chung. "Biogenic nanomaterials: Synthesis, characterization, growth mechanism, and biomedical applications." *Journal of Microbiological Methods* 157 (2019): 65–80.
2. Sharma, Deepali, Suvardhan Kanchi, and Krishna Bisetty. "Biogenic synthesis of nanoparticles: A review." *Arabian Journal of Chemistry* 12, no. 8 (2019): 3576–3600.
3. Mughal, Bilal, Syed Zohaib Javaid Zaidi, Xunli Zhang, and Sammer Ul Hassan. "Biogenicnanoparticles: Synthesis, characterisation and applications." *Applied Sciences* 11, no. 6 (2021): 2598.
4. Vaseghi, Zahra, Ali Nematollahzadeh, and Omid Tavakoli. "Green methods for the synthesis of metal nanoparticles using biogenic reducing agents: A review." *Reviews in Chemical Engineering* 34, no. 4 (2018): 529–559.
5. do Nascimento, Jéssica Mesquita, Nildo Duarte Cruz, Gabriel Rodrigues de Oliveira, Waldeemeson Silva Sá, Jorge Diniz de Oliveira, Paulo Roberto S. Ribeiro, and Selma GF Leite. "Evaluation of the kinetics of gold biosorption processes and consequent biogenic synthesis of AuNPs mediated by the fungus Trichoderma harzianum." *Environmental Technology & Innovation* 21 (2021): 101238.
6. Sundararajan, Niranjana, Heena Shabnam Habeebsheriff, Karthikkumar Dhanabalan, Vo Huu Cong, Ling Shing Wong, Ranjithkumar Rajamani, and Bablu Kumar Dhar. "Mitigating global challenges: Harnessing green synthesized nanomaterials for sustainable crop production systems." *Global Challenges* 8, no. 1 (December 25, 2023). https://doi.org/10.1002/gch2.202300187.
7. Avasthi, A., Caro, C., Pozo-Torres, E., Leal, M. P., and García-Martín, M. L. Magnetic Nanoparticles as MRI Contrast Agents. *Topics in current chemistry (Cham)*, 378, 3 (2020): 40. https://doi.org/10.1007/s41061-020-00302-w
8. Heath, George R., Ekaterina Kots, Janice L. Robertson, Shifra Lansky, George Khelashvili, Harel Weinstein, and Simon Scheuring. "Localization atomic force microscopy." *Nature* 594, no. 7863 (2021): 385–390.
9. Dinata, Agil Aditya, Abel Maaruf Rosyadi, Syukri Hamid, and Rahadian Zainul. 2018. "A review chemical vapor deposition: Process and application." *INA-Rxiv*. https://doi.org/10.31227/osf.io/yfeau.
10. Goel, Muskan, Anurag Sharma, and Bechan Sharma. "Recent advances in biogenic silver nanoparticles for their biomedical applications." *Sustainable Chemistry* 4, no. 1 (March 3, 2023): 61–94. https://doi.org/10.3390/suschem4010007.

11. He, Xiaojia, Hua Deng, and Huey-min Hwang. "The current application of nanotechnology in food and agriculture." *Journal of Food and Drug Analysis* 27, no. 1 (January 2019): 1–21. https://doi.org/10.1016/j.jfda.2018.12.002.
12. Chae, Byeongmin, Jinho Kim, and Do Gyun Lee. "Exploring partnering and joint ventures in the construction industry: A brief perspective." *Journal of Southwest Jiaotong University* 57, no. 1 (February 28, 2022): 706–17. https://doi.org/10.35741/issn.0258-2724.57.1.63.
13. Kulkarni, Uma, and Ashwini Desai. "Academic excellence through industry institute collaboration." *Journal of Engineering Education Transformations* 34 (January 31, 2021): 747. https://doi.org/10.16920/jeet/2021/v34i0/157177.
14. Zahin, Nuzhat, Raihanatul Anwar, Devesh Tewari, et al. "Nanoparticles and its biomedical applications in health and diseases: Special focus on drug delivery." *Environmental Science and Pollution Research* 27 (2020): 19151–19168.
15. He, Xiaojia, Hua Deng, and Huey-min Hwang. "The current application of nanotechnology in food and agriculture." *Journal of Food and Drug Analysis* 27, no. 1 (January 2019): 1–21. https://doi.org/10.1016/j.jfda.2018.12.002.
16. EFSA Panel on Food Additives and Nutrient Sources added to Food (ANS), Maged Younes, Peter Aggett, Fernando Aguilar, Riccardo Crebelli, Birgit Dusemund, Metka Filipič et al. "Re-evaluation of silicon dioxide (E 551) as a food additive." *EFSA Journal* 16, no. 1 (2018): e05088.
17. Sahoo, Dibakar, Abhishek Mandal, Tapas Mitra, Kaushik Chakraborty, Munmun Bardhan, and Anjan Kumar Dasgupta. "Nanosensing of pesticides by zinc oxide quantum dot: An optical and electrochemical approach for the detection of pesticides in water." *Journal of Agricultural and Food Chemistry* 66, no. 2 (2018): 414–423.
18. Zhang, Shenghe, Fukai Chu, Weizhao Hu, Bibo Wang, Richard K.K. Yuen, and Yuan Hu. "Investigation of nanocomposite system with long-chain phosphorus-containing intercalator-modified montmorillonite nanosheets." *Composites Part A: Applied Science and Manufacturing* 176 (2024): 107877.
19. Fan, Wenjiao, Junxiu Sun, Yunchuan Chen, Jian Qiu, Yan Zhang, and Yuanlong Chi. "Effects of chitosan coating on quality and shelf life of silver carp during frozen storage." *Food Chemistry* 115, no. 1 (July 1, 2009): 66–70. https://doi.org/10.1016/j.foodchem.2008.11.060.
20. Salama, Ahmed K., Khaled A. Osman, and Neama Abdel-Razeek Gouda. "Remediation of lead and cadmium-contaminated soils." *International Journal of Phytoremediation* 18, no. 4 (2016): 364–367.
21. Lee, S-K., and Andrew Mills. "Platinum and palladium in semiconductor photocatalytic systems." *Platinum Metals Review* 47, no. 2 (2003): 61–72.
22. Malkawi, Bashar, and Bashayer Almajed. "Treatment of intellectual property in the bankruptcy legal framework of the GCC states." *Queen Mary Journal of Intellectual Property* 14, no. 1 (March 12, 2024): 87–100. https://doi.org/10.4337/qmjip.2024.01.05.
23. Petryshak, Maria. "IP-court: A tool for intellectual property lawyer protection." *Law and Innovative Society*, no. 2 (15) (January 4, 2021). https://doi.org/10.37772/2309-9275-2020-2(15)-23.
24. de Wit-de Vries, Esther, Wilfred A. Dolfsma, Henny J. van der Windt, and M. P. Gerkema. "Knowledge transfer in university–industry research partnerships: A review." *The Journal of Technology Transfer* 44, no. 4 (March 28, 2018): 1236–1255. https://doi.org/10.1007/s10961-018-9660-x.
25. De Moortel, Kevin, and Thomas Crispeels. "International university-university technology transfer: Strategic management framework." *Technological Forecasting and Social Change* 135 (October 2018): 145–155. https://doi.org/10.1016/j.techfore.2018.05.002.
26. Patil, Sunita, and Rajkuberan Chandrasekaran. "Biogenic nanoparticles: A comprehensive perspective in synthesis, characterization, application and its challenges." *Journal of Genetic Engineering and Biotechnology* 18, no. 1 (2020): 67.
27. Ramanathan, Santheraleka, Subash CB Gopinath, M.K. Md. Arshad, Prabakaran Poopalan, and Veeradasan Perumal. "Nanoparticle synthetic methods: Atrength and limitations." In *Nanoparticles in Analytical and Medical Devices*, pp. 31–43. Elsevier, 2021. https://doi.org/10.1016/B978-0-12-821163-2.00002-9.
28. Raghuwanshi, Navdeep, Amit Kumar Srivastava, Tara Chand Yadav, Sonam Gupta, Kapil Khatri, Vikas Pruthi, and Vikas Pruthi. "Biogenic nanoparticles as theranostic agents: Prospects and challenges." In *Integrating Green Chemistry and Sustainable Engineering,* pp. 647–684. Wiley, 2019. https://doi.org/10.1002/9781119509868.ch19.
29. Ghosh, Sougata, Tanay Bhagwat, Balu A. Chopade, and Thomas J. Webster. "Patents, technology transfer, and commercialization aspects of biogenic nanoparticles." In *Nanobiotechnology*, pp. 323–339. Elsevier, 2021. https://doi.org/10.1016/B978-0-12-822878-4.00020-1.

30. Soni, Vatika, Pankaj Raizada, Pardeep Singh, Hoang Ngoc Cuong, S. Rangabhashiyam, Adesh Saini, Reena V. Saini et al. "Sustainable and green trends in using plant extracts for the synthesis of biogenic metal nanoparticles toward environmental and pharmaceutical advances: A review." *Environmental Research* 202 (2021): 111622.
31. Theivasanthi Thirugnanasambandan. *Patenting Issues in the Development of Nanodrugs*, 2020. https://www.taylorfrancis.com/chapters/edit/10.1201/9781003052104-16/patenting-issues-development-nanodrugs-theivasanthi-thirugnanasambandan.
32. Keswani, Chetan, ed. *Intellectual Property Issues in Nanotechnology*. CRC Press, Boca Raton, FL, 2020.
33. Badawy, Ali A., Nilly AH Abdelfattah, Salem S. Salem, Mohamed F. Awad, and Amr Fouda. "Efficacy assessment of biosynthesized copper oxide nanoparticles (CuO-NPs) on stored grain insects and their impacts on morphological and physiological traits of wheat (Triticum aestivum L.) plant." *Biology* 10, no. 3 (2021): 233.
34. Kumar, Ashwini, Shah, Sunny R, Jayeoye, Titilope John, Kumar, Akash, Parihar, Akshay, Prajapati, Bhupendra Gopalbhai, … & Kapoor, Devesh U. "Biogenic metallic nanoparticles: Biomedical, analytical, food preservation, and applications in other consumable products." *Frontiers in Nanotechnology* 5 (2023): 1175149.
35. Boomi, Pandi, Ramalingam Ganesan, Gurumallesh Prabu Poorani, Sonamuthu Jegatheeswaran, Chandrasekaran Balakumar, Halliah Gurumallesh Prabu, Krishnan Anand, Narayanasamy Marimuthu Prabhu, Jeyaraman Jeyakanthan, and Muthupandian Saravanan. "Phyto-engineered gold nanoparticles (AuNPs) with potential antibacterial, antioxidant, and wound healing activities under in vitro and in vivo conditions." *International Journal of Nanomedicine* 15 (2020): 7553–7568. https://doi.org/10.2147/ijn.s257499.
36. Mao, Bin-Hsu, Zi-Yu Chen, Ying-Jang Wang, and Shian-Jang Yan. "Silver nanoparticles have lethal and sublethal adverse effects on development and longevity by inducing ROS-mediated stress responses." *Scientific Reports* 8, no. 1 (2018): 2445.
37. Todaria, Moulika, Dipak Maity, and Rajendra Awasthi. "Biogenic metallic nanoparticles as game-changers in targeted cancer therapy: Recent innovations and prospects." *Future Journal of Pharmaceutical Sciences* 10, no. 1 (2024): 25.
38. Sharma, Deepali, Suvardhan Kanchi, and Krishna Bisetty. "Biogenic synthesis of nanoparticles: A review." *Arabian Journal of Chemistry* 12, no. 8 (2019): 3576–3600.

11 Microbes-Assisted Nanotechnology for Removal of Toxic Pollutants from Water

Shashwati Wankar, Onkar Kadam, Rutuja Gumathannavar, Rashmi Ranjan Guru, and Swayamprava Dalai

11.1 INTRODUCTION

The first major conference of the United Nations (UN) in the year 2023 was dedicated to water. UN World Water Development Report 2023 emphasized safeguarding water, food, and energy security through partnerships and cooperation for water. According to the United Nations Sustainable Development Goal 6, sustainable water management and sanitation, and access to clean water is a major target by the year 2030.

In the 21st century, water pollution remains a critical global issue with diverse implications for ecosystems, human health, and socio-economic development. This issue poses a threat to human well-being, agricultural productivity, the balance of ecosystems, and economic progress [1]. The decline in water quality directly contributes to a range of environmental, social, and economic complications. The world's already limited water resources face increasing strain due to the escalating anthropogenic and industrial activities [2]. This pollution stems from the discharge of significant volumes of inadequately treated or completely untreated wastewater into rivers, lakes, underground water reservoirs, and coastal areas [3]. Moreover, the emergence of new pollutants such as personal care items, pharmaceuticals, pesticides, household and industrial chemicals, combined with shifting climate patterns, presents a fresh set of challenges concerning water quality. These challenges carry uncertain, long-term implications for both human health and the stability of ecosystems.

In 2010, the United Nations General Assembly (RES/64/292) acknowledged the significance of "access to safe and clean drinking water and sanitation as a fundamental human right crucial for our existence and welfare" [4]. Even though better water sources do not always guarantee water suitable for direct human consumption, greater endeavours are necessary to ensure access to safe and pure water [5]. Enhanced accessibility to reliable water and sanitation services results in disease prevention, improved human health, equality across genders and incomes, better educational outcomes, and increased economic efficiency. In developing nations, the primary and most critical water issues revolve around poor water quality and pollution. These countries commonly face significant challenges with inadequate wastewater collection and treatment systems, alongside limited access to safe drinking water and sanitation for a substantial portion of their populations [1,6]. The persistence of water quality problems in these regions can be attributed to deficient wastewater management, a lack of governmental commitment, insufficient investment, ineffective water distribution, changes in land use, population growth, and a general unawareness among policymakers regarding the profound connections between water quality and other developmental facets such as health, poverty, gender disparities, environmental decline, and food security. As a result, there's an urgent requirement to fortify plans for managing wastewater, raise awareness among political leaders, revamp water distribution mechanisms, and meticulously plan rural and urban developments that impact water quality. Encouraging governments to take decisive steps in tackling pressing

DOI: 10.1201/9781003487142-11

water quality challenges and integrating comprehensive water quality policies into their national developmental agendas are imperative.

11.2 UNDERSTANDING WATER POLLUTION

Water pollution persists as a vital international challenge because it creates numerous impacts on both environment and human health as well as social economic advancement. The United Nations Environment Programme guidelines state that this issue threatens human well-being together with agricultural productivity and ecosystems balance and economic development [7,8]. Water quality deterioration leads directly to various environmental difficulties which combine with social and economic problems. Limited water resources worldwide continues to deteriorate as expanding human-made and industrial operations compound resource pressures [9,10]. Rivers along with lakes and underground reservoirs and coastal regions become polluted when major wastewater discharges originate from improperly treated or untreated sources [11,12]. Water quality faces emerging difficulties from new pollutants such as personal care items and pharmaceuticals and pesticides and household and industrial chemicals in addition to changing climate patterns. Studying this current situation leads to unpredictable effects that will continue indefinitely and threaten the health of ecosystems and people [13,14].

In 2010, the United Nations General Assembly (RES/64/292) acknowledged the significance of "access to safe and clean drinking water and sanitation as a fundamental human right crucial for our existence and welfare" . Beyond the substantial impact on human health and the environment, the absence of accessible, clean, and affordable drinking water and sanitation places an increasing strain on the economic and social advancement of nations [4, 5,15]. In developing nations, the primary and most critical water issues revolve around poor water quality and pollution [16].

The current water pollution scenario demands structured efforts to strengthen wastewater management plans and political leader education and improved water distribution systems alongside thorough urban and rural development planning that affects water quality. It's imperative to encourage governments to take decisive steps in tackling pressing water quality challenges and integrate comprehensive water quality policies into their national developmental agendas.

11.2.1 Sources of Toxic Pollutants in Water

Point source pollution refers to contamination originating from identifiable, single sources [17]. These sources include legally or illegally discharged wastewater from industrial facilities, oil refineries, or wastewater treatment plants, as well as septic system leaks, chemical spills, oil spills, and illegal dumping [18,19]. The Environmental Protection Agency (EPA) regulates point source pollution by setting limits on the discharge permitted from a facility directly into water bodies. Despite its specific source, point source pollution can affect extensive stretches of waterways and oceans. Polluted run-off from non-specific sources, on the other hand, arises from diffuse and less traceable origins [20]. Although nonpoint source pollution stands as the primary cause of water pollution in waters, its regulation proves challenging due to the absence of a singular, identifiable source [21]. Transboundary pollution illustrates how water pollution transcends geographical boundaries. This type of pollution occurs when contaminated water from one country enters the waters of another [22]. Such contamination can result from disasters like oil spills or the gradual movement downstream of industrial, agricultural, or municipal discharge [23].

Each year, power plants contribute over 2 billion pounds of pollutants into our water bodies, marking the largest source of toxic water pollution in the country. This wastewater carries heavy metals and harmful chemicals, such as lead, mercury, and arsenic, known to pose risks to human health [24,25]. Historically, these power plants lacked restrictions on the quantities of toxins they could release into waterways [26]. Worldwide, agriculture stands as the leading cause of water degradation, notably impacting rivers, streams, wetlands, lakes, estuaries, and groundwater [27,28].

Oil pollution isn't solely from large spills but primarily stems from everyday consumer actions, such as oil and gasoline leaks from vehicles [29]. Nearly half of the estimated 1 million tons of oil entering marine environments annually originates from land-based sources like factories, urban areas, and farms [30,31]. While tanker spills contribute about 10% of global oil in water, discharges from routine operations of the shipping industry, both legal and illegal, account for about one-third [32].

Radioactive waste, emitting radiation beyond natural environmental levels, is produced by uranium mining, nuclear power plants, military weapons production and testing, as well as research and medical usage in universities and hospitals [33]. Improperly disposed or accidentally released radioactive substances threaten groundwater, surface water, and marine ecosystems.

11.2.2 Types and Characteristics of Common Waterborne Pollutants

Waterborne contaminants encompass a diverse range of substances, both organic and inorganic contaminants, microbial and viral contaminants, pharmaceutical and personal care products, and soil and sediments that can jeopardize water quality. Understanding these diverse waterborne contaminants is crucial for monitoring, managing, and mitigating their impacts on human health and the environment [34].

Heavy metals such as lead (Pb), cadmium (Cd), chromium (Cr), nickel (Ni), arsenic (As), and mercury (Hg) are toxic metals often present due to industrial discharge or natural occurrences. Industrial discharge is the major contributor for heavy metal contaminants in aqueous ecosystems. Fertilizers and pesticides containing metals may leach into water sources. Arsenic contamination affects groundwater in regions like Bangladesh and parts of India [35].

Nitrate contamination in water originating from various sources, both natural and human-induced. They may result in "baby blue syndrome" of methemoglobinemia, a condition in which an infant's blood cannot carry enough oxygen [36]. High nitrate levels in adults have been linked to a number of health problems, including an increased risk of some cancers and other illnesses [37].

Organic contaminants include pesticides and herbicides, and industrial chemicals. Chemicals like atrazine and glyphosate from agriculture can leach into water sources, posing health threats. Atrazine contamination in surface water impacts aquatic life and potentially human health [38,39]. In addition, organic chemicals such as Polychlorinated Biphenyls (PCBs), used in various industries, persist in the environment and bioaccumulate [40]. Pharmaceuticals and Personal Care Products (PPCPs) are also emerging and persistent pollutants that prevail in aquatic matrices over a long period [41]. The overuse of medications, cosmetics and hygiene products result in the run-off and accumulation of such compounds in water supplies [42]. Traces of antibiotics and hormones detected in surface water, impacting aquatic organisms and potentially human health [43]. Compounds like bisphenol-A (BPA) and phthalates can act as endocrine-disrupting compounds (EDCs), chemicals that interfere with hormonal system [44,45]. Soil erosion and run-off can introduce sediments, silt, and suspended solids into waterways, affecting water clarity and aquatic habitats. Excessive sedimentation in rivers due to deforestation or construction, impact aquatic ecosystems [46].

11.2.3 Environmental and Human Health Impacts of Water Pollution

Water pollution has profound impacts on both the environment and human health, spanning various dimensions and causing far-reaching consequences. Water pollution negatively affects aquatic ecosystems, leading to decline in fish populations and disruption of food chains [47]. Pollutants like heavy metals, chemicals, and oil spills can poison aquatic organisms, affecting health and reproductive abilities of fish, leading to population declines, relocation, or local extinctions [48]. A decrease in biodiversity can lead to instability in ecosystems, making them more vulnerable to further disturbances like climate change, invasive species, and diseases [49,50].

Contaminants cause the degradation of water quality and significantly alter the natural habitats, impacting aquatic plants, invertebrates, and other species [51]. Additionally, the excessive nutrients from pollution lead to eutrophication, causing elevated algal blooms, reducing oxygen levels and causing "dead zones" in water bodies [52]. Pollutants like heavy metals and pesticides accumulate in water, affecting water quality and aquatic life [53].

Polluted water facilitates the breeding of disease-carrying vectors like mosquitoes, contributing to vector-borne diseases like malaria and dengue. Prolonged exposure to certain water pollutants, such as arsenic, benzene, and certain pesticides, has been linked to an increased risk of various cancers including bladder, liver, kidney, and others [54,55].

11.3 NANOTECHNOLOGY IN WATER TREATMENT

Globally, clean and safe water is a necessary resource due to the rapid increase in population, growing urbanization, industrialization, and rising agricultural practices. Current water decontamination and purification methods often involve the use of chemicals, making them energetically and operationally intensive. Consequently, these approaches demand substantial engineering expertise and infrastructure. In response to these challenges, there is a pressing need for the development of innovative wastewater treatment methodologies capable of efficiently removing contaminants. Nanotechnology emerges as a promising option, particularly in the realm of nanomaterial-based wastewater treatment [56,57]. Nanostructured materials boast unique characteristics, including high surface-to-volume ratios, heightened sensitivity and reactivity, substantial adsorption capacity, and facile functionalization. These distinctive properties position nanomaterials as potential solutions to address the limitations associated with traditional water treatment methods. Their ability to overcome these challenges makes them a compelling avenue for the advancement of wastewater treatment technologies.

11.3.1 Nanoparticles

Silver nanoparticles (Ag NPs) exhibit strong antimicrobial properties, exhibiting effective activity against a wide range of bacteria, viruses, and fungi. They are commonly used for disinfection purposes in water treatment. The size of silver nanoparticles allows for increased bioavailability for microorganisms, enhancing their biocidal activity [58]. Titanium dioxide nanoparticles (TiO_2 NPs) are known for their photocatalytic activity. TiO_2 nanoparticles produce reactive oxygen species in response to ultraviolet (UV) light, which can degrade organic dyes, pollutants, microbes, and other contaminants in water [59,60]. Iron oxide nanoparticles (Fe_3O_4 NPs), specifically magnetite (Fe_3O_4), can be used for the removal of heavy metals from water through adsorption. They can be easily separated from water using magnetic fields after the adsorption process, facilitating their recovery and reuse. Zinc oxide nanoparticles (ZnO NPs) exhibit photocatalytic properties similar to TiO_2. They can be used in water treatment processes for the degradation of organic pollutants under UV light. ZnO nanoparticles also show antimicrobial activity, contributing to disinfection [61]. Cerium oxide nanoparticles (CeO_2 NPs) have redox properties that make them suitable for catalytic applications. They can be employed for the removal of organic pollutants and heavy metals from water through adsorption and catalytic reactions. Alumina nanoparticles (Al_2O_3 NPs) can be employed in water treatment for adsorption of various contaminants, including fluoride, heavy metals, and organic compounds. The surface properties of alumina nanoparticles can be modified to enhance their adsorption capabilities [62].

Functionalized carbon nanotubes (CNTs) and graphene oxide nanoparticles (GO NPs) have enhanced adsorption capacity for specific pollutants, such as heavy metals, organic compounds, and even gases.

"Nanocomposites" refer to materials composed of at least two components: a polymer matrix and nanoparticles dispersed within it. In the context of water treatment, nanocomposites can be designed to address specific challenges such as water purification, contaminant removal, and enhanced performance of treatment processes. Nanocomposites offer versatile solutions in water treatment combining the unique properties of nanoparticles with the structural advantages of polymers. Polymer/clay nanocomposites consist of a polymer matrix and clay nanoparticles, such as montmorillonite. They are known for their improved mechanical strength, thermal stability, and enhanced adsorption capacity, making them suitable for water treatment applications. Graphene-based nanocomposites exhibit excellent electrical conductivity and large surface area, and are effective for the removal of contaminants through adsorption and membrane filtration. Carbon nanotube/polymer nanocomposites have applications in water purification and sensing [63]. Metal-organic framework (MOF) nanocomposites can be integrated into polymer matrices to form nanocomposites with high porosity and large surface area [64]. Metal nanocomposites incorporated into polymeric matrices such as dendrimers, chitosan, cellulose and resins prove effective in adsorbing gases, organic pollutants, and heavy metals from water. [65]. The specific choice of nanocomposite depends on the targeted application and the contaminants present in the water source.

11.3.2 Mechanisms Involved in Water Treatment

As summarized in Figure 11.1, the mechanisms involved in water treatment by nanoparticles can differ depending on the functionality and type of nanoparticles used and the specific contaminants targeted. It's important to note that the effectiveness of nanoparticles in water treatment depends on various factors, including the type of nanoparticles, their concentration, the specific contaminants present, and water conditions.

Nanoparticles, especially those with high surface area and functional groups, can adsorb contaminants onto their surfaces. The porous structures and surface functionalities make them effective adsorbents for gases, organic pollutants, and heavy metals. This is particularly effective for removing organic pollutants and heavy metals. Graphene oxides, carbon nanotubes, nano clay, polymeric nanoparticles, and biochar nanomaterials follow this mechanism [66]. Some nanoparticles possess catalytic properties such as photocatalytic activity, Fenton reactions, redox reactions, ozonation, hydrolysis, and electrocatalysis that can enhance the breakdown of pollutants. Metal oxides like manganese oxide (MnOx) nanoparticles can catalyse the decomposition of ozone (O_3) into reactive oxygen species [67]. Metal or metal oxide nanoparticles, such as gold (Au) or zinc oxide (ZnO), can catalyse the hydrolysis of specific organic pollutants, breaking them down into less harmful by-products [28].

Nanoparticles can be used as coagulants or flocculants to destabilize and aggregate suspended particles in water. Because of their great surface area and reactivity, nanoparticles can function as

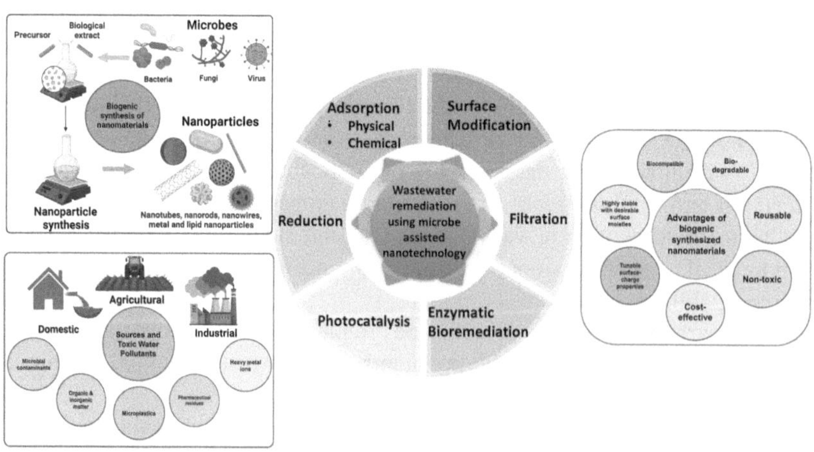

FIGURE 11.1 Schematic showing wastewater remediation using microbe-assisted nanotechnology.

efficient coagulants. Iron oxide nanoparticles, for instance, are being reported to promote coagulation by neutralising the surface charges of suspended particles and encouraging their aggregation [68]. Furthermore, coagulation potential has been demonstrated by other metallic nanoparticles, such as aluminium and titanium dioxide nanoparticles, via mechanisms including adsorption and charge neutralization. Nanoparticles can be very effective flocculants in flocculation because they facilitate the bridging and binding of destabilized particles into bigger flocs. For example, the potential of silica nanoparticles to improve flocculation through surface modification and electrostatic interactions has been investigated.

Rheological properties of polymer nanocomposite membranes such as mechanical strength, water permeability, separation efficiency, and fouling reduction are all enhanced by the use of nanoparticles. Researchers have documented enhancements in membranes with the incorporation of several nanomaterials, including nanoparticles, two-dimensional (2-D) layer materials, nanofibers, nanosheets, nanobeads, and other structural materials for nanocomposite. Nanoparticles can be incorporated into membrane materials to enhance their filtration properties. This can increase the mechanical strength, selectivity and efficiency of removing contaminants by creating additional barriers for particles or ions to pass through [9]. Incorporating materials like carbon nanotubes and graphene oxide, these membranes enhance the efficiency of water purification, addressing challenges in both municipal and industrial settings. The integration of nanosensors in water treatment systems enables real-time monitoring of water quality. Nanotechnology-based sensors offer high sensitivity and selectivity, allowing for the detection of environmental contaminants. As a result, nanomaterials open new possibilities for incredibly quick and precise water filtration membranes.

11.4 MICROBES IN WATER TREATMENT

Microbes play a vital role in water treatment, particularly in processes like wastewater treatment and purification of drinking water. One of the key processes where microbes are instrumental is in bioremediation, where they help break down and remove contaminants from water.

Microbial-driven bioremediation is a powerful and environmentally friendly approach to cleaning up pollutants in various ecosystems. This process harnesses the metabolic activities of microorganisms to transform or degrade contaminants into less harmful or non-toxic substances.

Microbial bioremediation is a natural and sustainable solution that utilizes the inherent capabilities of certain bacteria and fungi to degrade hydrocarbons present in waste material. Microbial enzymes, such as hydroxylases and oxygenases, enable them to break down complex hydrocarbon molecules into simpler compounds that can be utilized as carbon and energy sources.

11.4.1 Synergizing Microbes and Nanotechnology

The combination of biotechnological and nanotechnological approaches to synthesize nanoparticles using microorganisms is called microbial-assisted nanotechnology. In this approach, microorganisms are used with or without any modification to synthesize nanoparticles. Microorganisms are cultured on a large scale in fermenters. Nanoparticles are produced either by microbes themselves or by using the biomolecules extracted from microbes [69,70].

11.4.1.1 Algae-Mediated Synthesis of Nanoparticles

Ismail et al. mixed $AgNO_3$ solution with homogenized cultures of algae like *Arthrospira platensis* (formerly known as *Spirulina platensis*) and *Anabaena variabilis* separately and synthesized silver nanoparticles from both algal species. Nanoparticles from both algal species showed capacity to adsorb dyes like malachite green and remove them from water. The authors also tested nanoparticle-treated water on wheat seedlings and compared their growth with controls grown in distilled water and malachite green. Wheat seedlings in nanoparticle-treated water showed similar growth patterns as those in distilled water showing that the nanoparticles are not toxic [71].

Borah et al. synthesized silver nanoparticles using silver nitrate and dried biomass of unicellular algae *Chlorella ellipsoidea*. The synthesized nanoparticles catalysed degradation of dyes like methyl orange and methylene blue by $NaBH_4$ under UV light. Furthermore, these nanoparticles were bactericidal against Gram-positive bacteria like *Staphylococcus aureus* and Gram-negative bacteria like *Pseudomonas aeruginosa*, *Klebsiella pneumoniae* and *Escherichia coli* [72].

Goverthanan et al. synthesized Fe_3O_4 nanoparticles encapsulated with exopolysaccharides from algae *Chlorella vulgaris*. These nanoparticles were able to remove phosphate and ammonium ions from wastewater [73].

11.4.1.2 Bacteria-Mediated Synthesis of Nanoparticles

Zhu et al. genetically engineered *E. coli* bacteria to produce proteins from inserted genes, namely, SynHMB and T6SS (Type 6 secretory system which is isolated from *Pseudomonas putida*). These modified bacteria produced SynHMB proteins on their surface. The authors also synthesized silica nanoparticles with polyethyleneimine (PEI) and diethylenetriaminepentaacetic acid (DTPA). The engineered bacteria and silica nanoparticles got bound to each other to form aggregates. These aggregates had SynHMB, PEI and DTPA molecules on their surface which trapped heavy metal ions like Cd^{2+} and Pb^{2+} in water by complexing with them [74].

Qunying Yuan et al. inserted the metallothionein gene from *Candida albicans* into *E. coli* bacteria. These modified *E. coli* bacteria produced metallothionein protein which reduced and capped silver ions to produce silver nanoparticles. These nanoparticles which were produced inside the bacterial cells were extracted using repeated cycles of urea treatment, sonication and centrifugation. However, it is yet to be tested if it can be used to treat wastewater or not [75].

Kamal et al. synthesized copper nanoparticles using copper sulphate, carboxymethylcellulose (CMC) and hydrazine hydrate. Copper ions (Cu^{2+}) were stabilized by CMC to form copper nanoparticles. These copper nanoparticles were adsorbed onto a nanofiber sheet made of bacterial cellulose synthesized by *Gluconacetobacter xylinus* bacteria, leading to formation of an immobilized catalyst. This catalyst was able to reduce dyes like methylene blue and 4-nitrophenol in the presence of sodium borohydride [76].

11.4.1.3 Fungi-Mediated Synthesis of Nanoparticles

Mahanty et al. [77] synthesized iron oxide nanoparticles by using the fungus *Aspergillus tubingensis* STSP25 isolated from the rhizosphere of the mangrove plant *Avicenna officinalis*. The fungi were cultured in a liquid medium and after achieving sufficient growth they filtered the culture. The filtrate (also called as fungal cell filtrate) containing extracellular biomolecules was mixed with $Fe(NO_3)_3$ salt solution to produce nanoparticles. The biomolecules in the fungal cell filtrate capped the reduced Fe atoms.

These nanoparticles successfully removed heavy metals like nickel, lead, zinc and copper from wastewater by the process of adsorption of these heavy metal ions onto the nanoparticle surface [77]. Sharma et al. synthesized zinc oxide (ZnO) nanoparticles by mixing cell-free extracts of the fungus *Phanerochaete chrysosporium* with zinc sulphate salt. These nanoparticles were stabilized using cellulose from sugarcane bagasse. The ZnO-cellulose nanocomposites thus formed showed antibacterial activity against pathogenic bacteria like *E. coli* and *S. aureus* which are found in contaminated water [78].

Kalpana et al. synthesized zinc oxide (ZnO) nanoparticles from cell-free extracts of the fungus *Aspergillus niger*. They used zinc nitrate as a source of zinc ions. These ZnO nanoparticles showed catalytic activity in the reduction of the dye Bismarck Brown. These nanoparticles also showed antibacterial activity against waterborne disease-causing bacteria like *E. coli* and *S. aureus*. Moreover, these nanoparticles are not phytotoxic as they showed no inhibition of germination of pea seeds in phytotoxicity studies [79].

11.4.2 Benefits and Challenges of Integrating Microbes and Nanomaterials

Integrating microbes with nanoparticle synthesis has many advantages over traditional physical and chemical methods. Most traditional methods involve the use of reagents that are toxic to humans and can pollute the environment. In microbial nanoparticle synthesis, these reagents are completely avoided, thereby making this method safer and more ecofriendly over traditional methods. Microorganisms divide very rapidly, and thus, one can maintain a stock culture of microbes which can be used again and again for batch-to-batch synthesis of nanoparticles. Due to this, microbial methods are cheaper and more feasible than traditional methods wherein the chemicals used can be expensive [70].

However, microbial method of nanoparticle synthesis comes with challenges. These challenges include selecting the best strains of microbial species for nanoparticle production; optimizing conditions like pH, temperature, and salt concentration for the growth of these microbes; and extracting nanoparticles from microbial cells. Moreover, controlling the size and shape of nanoparticles in this method is difficult. The cellular mechanisms involved in the synthesis of metal nanoparticles are not yet fully understood. Scaling up nanoparticle production to industrial levels can be a challenge in this method [80].

11.5 ENVIRONMENTAL IMPACTS AND SAFETY

Nanotechnology not only offers promising approaches for water and wastewater remediation due to its unique properties, but it also raises environmental concerns and safety considerations. While nanotechnology shows promise in water remediation, its environmental impacts and safety concerns demand comprehensive risk assessments, proper regulation, and ethical considerations. Understanding and mitigating potential risks associated with nanomaterials are crucial for ensuring their safe and responsible use in water and wastewater treatment.

Engineered nanoparticles used in remediation processes might have high mobility in aquatic environments, potentially leading to unintended ecological exposure. Nanoparticles might accumulate in organisms, potentially affecting the food chain and ecosystem dynamics. Some nanoparticles have also shown adverse effects on aquatic organisms, raising concerns about their ecological impacts. Interactions between nanoparticles and biota can lead to changes in biological processes and ecosystem functioning.

11.5.1 Assessing the Potential Environmental Impacts of Microbial-Assisted Nanotechnology

Microbial-assisted nanotechnology, while holding promise for various environmental applications, requires a thorough assessment of its potential impacts to ensure responsible and safe implementation. Microbes can interact with nanoparticles, potentially altering their behaviour, mobility, and fate in the environment. Microbially mediated processes might influence the movement of nanoparticles in soil, water, and air, affecting their environmental distribution [81]. Microbes play vital roles in ecosystems, and alterations due to nanoparticle exposure might impact ecological processes. Changes in microbial communities due to nanoparticle interactions could have cascading effects on ecosystem health and function [82]. Workers involved in microbial-assisted nanotechnology production and application may face exposure risks, necessitating proper safety protocols. Nanomaterials produced or manipulated using microbial systems might pose risks via inhalation, dermal contact, or ingestion. Current regulatory frameworks might not sufficiently tackle the distinctive challenges posed by microbial-assisted nanotechnology. The environmental risks associated with the release of nanomaterials into the environment are also essential aspects. These include examining their potential toxicity to ecosystems, their persistence in various environments, and their likelihood of bioaccumulation in living organisms [48]. Balancing the potential benefits of microbial-assisted nanotechnology against the risks is crucial. While these technologies hold promise for advancements, understanding and mitigating their environmental implications are essential to ensure their sustainable and safe deployment [83]. Discussions on risk communication, public perception, and ethical implications are crucial for responsible adoption. Microbial-assisted nanotechnology holds immense potential but demands careful evaluation of environmental impacts, ecosystem interactions, and human safety considerations. Comprehensive risk assessments, regulatory updates, and ethical discussions are imperative to ensure its safe and responsible deployment in diverse environmental applications.

11.5.2 Safety Considerations and Regulatory Aspects

Microbial-assisted nanotechnology for water and wastewater remediation presents both innovative solutions and potential safety concerns, prompting a comprehensive assessment of its safety considerations and regulatory aspects. Those involved in the production, handling, and application of microbial-assisted nanotechnologies may face exposure risks, necessitating stringent safety protocols [84].

Inhalation, dermal contact, and ingestion could serve as potential exposure routes for workers dealing with these nanomaterials. Comprehensive studies on nanomaterial toxicity and their potential effects on human health are crucial for risk assessment and management [85]. Understanding how these nanomaterials interact within biological systems is essential to assess their bioavailability and potential for accumulation in living organisms. Existing regulations might have certain shortcomings while addressing the unique risks associated with microbial-assisted nanotechnologies, necessitating updates [83]. Development of specific guidelines and frameworks addressing the safety and environmental aspects of these technologies is imperative for their responsible use. Organizations such as ISO (International Organization for Standardization) develop standards to safeguard the use of nanotechnology. These standards may cover aspects like risk assessment methodologies, handling protocols, and environmental impact assessments [86]. Open dialogues addressing ethical concerns, risk communication, and societal perceptions are essential for the responsible adoption of these technologies. Engaging stakeholders in discussions regarding the risks and benefits of these technologies can lead to informed decision-making and regulatory frameworks. Microbial-assisted nanotechnology holds promise for water and wastewater remediation but demands rigorous safety assessments, updated regulatory frameworks, and ethical deliberations. Balancing innovation with safety is crucial for ensuring the responsible deployment of these technologies while safeguarding human health and the environment.

11.6 FUTURE DIRECTIONS AND CHALLENGES

The future of microbial-assisted nanotechnology presents a promising landscape filled with both opportunities and challenges. One emerging trend lies in harnessing the power of microbial communities to drive nanomaterial synthesis and manipulation. This innovative approach offers a sustainable and ecofriendly alternative to traditional methods and enables precise control over nanoparticle properties, opening new avenues in fields like medicine, environmental remediation, and electronics. However, as with any burgeoning field, significant technological challenges and research gaps persist. One such challenge involves elucidating the intricate mechanisms underlying microbial-mediated nanomaterial synthesis and optimizing the process for scalability and reproducibility. Additionally, there is a pressing need to enhance our understanding of microbial interactions within nanomaterial matrices to mitigate potential environmental and biological risks. Moreover, bridging the gap between laboratory-scale demonstrations and large-scale industrial applications remains a formidable obstacle. Nevertheless, the potential for scale-up and commercialization is vast. With advancements in bioreactor design, process engineering, and automation, the translation of microbial-assisted nanotechnology from bench to market is becoming increasingly feasible.

Moreover, strategic collaborations between academia, industry, and regulatory bodies will be paramount in navigating the complex landscape of intellectual property rights, safety regulations, and market dynamics. Despite these challenges, the transformative impact of microbial-assisted nanotechnology on various sectors cannot be overstated. From revolutionizing drug delivery systems to enabling sustainable energy production, the convergence of microbiology and nanotechnology holds immense promise for addressing pressing global challenges. By fostering interdisciplinary collaboration, investing in foundational research, and fostering an innovation-friendly ecosystem, we can propel this nascent field towards realizing its full potential and usher in a new era of technological advancement and societal benefit.

11.7 CONCLUSION

In conclusion, the potential of microbial-assisted nanotechnology represents a ground-breaking frontier in scientific innovation, offering sustainable solutions to pressing global challenges. By harnessing the unique capabilities of microbial communities to synthesize and manipulate nanomaterials, we have unlocked unprecedented opportunities across diverse fields, from medicine to environmental remediation. Particularly, in the realm of water pollution, microbial-assisted nanotechnology emerges as a sustainable solution, offering efficient and cost-effective methods for contaminant removal and water purification. As we confront escalating global water challenges, characterized by pollution, scarcity, and inequitable access, collaborative efforts and interdisciplinary research are imperative. By fostering synergistic collaborations among scientists, engineers, policymakers, and stakeholders, we can accelerate the translation of innovative technologies from the lab to real-world applications. Moreover, the role of policy and public awareness cannot be overstated. Strong regulatory frameworks that are based on ethical and scientific principles are necessary to guarantee the appropriate and safe application of microbially aided nanotechnology. Concurrently, raising public awareness about the potential benefits and risks associated with these emerging technologies fosters informed decision-making and societal engagement. As we navigate towards a future defined by sustainability and resilience, it is imperative that we leverage the transformative potential of microbial-assisted nanotechnology to address the complex water challenges facing our planet. Through concerted efforts and a shared commitment to scientific excellence and ethical stewardship, we can harness the power of innovation to safeguard water resources, protect human health, and advance the welfare of current and future generations. In doing so, we chart a path towards a more equitable, prosperous, and sustainable world.

REFERENCES

1. Bashir I, Lone FA, Bhat RA, et al. Concerns and threats of contamination on aquatic ecosystems. In: Hakeem KR, Bhat RA, Qadri H, editors. *Bioremediation and Biotechnology: Sustainable Approaches to Pollution Degradation*. Cham: Springer International Publishing; 2020. pp. 1–26.
2. Ingrao C, Strippoli R, Lagioia G, et al. Water scarcity in agriculture: An overview of causes, impacts and approaches for reducing the risks. *Heliyon*. 2023;9(8):e18507.
3. Adla K, Dejan K, Neira D, et al. Chapter 9: Degradation of ecosystems and loss of ecosystem services. In: Prata JC, Ribeiro AI, Rocha-Santos T, editors. *One Health*. Cambridge, MA: Academic Press; 2022. pp. 281–327.
4. Tignino M, Bréthaut C. *Research Handbook on Freshwater Law and International Relations*. Cheltenham: Edward Elgar Publishing; 28 Dec. 2018.
5. Jianping L, Minrong L, Jinnan W, et al. Global environmental issues and human wellbeing. In: Jianping L, Minrong L, Jinnan W, et al., editors. *Report on Global Environmental Competitiveness*. Berlin: Springer; 2014. pp. 3–21.
6. Vijayaraghavan P, Veeramanikandan V, Pradeep BV, et al. Enhanced aerobic naphthalene degradation utilizing indigenous microbial flora as a biocatalyst in oil-contaminated wastewater. *Topics in Catalysis*. 2024; 67:1–11.

7. Desai BH. 14. United nations environment programme (UNEP). *Yearbook of International Environmental Law.* 2020;31(1):319–325.
8. Panda J, Mishra AK, Mohanta YK, et al. Exploring biopolymer for food and pharmaceuticals application in the circular bioeconomy: An agro-food waste-to-wealth approach. *Waste and Biomass Valorization.* 2024;15:1–31.
9. Li X, Liu E, Zhang E, et al. Spatio-temporal variations of sedimentary metals in a large suburban lake in southwest China and the implications for anthropogenic processes. *Science of the Total Environment.* 2020;707:135650.
10. Rawindran H, Khoo KS, Ethiraj B, et al. Environmental impact assessment via life cycle analysis on ultrafiltration membrane fabricated from polyethylene terephthalate waste to treat microalgal cultivation wastewater for reusability. *Environmental Research.* 2024;251:118687.
11. Fritsch O, Benson D. *Handbook on the Governance and Politics of Water Resources.* Cheltenham: Edward Elgar Publishing; 04 Jun. 2024.
12. Zacharias I, Liakou P, Biliani I. A review of the status of surface European waters twenty years after WFD introduction. *Environmental Processes.* 2020;7(4):1023–1039.
13. Zango ZU, Khoo KS, Garba A, et al. A review on carbon-based biowaste and organic polymer materials for sustainable treatment of sulfonamides from pharmaceutical wastewater. *Environmental Geochemistry and Health.* 2024;46(4):145.
14. François M, Lin K-S, Rachmadona N, et al. Utilization of carbon-based nanomaterials for wastewater treatment and biogas enhancement: A state-of-the-art review. *Chemosphere.* 2023;350:141008.
15. Gao W, Du Y, Gao S, et al. Heavy metal accumulation reflecting natural sedimentary processes and anthropogenic activities in two contrasting coastal wetland ecosystems, eastern China. *Journal of Soils and Sediments.* 2016;16(3):1093–1108.
16. Xiao L, Liu J, Ge J. Dynamic game in agriculture and industry cross-sectoral water pollution governance in developing countries. *Agricultural Water Management.* 2021;243:106417.
17. Gupta A. *Water Pollution-Sources, Effects and Control.* Jaipur: Pointer Publishers; 2016.
18. Schweitzer L, Noblet J. Chapter 3.6-Water contamination and pollution. In: Török B, Dransfield T, editors. *Green Chemistry.* Elsevier; 2018. pp. 261–290. https://doi.org/10.1016/B978-0-12-809270-5.00011-X.
19. Ng CA, Chew SN, Bashir MJ, et al. Enhancing microbial fuel cell performance for sustainable treatment of palm oil mill wastewater using carbon cloth anode coated with activated carbon. *International Journal of Hydrogen Energy.* 2024;52:1092–1104.
20. Adu JT, Kumarasamy MV. Assessing non-point source pollution models: A review. *Polish Journal of Environmental Studies.* 2018;27(5):1913–1922.
21. Wang R, Wang Q, Dong L, et al. Cleaner agricultural production in drinking-water source areas for the control of non-point source pollution in China. *Journal of Environmental Management.* 2021;285:112096.
22. Lu J. Turnover of environmental protection officials and transboundary water pollution. *Environmental Science and Pollution Research.* 2021;28(8):10207–10223.
23. Parven A, Hasan Md.S. Trans-boundary water conflicts between Bangladesh and India: water governance practice for conflict resolution. *International Journal of Agricultural Research, Innovation and Technology.* 2018;8:79.
24. Verma C, Madan S, Hussain A. Heavy metal contamination of groundwater due to fly ash disposal of coal-fired thermal power plant, Parichha, Jhansi, India. *Cogent Engineering.* 2016;3(1):1179243.
25. Miara A, Macknick JE, Vörösmarty CJ, et al. Climate and water resource change impacts and adaptation potential for US power supply. *Nature Climate Change.* 2017;7(11):793–798.
26. Dmitrienko MA, Strizhak PA. Coal-water slurries containing petrochemicals to solve problems of air pollution by coal thermal power stations and boiler plants: An introductory review. *Science of the Total Environment.* 2018;613–614:1117–1129.
27. Pericherla S, Karnena MK, Vara S. A review on impacts of agricultural runoff on freshwater resources. *International Journal on Emerging Technologies.* 2020;11:829–833.
28. Ray MK, Panda J, Panda BP, et al. Mycoremediation of heavy metals and/or metalloids in soil. In: Sarma H, Joshi S, editors. *Land Remediation and Management: Bioengineering Strategies.* Singapore: Springer Nature; 2023. pp. 161–190.
29. Nambi IM, Rajasekhar B, Loganathan V, et al. An assessment of subsurface contamination of an urban coastal aquifer due to oil spill. *Environmental Monitoring and Assessment.* 2017;189(4):148.
30. Kankara RS, Arockiaraj S, Prabhu K. Environmental sensitivity mapping and risk assessment for oil spill along the Chennai Coast in India. *Marine Pollution Bulletin.* 2016;106(1):95–103.

31. Pham MN, Nishimura F, Lan JCW, et al. Recent advancement of eliminating antibiotic resistance bacteria and antibiotic resistance genes in livestock waste: A review. *Environmental Technology & Innovation*. 2024;36:103751.
32. Chen J, Di Z, Shi J, et al. Marine oil spill pollution causes and governance: A case study of Sanchi tanker collision and explosion. *Journal of Cleaner Production*. 2020;273:122978.
33. Bajwa BS. Uranium and other heavy toxic elements distribution in the drinking water samples of SW-Punjab, India – A potentially dangerous trend. India: Hemvati Nandan Bahuguna Garhwal University; 2015.
34. Ahamad A, Madhav S, Singh AK, et al. Types of water pollutants: Conventional and emerging. In: Pooja D, Kumar P, Singh P, et al., editors. *Sensors in Water Pollutants Monitoring: Role of Material*. Singapore: Springer; 2020. pp. 21–41.
35. Shaji E, Santosh M, Sarath KV, et al. Arsenic contamination of groundwater: A global synopsis with focus on the Indian Peninsula. *Geoscience Frontiers*. 2021;12(3):101079.
36. Shaban J, Al-Najar H, Kocadal K, et al. The effect of nitrate-contaminated drinking water and vegetables on the prevalence of acquired methemoglobinemia in Beit Lahia City in Palestine. *Water*. 2023. https://doi.org/10.3390/w15111989.
37. Picetti R, Deeney M, Pastorino S, et al. Nitrate and nitrite contamination in drinking water and cancer risk: A systematic review with meta-analysis. *Environmental Research*. 2022;210:112988.
38. Urseler N, Bachetti R, Biolé F, et al. Atrazine pollution in groundwater and raw bovine milk: Water quality, bioaccumulation and human risk assessment. *Science of the Total Environment*. 2022;852:158498.
39. Ardo FM, Khoo KS, Sobri MZA, et al. Modelling photoperiod in enhancing hydrogen production from Chlorella vulgaris sp. while bioremediating ammonium and organic pollutants in municipal wastewater. *Environmental Pollution*. 2024;346:123648.
40. Mikolajczyk S, Warenik-Bany M, Maszewski S, et al. Dioxins and PCBs – Environment impact on freshwater fish contamination and risk to consumers. *Environmental Pollution*. 2020;263:114611.
41. Ebele AJ, Abou-Elwafa Abdallah M, Harrad S. Pharmaceuticals and personal care products (PPCPs) in the freshwater aquatic environment. *Emerging Contaminants*. 2017;3(1):1–16.
42. Narayanan M, El-sheekh M, Ma Y, et al. Current status of microbes involved in the degradation of pharmaceutical and personal care products (PPCPs) pollutants in the aquatic ecosystem. *Environmental Pollution*. 2022;300:118922.
43. Galindo-Miranda JM, Guízar-González C, Becerril-Bravo EJ, et al. Occurrence of emerging contaminants in environmental surface waters and their analytical methodology – A review. *Water Supply*. 2019;19(7):1871–1884.
44. Erkekoglu P, Kocer-Gumusel B. Environmental effects of endocrine-disrupting chemicals: A special focus on phthalates and bisphenol A. In: Larramendy ML, Soloneski S, editors. *Environmental Health Risk – Hazardous Factors to Living Species*. Intech Open; 2016. pp. 155–90. https://doi.org/10.5772/62455.
45. Kawa IA, Akbar M, Fatima Q, et al. Endocrine disrupting chemical Bisphenol A and its potential effects on female health. *Diabetes & Metabolic Syndrome: Clinical Research & Reviews*. 2021;15(3):803–811. https://doi.org/10.1016/j.dsx.2021.03.031.
46. Gharibreza M, Zaman M, Arabkhedri M, et al. The off-site implications of deforestation on sedimentation rates and pollution in Abkenar open water (Anzali Lagoon, Caspian Sea) using radionuclide techniques and sediment quality indices. *International Journal of Sediment Research*. 2022;37(3):370–382.
47. Driscoll CT, Mason RP, Chan HM, et al. Mercury as a global pollutant: Sources, pathways, and effects. *Environmental Science & Technology*. 2013;47(10):4967–4983.
48. Islam FS, Gault AG, Boothman C, et al. Role of metal-reducing bacteria in arsenic release from Bengal delta sediments. *Nature*. 2004;430(6995):68–71.
49. Dudgeon D, Arthington AH, Gessner MO, et al. Freshwater biodiversity: Importance, threats, status and conservation challenges. *Biological Reviews*. 2006;81(2):163–182.
50. Bogardi JJ, Leentvaar J, Sebesvári Z. *Biologia Futura*: Integrating freshwater ecosystem health in water resources management. *Biologia Futura*. 2020;71(4):337–358.
51. Posthuma L, Brack W, van Gils J, et al. Mixtures of chemicals are important drivers of impacts on ecological status in European surface waters. *Environmental Sciences Europe*. 2019;31(1):71.
52. Kakade A, Salama E-S, Han H, et al. World eutrophic pollution of lake and river: Biotreatment potential and future perspectives. *Environmental Technology & Innovation*. 2021;23:101604.
53. Rai PK. Heavy metal pollution in aquatic ecosystems and its phytoremediation using wetland plants: An ecosustainable approach. *International Journal of Phytoremediation*. 2008;10(2):133–160.

54. Bellinger DC. Lead neurotoxicity and socioeconomic status: Conceptual and analytical issues. *NeuroToxicology.* 2008;29(5):828–832.
55. Bräuner EV, Andersen ZJ, Frederiksen M, et al. Health effects of PCBs in residences and schools (HESPERUS): PCB – health cohort profile. *Scientific Reports.* 2016;6(1):24571.
56. Yadav S, Chander S, Gupta A, et al. Biogenic engineered zinc oxide nanoparticle for sulfur black dye removal from contaminated wastewater: Comparative optimization, simulation modeling, and isotherms. *Bioengineered.* 2024;15(1):2325721.
57. Usman A, Khoo KS, Ariffin MM, et al. Adsorption of terbutaline β-agonists from wastewater by mechano-synthesized iron oxide nanoparticles modified copper (II) isonicotinate metal-organic framework. Available at SSRN 4850341.2024.
58. Rasheed A, Hussain S, Mushtaq W, et al. Application of silver nanoparticles synthesized through varying biogenic and chemical methods for wastewater treatment and health aspects. *Environmental Science and Pollution Research.* 2023. https://doi.org/10.1007/s11356-022-24761-4.
59. Riaz S, Park S-J. An overview of TiO_2-based photocatalytic membrane reactors for water and wastewater treatments. *Journal of Industrial and Engineering Chemistry.* 2019;84:23–41.
60. Zango ZU, Khoo KS, Garba A, et al. A review on titanium oxide nanoparticles modified metal-organic frameworks for effective CO_2 conversion and efficient wastewater remediation. *Environmental Research.* 2024;252:119024.
61. Sheikh M, Pazirofteh M, Dehghani M, et al. Application of ZnO nanostructures in ceramic and polymeric membranes for water and wastewater technologies: A review. *Chemical Engineering Journal.* 2020;391:123475.
62. Banerjee S, Dubey S, Gautam RK, et al. Adsorption characteristics of alumina nanoparticles for the removal of hazardous dye, Orange G from aqueous solutions. *Arabian Journal of Chemistry.* 2019;12(8):5339–5354.
63. Nasrollahzadeh M, Sajjadi M, Iravani S, et al. Carbon-based sustainable nanomaterials for water treatment: State-of-art and future perspectives. *Chemosphere.* 2021;263:128005.
64. Yang F, Du M, Yin K, et al. Applications of metal-organic frameworks in water treatment: A review. *Small.* 2022;18(11):2105715.
65. Fu S, Sun Z, Huang P, et al. Some basic aspects of polymer nanocomposites: A critical review. *Nano Materials Science.* 2019;1(1):2–30.
66. Neeti K, Singh R, Ahmad S. The role of green nanomaterials as effective adsorbents and applications in wastewater treatment. *Materials Today: Proceedings.* 2023;77:269–276.
67. Fan K, Chen Q, Zhao J, et al. Preparation of MnO_2-carbon materials and their applications in photocatalytic water treatment. *Nanomaterials.* 2023. https://doi.org/10.3390/nano13030541.
68. Gupte T, Pradeep T. Chapter 3- Nanosensors for water quality monitoring. In: Ahuja S, editor. *Separation Science and Technology.* Vol. 15: Academic Press; 2022. pp. 37–53. https://doi.org/10.1016/B978-0-323-90763-7.00010-X.
69. Dhanker R, Hussain T, Tyagi P, et al. The emerging trend of bio-engineering approaches for microbial nanomaterial synthesis and its applications [Review]. *Frontiers in Microbiology.* 2021;12:638003.
70. Koul B, Poonia AK, Yadav D, et al. Microbe-mediated biosynthesis of nanoparticles: Applications and future prospects. *Biomolecules.* 2021;11(6):886.
71. Ismail GA, Allam NG, El-Gemizy WM, et al. The role of silver nanoparticles biosynthesized by Anabaena variabilis and Spirulina platensis cyanobacteria for malachite green removal from wastewater. *Environmental Technology.* 2021;42(28):4475–4489.
72. Borah D, Das N, Das N, et al. Alga-mediated facile green synthesis of silver nanoparticles: Photophysical, catalytic and antibacterial activity. *Applied Organometallic Chemistry.* 2020;34(5):e5597.
73. Govarthanan M, Jeon C-H, Jeon Y-H, et al. Non-toxic nano approach for wastewater treatment using Chlorella vulgaris exopolysaccharides immobilized in iron-magnetic nanoparticles. *International Journal of Biological Macromolecules.* 2020;162:1241–1249.
74. Zhu N, Zhang B, Yu Q. Genetic engineering-facilitated coassembly of synthetic bacterial cells and magnetic nanoparticles for efficient heavy metal removal. *ACS Applied Materials & Interfaces.* 2020;12(20):22948–22957.
75. Yuan QA-O, Bomma MA-O, Xiao Z. Enhanced silver nanoparticle synthesis by *Escherichia coli* transformed with *Candida albicans* metallothionein gene. *Materials.* 2019;12:4180. https://doi.org/10.3390/ma12244180.
76. Kamal T, Ahmad I, Khan SB, et al. Microwave assisted synthesis and carboxymethyl cellulose stabilized copper nanoparticles on bacterial cellulose nanofibers support for pollutants degradation. *Journal of Polymers and the Environment.* 2019;27(12):2867–2877.

77. Mahanty S, Chatterjee S, Ghosh S, et al. Synergistic approach towards the sustainable management of heavy metals in wastewater using mycosynthesized iron oxide nanoparticles: Biofabrication, adsorptive dynamics and chemometric modeling study. *Journal of Water Process Engineering.* 2020;37:101426.
78. Sharma JL, Dhayal V, Sharma RK. White-rot fungus mediated green synthesis of zinc oxide nanoparticles and their impregnation on cellulose to develop environmental friendly antimicrobial fibers. *3 Biotech.* 2021;11(6):269.
79. Kalpana VN, Kataru BAS, Sravani N, et al. Biosynthesis of zinc oxide nanoparticles using culture filtrates of Aspergillus niger: Antimicrobial textiles and dye degradation studies. *OpenNano.* 2018;3:48–55.
80. Kapoor RT, Salvadori MR, Rafatullah M, et al. Exploration of microbial factories for synthesis of nanoparticles – A sustainable approach for bioremediation of environmental contaminants. *Front Microbiol.* 2021;12:658294. https://doi.org/10.3389/fmicb.2021.658294.
81. Shende, S.S. *et al.* Interaction of Nanoparticles with Microbes. In: Singh, P., Singh, R., Verma, P., Bhadouria, R., Kumar, A., Kaushik, M. (eds) *Plant-Microbes-Engineered Nano-particles (PM-ENPs) Nexus in Agro-Ecosystems. Advances in Science, Technology & Innovation.* Springer, Cham; 2021. https://doi.org/10.1007/978-3-030-66956-0_12
82. Rico CM, Majumdar S, Duarte-Gardea M, et al. Interaction of nanoparticles with edible plants and their possible implications in the food chain. *Journal of Agricultural and Food Chemistry.* 2011;59(8):3485–3498.
83. Hull M, Bowman D. *Nanotechnology Environmental Health and Safety: Risks, Regulation, and Management.* Norwich: William Andrew; 2014.
84. Fadeel B, Farcal L, Hardy B, et al. Advanced tools for the safety assessment of nanomaterials. *Nature Nanotechnology.* 2018;13(7):537–543.
85. Avramescu SM, Fierascu I, Akhtar K, et al. *Engineered Nanomaterials: Health and Safety.* Norderstedt: BoD–Books on Demand; 2020.
86. Lee JH, Kuk WK, Kwon M, Lee JH, et al. Evaluation of information in nanomaterial safety data sheets and development of international standard for guidance on preparation of nanomaterial safety data sheets. *Nanotoxicology.* 2013;7(3):338-345. https://doi.org/10.3109/17435390.2012.658095.

12 Microbial Polysaccharide–Based Nanomaterials for Efficient Removal of Pollutants from Wastewater

Anita Mandal, Siddharth Solanki, Akbar Husen Sunasara, Kalyani Vikhe, Amilia Nongbet, Bibhu Prasad Panda, and Neelam Amit Kungwani

12.1 INTRODUCTION

Water is the most valued natural resource for living beings but is vulnerable to pollution due to urban development and rapid industrialization [1]. Freshwater resources are being compromised due to the discharge of untreated wastewater contaminated with various pollutants including heavy metals, oils, personal care products, and pharmaceuticals at an alarming rate. Many of these pollutants are non-biodegradable and tend to accumulate in the ecosystem, affecting different aquatic biota [2,3]. Removal of pollutants from wastewater is traditionally done by techniques such as ion exchange, flotation, coagulation, flocculation, reverse osmosis, oxidation, adsorption, ultrafiltration, sedimentation, and electro-precipitation [1]. However, it also adds to high operational costs, energy consumption, and use of hazardous chemicals, which leads to secondary pollution. In the search for sustainable solutions to mitigate environmental hazards caused by xenobiotics, the use of nanomaterials has opened many possibilities for enhancing the efficiency of traditional wastewater treatment technologies [4,5]. Polysaccharide-based nanomaterials hold significant potential among various nanomaterials due to their biodegradable, biocompatible, and eco-friendly nature. Moreover, they can be derived from renewable resources and offer sustainable solutions for environmental applications [6,7].

Polysaccharides sourced from animals, plants, algae, and microbes have gained much interest due to their varied structural and functional diversity [8,9]. Among different sources of biopolymer, microbial polysaccharides are currently receiving much attention due to their several key advantages. Microbial production is independent of climatic conditions, they exhibit rapid growth, and their production process is much simpler. Furthermore, they can be genetically modified for the synthesis of specific polysaccharides [8]. Microbial polysaccharides derived from bacteria, fungi, yeast, and microalgae have been widely studied for their biomedical and environmental applications [10,11]. They can be categorized into three different types based on their localization within a microbial cell, i.e. intracellular, extracellular, and cell wall polysaccharides. Furthermore, they are divided into homo and heteropolysaccharides based on monomer units. Structurally, they have distinctive hydrophilic groups and highly branched architecture which enable them to interact with various pollutants through various mechanisms [12]. Some examples of microbial polysaccharides include dextran, pullulan, bacterial cellulose, xanthan, gellan, curdlan, and levan [13] (Table 12.1, Figure 12.1). These polysaccharides can be transformed into nanomaterials by using various techniques described further in this chapter.

DOI: 10.1201/9781003487142-12

TABLE 12.1
Sources and Application of Microbial Polysaccharides

Polysaccharide	Organism	Environmental Application	References
Xanthan gum	*Xanthomonas campestris*	Oil recovery, adsorption of heavy metals such as lead	[21,22]
Curdlan	*Alcaligenes faecalis*	Heavy metal removal	[23–26]
Pullulan	*Aureobasidium pullulans*	Dye removal, antibiotics removal	[27,28]
Dextran	*Leuconostoc mesenteroides, Weisella, Streptococcus*	Degradation and removal of antibiotics, helps to remove endocrine-disrupting compounds	[29–31]
Bacterial cellulose	*Acetobacter, Aerobacter, Agrobacterium, Pseudomonas*	Heavy metal adsorption, removal of oil and organic solvents, pesticides	[32,33]
Alginate	*Pseudomonas aeruginosa* and *Azotobacter* sp.	Dye removal, antibiotic removal, heavy metal removal	[34]
Glucan	*Leuconostoc dextranicum*	Heavy metal adsorption	[35]
Levan	*Alcaligenes viscosus, Halomonas eurihalina, Zymomonas mobilis*	Antimicrobial application	[36]
Gellan	*Pseudomonas elodea, Sphingomonas elodea, Sphingomonas paucimobilis*	Adsorption of heavy metal ions	[37]
Scleroglucan	*Sclerotium rolfsii, Epicoccum nigrum, Botrytis cinerea*	Oil recovery	[38]

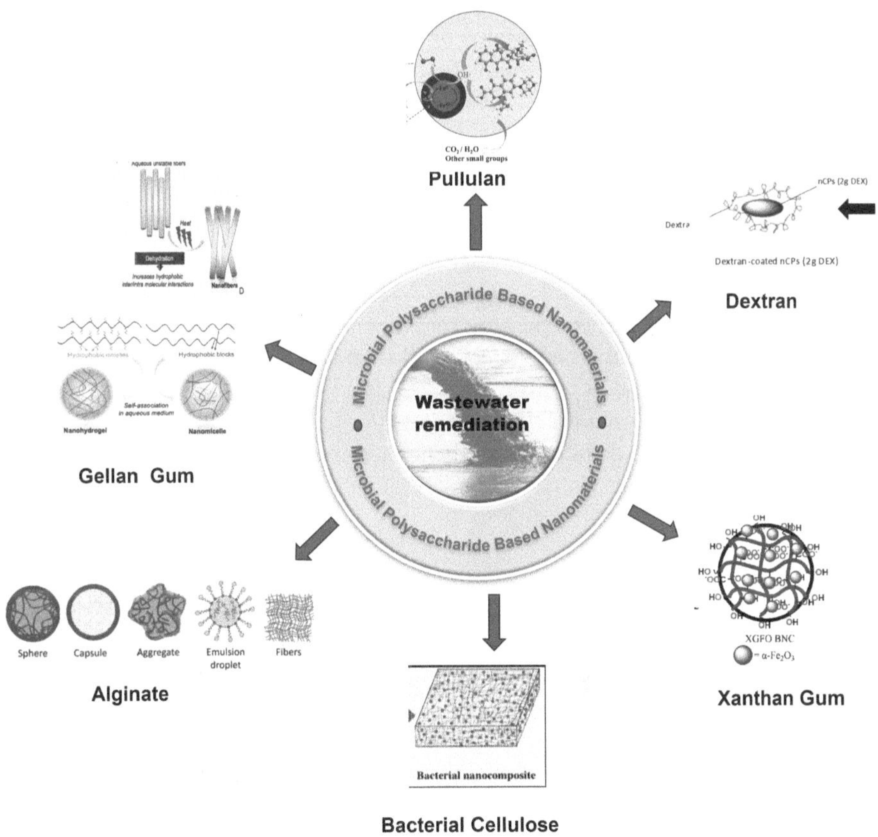

FIGURE 12.1 Microbial polysaccharides and their potential applications as nanomaterials.

Nanomaterials (sizes from 1 to 100 nm) are nanosized structures that are used for various applications including wastewater treatment, biomedical, sensors, cosmetics, food packaging, and coatings [14–16]. In this context, microbial-based nanomaterials offer several advantages due to their large surface area to volume ratio, and optical, magnetic, hydrophilic, hydrophobic, and antimicrobial properties. Some examples of nanomaterials are nanotubes, nanorods, nanofibers, quantum dots, etc. [17,18]. Their ability to form colloidal suspensions and gels in an aqueous environment makes them an effective candidate for capturing and immobilizing pollutants such as heavy metals, dyes, inorganic and organic pollutants, and oil. Microbial polysaccharides can serve as matrices for the incorporation of nanomaterials such as carbon nanotubes, magnetic nanoparticles, and metal and metal oxide nanoparticles, enhancing their functionality in various applications [19,20]. Since these polysaccharides are biodegradable, they do not cause much harm to the environment and can be effectively used in wastewater remediation [12].

This chapter provides in-depth knowledge of properties, synthesis, and applications of nanomaterials derived from microbial polysaccharides for wastewater remediation. It further discusses the methods by which these materials are created and their effectiveness in removing a variety of pollutants. Additionally, this chapter highlights the role of microbial polysaccharides in eco-friendly and sustainable nanotechnologies, contributing to the creation of efficient, low-cost solutions for wastewater purification.

12.2 MICROBIAL POLYSACCHARIDE: SOURCES AND PRODUCTION

12.2.1 Xanthan Gum

Xanthan gum (XG) a natural, branched microbial exopolysaccharide produced by gram-negative bacteria genus *Xanthomonas*. Different strains of *Xanthomonas* that are capable of producing XG are *X. campestris*, *X. citri*, *X. vasculorium*, *X. fragaria*, *X. axonopodis*, and *X. arboricola* [39,40]. Among these, the most commonly used strains for XG production is *Xanthomonas campestris* [21]. Its structure includes a cellulose backbone connected to a side chain through α-1,3 linkage (Figure 12.2). It has a molecular weight in the range of 1×10^6 to 20×10^6 Daltons. The basic repeating units of xanthan is a Penta saccharide containing D-glucose (Glc), D-mannose (Man), and D-glucuronic acid (GlcA) with acetate (Ac) and pyruvate (Pyr) group [40]. In industrial applications, it is mainly used as a thickener and emulsion stabilizer.

For the synthesis of XG, the bacterium is grown in a nutrient-rich medium containing macronutrients such as carbon and nitrogen and micronutrients such potassium, iron, and calcium [41]. Most frequently used carbon sources are glucose and sucrose. During the fermentation process, the bacteria metabolize these nutrients and produce XG as an extracellular polysaccharide [42]. The structure of XG is formed through polymerization. After fermentation process, the gum is recovered by precipitating it with alcohols such as ethanol and isopropanol, followed by the purification step. The purified gum is then dried and milled to form powder [21].

12.2.2 Dextran

Dextran is a linear polymer of glucose containing α-1,6 glycosidic linkages. It is an exopolysaccharide commonly secreted by *Leuconostoc mesenteroides* and other lactic acid bacteria such as *Lactobacillus*, *Weisella*, and *Streptococcus* [29]. These bacteria have been isolated from various plant sources such as *Agave salmiana* and pummelo, as well as from fermented products like rice batter, cabbage, idli batter, and pickles. Some dextrans have α (1,2), α (1,3), and α (1,4) linkages [30]. The structure and size of dextrans vary depending on the specific enzymes produced by the bacteria.

Commercially dextran is produced by culturing *L. mesenteroides*, a bacterium that is considered safe and stable. In the fermentation process, the bacteria utilizes sucrose as the main carbon source [42]. Dextran can be synthesized enzymatically using dextransucrases enzyme, which polymerizes

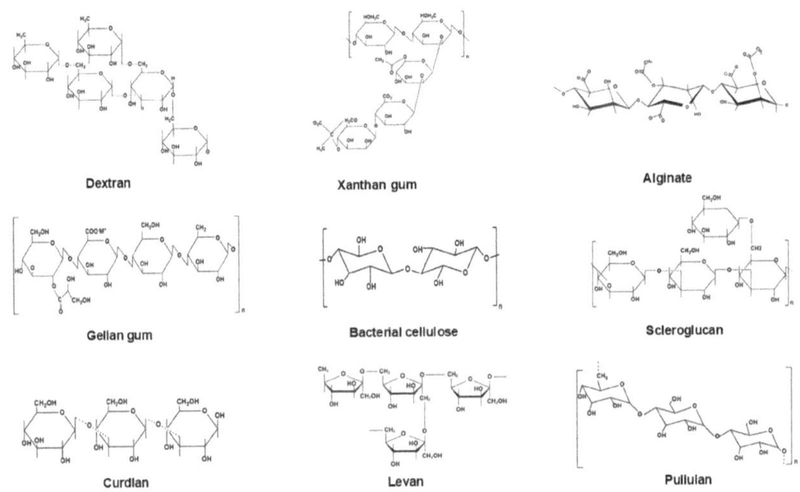

FIGURE 12.2 Structure of various microbial polysaccharides.

glucose from sucrose into dextran. During the fermentation process, dextran gets accumulated in the surrounding medium, which is then precipitated using ethanol or methanol. The precipitated dextran is purified by washing away impurities, including bacterial cells and other by-products. The purified dextran is dried and then used for various environmental applications [43].

12.2.3 Pullulan

Pullulan is a linear polysaccharide composed of malt triose repeating units which are linked together by α-1,6 glycosidic linkages. The internal malt triose is linked together by α-1,4 glycosidic linkages [22]. The molecular weight of pullulan ranges from 4.5×10^4 to 6×10^5 Dalton. It is a white-colored tasteless powder with a thermal stability of 250°C–280°C with the ability to form fibers. Pullulan is mainly produced by a fungus *Aureobasidium pullulans* through fermentation process [27]. The production of pullulan occurs intracellularly where glucose or other carbon sources are converted into pullulan precursors like uridine diphosphate glucose (UDPG). Enzymes like UDPG-pyrophosphorylase and glucosyltransferase are necessary for the conversion of glucose into pullulan.

12.2.4 Bacterial Cellulose

Bacterial cellulose is a mono-polysaccharide consisting of β-D-glucose units which are linked together by β-(1,4)-glycosidic linkage [44]. These are basically synthesized by bacteria and are considered much purer than plant cellulose as plant cellulose is generally associated with lignin, pectin, hemicellulose, etc. Bacterial cellulose and plant cellulose differ in physical and chemical properties but share a similar structure, consisting of two cellulosic subunits I and II and has the degree of polymerization ranging from 2,000 to 6,000 in bacterial cellulose [32].

There are very few bacterial species that produce bacterial cellulose such as *Acetobacter* spp., *Azotobacter* spp., *Agrobacterium* spp., and *Rhizobium* spp. Production of bacterial cellulose depends on the carbon source provided such as glucose, fructose, xylose, arabinose, sucrose, glycerol, and starch. These compounds are then converted into intermediate compounds which enter metabolic pathways for further conversion. During the process, uridine diphosphoglucose (UDP-glucose) serves as a precursor for cellulose synthesis. At first, glucose is converted into glucose-6-phosphate and then to glucose-1-phosphate by the enzyme phosphoglucomutase (PGM). Glucose-1-phosphate is then converted into UDP-glucose by the enzyme glucose-pyrophosphorylase (UGP). At last, UDP-glucose is synthesized into cellulose with the help of cellulose synthase enzyme [45].

12.2.5 Fungal Chitosan

Chitosan (CS) is a natural, biodegradable biopolymer of linear heteroglycan composed of β-(1,4)-linked D-glucosamine units which are typically produced through the deacetylation of chitin [46]. A substance is classified as chitosan when the concentration of acetylglucosamines is less than 50%, whereas it is considered chitin if the concentration is higher than 50%. Chitosan is commercially manufactured from crab shells and shrimp, but there are some fungi, from phylum *Zygomycetes*, e.g. *Mucor rouxii*, *Rhizopus oryzae*, and *Aspergillus niger*, from which chitosan can be efficiently produce [47]. The molecular weight of chitosan ranges from 20 to 1,200 kDa [48]. CS possess a positive charge due to its free amino group and can form electrostatic attractions with negatively charged molecules.

Fungal chitosan is produced by several key steps starting with the cultivation of fungi like *Rhizopus*, *Aspergillus*, and *Mucor*, which are rich in chitin. After harvesting the fungal biomass, chitin is extracted first by treating the biomass with an alkali solution, usually sodium hydroxide (NaOH) to remove proteins in a process called deproteinization [49]. The extracted chitin may also undergo decolorization if necessary to remove pigments. This chitin is converted into chitosan through deacetylation, where it is treated with concentrated alkali at high temperatures to remove acetyl groups, yielding chitosan with free amine groups [46]. This crude chitosan undergoes purification by washing with ethanol or acetone [50], neutralization and drying to eliminate impurities and residual chemicals. The final product is characterized by its degree of deacetylation and molecular weight, which determine its functional properties like solubility and bioactivity. Chitosan derived from microbes offers a sustainable and lower environmental footprint and more consistent quality [20,48].

12.2.6 Microbial Hyaluronic Acid

Hyaluronic acid (HA) is a natural polysaccharide produced from microorganisms such as yeast and bacteria like *Streptococcus zooepidemicus*, *Bacillus subtilis*, and *Escherichia coli* [51]. It consists of repeating units of N-acetylglucosamine and glucuronic acid which are linked by β-(1,3) and β-(1,4) glycosidic bonds. The molecular weight of HA is 4.0×10^3 kDa [52,53]. Nanomaterials synthesized from HA have great potential as antibacterial agents [6].

HA is primarily produced through microbial fermentation. The process begins by culturing HA-producing strains in an enriched nutrient medium under controlled conditions where they naturally secrete hyaluronic acid as a metabolic by-product. After sufficient growth, the bacterial cells are separated from the fermentation broth and then the HA is extracted [52]. The extraction through various steps includes filtration, precipitation, and centrifugation to remove other impurities. The resulting HA is further purified, through ultrafiltration and chromatography to achieve the desired molecular weight and purity. The purified product is then dried and used for commercial use. This biotechnological production method offers a sustainable and efficient way to produce HA.

12.3 FABRICATION AND SYNTHESIS OF POLYSACCHARIDES-BASED NANOMATERIALS AND THEIR COMPOSITES

Polysaccharide-based nanomaterials are renewable, biocompatible, and biodegradable materials that are derived from naturally occurring polysaccharides. These nanomaterials are gaining importance due to their environment friendly nature and their potential to replace synthetic polymers. Polysaccharides such as XG, dextran, bacterial cellulose, pullulan, and alginate are commonly used

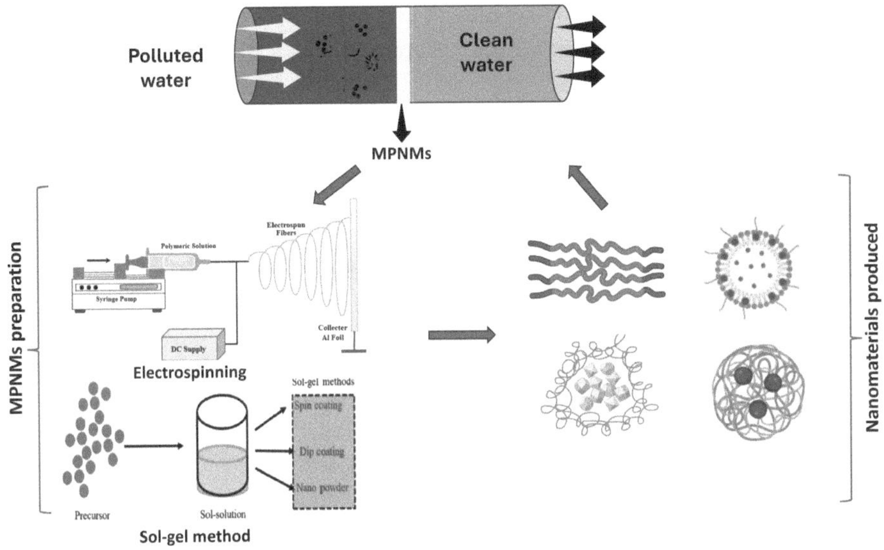

FIGURE 12.3 Process for the fabrication and synthesis of polysaccharides-based nanomaterials suitable for wastewater treatment.

for the synthesis of nanomaterials. They can be easily modified to enhance different properties like mechanical strength, flexibility, adsorption strength, and other various properties, making them useful for a wide range of applications [54,55]. The synthesis of microbial polysaccharides–based nanomaterials includes transforming microbial polysaccharides into nanosized structures through various physicochemical methods. These polysaccharides derived from bacteria, fungi, etc. offer unique and functional characteristics for the synthesis of nanomaterials [4,56]. The physicochemical methods used for their preparation tailor them to produce particular nanostructures that enhance their reactivity, surface area, and pollutant-binding capacity, making them efficient for wastewater treatment (Figure 12.3). The synthesis of nanomaterial is usually divided into two categories: top-down and bottom-up approaches [17,57]. In the top-down approach, larger materials are broken down into nanosized structures through mechanical, chemical, and physical process, and in the bottom-up approach, small, tiny molecules and atoms are combined to form nanostructured particles [58]. Various methods for the synthesis of these nanomaterials are discussed below.

12.3.1 Electrospinning

Electrospinning is a top-down method used for the development of nanomaterials. It generally produces nanofibers from a variety of materials, typically polymers. In this process, a polymer solution is pushed through a syringe needle under high electric field which stretches polymers into nanofibers [58,59]. The fibers then formed get collected on a collection plate. There are various types of electrospinning such as wet-dry electrospinning, wet-wet electrospinning, and coaxial electrospinning. In wet-dry electrospinning, a solvent evaporates as the fibers are getting formed, while wet-wet electrospinning involves a non-volatile solvent which is spun into a second solvent contained in a dish between the spinneret and the collection plate [60]. Coaxial electrospinning consists of two coaxial capillaries that allow for the use of two liquids and can fabricate two different fibers simultaneously [19], and these fibers particularly have a core-sheath structure. Polysaccharides such as chitosan and cellulose are often used in wet-dry electrospinning. Crystalline polysaccharides like cellulose and XG are harder to dissolve, and thus, they are less commonly used for this process. Microbial polysaccharides can also be co-spun with other polymers to achieve more stable nanomaterial. For example, cellulose and heparin can be spun together along with some nanoparticles to form nanocomposite fibers [61]. Pullulan combined with carbon nanodots can be an effective nanomaterial for the removal of various antibiotics from wastewater [22,62]. Nanofibers produced through the electrospinning process, especially those with core-sheath structures, offer significant potential in wastewater treatment [61]. Their high surface area, porosity, ability to incorporate different metals, and magnetic nanoparticles make them beneficial for removing pollutants. These nanofibers can adsorb heavy metals, antibiotics, and organic and inorganic contaminants, and capture pathogens. Additionally, they can be modified with antibacterial agents, enhancing their effectiveness in treating wastewater while ensuring reusability and stability [42].

12.3.2 Co-precipitation

Co-precipitation is an effective method for the synthesis of nanomaterials from polysaccharides such as dextran, bacterial cellulose, and alginate, offering a green and sustainable approach to nanomaterial production [58,63]. These polysaccharides can act as a stabilizing or capping agent for controlling the shape, size, and dispersion of nanoparticles. When metal ions are introduced into the solvent containing these polysaccharides, a precipitating agent like sodium hydroxide (NaOH) is added, which triggers the co-precipitation of metal ions in the presence of the biopolymer. This results in the formation of nanoparticles coated with polysaccharides, which provides stability and prevents their aggregation. Each polysaccharide brings specific properties to the co-precipitated nanomaterial. Dextran is known for its ability to form highly stable metal oxide nanoparticles using this co-precipitation method [64]. Bacterial cellulose, due to its fibrillar network, enhances the structural integrity of composite nanomaterials, while alginate provides excellent chelating properties that allow it to bind with metal ions, making it ideal for trapping heavy metals in wastewater. Due to the high surface area of the nanoparticles and the functional groups provided by the polysaccharides, these nanocomposites exhibit enhanced adsorption capacity and catalytic activity [54].

12.3.3 Sol-Gel Method

Sol-gel method is a chemical technique for synthesizing nanomaterials. It involves transformation of the liquid precursor into a sol, which subsequently forms a network structure known as gel. This method is employed for creating high-quality metal oxide nanomaterials [58]. In this process, the synthesis of nanomaterials happens through multiple stages. First, the precursor undergoes hydrolysis in the presence of water or alcohol, to generate the sol and then followed by the condensation phase where the viscosity of the solvent increases, thus forming porous structures. During the condensation stage, molecular bridges (such as polysaccharides-hydroxo or polysaccharide-oxo linkages) form, leading to polymeric structures within the solution [17]. As the aging process continues, the material's structure properties and porosity evolve. Drying process removes excess water and remaining solvents. Furthermore, calcination can be applied to achieve the desired nanomaterial structure. The sol-gel method is considered cost-effective and offers various benefits such as low processing temperatures, ease in producing composites, and complex nanostructures, thereby making it an excellent choice for the development of polysaccharide-based nanomaterials for removing pollutants from wastewater [17].

12.3.4 Physical and Chemical Cross-linking

Polysaccharide-based nanoparticles can be synthesized through cross-linking process, where the chains are interconnected to form three-dimensional network, thereby increasing their stability and functionality. There are usually two methods for cross-linking: covalent and ionic cross-linking [65,66]. In covalent cross-linking, permanent chemical bonds are formed between polymer chains, resulting in a rigid stable network. These covalent bonds ensure that NPs can adsorb different substances like water or other contaminants present in water without dissolving even under different pH changes. NPs formed from this process are highly durable due to the presence of common cross-linkers such as glutaraldehyde and genipin [67,68]. On the other hand, ionic cross-linking offers a simpler and more biocompatible method which involves electrostatic attraction between charged polysaccharides and oppositely charged ions. For example, calcium (Ca^{2+}) ions are used for cross-linking of alginate. However, it offers a simpler alternative to covalent methods but can lead to less stable NPs [67,69].

12.4 APPLICATION OF MICROBIAL POLYSACCHARIDE–BASED NANOMATERIALS AND THEIR COMPOSITES IN WASTEWATER TREATMENT

Microbial polysaccharide–based nanomaterials have emerged as promising tools in various environmental applications (Table 12.2). These nanomaterials can be explored for wastewater remediation, as they can exhibit excellent capabilities in adsorbing dyes, heavy metals, and other organic and inorganic pollutants (Figure 12.4). Their ability to be functionalized and modified enhances their efficiency in capturing a wide range of contaminants [12].

12.4.1 Treatment of Heavy Metals

Heavy metal contamination arises from both direct and indirect anthropogenic activities. Direct sources are effluents discharged from industries such as refineries, manufacturing plants, and wastewater treatment facilities. Indirect sources include leaching through soil and groundwater systems. Heavy metals pose significant risks due to their accumulation and toxicity in the

TABLE 12.2
Microbial Polysaccharide Based Nanomaterials and Their Application

Nanomaterials	Polysaccharides	Mode of Synthesis	Environmental Applications	References
XG-anchored α-Fe_2O_3 nanoparticles	Xanthan gum	Monowave synthesis	Lead (Pb) removal	[70]
Polyaniline-coated bacterial cellulose	Bacterial cellulose	Electrospinning	Chromium (Cr) removal	[71]
Magnetic cellulose/graphene oxide composite	Bacterial cellulose	Co-precipitation	Methylene blue adsorption	[72]
Chitosan-based iron@silver nanocomposite	Chitosan	Cross-linking	Industrial wastewater treatment	[73]
Cellulose nanofiber/graphene oxide/ silica nanocomposite	Bacterial cellulose	Sol-gel method	Methylene blue adsorption	[74]
Carbon nanodots (CNDs)–embedded pullulan nanofibers	Pullulan	Electrospinning	Antibiotic (sulphathiazole) removal	[22]
Dextran-coated iron oxide nanoparticles	Dextran	Co-precipitation	Brackish water desalination	[75]
Chitosan-supported TiO_2 and polyvinyl alcohol/chitosan blend–supported TiO_2	Chitosan	Co-precipitation	Removal of metronidzole	[76]
Polydimethylsiloxane magnetic cellulose nanofibers	Bacterial cellulose	Electroless deposition	Removal of oil spills	[77]
Hybrid chitosan-TiO_2/ZnS	Chitosan	Cross-linking and hydrothermal process	Aromatic amines, carboxylic acids	[78]

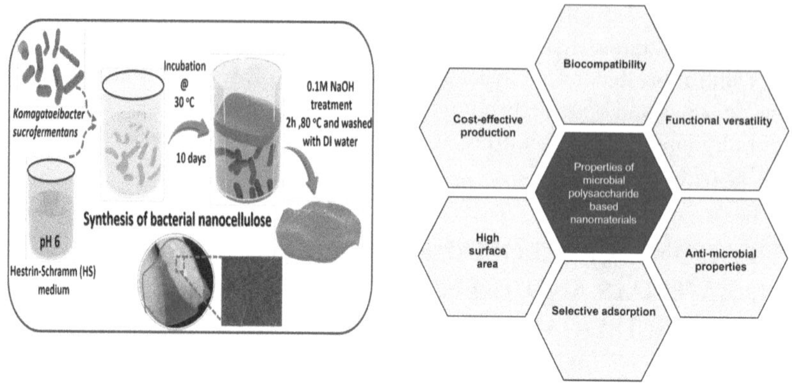

FIGURE 12.4 Synthesis and properties of polysaccharide nanoparticles.

aquatic environment. Chronic exposure to elevated levels of heavy metals leads to various disorders in humans and other organisms.

Polysaccharide NPs from cellulose, alginate, and XG are highly efficient in heavy metal adsorption due to their abundant functional groups (-COOH and -OH) [12]. These groups provide high affinity and selective binding sites for various heavy metals such as lead (Pb), chromium (Cr), arsenic (As), and cobalt (Co). The adsorption mechanism involves electrostatic interactions as the negatively charged functional group binds to the positive ions of the metals. More functional groups create chelation complexes with the metal ions, thereby enhancing their adsorption capacity. To enhance the adsorption capacity of polysaccharides-based nanomaterials, surface modification can be done by adding nanoparticles, with especially metal oxide or magnetic nanoparticles. For instance, XG-anchored α-Fe_2O_3 (magnetic nanoparticles) demonstrates a higher affinity for heavy metals such as lead (Pb) [70]. This magnetic nanoparticle, along with XG, shows improved performance by enhancing the metal ion binding. Bacterial cellulose coated with polyaniline is used as an adsorbent material to eliminate hazardous heavy metals like chromium (Cr) from wastewater [71]. Similarly, carbon nanotubes–alginate beads work efficiently to enhance heavy metal removal from wastewater. The incorporation of carbon nanotubes into alginate significantly boosts the physicochemical characteristics of the alginate-based composites, leading to improved structural stability, surface area, and adsorption sites. These enhanced properties contribute to a greater capacity for adsorbing heavy metals [79].

12.4.2 Removal of Dyes

Among various contaminants present in wastewater, synthetic dyes from multiple industries, like food, paper, plastics, cosmetics, printing, and textiles, are released into the water bodies in large quantities every year. These dyes are non-degradable and cannot be removed down by sludge treatment plants. They are completely soluble in water, which makes them very difficult to be removed. Even a small amount of dye can cause significant harm to the environment and contribute to diseases in animals, humans, and even aquatic life. Polysaccharide-based nanomaterials emerged as promising candidates for dye removal due to their high adsorption capacity. Dyes are organic compounds having complex aromatic structures and also include functional groups such as COOH, NH_2, NR_2, NO, N_2, and NO_2 [80]. These groups impart color to dyes. Dyes can be cationic or anionic, so polysaccharides that are positively charged exhibit strong affinity for anionic dyes (methyl orange, Congo red), while polysaccharides that are positively charged can bind to cationic dyes (methylene blue). Functional groups on polysaccharides form hydrogen bonds with the dye molecules, further enhancing their dye absorption capacity [80]. Polysaccharide nanomaterials with a high surface area and porous structure, such as cellulose nanofibers and pullulan nanofibers, show higher adsorption efficiency.

Microspheres grafted using pullulan show high removal efficiency of organic dye contaminants, e.g., adsorption of Azocarmine B, methyl orange, and methylene blue. The cross-linked pullulan microspheres have the ability to absorb water. The hydrophilic pullulan allows fast movement of adsorbates, and its grafting with pAPTAC (poly-3-acrylamidopropyltrimethylammonium chloride) adds hydrophobic properties for trapping dyes [81]. Chitosan/graphene oxide nanocomposites also exhibit superior adsorption performance due to their increased surface area and enhanced hydrophilicity. Graphene oxide introduces more functional groups like carboxyl and hydroxyl that act as additional adsorption sites for dye molecules [74]. Combining alginate with magnetic nanoparticles also shows enhanced adsorption capacity and also allows for easy removal from solution after treatment using magnetic field [82].

12.4.3 Oil Spill Treatment

Recurrent oil spills, industrial chemical leaks, and oil-laden domestic sewage pose significant risks to marine economies and ecosystems. Available solutions under the National Contingency Plan use toxic synthetic chemicals that may cause harm to the marine ecosystem. As an alternative to synthetic chemicals, nanomaterials had been tested effectively for the removal of oil spills. Modified cellulose-based nanomaterials in the form of sponges and nanocrystals modified with octadecanethiol and silicon dioxide show potential in removing oil spills from water. Due to the simplicity of operation, these biobased nanomaterials could be an ideal method for wastewater remediation, effectively separating oil and organic solvent with high efficiency at low cost. Additionally, a water-repellent porous nanometallic composite with a hydrophobic surface has also been tested for efficient oil–water separation. The research finding of an earlier study [77] shows promising application of a polydimethylsiloxane-decorated magnetic cellulose nanofiber composite for effective oil–water separation. XG functionalized with silica nanoparticles is also studied as a novel dispersant for marine oil spill remediation. The process utilizes the electrostatic attraction between XG and oppositely charged silica nanoparticles to promote synergistic emulsification [83].

12.4.4 Wastewater Treatment

Effluents from domestic wastewater contains contaminants like organic pollutants, laundry detergents, pharmaceuticals and personal care products, microplastics, and oils. Effective treatment of this wastewater is crucial for protecting the aquatic ecosystems and human health. Biobased nanomaterials such as bacterial cellulose nanocrystals, cellulose nanofibrils, and nanocomposites have adaptable surface chemistry and have the ability to make membranes for adsorption of various contaminants. Additionally metal oxides like TiO_2 and organosilanes when conjugated with nanocellulose, it provides more efficiency nanocellulose based membranes by adding hydrophilic and oleophilic property.. Thus, they could adsorb kitchen oils and other organic solvents from the water surface [54]. On the other hand, chitosan is one of the potent polysaccharides that are found to adsorb pesticides such as oxadiazon [84] and atrazine [85]. Nanosorbents or nanomaterials formed in combination with polysaccharides show high performance in removing pesticides from agricultural wastewater. Chitosan and chitosan-modified sepiolite show high binding affinity toward different pesticides, thereby removing them from aqueous solution. Chitosan nanomaterials made along with alginate, chloroacetic acid, polyethylenimine, polyvinyl alcohol, and Fe_3O_4 nanoparticles display greater performance in absorbing anti-inflammatory drugs and antibiotics, and also exhibits antimicrobial activity [76,86].

12.5 CONCLUSION

Microorganisms like bacteria, fungi, etc. produce various polysaccharides including HA, dextran, pullulan, XG, and chitosan, which can be used as efficient sources for nano-biomaterial for the purification of wastewater. Microbial polysaccharides–based nanomaterials represent a sustainable approach for the removal of pollutants from wastewater due to their biodegradable, eco-friendly, renewable, and versatile nature. For the fabrication of nanomaterials, methods like cross-linking, electrospinning, sol gel, and co-precipitation are used, which results in the formation of nanofibers and cross-linked nanoparticles. They emerge as potential candidates that can be used as membranes, adsorbents, photocatalysts, etc. As research in this field progresses, they have very much potential to play a major role in developing eco-friendly, sustainable, scalable, and low-cost wastewater treatment technologies for water resource management.

12.6 CHALLENGES AND OPPORTUNITIES

Although microbial polysaccharide–based nanomaterials exhibit remarkable prospects in wastewater treatment, but scaling up their production can be a major challenge. Large-scale production demands significant resources like advanced equipment and fermentation systems, which can increase production costs. Although these nanomaterials are biodegradable, their effectiveness may decrease under high pH, salinity, and temperature conditions, which are commonly found in industrial wastewater. The long-term consequences of these nanomaterials, particularly in the form of nanoparticles, are not yet fully investigated so ecotoxicity studies are essential to ensure their safe use and application. The production techniques, such as electrospinning and sol-gel process, may require high energy expenditure. This factor could diminish the advantages offered by nanomaterials. On the other hand, nanomaterials derived from polysaccharides can show immense potential in adsorbing specific pollutants such as heavy metals, pharmaceuticals dyes, and other pollutants. This can be achieved by enhancing their surface chemistry through the incorporation of functional groups designed to selectively bind to these specific pollutants. They can also be the more advanced version of the present wastewater treatment technologies by enhancing their overall system efficiency. Microbial polysaccharide–based nanomaterials minimize the dependency on synthetic materials, thus providing green technologies in wastewater treatment, reducing overall environment footprint of treatment processes, and promoting circular economy.

REFERENCES

1. Blaga, A.C.; Zaharia, C.; Suteu, D. Polysaccharides as support for microbial biomass-based adsorbents with applications in removal of heavy metals and dyes. *Polymers* 2021, *13*, 2893.
2. Ghiorghita, C.-A.; Dinu, M.V.; Lazar, M.M.; Dragan, E.S. Polysaccharide-based composite hydrogels as sustainable materials for removal of pollutants from wastewater. *Molecules* 2022, *27*, 8574.
3. Yadav, S.; Chander, S.; Gupta, A.; Kataria, N.; Khoo, K.S. Biogenic engineered zinc oxide nanoparticle for sulfur black dye removal from contaminated wastewater: Comparative optimization, simulation modeling, and isotherms. *Bioengineered* 2024, *15*, 2325721, https://doi.org/10.1080/21655979.2024.2325721.
4. Saud, A.; Gupta, S.; Allal, A.; Preud'homme, H.; Shomar, B.; Zaidi, S.J. Progress in the sustainable development of biobased (nano)materials for application in water treatment technologies. *ACS Omega* 2024, *9*, 29088–29113.
5. Poonia, K.; Singh, P.; Ahamad, T.; Le, Q. Van; Phan Quang, H.H.; Thakur, S.; Mishra, A.K.; Selvasembian, R.; Hussain, C.M.; Nguyen, V.-H.; et al. Sustainability, performance, and production perspectives of waste-derived functional carbon nanomaterials towards a sustainable environment: A review. *Chemosphere* 2024, *352*, 141419, https://doi.org/10.1016/j.chemosphere.2024.141419.
6. Arora, D.; Sharma, N.; Sharma, V.; Abrol, V.; Shankar, R.; Jaglan, S. An update on polysaccharide-based nanomaterials for antimicrobial applications. *Appl. Microbiol. Biotechnol.* 2016, *100*, 2603–2615.
7. François, M.; Lin, K.-S.; Rachmadona, N.; Khoo, K.S. Utilization of carbon-based nanomaterials for wastewater treatment and biogas enhancement: A state-of-the-art review. *Chemosphere* 2024, *350*, 141008, https://doi.org/10.1016/j.chemosphere.2023.141008.
8. Qi, X.; Tong, X.; Pan, W.; Zeng, Q.; You, S.; Shen, J. Recent advances in polysaccharide-based adsorbents for wastewater treatment. *J. Clean. Prod.* 2021, *315*, 128221, https://doi.org/10.1016/j.jclepro.2021.128221.

9. Rawindran, H.; Khoo, K.S.; Ethiraj, B.; Lim, J.W.; Liew, C.S.; Goh, P.S.; Raksasat, R.; Leong, W.H.; Rajarathinam, R.; Ng, H.-S.; et al. Environmental impact assessment via life cycle analysis on ultrafiltration membrane fabricated from polyethylene terephthalate waste to treat microalgal cultivation wastewater for reusability. *Environ. Res.* 2024, *251*, 118687, https://doi.org/10.1016/j.envres.2024.118687.
10. Zango, Z.U.; Khoo, K.S.; Garba, A.; Garba, Z.N.; Danmallam, U.N.; Aldaghri, O.; Ibnaouf, K.H.; Ahmad, N.M.; Binzowaimil, A.M.; Lim, J.W.; et al. A review on titanium oxide nanoparticles modified metal-organic frameworks for effective CO_2 conversion and efficient wastewater remediation. *Environ. Res.* 2024, *252*, 119024, https://doi.org/10.1016/j.envres.2024.119024.
11. Ardo, F.M.; Khoo, K.S.; Ahmad Sobri, M.Z.; Suparmaniam, U.; Ethiraj, B.; Anwar, A.F.; Lam, S.M.; Sin, J.C.; Shahid, M.K.; Ansar, S.; et al. Modelling photoperiod in enhancing hydrogen production from Chlorella vulgaris sp. while bioremediating ammonium and organic pollutants in municipal wastewater. *Environ. Pollut.* 2024, 346, 123648, https://doi.org/10.1016/j.envpol.2024.123648.
12. Al-Hazmi, H.E.; Łuczak, J.; Habibzadeh, S.; Hasanin, M.S.; Mohammadi, A.; Esmaeili, A.; Kim, S.-J.; Khodadadi Yazdi, M.; Rabiee, N.; Badawi, M.; et al. Polysaccharide nanocomposites in wastewater treatment: A review. *Chemosphere* 2024, *347*, 140578.
13. Mohammadi, A.A.; Zarei, A.; Esmaeilzadeh, M.; Taghavi, M.; Yousefi, M.; Yousefi, Z.; Sedighi, F.; Javan, S. Assessment of heavy metal pollution and human health risks assessment in soils around an industrial zone in Neyshabur, Iran. *Biol. Trace Elem. Res.* 2020, *195*, 343–352, https://doi.org/10.1007/s12011-019-01816-1.
14. Aun Ng, C.; Chew, S.N.; Bashir, M.J.K.; Abunada, Z.; Wong, J.W.C.; Habila, M.A.; Khoo, K.S. Enhancing microbial fuel cell performance for sustainable treatment of palm oil mill wastewater using carbon cloth anode coated with activated carbon. *Int. J. Hydrogen Energy* 2024, *52*, 1092–1104, https://doi.org/10.1016/j.ijhydene.2023.09.162.
15. Mohanta, Y.K.; Mishra, A.K.; Panda, J.; Chakrabartty, I.; Sarma, B.; Panda, S.K.; Chopra, H.; Zengin, G.; Moloney, M.G.; Sharifi-Rad, M. Promising applications of phyto-fabricated silver nanoparticles: Recent trends in biomedicine. *Biochem. Biophys. Res. Commun.* 2023, *688*, 149126, https://doi.org/10.1016/j.bbrc.2023.149126.
16. Ghotekar, S.; Ravikumar, C.R.; Chauhan, A.; Hikku, G.S.; Lin, K.-Y.A.; Rahdar, A.; Hitler, L.; Jabir, M.S.; Marzban, A.; Oza, R. Eco-friendly fabrication of CdO nanoparticles using polyalthia longifolia leaves extract for antibacterial and electrochemical sensing studies. *J. Sol-Gel Sci. Technol.* 2024, *110*, 221–232, https://doi.org/10.1007/s10971-024-06352-6.
17. Paramasivam, G.; Palem, V.V.; Sundaram, T.; Sundaram, V.; Kishore, S.C.; Bellucci, S. Nanomaterials: Synthesis and applications in theranostics. *Nanomaterials* 2021, *11*, https://doi.org/10.3390/nano11123228.
18. Mohanta, Y.K.; Biswas, K.; Mishra, A.K.; Patra, B.; Mishra, B.; Panda, J.; Avula, S.K.; Varma, R.S.; Panda, B.P.; Nayak, D. Amelioration of gold nanoparticles mediated through ocimum oil extracts induces reactive oxygen species and mitochondrial instability against MCF-7 breast carcinoma. *RSC Adv.* 2024, *14*, 27816–27830, https://doi.org/10.1039/D4RA04807E.
19. Altammar, K.A. A review on nanoparticles: Characteristics, synthesis, applications, and challenges. *Front. Microbiol.* 2023, *14*, 1155622.
20. Nayak, D.; Mishra, A.K.; Malla, S.; Swain, U.; Panigrahi, K.; Panda, J.; Rustagi, S.; Rauta, P.R.; Mohanta, Y.K. Eco-friendly mycogenic synthesis of iron oxide nanoparticles using volvariella volvacea and unveiling its potential as antibiofilm and anticancer agents. *ECS J. Solid State Sci. Technol.* 2024, *13*, 087003, https://doi.org/10.1149/2162-8777/ad658b.
21. García-Ochoa, F.; Santos, V.E.; Casas, J.A.; Gómez, E. Xanthan gum: Production, recovery, and properties. *Biotechnol. Adv.* 2000, *18*, 549–579.
22. Aijaz, M.O.; Ahmad, M.; Al-Wabel, M.I.; Karim, M.R.; Usman, A.R.A.; Assaifan, A.K. Carbon nanodots-embedded pullulan nanofibers for sulfathiazole removal from wastewater streams. *Membranes* 2022, *12*, 228.
23. Aquinas, N.; Bhat, MR.; Selvaraj, S. A review presenting production, characterization, and applications of biopolymer curdlan in food and pharmaceutical sectors. *Polym. Bull. (Berl.)* 2022, *79*, 6905–6927.
24. Moon, C.J.; Lee, J.H. Use of curdlan and activated carbon composed adsorbents for heavy metal removal. *Process Biochem.* 2005, *40*, 1279–1283, https://doi.org/10.1016/j.procbio.2004.05.009.
25. Kanmani, P.; Lim, S.T. Synthesis and characterization of pullulan-mediated silver nanoparticles and its antimicrobial activities. *Carbohydr. Polym.* 2013, *97*, 421–428, https://doi.org/10.1016/j.carbpol.2013.04.048.
26. Al-Rmedh, Y.S.S.; Ali, H.I.; Al-Sahlany, S.T.G. Curdlan gum, properties, benefits and applications. *IOP Conf. Ser. Earth Environ. Sci.* 2023, 1158, 112011.

27. Cheng, K.-C.; Demirci, A.; Catchmark, J.M. Pullulan: Biosynthesis, production, and applications. *Appl. Microbiol. Biotechnol.* 2011, *92*, 29–44.
28. Leathers, T.D. Biotechnological production and applications of pullulan. *Appl. Microbiol. Biotechnol.* 2003, *62*, 468–473.
29. Diaz-Montes, E. Dextran: Sources, structures, and properties. *Polysaccharides* 2021, *2*, 554–565.
30. Sarwat, F.; Ul Qader, S.A.; Aman, A.; Ahmed, N. Production characterization of a unique dextran from an indigenous Leuconostoc mesenteroides CMG713. *Int. J. Biol. Sci.* 2008, *4*, 379–386.
31. Santoshi Kumari, A.; Ayodhya, D.; Pradeep Kumar, D.; Maragoni, V. Xanthan gum mediated monowave synthesis of silver nanoparticles: Characterization, anticancer, antimicrobial, antioxidant and catalytic studies. *Results Chem.* 2023, *5*, 100930.
32. Lahiri, D.; Nag, M.; Dutta, B.; Dey, A.; Sarkar, T.; Pati, S.; Edinur, H.A.; Abdul Kari, Z.; Mohd Noor, N.H.; Ray, R.R. Bacterial cellulose: Production, characterization, and application as antimicrobial agent. *Int. J. Mol. Sci.* 2021, *22*, 12984.
33. Amerhaider Nuar, N.N.; Siti Nurul, S.N.A.; Li, F.; Mat Azmi, I.D.; Chiang, P.C.; Choong, T.S.Y. Synthesis of controlled-release calcium peroxide nanoparticles coated with dextran for removal of doxycycline from aqueous system. *Polymers (Basel).* 2022, *14*, https://doi.org/10.3390/polym14183866.
34. Hay, I.D.; Ur Rehman, Z.; Moradali, M.F.; Wang, Y.; Rehm, B.H.A. Microbial alginate production, modification and its applications. *Microb. Biotechnol.* 2013, *6*, 637–650.
35. Majumder, A.; Goyal, A. Rheological and gelling properties of a novel glucan from Leuconostoc dextranicum NRRL B-1146. *Food Res. Int.* 2009, *42*, 525–528.
36. Domżał-Kedzia, M.; Ostrowska, M.; Lewińska, A.; Łukaszewicz, M. Recent developments and applications of microbial levan, a versatile polysaccharide-based biopolymer. *Molecules* 2023, *28*, 5407.
37. Wang, X.; Zhao, C.; Zhao, P.; Dou, P.; Ding, Y.; Xu, P. Gellan gel beads containing magnetic nanoparticles: An effective biosorbent for the removal of heavy metals from aqueous system. *Bioresour. Technol.* 2009, *100*, 2301–2304.
38. Castro, R.H.; Corredor, L.M.; Llanos, S.; Causil, M.A.; Arias, A.; Pérez, E.; Quintero, H.I.; Romero Bohórquez, A.R.; Franco, C.A.; Cortés, F.B. Experimental investigation of the viscosity and stability of scleroglucan-based nanofluids for enhanced oil recovery. *Nanomaterials* 2024, *14*, 156.
39. de Monaco Lopes, B.; Lessa, V.L.; Silva, B.M.; da Silva Carvalho Filho, M.A.; Schnitzler, E.; Lacerda, L.G. Xanthan gum: Properties, production conditions, quality and economic perspective. *J. Food Nutr. Res.* 2015, *54*, 185–194.
40. Jadav, M.; Pooja, D.; Adams, D.J.; Kulhari, H. Advances in xanthan gum-based systems for the delivery of therapeutic agents. *Pharmaceutics* 2023, *15*, 1–27, https://doi.org/10.3390/pharmaceutics15020402.
41. Silva, M.F.; Fornari, R.C.G.; Mazutti, M.A.; de Oliveira, D.; Padilha, F.F.; Cichoski, A.J.; Cansian, R.L.; Di Luccio, M.; Treichel, H. Production and characterization of xantham gum by Xanthomonas campestris using cheese whey as sole carbon source. *J. Food Eng.* 2009, *90*, 119–123, https://doi.org/10.1016/j.jfoodeng.2008.06.010.
42. Ahmad, N.H.; Mustafa, S.; Che Man, Y.B. Microbial polysaccharides and their modification approaches: A review. *Int. J. Food Prop.* 2015, *18*, 332–347.
43. Bhatia, S. Microbial polysaccharides as advance nanomaterials. In *Systems for Drug Delivery*; Springer International Publishing: Cham, 2016; pp. 29–54. https://doi.org/10.1007/978-3-319-41926-8_2.
44. Choi, S.M.; Rao, K.M.; Zo, S.M.; Shin, E.J.; Han, S.S. Bacterial cellulose and its applications. *Polymers* 2022, *14*, 1080.
45. Zhong, C. Industrial-scale production and applications of bacterial cellulose. *Front. Bioeng. Biotechnol.* 2020, *8*, 1–19, https://doi.org/10.3389/fbioe.2020.605374.
46. Crognale, S.; Russo, C.; Petruccioli, M.; D'Annibale, A. Chitosan production by fungi: Current state of knowledge, future opportunities and constraints. *Fermentation* 2022, *8*, 76.
47. Pochanavanich, P.; Suntornsuk, W. Fungal chitosan production and its characterization. *Lett. Appl. Microbiol.* 2002, *35*, 17–21.
48. Román-Doval, R.; Torres-Arellanes, S.P.; Tenorio-Barajas, A.Y.; Gómez-Sánchez, A.; Valencia-Lazcano, A.A. Chitosan: Properties and its application in agriculture in context of molecular weight. *Polymers (Basel)* 2023, *15*, 1–26, https://doi.org/10.3390/polym15132867.
49. Huq, T.; Khan, A.; Brown, D.; Dhayagude, N.; He, Z.; Ni, Y. Sources, production and commercial applications of fungal chitosan: A review. *J. Bioresour. Bioprod.* 2022, *7*, 85–98.
50. Tasar, O.C.; Erdal, S.; Taskin, M. Chitosan production by psychrotolerant rhizopus oryzae in non-sterile open fermentation conditions. *Int. J. Biol. Macromol.* 2016, *89*, 428–433.

51. Kaur, M.; Jayaraman, G. hyaluronan production and molecular weight is enhanced in pathway-engineered strains of lactate dehydrogenase-deficient lactococcus lactis. *Metab. Eng. Commun.* 2016, *3*, 15–23, https://doi.org/10.1016/j.meteno.2016.01.003.
52. Serra, M.; Casas, A.; Toubarro, D.; Barros, A.N.; Teixeira, J.A. Microbial hyaluronic acid production: A review. *Molecules* 2023, *28*, 2084.
53. Necas, J.; Bartosikova, L.; Brauner, P.; Kolar, J. Hyaluronic acid (Hyaluronan): A review. *Vet. Med. (Praha).* 2008, *53*, 397–411, https://doi.org/10.17221/1930-VETMED.
54. Nasrollahzadeh, M.; Sajjadi, M.; Iravani, S.; Varma, R.S. Starch, cellulose, pectin, gum, alginate, chitin and chitosan derived (nano)materials for sustainable water treatment: A review. *Carbohydr. Polym.* 2021, *251*, 116986, https://doi.org/10.1016/j.carbpol.2020.116986.
55. Ray, M.K.; Panda, J.; Panda, B.P.; Mohanta, T.K.; Mohanta, Y.K. Mycoremediation of heavy metals and/or metalloids in soil. In *Land Remediation and Management: Bioengineering Strategies.* Sarma, H., Joshi, S., Eds.; Springer Nature: Singapore, 2023; pp. 161–190. ISBN 978-981-99-4221-3.
56. Behzadmehr, R.; Fathi-karkan, S.; Razzaq, S.; Moafian, Z.; Rahdar, A.; Ghotekar, S. Unleashing the potential of ibrutinib-loaded nanoparticles for cancer treatment—A comprehensive review. *Bionanoscience* 2024, *14*, 3784–3797, https://doi.org/10.1007/s12668-024-01445-6.
57. Wang, C.; Gao, X.; Chen, Z.; Chen, Y.; Chen, H. Preparation, characterization and application of polysaccharide-based metallic nanoparticles: A review. *Polymers (Basel).* 2017, *9*, https://doi.org/10.3390/polym9120689.
58. Baig, N.; Kammakakam, I.; Falath, W. Nanomaterials: A review of synthesis methods, properties, recent progress, and challenges. *Mater. Adv.* 2021, *2*, 1821–1871.
59. Sanap, D.; Avhad, L.; Ahire, S.; Mirzaei, M.; Kumar, D.; Ghotekar, S.; Gaikwad, N.D. Insights into the promising heterogeneous catalysis of eco-friendly synthesized spinel $CuFe_2O_4$ nanoparticles for biginelli reaction. *Res. Chem. Intermed.* 2024, *50*, 3687–3711, https://doi.org/10.1007/s11164-024-05322-5.
60. Zheng, Y.; Monty, J.; Linhardt, R.J. Polysaccharide-based nanocomposites and their applications. *Carbohydr. Res.* 2015, *405*, 23–32.
61. Hou, L.; Udangawa, W.M.R.N.; Pochiraju, A.; Dong, W.; Zheng, Y.; Linhardt, R.J.; Simmons, T.J. Synthesis of heparin-immobilized, magnetically addressable cellulose nanofibers for biomedical applications. *ACS Biomater. Sci. Eng.* 2016, *2*, 1905–1913.
62. Rani, S.; Sabharwal, H.; Kumar, P.; Chauhan, A.K.; Khoo, K.S.; Kataria, N. Comparative behavior of carbon-based materials for the removal of emerging Bisphenol A from water: Adsorption modelling and mechanism. *Groundw. Sustain. Dev.* 2024, *25*, 101121, https://doi.org/10.1016/j.gsd.2024.101121.
63. Chawla, A.; Sudhaik, A.; Sonu; Kumar, R.; Raizada, P.; Khan, A.A.P.; Ahamad, T.; Nguyen, V.-H.; Selvasembian, R.; Kaushik, A.; et al. Recent advances in synthesis methods and surface structure manipulating strategies of copper selenide (CuSe) nanoparticles for photocatalytic environmental and energy applications. *J. Environ. Chem. Eng.* 2024, *12*, 113125, https://doi.org/10.1016/j.jece.2024.113125.
64. Predescu, A.M.; Matei, E.; Berbecaru, A.C.; Pantilimon, C.; Drăgan, C.; Vidu, R.; Predescu, C.; Kuncser, V. Synthesis and characterization of dextran-coated iron oxide nanoparticles. *R. Soc. Open Sci.* 2018, *5*, https://doi.org/10.1098/rsos.171525.
65. Mizrahy, S.; Peer, D. Polysaccharides as building blocks for nanotherapeutics. *Chem. Soc. Rev.* 2012, *41*, 2623–2640, https://doi.org/10.1039/c1cs15239d.
66. Vijayaraghavan, P.; Veeramanikandan, V.; Pradeep, B.V.; Pothiraj, C.; Alarjani, K.M.; Al Farraj, D.A.; Nguyen, V.-H.; Balaji, P. Enhanced aerobic naphthalene degradation utilizing indigenous microbial flora as a biocatalyst in oil-contaminated wastewater. *Top. Catal.* 2024, *67*, 1183–1193, https://doi.org/10.1007/s11244-024-01953-5.
67. Plucinski, A.; Lyu, Z.; Schmidt, B.V.K.J. Polysaccharide nanoparticles: From fabrication to applications. *J. Mater. Chem. B Mater. Biol. Med.* 2021, *9*, 7030–7062.
68. Kimta, N.; Chauhan, A.; Puri, S.; Kumari, A.; Sharma, R.; Kumar, A.; Kapoor, D. Investigation on the influence of silver doping on structural and biomedical properties of zinc oxide nanoparticles synthesized using crude ethanolic extract of adiantum venustum D. Don. *Emergent Mater.* 2024, *7*, 1619–1645, https://doi.org/10.1007/s42247-024-00715-z.
69. Barwant, M.; Karande, V.; Basnet, P.; Kumar, D.; Sargazi, S.; Mirzaei, M.; Jabir, M.S.; Sanap, D.; Ghotekar, S. Insights into the antioxidant and anticancer properties of novel biologically synthesized NiO/Ni_2O_3 nanoparticles using sargassum tenerrimum. *J. Sol-Gel Sci. Technol.* 2024, *111*, 409–420, https://doi.org/10.1007/s10971-024-06446-1.
70. Alharthi, F.A.; Alshammari, R.H.; Hasan, I. Synthesis of xanthan gum anchored α-Fe_2O_3 bionanocomposite material for remediation of Pb (II) contaminated aquatic system. *Polymers.* 2023, *15*, 1134.

71. Jahan, K.; Kumar, N.; Verma, V. Removal of hexavalent chromium from potable drinking using a polyaniline-coated bacterial cellulose mat. *Environ. Sci.* 2018, *4*, 1589–1603.
72. Shi, H.; Li, W.; Zhong, L.; Xu, C. Methylene blue adsorption from aqueous solution by magnetic cellulose/graphene oxide composite: Equilibrium, kinetics, and thermodynamics. *Ind. Eng. Chem. Res.* 2014, *53*, 1108–1118.
73. Olajire, A.A.; Bamigbade, L.A. Green synthesis of chitosan-based iron@silver nanocomposite as adsorbent for wastewater treatment. *Water Resour. Ind.* 2021, *26*, 100158, https://doi.org/10.1016/j.wri.2021.100158.
74. Abouzeid, R.E.; Owda, M.E.; Dacrory, S. Effective adsorption of cationic methylene blue dye on cellulose nanofiber/graphene oxide/silica nanocomposite: Kinetics and equilibrium. *J. Appl. Polym. Sci.* 2022, *139*, e52377, https://doi.org/10.1002/app.52377.
75. Bai, H.; Liu, Z.; Sun, D.D. Highly water soluble and recovered dextran coated Fe_3O_4 magnetic nanoparticles for brackish water desalination. *Sep. Purif. Technol.* 2011, *81*, 392–399.
76. Neghi, N.; Kumar, M.; Burkhalov, D. Synthesis and application of stable, reusable TiO_2 polymeric composites for photocatalytic removal of metronidazole: Removal kinetics and density functional analysis. *Chem. Eng. J.* 2019, *359*, 963–975.
77. Chu, Z.; Li, Y.; Zhou, A.; Zhang, L.; Zhang, X.; Yang, Y.; Yang, Z. Polydimethylsiloxane-decorated magnetic cellulose nanofiber composite for highly efficient oil-water separation. *Carbohydr. Polym.* 2022, *277*, 118787, https://doi.org/10.1016/j.carbpol.2021.118787.
78. Jbeli, A.; Ferraria, A.M.; Botelho do Rego, A.M.; Boufi, S.; Bouattour, S. Hybrid chitosan-TiO_2/ZnS prepared under mild conditions with visible-light driven photocatalytic activity. *Int. J. Biol. Macromol.* 2018, *116*, 1098–1104, https://doi.org/10.1016/j.ijbiomac.2018.05.141.
79. Wang, B.; Wan, Y.; Zheng, Y.; Lee, X.; Liu, T.; Yu, Z.; Huang, J.; Ok, Y.S.; Chen, J.; Gao, B. Alginate-based composites for environmental applications: A critical review. *Crit. Rev. Environ. Sci. Technol.* 2018, *49*, 318–356.
80. Saya, L.; Gautam, D.; Malik, V.; Singh, W.R.; Hooda, S. Natural polysaccharide based graphene oxide nanocomposites for removal of dyes from wastewater: A review. *J. Chem. Eng. Data* 2021, *66*, 11–37.
81. Ghimici, L.; Constantin, M. A review of the use of pullulan derivatives in wastewater purification. *React. Funct. Polym.* 2020, *149*, 104510, https://doi.org/10.1016/j.reactfunctpolym.2020.104510.
82. Rocher, V.; Siaugue, J.-M.; Cabuil, V.; Bee, A. Removal of organic dyes by magnetic alginate beads. *Water Res.* 2008, *42*, 1290–1298, https://doi.org/10.1016/j.watres.2007.09.024.
83. Pi, G.; Li, Y.; Bao, M.; Mao, L.; Gong, H.; Wang, Z. Novel and environmentally friendly oil spill dispersant based on the synergy of biopolymer xanthan gum and silica nanoparticles. *ACS Sustain. Chem. Eng.* 2016, *4*, 3095–3102, https://doi.org/10.1021/acssuschemeng.6b00063.
84. Arvand, M.; Latify, L.; Tajmehri, H.; Yagubov, A.I.; Nuriyev, A.N.; Pourhabib, A.; Mousavi, S.J.; Abolhassani, M.R. Comparative study for the removal of oxadiazon from aqueous solutions by adsorption on chitosan and activated carbon. *Anal. Lett.* 2009, *42*, 856–869.
85. Liu, H.-C.; Chen, W.; Cui, B.; Liu, C. Enhanced atrazine adsorption from aqueous solution using chitosan-modified sepiolite. *J. Cent. S. Univ.* 2015, *22*, 4168–4176.
86. Niu, H.; Dizhang; Meng, Z.; Cai, Y. Fast defluorination and removal of norfloxacin by alginate/Fe@Fe_3O_4 core/shell structured nanoparticles. *J. Hazard. Mater.* 2012, *227–228*, 195–203.

13 Bioremediation of Various Industrial Effluents Utilizing Microorganism-Assisted Nanotechnology

S. Shanmugavani and M. Thenmozhi

13.1 INTRODUCTION

One of the biggest global concerns is how to handle industrial sludge in a way that is both environmentally and economically feasible. The methods used for disposing industrial sludge includes burning, composting, drying at high temperature, dispersing the sludge on the ground, lime stabilization, and landfilling. The bulk of India's textile industry discharges its waste into agriculture fields, waste yard, fallow ground, poorly managed wasteland, and along railroad line due to exorbitant sludge management costs, which further contaminates groundwater, creating a serious threat to human health. Meanwhile, landfilling and waste disposal have become increasingly expensive and impractical due to limited available landfill space, stringent national waste disposal regulations, and heightened local awareness. In the majority of developing nations, sludge management procedures are inadequate and not well developed. Nowadays, several factories are practicing ecofriendly and low-cost industrial effluent treatment processes [1]. The improper handling and subsequent discharge of dangerous effluents into groundwater and other large water bodies through sewage drains adversely affect animal health and aquatic life. Additionally, inadequately treated effluents may lead to contamination of air, land surface, soil, and other elements of the ecosystem [2].

Because of inadequate wastewater treatment facilities, including proper electric power supply, inadequate sustenance, and shortage of skilled and experienced labour, industrial effluents are very often discharged into water bodies, which has negative impact on underwater life. Contamination of groundwater and other water sources can harm both human health and the ecosystem [3]. In order to meet the growing demand for metals, society has been forced to increase mining operations, largely due to the overuse of natural resources. Consequently, expansion of industries results in increased resource and water consumption, which leads to the release of a significant amount of heavy metals as effluents. As a result, industrial wastewater management has gained international attention. Several research studies and experiments have been carried out to recycle wastewater and utilize it for various processes [4]. Nevertheless, each of these techniques still has some imperfections such as low productivity, the enormous amounts of solvents and reagents required, and the production of secondary pollutants such as dangerous chemicals, sludge, waste residues, and effluents [5]. Therefore, it is now essential everywhere to find affordable, environmentally friendly ways to treat wastewater more effectively. The objective of this study is to give a clear view to researchers and industry stakeholders about the present scenario, available treatments, difficulties, and possibilities related to wastewater treatment techniques. This study additionally uncovers feasible methods for treating wastewater and turning it into readily available and highly valuable supplies of clean water, all of which help to fulfil sustainable development goals.

DOI: 10.1201/9781003487142-13

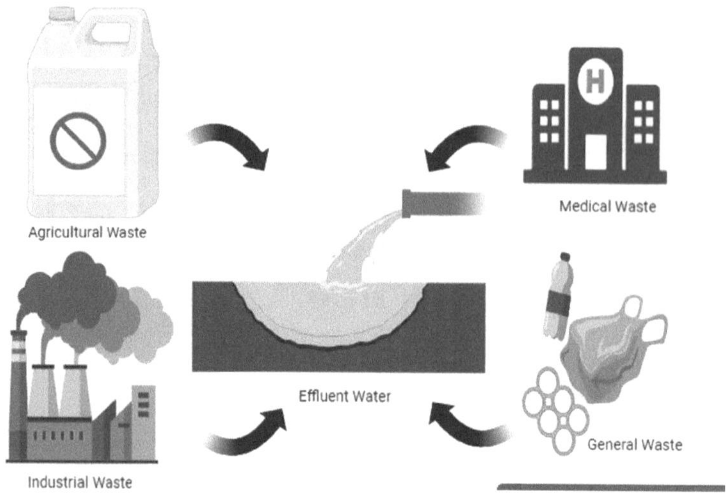

FIGURE 13.1 Various sources of effluent water.

13.2 IMPACT OF INDUSTRIAL EFFLUENT

A country's economic progress is increasingly dependent on rapid industrialization and urbanization, which are also essential for human welfare. However, it is also turning into a growing source of water pollution and a potent, unavoidable danger to upset the environment's balance and ecosystem. Numerous hazardous industrial pollutants have either directly or indirectly poisoned various biological forms after being poorly processed and released into water sources. Heavy metals are one of the nature's primary, persistent, and non-biodegradable water pollutants. Certain dangerous heavy metals that aquatic animals ingest can harm the health of other species and ultimately humans through the food chain. They may result in diminished growth and development, oxidative stress, organ damage, nervous system abnormalities, and cancer. The tannery effluent ranked highest among all the industrial wastes [6]. The various sources of industrial effluent are listed in Figure 13.1.

Phenolic compounds are another common chemical contaminants released by industry. They affect the processes of biological treatment by preventing normal microbial activity, which is how they demonstrate toxicity. Additionally, they may result in cyanosis, sweating, low body temperature, reflex loss, reduced breathing, and respiratory failure. Genotoxicity and mutagenicity can be caused by the main effluent elements of the paper and pulp industries, such as tannins, resins, and chlorinated organic compounds. In addition, they can result in a decrease in reflexes, perspiration, lowered body temperature, cyanosis, and difficulty breathing. The tannins, resins, and chlorinated organic compounds from the pulp and paper industries can cause genotoxicity and mutagenicity. Lignin is an important industrial waste from the pulp industry. During biological treatment, they may transform into hazardous chemicals and are partially biodegradable.

This, in turn, affects the hormonal balance of aquatic animals. Certain essential sterols, such stigmasterol and β-sitosterol, can bind to fish oestrogen receptors and cause reproductive disorders. Chromium compounds and oil foam form a colloidal substance that hinders sunlight from entering water bodies and reduces the amount of dissolved oxygen. Many textile industries use organic dyes related to chlorine, which are carcinogenic. The tannery and textile effluents can increase the level of Chemical Oxygen Demand and Biochemical Oxygen Demand. Sodium and chromium adversely affect the plant growth, soil quality and water bodies. For a clean and healthy environment, it is crucial to use efficient methods to eliminate these harmful contaminants before they reach different water bodies. The effects of industrial effluents are shown in Figure 13.2.

13.3 METHODS OF TREATING INDUSTRIAL EFFLUENTS

The primary industries that contribute to the presence of metals in wastewater include chemical, petrochemical, sugar, textile and dyeing, bleaching, paper and pulp, mining and quarries, organic chemical, nuclear power, leather/tannery, iron and steel, electric power plants, metal refining, soap and detergent, pharmaceutical, pesticide and biocide metal processing, and electroplating industries [7].

The advancement of technology and research has improved the research ideas to extract valuable raw materials from industrial effluents. In order to lessen their harmful effect on human health and preserve the aesthetic value of the environment, many techniques have been developed. The industrial effluent water was treated using chemical, biological, and physical processes. Several advances and breakthroughs in hybrid technology, which combines two and more processes, have increased the recovery rate and efficacy [8]. The various effluent treatment processes are listed in Figure 13.3.

Most commonly, the untreated industrial wastewater effluents are discharged into water bodies, damaging normal underwater life, due to poor wastewater treatment facilities, including malfunctioning of electricity, poor sustenance, and a shortage of skilled and experienced labour. Water

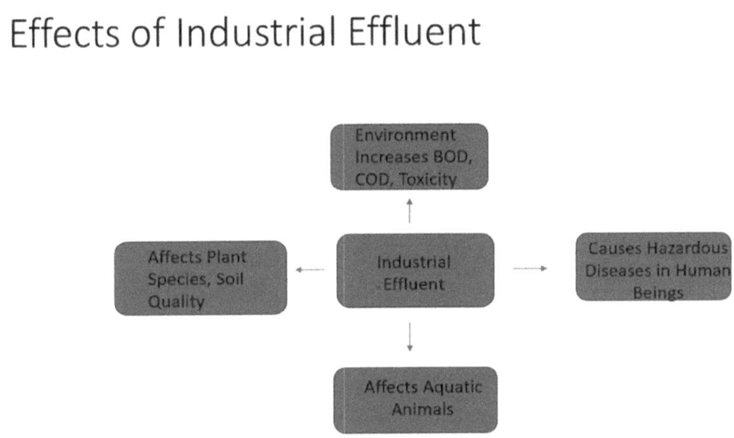

FIGURE 13.2 Effects of industrial effluents.

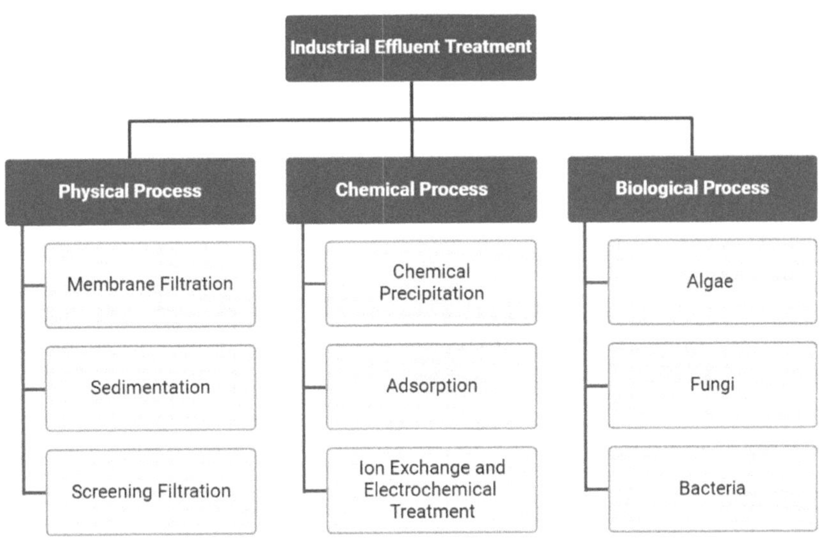

FIGURE 13.3 Methods of industrial effluent treatment.

bodies, including groundwater, become polluted, leading to both environmental degradation and health risks [9]. A great deal of research has been carried out by utilizing biological, chemical, and physical processes. There have also been several advancements and innovations in hybrid techniques, which combine two or three methods to increase accuracy and recovery rates.

13.3.1 Physical Processes

This is the first step in the treatment process to remove bigger or more solid and stable particles from the industrial wastewater. This procedure consists of membrane filtration, sedimentation, filtration, and screening [10].

13.3.1.1 Membrane Filtration

The technique of removing metal from mixed metal wastes by filtering via a porous or non-porous membrane is called membrane filtration. In simple terms, a membrane is a barrier that divides two phases apart by limiting the selective passage of constituents across it. The wastewater composition, the size of the membrane pore, and the material utilized are some of the variables that might impact the membrane's performance. This technique includes nanofiltration (NF), reverse osmosis (RO), microfiltration (MF), and ultrafiltration (UF). There is considerable similarity between UF and MF. The primary function of MF is pre-treatment, and it functions well at low pressures. The suspended or colloidal particles are separated from wastewater. Typically, the surface coating has a thickness of 10–150 μm. This membrane removes oil, grease, particles, germs, and organic materials [10,11]. Track-etching, phase inversion, stretching, and sintering are the processes used to produce the membranes for MF methods. The surface layer thickness in UF is typically 150–250 μm. Proteins, viruses, polysaccharides, and macromolecules are eliminated using this membrane. It is more efficient and uses less energy. High pressure is necessary to flow water through NF, which effectively separates substances with low molecular weights at a pore size of 0.001 μm. Typically, the surface coating has a thickness of 150 μm. High molecular weight substances, like mono-, di-, and oligosaccharides, are removed by this membrane. NF is capable of efficiently removing heavy metals from industrial effluents [10]. RO is frequently used with a semi-permeable membrane to extract salt from salty water. It is capable of successfully eliminating the multivalent ions from industrial wastewater. The surface coating typically has a thickness of 150 μm. Amino acids, glucose, proteins, sodium chloride, and high- and low-molecular-weight substances are all eliminated using this membrane [10].

13.3.1.2 Sedimentation

Sedimentation is the process of making the suspended solid particles in wastewater to settle down. Wastewater treatment facilities employ sedimentation basins to separate water from suspended particles. Secondary clarifiers are employed in biological processes, such as the activated sludge process, to eliminate settleable particles that are produced as an active component of the biological process. To accelerate the process, chemicals can be added externally to the aqueous media. The flow speed, detention duration, and solid loading can all have a significant impact on sedimentation efficiency [12,13].

13.3.1.3 Screening and Filtration

In this process, the suspended or floating items are removed from the wastewater. Screens of different sizes are used for different particles. To remove oils, grease, germs, and other materials through filtering, pore sizes between 0.1 and 0.5 mm are utilized [12].

13.3.2 Chemical Processes

13.3.2.1 Chemical Precipitation

One popular conventional method for removing metals from wastewater is chemical precipitation, which involves adding various chemicals. Chemical precipitation falls into several types, including hydroxide precipitation, chelating precipitation, and sulphide precipitation. The pH shift is the crucial factor in the process [14]. However, during the precipitation process, sludge gets produced in large quantities, which raises the expense of management and disposal [14]. The concentration of wastewater, the quantity of metals in the wastewater, the reaction's parameters, and the presence of substances that might inhibit reactions all have an impact on the efficiency of the process.

13.3.2.2 Adsorption

The conventional technique for extracting metals from wastewater is adsorption. This method works well and offers a lot of surface area for the metals to be adsorbed. This technique involves the passing of metal-containing wastewater through a solid phase that is called as an adsorbent. On the surface, the specific metals that are adsorbed may then be gathered and separated. After the removal of metal from the adsorbent, desorption method was used to replenish it. Natural materials, industrial by-products, and agricultural waste can all be sources of the adsorbent. The most frequently used adsorbent is activated carbon. Other adsorbents are sawdust, fly ash, rice husk, carbon nanotubes, zeolite, water hyacinth, orange peel pith, coconut coir pith, sunflower, etc. The selectivity of the adsorbents influences the cost of activated carbon throughout the adsorption process. Certain adsorbents are mechanically weak, have poor adsorption capacity, and may disintegrate in acidic environments. This procedure involves choosing the right adsorbent and waste products produced, which requires more consideration [15].

13.3.2.3 Ion Exchange Treatment

Ion exchange is the process of exchanging ions between the liquid and solid phases. The exchanging of ions is influenced by a number of factors such as temperature, pH, anions, adsorbent concentrations, initial sorbate, and contact duration. Typically, metals from wastewater are removed using natural or synthetic resins. Metals from wastewater are often removed using synthetic or natural polymers. Unfortunately, there are several drawbacks with this method. For example, it is not that much cost-effective because the resins required are very sensitive to the particles being removed and expensive, which results in reduced efficiency of metal extraction. Additionally, it is necessary to change the resins periodically because repurposed resins eventually begin to deteriorate and smell nasty [14]. The ion exchange wastewater treatment was illustrated in Figure 13.4.

13.3.2.4 Electrochemical Treatment

It is possible to clean municipal and industrial wastewater using electrochemical technologies. Preliminary removal of bigger particles and other physical inclusions from water is accomplished by conventional treatment. Once this requirement exists, any type of pollution, including ions, bacteria, and organic and inorganic components, may be eliminated using electrochemical techniques,

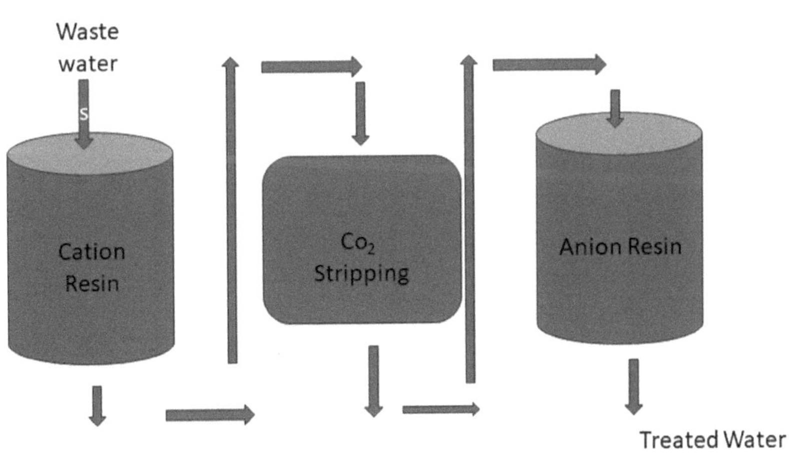

FIGURE 13.4 Ion exchange wastewater treatment.

producing pure water with distilled water quality. The most popular electrochemical water treatment techniques are electrochemical reduction, which recovers metals and changes persistent organic compounds into less toxic forms; electrocoagulation and electro flotation, which remove suspended particles; electrodialysis; and electrochemical oxidation, which mineralizes organic pollutants, disinfects water, and removes cyanides and sulphides. Electrodialysis is also used to desalinate water, while electroflotation is used to remove suspended particles [16].

13.3.3 Biological Process

The alternative process to conventional metal recovery procedure is the biological process. Microorganisms are crucial to the biological treatment of wastewater. Bacteria are used to break down metallic compounds into simpler forms so that they may be more easily separated from mixed metal solutions [17].

13.4 BIOREMEDIATION

Bioremediation is a prominent and inventive technology because of its eco-friendliness and economic expediency. In such technology, the naturally growing microbial strains degrade toxic substances into nontoxic ones. This degradation is mainly done by enzymes. Hazardous materials can be broken down or detoxified by the use of bioremediation, which involves giving the organisms the nutrition and other elements they require to thrive. Every phase of the metabolic process depends on enzymes [18]. It belongs to a family that also includes hydrolases, lyases, transferases, and oxidoreductases. Numerous enzymes may break down a broad variety of substrates due to their specific and non-specific substrate affinities. The effectiveness of bioremediation depends on the contaminants' enzymatic activity. During bioremediation, it is frequently necessary to adjust ambient factors in order to hasten microbial growth and breakdown [19]. The most commonly used methods of bioremediation are listed in Figure 13.5.

13.5 ADVANTAGES OF BIOREMEDIATION

Environmental pollutants can be effectively and economically removed by the process of bioremediation [20]. Promoting plant growth, controlling insects, protecting soil, recycling nutrients, and lowering pollutants are some of the crucial roles played by soil microorganisms [21]. Bioremediation had become more significant because of its cost-effectiveness, societal acceptability, and efficiency.

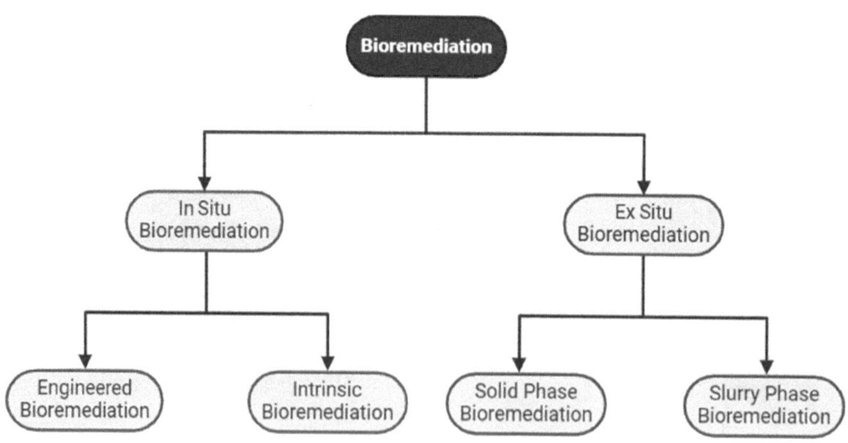

FIGURE 13.5 Types of bioremediations.

Bacteria play a key role in bioremediation. Archaea have been identified to have a role in bioremediation in a variety of bacterial-based applications. Microbes can also aid in the removal of impurities from industrial waste that is basic, acidic, hypersaline, or hyperthermal [22]. Based on current research studies, employing many live organisms in bioremediation can lead to better outcomes and efficiency, as well as more microbial diversity. The bioremediation technique was utilized by several researchers to eliminate the inorganic and organic contaminants [23,24]. In a research study, *Aspergillus sydowii* was used to treat a variety of contaminants using bioremediation technology, including endophytic fungus and organophosphate pesticides including profenofos, methyl parathion, and chlorpyrifos [25].

Microbial enzymes are employed in the bioremediation process in order to convert hydrocarbons into less harmful substances. The widespread application of genetically modified microorganisms that may also help to remove xenobiotic substances, including petroleum, naphthalene, toluene, and benzene, is currently being investigated [24]. The ambient temperature, aerobic or anaerobic conditions, and nutrient availability are some of the factors that influence bioremediation for improved outcomes. Emerging environmental pollutants, which are mostly brought into ecosystems by human activity, include poisons, heavy metals, persistent organic compounds, and air pollutants that can be either natural or artificial. These pollutants represent a hazard to all living things, including humans, animals, and plants. One of the most economical and environmentally advantageous biotechnological developments is bioremediation. Waste management is mainly based on bioremediation. Because of their difficulty in breaking down, persistent organic pollutants are considered as heterologous biological substances [26].

13.6 ROLE OF NANOTECHNOLOGY IN WASTEWATER TREATMENT

A robust economy and the development of society mainly depend on access to safe and clean water, which is a basic human necessity. There are steps being taken to lessen the current circumstances. The ecology is adversely affected by wastewater produced from both industrial and residential sources [27]. Generally, chemicals are used to tackle this problem, which may contribute to the rising concern over wastewater treatment. The treatment process needs to be easy, economical, and ecologically benign. It has been proven that nanomaterials may effectively remove a variety of contaminants from wastewater [28]. Nanotechnology has developed into a viable and environmentally beneficial method for treating water and other environmental clean-ups in recent years. Toxic contaminants have been effectively adsorbed, degraded, and removed from water bodies by means of this technology through the structural and physiochemical alteration of frequently used materials [29].

The efficacy of water treatment systems has been increased by a variety of engineered nanomaterials and tools that have been created at the molecular and atomic level, including photocatalysis, dendrimers, biosorption, and nano-filters [30].

In addition, various kinds of integrated and composite treatment filters or system, such as those made of mixed oxides, zeolites, metallic and bimetallic nanoparticles, and carbon nanotubes and wires, are made of chemical and biogenic nanoparticles (NPs) and are intended to improve the surface-volume ratio, have better reaction, and neutralize the toxins with a notable recovery rate [31]. In nanotechnology, the use of nanomembranes to remove physical particles and molecules, as well as biological and chemical pollutants, softens water. Consequently, the field of nanobiotechnology has broadened to encompass the production and entrapment of nanoparticles (NPs) for bioremediation using microbes and their active catalytic molecules as bio-nanomachinery.

Microorganisms and their enzymes are used for removing the hazardous pollutants in wastewater by the process of bioremediation. Microbially assisted nanotechnology is economical and environmentally beneficial [32]. Hazardous dyes, pharmaceuticals, and other developing toxins emitted by industry, municipal, and agriculture waste in water bodies is efficiently broken down and

TABLE 13.1
Current Nanotechniques Used for Wastewater Treatment

S. No.	Nanotechnique Methods for Wastewater Treatment	Nanomaterials Involved	Description
1	Adsorption	Carbon-based nanotubes, carbon-based nanoadsorbents, and nano metal oxide are used	Through absorption, wastewater is cleaned by removing organic and Inorganic substances from it [30].
2	Membrane and membrane process	Nano-TiO_2 nano-Ag, nanozeolites	This method efficiently purifies wastewater with less usage of chemicals [31].
3	Photocatalysis	Fullerene derivatives, TiO_2 photocatalyst	Photocatalysis process eliminates hazardous pollutant and microorganisms [30].
4	Sensing and monitoring	Quantum dots, CNTs, and nanoparticles	These nanosensors detect pathogens and remaining residues in water [31].
5	Microbial-assisted nanotechnology	Example: *Pseudomonas aeruginosa* synthesizes zirconia nanoparticles which was involved in bioremediation [34]	The microbial-assisted nanotechnology is ecofriendly and cost-effective [32]. Pharmaceuticals, toxic dyes, and other emerging toxins released by municipal, agricultural, and industrial wastes into water bodies are efficiently broken down and removed by the nanoparticles generated by bacteria or biogenic nanoparticles.

removed by the nanoparticles generated by bacteria or biogenic nanoparticles. These nanoparticles are not only used to remove chemical and heavy metal contaminants from wastewater, but they also show promise in terms of detecting pathogens, toxicants, and disinfection by-products using sensing technologies [33]. Some of the nanotechniques currently used for wastewater treatment are listed in Table 13.1.

13.6.1 Significance of Nano-bioremediation

Nanotechnology has become a promising and useful technology that offers several important advantages for enhancing environmental technologies that use nanomaterials with effective performance in the bioremediation process [35]. Nano-bioremediation uses nanoparticles to speed up bioremediation, with the goal of cleansing the environment [36]. There are two subgroups under nano-bioremediation: microbial nano-remediation and nano-phytoremediation of nanoparticle [37,38].

The utilization of microorganisms to remediate contaminants is known as bioremediation. After nanotechnology and bioremediation were combined, the way nanoparticles interact with live organisms largely determines the effectiveness of nano-bioremediation. A small number of instances where biotic components and nanoparticles interacted produced biocidal effects that were shown to be detrimental to the organisms engaged in the bioremediation process [39]. Therefore, evaluating the interaction between nanoparticles and biotic components is a prerequisite for the nano-bioremediation process. The success of nano-bioremediation can be influenced by a number of variables, including the size, shape, and chemical makeup of nanoparticles, the physiological traits of the organism, the pH and temperature of the soil, and the kind of pollution [40]. These elements operate both directly and indirectly. The ideal growth of biological organisms depends on factors such as temperature, pH, and medium, and the interaction of nanoparticles with organisms controls a range of processes, including absorption, dissolution, and biotransformation. The procedure of nano-bioremediation involves two steps. First, pollutants are broken down by nanoparticles to a level that is suitable for bioremediation. Next, pollutants undergo biodegradation [41].

13.6.2 Microorganisms-Assisted Nanotechnology

The employment of nanomaterials and microorganisms in tandem makes nanotechnology eco-friendlier and sustainable. The chemically produced nanoparticles may have certain disadvantages in terms of self-agglomeration and chemical consumption. Thus, the green production of nanoparticles employing enzymes from plant extracts and bacterial and fungal extracts was used as an alternative. They act as reductive agents for the metal complex salt and synthesize metallic nanoparticles [42].

This microbial nanotechnology uses soil bacteria and nanoparticles to speed up the biodegradation process. The microorganisms take up metal ions, undergo a reduction process, and synthesize nanoparticles. Metals and microbial enzymes together help in the formation of nanoparticles for nano-bioremediation [43]. Microbial-assisted nanotechnology is a two-phase process [44]. In the initial stage, nanoparticles are introduced into the system, and the contaminant particles endure processes such as absorption adsorption, photocatalysis, and dissolution. In the second phase, a variety of biotic processes, such as biostimulation and biotransformation, remove these particles from the system [45]. The biotic phase, which is the second stage, is the important phase for the bioremediation of contaminants. The pollutants, including both organic and inorganic, have been treated via microbial nano-bioremediation [46]. By adding protein-based bioactive components, these nanosized particles achieve solidity in a liquid solution by co-precipitation [42]. Figure 13.6 briefly describes the methods of bioremediation.

13.6.3 Nanotechnology Methods Combined with Microbial Assistance to Facilitate Bioremediation of Wastewater

Aspergillus tubingensis, which is obtained from the rhizosphere of *Avicennia officinalis* in the Indian Sundarbans, was used to extract iron oxide nanoparticles. The produced nanoparticles removed more than 90% of the heavy metals [Zn (II), Cu (II), Ni (II), and Pb (II)] from wastewater after five cycles of regeneration capacity. The metal ions were chemically adsorbed on the surface of nanoparticles by endothermic reaction [42]. Iron oxide and exopolysaccharides (EPS) from *Chlorella vulgaris* were co-precipitated in a different investigation. EPS functional groups were shown to effectively modify the nanoparticles and were discovered using Fourier-transform infrared spectroscopy (FT-IR) research. Furthermore, it was demonstrated that the nanocomposite could remove 91% of PO_4^{3-} and 85% of $NH4^+$ [47].

The copper-resistant bacteria *Escherichia* sp. SINT7 were used for the synthesis of copper nanoparticles. It was exemplified that the biogenic nanoparticles broke down both and azo and textile effluent dyes. Dyes like reactive black-5, direct blue-1, Congo red, and malachite green were lowered to 83.61%, 88.42%, 97.07%, and 90.55%, respectively, at a low concentration of 25 mg/L.

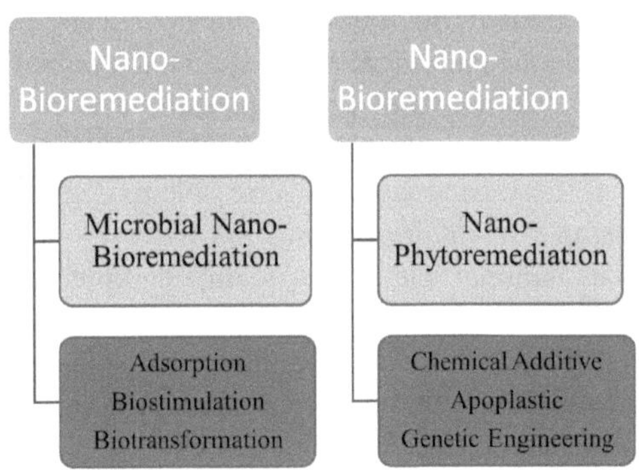

FIGURE 13.6 Nano-bioremediation.

TABLE 13.2
Microbial-Assisted Nanotechnology

S. No	Nanoparticles Synthesized	Microorganism Assisted	Degradation Capacity	Important Features	Reference
1	Zirconia (ZrO_2 NPs) nanoparticles	*P. aeruginosa*	Degradation of tetracycline 526.32 mg/g	Involved in bioremediation and green synthesis of nanoparticles.	[34]
2	Silica nanoparticles	*Actinomycetes*	80% decolouration of industrial dye	Economical and long lasting	[49]
3	Enzyme immobilized nanoparticle	*Pleurotus ostreatus*	Degradation of 90% of bisphenol-A and 10% of carbamazepine	Economical, reusable enzyme	[50]
4	Cobalt oxide (Co_3O_4) and cobalt nanoparticles		Removal of murexide dye 39.4% and 43.6%	More economical, simpler, easier to produce, and more efficient in photocatalysis degradation	[51]
5	Cyclodextrin fibres	*Lysinibacillus sphaericus*	Chromium degradation (VI) = 58% ± 1.4% Nickel (II) = 70% ± 0.2%,	Cyclodextrin offers an additional source of carbon for the growth of bacteria.	[52]
6	Iron oxide nanoparticles	*Aspergillus niger* BSC-1	Degradation of 99% of cr VI		[53]
7	Cadmium sulphide nanoparticles	*Pseudomonas aeruginosa* JP-11	Degradation of 88.66% Cd (II)		[28]
8	Palladium nanoparticles	*Spirulina platensis*	Degradation of 90% Pb		[54]
9	Iron oxide nanoparticles	*Aspergillus tubingensis* STSP 25	Degradation of 98% of Pb(II), 96.45% of Ni(II), 92.19% of Cu(II), 93.99% of Zn(II)		[53]

At a concentration of 100 mg/L, these were further reduced to 76.84%, 62.32%, 83.90%, and 31.08%, respectively. The suspended particles, chloride, and phosphate ions in treated samples decreased, and industrial effluent was also treated. Such biogenic nanoparticles' performance supports industry's efforts to produce goods in an economical and sustainable manner [48]. Some of the microorganisms associated with removal of industrial waste pollutants are listed in Table 13.2.

In another study, Cheng et al. [55] synthesized iron–sulphur nanoparticle. Through the extracellular transport of electrons, these nanoparticles were able to degrade napthal green B dye. The synthesis of nanoparticles using *Pseudoalteromonas* sp. CF10-13 provides a sustainable method of biodegradation. Endogenous nanoparticle synthesis impeded the production of hazardous gases and metal complexes. Utilizing biogenic particles is an excellent approach for cleaning up industrial effluents. There are several ways that microorganisms might contribute to the development of nanotechnology, in addition to their direct ability to produce nanoparticles. For example, microbes when combined with nanoparticles produce catalytic enzymes that aid in the cleanup of effluents. Figure 13.7 explains the various approaches of industrial wastewater treatment using nanotechnology in combination with microorganisms.

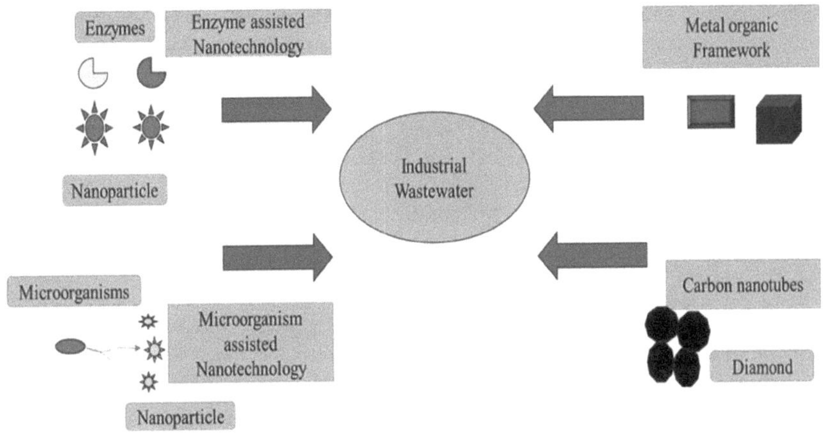

FIGURE 13.7 Different nanotechnology methods combined with microorganisms.

13.7 FUTURE PERSPECTIVES AND CONCLUSION

Researchers are interested in nanotechnology because of its numerous positive impacts, including its vast surface area, adaptability, durability in difficult environments, ease of handling materials, greater interaction, and several more. The combination of nanotechnology with microbes and enzymes has enabled an environmentally friendly method of treating industrial wastewater. Using microorganism can reduce the risks associated with chemically generated nanoparticles. The residues are separated by precipitation method and simple filtration technique, and they are found to be biocompatible. The commercialization of various components of nanotechnology presents a bigger hurdle. One percent of these aspects of nanotechnology have been introduced to the market. Thus, the large-scale application of these simple and effective microorganism combined with nanotechnology techniques is found to be a springboard for industry. An environmentally sustainable method for bioremediation of industrial wastewater has been made possible by microbial-assisted nanotechnology. The synthesis of nanoparticles from microbes offers better options for sustainable and affordable industrial wastewater treatment. To fully realize the potential of this method, further commercial research and development should be done. This will generate more green energy and strengthen the industrial economy.

REFERENCES

1. K. Kümmerer, D.D. Dionysiou, O. Olsson and D. Fatta-Kassinos, "Reducing aquatic micropollutants – Increasing the focus on input prevention and integrated emission management", *Sci Total Environ* 652 (2019):836–850.
2. J. Ahmed, A. Thakur and A. Goyal, "Industrial wastewater and its toxic effects", in *Biological Treatment of Industrial Wastewater*, The Royal Society of Chemistry (2021):1–14. https://doi.org/10.1039/9781839165399-00001.
3. A.J. Bora and R.K. Dutta, "Removal of metals (Pb, Cd, Cu, Cr, Ni, and Co) from drinking water by oxidation-coagulation-absorption at optimized pH", *J Water Process Eng* 31 (2019):100839.
4. H.A. Alalwan, M.A. Kadhom and A.H. Alminshid, "Removal of heavy metals from wastewater using agricultural byproducts", *J Water Supply Res Technol Aqua* 69 (2020):99–112.
5. H. Gebretsadik, A. Gebrekidan and L. Demlie, "Removal of heavy metals from aqueous solutions using Eucalyptus Camaldulensis: An alternate low cost adsorbent", *Cogent Chem* 6 (2020):1720892.
6. M. Bilal, J.A. Shah, T. Ashfaq, S.M.H. Gardazi, A.A. Tahir, A. Pervez, et al., "Waste biomass adsorbents for copper removal from industrial wastewater—A review", *J Hazard Mater* 263 (2013):322–333.
7. C.G. Awuchi, H. Twinomuhwezi, C.G. Awuchi, I. Victory and A.O. Ikechukwu, "Industrial waste management, treatment, and health issues: Wastewater, solid, and electronic wastes", *Eur Acad Res* 8 (2020):1081–1119.

8. M. Hemalatha, J.S. Sravan, B. Min and S. Venkata Mohan, "Microalgae-biorefinery with cascading resource recovery design associated to dairy wastewater treatment", *Bioresour Technol* 284 (2019):424–429.
9. Y. Wu, J. Luo, Q. Zhang, M. Aleem, F. Fang, Z. Xue, et al., "Potentials and challenges of phosphorus recovery as vivianite from wastewater: A review", *Chemosphere* 226 (2019):246–258.
10. M.A. Abdel-Fatah, "Nanofiltration systems and applications in wastewater treatment: Review article", *Ain Shams Eng J* 9 (2018):3077–3092.
11. E. Obotey Ezugbe and S. Rathilal, "Membrane technologies in wastewater treatment: A review", *Membranes (Basel)* 10 (2020):89.
12. M. Ince and O. Kaplan Ince, "Heavy metal removal techniques using response surface methodology: Water/wastewater treatment", in *Biochemical Toxicology - Heavy Metals and Nanomaterials*, Intech Open (2020). https://doi.org/10.5772/intechopen.88915.
13. S. Conserva, F. Tatti, V. Torretta, N. Ferronato and P. Viotti, "An integrated approach to the biological reactor–sedimentation tank system", *Resources* 8 (2019):94.
14. G. Crini and E. Lichtfouse, "Advantages and disadvantages of techniques used for wastewater treatment", *Environ Chem Lett* 17 (2019):145–155.
15. A. Pohl, "Removal of heavy metal ions from water and wastewaters by sulfur-containing precipitation agents", *Water Air Soil Pollut* 231 (2020):503.
16. M. Sillanpää and M. Shestakova, Electrochemical water treatment methods, *Science Direct* (2017):47–130. https://doi.org/10.1016/B978-0-12-811462-9.00002-5.
17. S.L.R.K. Kanamarlapudi, V.K. Chintalpudi and S. Muddada, "Application of biosorption for removal of heavy metals from wastewater", in *Biosorption*, InTech (2018). https://doi.org/10.5772/intechopen.77315.
18. S. Malik, A. Dhasmana, S. Kishore and M. Kumari, "Microbes and microbial enzymes for degradation of pesticides", in *Bioremediation and Phytoremediation Technologies in Sustainable Soil Management*, Apple Academic Press, Boca Raton, FL (2022):95–127. https://doi.org/10.1201/9781003281207-5.
19. X. Ren, G. Zeng, L. Tang, J. Wang, J. Wan, J. Wang, et al., "The potential impact on the biodegradation of organic pollutants from composting technology for soil remediation", *Waste Manag* 72 (2018):138–149.
20. M. Tripathi and R. Gaur, "Bioactivity of soil microorganisms for agriculture development", in *Microbes in Land Use Change Management*, Elsevier (2021):197–220. https://doi.org/10.1016/B978-0-12-824448-7.00012-7.
21. S. Alaira, C. Padilla, E. Alcantara and N. Aggangan, "Social acceptability of the bioremediation technology for the rehabilitation of an abandoned mined-out area in Mogpog, Marinduque, Philippines", *J Environ Sci Manag* 24 (2021):77–91.
22. M.J. Krzmarzick, D.K. Taylor, X. Fu and A.L. McCutchan, "Diversity and niche of archaea in bioremediation", *Archaea* 2018 (2018):1–17.
23. D. Kour, T. Kaur, R. Devi, A. Yadav, M. Singh, D. Joshi, et al., "Beneficial microbiomes for bioremediation of diverse contaminated environments for environmental sustainability: Present status and future challenges", *Environ Sci Pollut Res* 28 (2021):24917–24939.
24. P. Sharma, A.K. Pandey, S.-H. Kim, S.P. Singh, P. Chaturvedi and S. Varjani, "Critical review on microbial community during in-situ bioremediation of heavy metals from industrial wastewater", *Environ Technol Innov* 24 (2021):101826.
25. P.R.S. Soares, W.G. Birolli, I.M. Ferreira and A.L.M. Porto, "Biodegradation pathway of the organophosphate pesticides chlorpyrifos, methyl parathion and profenofos by the marine-derived fungus Aspergillus sydowii CBMAI 935 and its potential for methylation reactions of phenolic compounds", *Mar Pollut Bull* 166 (2021):112185.
26. S. Singh, S. Singh and R.K. Kushwaha, "Bioremediation of hydrocarbons and xenobiotic compound", in S. Singh, S. Singh and R.K. Kushwaha, eds. *Bioremediation: Challenges and Advancements*, Bentham Science Publishers, (2022):1–48. https://doi.org/10.2174/9789815036039122010004.
27. G.K. Teye, W.K. Darkwah, H. Jingyu, L. Ke and Y. Li, "Photodegradation of pharmaceutical and personal care products (PPCPs) and antibacterial activity in water by transition metals", *Rev Environ Contam Toxicol* 254 (2021):131–162.
28. R.K. Raja, S. Hazir, G. Balasubramani, G. Sivaprakash, E.S.J. Obeth, T. Boobalan, et al., "Green nanotechnology for the environment", in *Handbook of Microbial Nanotechnology*, Elsevier (2022):461–478. https://doi.org/10.1016/B978-0-12-823426-6.00006-1.
29. K. Jain, A.S. Patel, V.P. Pardhi and S.J.S. Flora, "Nanotechnology in wastewater management: A new paradigm towards wastewater treatment", *Molecules* 26 (2021):1797.
30. P. Ciambelli, G. La Guardia and L. Vitale, "Nanotechnology for green materials and processes", in *Studies in Surface Science and Catalysis* (2020):97–116. https://doi.org/10.1016/B978-0-444-64337-7.00007-0.

31. Y. Han, F. Lian, Z. Xiao, S. Gu, X. Cao, Z. Wang, et al., "Potential toxicity of nanoplastics to fish and aquatic invertebrates: Current understanding, mechanistic interpretation, and meta-analysis", *J Hazard Mater* 427 (2022):127870.
32. H. Ben Slama, A. Chenari Bouket, Z. Pourhassan, F.N. Alenezi, A. Silini, H. Cherif-Silini, et al., "Diversity of synthetic dyes from textile industries, discharge impacts and treatment methods", *Appl Sci* 11 (2021):6255.
33. V. Yadav, J. Ali and M.C. Garg, "Biosorption of methylene blue dye from textile-industry wastewater onto sugarcane bagasse: Response surface modeling, isotherms, kinetic and thermodynamic modeling", *J Hazard Toxic Radioact Waste* 25 (2021). https://doi.org/10.1061/(ASCE)HZ.2153-5515.000057.
34. B. Debnath, M. Majumdar, M. Bhowmik, K.L. Bhowmik, A. Debnath and D.N. Roy, "The effective adsorption of tetracycline onto zirconia nanoparticles synthesized by novel microbial green technology", *J Environ Manage* 261 (2020):110235.
35. Y. Singh and M.K. Saxena, "Insights into the recent advances in nano-bioremediation of pesticides from the contaminated soil", *Front Microbiol* 13 (2022):982611.
36. P. Bhatt, S.C. Pandey, S. Joshi, P. Chaudhary, V.M. Pathak, Y. Huang, et al., "Nanobioremediation: A sustainable approach for the removal of toxic pollutants from the environment", *J Hazard Mater* 427 (2022):128033.
37. R. Singh, M. Behera and S. Kumar, "Nano-bioremediation: An innovative remediation technology for treatment and management of contaminated sites", in G. Saxena, R.N. Bharagava, eds. *Bioremediation of Industrial Waste for Environmental Safety*, Springer: Singapore (2020):165–182.
38. A. Kumari, P. Kumari, V.D. Rajput, S.N. Sushkova and T. Minkina, "Metal(loid) nanosorbents in restoration of polluted soils: Geochemical, ecotoxicological, and remediation perspectives", *Environ Geochem Health* 44 (2022):235–246.
39. A. Juárez-Maldonado, H. Ortega-Ortíz, A.B. Morales-Díaz, S. González-Morales, Á. Morelos-Moreno, M. Cabrera-De la Fuente, et al., "Nanoparticles and nanomaterials as plant biostimulants", *Int J Mol Sci* 20 (2019):162.
40. W. Tan, J.R. Peralta-Videa and J.L. Gardea-Torresdey, "Interaction of titanium dioxide nanoparticles with soil components and plants: current knowledge and future research needs – a critical review", *Environ Sci Nano* 5 (2018):257–278.
41. E. Kranjc and D. Drobne, "Nanomaterials in plants: A review of hazard and applications in the agri-food sector", *Nanomaterials* 9 (2019):1094.
42. S. Mahanty, S. Chatterjee, S. Ghosh, P. Tudu, T. Gaine, M. Bakshi, et al., "Synergistic approach towards the sustainable management of heavy metals in wastewater using mycosynthesized iron oxide nanoparticles: Biofabrication, adsorptive dynamics and chemometric modeling study", *J Water Process Eng* 37 (2020):101426.
43. G. Pandey, "Prospects of nanobioremediation in environmental cleanup", *Orient J Chem* 34 (2018):2838–2850.
44. M. Usman, M. Farooq, A. Wakeel, A. Nawaz, S.A. Cheema, H. ur Rehman, et al., "Nanotechnology in agriculture: Current status, challenges and future opportunities", *Sci Total Environ* 721 (2020):137778.
45. B. Abebe, H.C.A. Murthy and E. Amare, "Summary on adsorption and photocatalysis for pollutant remediation: Mini review", *J Encapsulation Adsorpt Sci* 08 (2018):225–255.
46. W.L. Desiante, N.S. Minas and K. Fenner, "Micropollutant biotransformation and bioaccumulation in natural stream biofilms", *Water Res* 193 (2021):116846.
47. M. Govarthanan, C.-H. Jeon, Y.-H. Jeon, J.-H. Kwon, H. Bae and W. Kim, "Non-toxic nano approach for wastewater treatment using Chlorella vulgaris exopolysaccharides immobilized in iron-magnetic nanoparticles", *Int J Biol Macromol* 162 (2020):1241–1249.
48. M. Noman, T. Ahmed, S. Hussain, M.B.K. Niazi, M. Shahid and F. Song, "Biogenic copper nanoparticles synthesized by using a copper-resistant strain Shigella flexneri SNT22 reduced the translocation of cadmium from soil to wheat plants", *J Hazard Mater* 398 (2020), pp123175.
49. R. Mohanraj, B.M. Gnanamangai, S. Poornima, V. Oviyaa, K. Ramesh, G. Vijayalakshmi, et al., "Decolourisation efficiency of immobilized silica nanoparticles synthesized by actinomycetes", *Mater Today Proc* 48 (2022):129–135.
50. C. Ji, L.N. Nguyen, J. Hou, F.I. Hai and V. Chen, "Direct immobilization of laccase on titania nanoparticles from crude enzyme extracts of P. ostreatus culture for micro-pollutant degradation", *Sep Purif Technol* 178 (2017):215–223.
51. A.S. Adekunle, J.A.O. Oyekunle, L.M. Durosinmi, O.S. Oluwafemi, D.S. Olayanju, A.S. Akinola, et al., "Potential of cobalt and cobalt oxide nanoparticles as nanocatalyst towards dyes degradation in wastewater", *Nano-Struct Nano-Objects* 21 (2020):100405.

52. N.O. San Keskin, A. Celebioglu, O.F. Sarioglu, T. Uyar and T. Tekinay, "Encapsulation of living bacteria in electrospun cyclodextrin ultrathin fibers for bioremediation of heavy metals and reactive dye from wastewater", *Colloids Surf B Biointerfaces* 161 (2018):169–176.
53. S. Mahanty, M. Bakshi, S. Ghosh, S. Chatterjee, S. Bhattacharyya, P. Das, et al., "Green synthesis of iron oxide nanoparticles mediated by filamentous fungi isolated from sundarban mangrove ecosystem, India", *Bionanoscience* 9 (2019):637–651.
54. M.H. Sayadi, N. Salmani, A. Heidari and M.R. Rezaei, "Bio-synthesis of palladium nanoparticle using Spirulina platensis alga extract and its application as adsorbent", *Surf Interfac* 10 (2018):136–143.
55. S. Cheng, N. Li, L. Jiang, Y. Li, B. Xu and W. Zhou, "Biodegradation of metal complex Naphthol Green B and formation of iron–sulfur nanoparticles by marine bacterium Pseudoalteromonas sp CF10-13", *Bioresour Technol* 273 (2019):49–55.

14 Microbial Nanotechnology in Medicine and Healthcare Industries

Mwalimu Raphael Jilani, Azhagu Saravana Babu Packirisamy, Maheswari Vinodkumar, and Abinaya Krishnan

14.1 INTRODUCTION

Nanomaterial production is increasingly emphasizing environmentally friendly approaches, with microbial nanotechnology emerging as a sustainable solution for producing metallic nanoparticles [1]. Nanoparticles are defined as particles of not more than 100 nm with special properties that are useful in diverse applications. These nanomaterials exhibit distinct optical, electrical, and magnetic characteristics at the nanoscale, stemming from their large surface area-to-volume ratio governed by quantum mechanics. The environmental impact of traditional nano production is hampered by its high energy use, limited yields, and risks. Because of their effectiveness and environmental friendliness, bio-based methods—which use live things or their derivatives—are becoming more and more popular [2,3]. Microbial synthesis of nanoparticles, in particular, offers advantages such as greater stability and reduced reliance on stringent conditions and pure starting materials [4,5]. Various biological entities, including algae, fungi, bacteria, and plants, serve as both reducing and capping agents in nanoparticle generation, resulting in a relatively simple, reproducible, and stable synthesis process. Among these, bacteria-mediated biosynthesis stands out as a promising bottom-up approach [6].

Nanobiomedicine, particularly the microbial fabrication of nanoparticles, presents a burgeoning field with extensive possibilities for biotechnological applications. The utilization of microorganisms in nanoparticle synthesis aligns with the principles of green chemistry, aiming to minimize toxicity associated with conventional chemical synthesis routes [7]. Consequently, scientists are actively exploring the ecological benefits and diverse applications of biogenic nanomaterials especially in medicine and pharmaceutics where nanotechnology finds applications in drug development, infectious agent detection, biosensors, and bioassays, offering novel strategies for managing diseases and improving diagnostics [8,9].

Microbial nanoparticles show potential in medicine and healthcare for cancer therapy, infectious disease mitigation, and antimicrobial drug resistance [10]. They have high carrier capacity and can bind to hydrophilic and hydrophobic compounds, but their potential side effects and toxicity should be considered [11,12]. Nanotechnology is transforming medical treatments across various fields, including pharmacology and imaging. Techniques like chemical and physical deposition, laser ablation, and melt mixing are used to produce nanoparticles, but they can lead to dangerous chemicals, unstable nanoparticles, and harmful by-products [13]. Biogenic nanoparticle synthesis, based on environmentally friendly methods, produces stable nanoparticles using bacteria and fungi, which can produce reductase enzymes that promote biosorption [14]. These enzymes are crucial in large-scale manufacturing, and microorganisms are rich in bioactive compounds that facilitate microbial biomineralization through biologically induced and intracellular processes [7]. This book chapter discusses the production, application, and advancement of microbial nanoparticles and nanotechnologies in the medicine and healthcare industries.

DOI: 10.1201/9781003487142-14

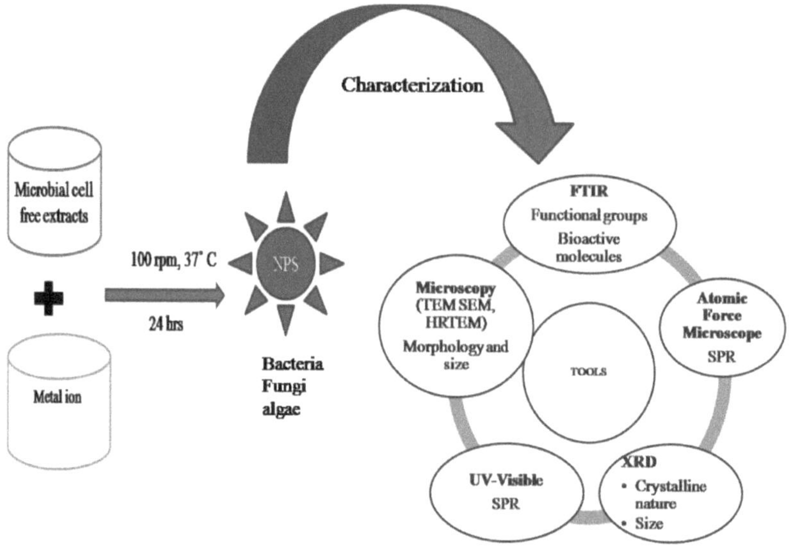

FIGURE 14.1 Illustration of microbial synthesis and characterization techniques.

14.2 MICROBIAL PRODUCTION OF NANOPARTICLES AND THEIR CHARACTERIZATION

The synthesis routes, methods of characterization, and applications of nanoparticles have advanced the field of nanotechnology. However, despite the efficiency and success of physical and chemical synthesis methods for nanoparticle synthesis, concerns have surfaced over the ecotoxicological effects of the rising production of metal nanoparticles and oxides upon their release into the environment. Consequently, attention has shifted toward "green synthesis" or "biosynthesis" of nanoparticles using microorganisms such as bacteria, fungi, viruses, yeasts, and algae. Various microorganisms utilize distinct mechanisms to form nanoparticles, often entailing the sequestration of metal ions on microbial surfaces or within cells, succeeded by their reduction to nanoparticles facilitated by enzymes. Microbial synthesis produces precise nanoparticles with vast applications in drug delivery, cancer therapy, antimicrobial agents, and biosensors [9,10]. In this section, we spotlight recent microbial-synthesized nanoparticles by various researchers, emphasizing their characterization, including shape, size, and other relevant properties. Figure 14.1 illustrates the microbial synthesis of nanoparticles and the tools used for their characterization.

14.2.1 Bacterial Biosynthesis of Nanoparticles

Bacterial-synthesized nanoparticles, alongside their investigated biological activities, are shown in Table 14.1. Abdollahnia et al. [5] produced silver, selenium, and bimetallic Ag-Se nanoparticles with uniform size and stability using exceptional strains E106 and E108 from the *Haloferax* and *Halogemetricum* genera. El-Behery et al. employed *Bacillus paramycoides* filtrates, along with gamma rays, to synthesize nanoparticles (AgNPs, SeNPs, and Ag-SeNPs) of varied shapes. The highest optical density and optimal AgNP generation were achieved at 20.0 kGy, while SeNPs and bimetallic Ag-SeNPs were most effectively synthesized at 15.0 kGy. Synthesis procedures involved radiolysis, hydrolysis, reduction, and complexation. Particle size dispersion increased with higher gamma radiation dosage [15].

Mona and Syoof [16] synthesized AgNPs using *Klebsiella pneumoniae*, observing nanoparticle sizes ranging from 26.84 to 44.42 nm. Notably, they found that the synthesis process was accelerated

TABLE 14.1
Bacterial Nanoparticles and Their Biological Actions

Bacteria	Nanoparticle	Average Size (nm)	Biological Action	References
Haloferax and *Halogemetricum* strains	Ag, Se, Ag-Se		Antibacterial	[5]
Bacillus paramycoides	Ag	38.35	Antifungal	[15]
	Se	60.25	Anti-biofilm	
	Ag-Se	48.75	Antioxidant	
Klebsiella pneumoniae	Ag	35.52	–	[16]
Lactobacillus paracasei	Se	26.5		[17]
Rothia endophytica	Ag	59.5	Anti-Candidal	[18]
Acinetobacter sp. SW 30	Au	10	–	[19]
Acinetobacter sp. SW 30	Se	100	–	

at 90°C under alkaline conditions with a pH of 10, enhancing the reduction and stabilizing capacities of reductase enzymes.

El-Saadony et al. [17] utilized *Lactobacillus paracasei* to biosynthesize 3.0–50 nm SeNPs exhibiting a distinct absorption peak at 300 nm in UV–Vis spectroscopy, and Elbahnasawy et al. [18] used endophytic bacteria (*Rothia endophytica*) to fabricate Ag ions. Wadhwani et al. [19] illustrated the generation of AuNPs and SeNPs leading to distinct color alterations to purple and orange, respectively, through the utilization of purified lignin peroxidase (97.4 KDa) derived from *Acinetobacter* sp. SW 30. This enzymatic approach yielded spherical, crystalline gold nanoparticles alongside amorphous selenium nanoparticles.

14.2.2 Fungal Biosynthesis of Nanoparticles

Fungi are prioritized over bacteria for large-scale nanoparticle production due to their facile growth, high yield, and possession of metal-reducing enzymes, rendering them a critical organism in nanoparticle manufacturing [20]. Table 14.2 displays a range of fungal-synthesized nanoparticles alongside their corresponding biological activities.

14.2.2.1 Fungal Silver Nanoparticles

A number of studies have been conducted for the biosynthesis of silver nanoparticles (AgNPs), leveraging diverse biological sources. Among these, *Talaromyces purpureogenus*, a fungus known for producing various secondary metabolites like terpenoids and phenols, has attracted attention for nanoparticle production. Sharma et al. [21] synthesized crystalline AgNPs. Bhatnagar et al. [22] produced AgNPs of varied morphologies and sizes with a surface plasmon resonance around 410 nm from the extracellular pigment of *T. purpureogenus*. Hu et al. [23] utilized the mycelial extract from *T. purpureogenus* to synthesize AgNPs below 50 nm.

Gemishev et al. [24] demonstrated the extracellular biosynthesis of AgNPs from a silver nitrate solution using the cell-free water extract of *Trichoderma reesei* biomass, which was confirmed by HRTEM micrographs showing crystalline AgNPs.

AgNPs were also synthesized from other fungal species such as *Penicillium verrucosum* [25], *Botryosphaeria rhodina* [26], *Aspergillus terreus* [27], *Beauveria bassiana* [28], and *Aspergillus tamari*, [29] revealing diverse morphologies and sizes in transmission electron microscopy (TEM) micrographs [28] and exhibiting strong surface plasmon resonance (SPR) and nanoparticle sizes determined through UV–Vis spectroscopy and atomic force microscopy (AFM) [29]. Majeed et al. [30] utilized *Penicillium italicum*. Janakiraman and Govindarajan [31] employed *Botryodiplodia theobromae*. Othman et al. [32] utilized the cell-free extract of *Aspergillus fumigatus*, revealing

TABLE 14.2
Fungal Nanoparticles and Their Biological Applications

Nanoparticles	Fungi	Size (nm)	Biological Action	References
AgNPs	*Talaromyces purpureogenus*	49.3	Antibacterial, antioxidant	[21]
		4–41	Antibacterial, anticancer	[22]
		25	Antibacterial, anticancer, wound healing	[23]
	Trichoderma reesei	15–25	–	[24]
	Penicillium verrucosum	10–12	Antifungal	[25]
	Botryosphaeria rhodina		Antioxidant, anticancer	[26]
	Aspergillus terreus BA6	7–23	Antimicrobial, antitumor	[27]
	Beauveria bassiana	10–50	Antimicrobial	[28]
	Aspergillus tamari	40	Antimicrobial	[29]
	Penicillium italicum	14.5–23.3	Antibacterial, anticancer	[30]
	Botryodiplodia theobromae	62–111	Anticancer	[31]
	Aspergillus fumigatus	10–34	Antimicrobial, antitumor	[32]
	Tritirachium oryzae	7–75	Antibacterial	[33]
	Punctularia atropurpurascens, Penicillium expansum, Phanerochaete chrysosporium	10–30	Antimicrobial	[34]
AuNPs	*Fusarium solani*	40–45	Anticancer	[36]
	Pleurotus ostreatus	10–30	Anticancer, antimicrobial	[37]
Ag-AuNPs	*Aspergillus niger*	10.12–52.51	Antimicrobial, antioxidant, anticoagulant, thrombolytic	[38]
	Trichoderma longibrachiatum	6.98–25.20	Antimicrobial, antioxidant, anticoagulant, thrombolytic	[38]
ZnONPs	*Alternaria tenuissima*	46.58	Antioxidant, anticancer, antimicrobial	[2]
	Fusarium solani	8.55–21.76	Antioxidant, antimicrobial	[3]
	Periconium sp.	16–78	Antimicrobial, antioxidant	[4]
	Fusarium keratoplasticum	10–42	Antibacterial, UV protection	[35]
ZnONPs	*Aspergillus niger*	8–38	Antibacterial, UV protection	[35]
CuNPs	*Fusarium solani*	9.97–19.49	Antimicrobial	[3]
FeNPs	*Aspergillus flavus*	28–33	Antimicrobial	[39]
FeONPs	*Trichoderma asperellum, Phialemoniopsis ocularis, Fusarium incarnatum*	13–31	–	[40]
SeNPs	*Nematospora coryli*	50–250	Anti-Candida, antioxidant	[41]
	Aspergillus quadrilineatus, Aspergillus terreus, Fusarium equiseti		Antimicrobial, antioxidant	[42]
	Mariannaea sp. HJ	45.19–212.65	–	[43]

strong absorption peaks and morphology. Al-limoun et al. [33] synthesized silver nanoparticles (AgNPs) from mycelia-free *Tritirachium oryzae* filtrate at various conditions such as pH, temperature, and different concentration of silver nitrate salt. The nanoparticles obtained were monodispersed and exhibited spherical to ovular shapes with diameters ranging from 7 to 75 nm, as confirmed by TEM analysis. Additionally, Sanguiñedo et al. [34] synthesized AgNPs using the fungi *Punctularia atropurpurascens*, *Penicillium expansum*, and *Phanerochaete chrysosporium*, ranging from 10 to 30 nm in size with a spherical shape.

14.2.2.2 Fungal Zinc Oxide Nanoparticles

Abdelhakim et al. [2] presented an eco-friendly method for producing zinc oxide nanoparticles (ZnONPs) using the endophytic fungus *Alternaria tenuissima*. The synthesized ZnONPs displayed an SPR absorption peak at 369 nm in the UV–Vis spectrum. TEM images depicted spherical particles ranging from 10 to 30 nm in size, with a mean particle size of 15.62 ± 4.51 nm. The SAED pattern confirmed distinct lattice planes, corroborating the particle size findings.

Similarly, El-Sayed and El-Sayed [3] synthesized ZnONPs from *Fusarium solani* alongside other metal nanoparticles. Ganesan et al. [4] employed the endophytic fungus *Periconium* sp. extract in a sol-gel process for ZnONP synthesis, attributing the mycochemical constituents present in the fungal extract as crucial in the process. Additionally, the cell-free filtrates of *Fusarium keratoplasticum* and *Aspergillus niger* were utilized as biocatalysts for ZnONP synthesis by exploiting metabolites such as enzymes and proteins present in the filtrates, resulting in the production of hexagonal-shaped nanoparticles by *A. niger* and nanorods by *F. keratoplasticum* with sizes ranging from 8 to 42 nm [35].

14.2.2.3 Fungal Gold Nanoparticles

Clarance et al. [36] synthesized gold nanoparticles (AuNPs) using *F. solani* from *Chonemorpha fragrans*, ranging from 40 to 45 nm in size, exhibiting needle- and flower-like structures with a spindle shape, while El Domany et al. [37] utilized *Pleurotus ostreatus* fungus extracellular fluid to produce stable AuNPs in solution, ranging from 10 to 30 nm in size, with an uneven spherical shape and a crystalline nature, stabilized by an amino acid residue as a capping agent.

Elegbede et al. [38] employed xylanase extracted from *A. niger* and *Trichoderma longibrachiatum* for the biosynthesis of bimetallic Ag-AuNPs. The SPR band of these nanoparticles was identified by UV–Vis spectroscopy examination as having intensity peaks at 520 and 534 nm, with varying shapes (spherical, oval, and regular) of different sizes ranging from 6.98 to 52.51 nm as identified by field emission scanning electron microscopy and TEM.

14.2.2.4 Fungal Iron Nanoparticles

Amanyamadan et al. [39] synthesized iron nanoparticles using *Aspergillus flavus*, resulting in spherical and irregular nanoparticles with an average diameter of 30 ± 3 nm, while Mahanty et al. [40] employed *Trichoderma asperellum*, *Phialemoniopsis ocularis*, and *Fusarium incarnatum* to synthesize iron oxide nanoparticles with varying sizes and shapes, facilitated by bioactive compounds in the fungal cell-free filtrate.

14.2.2.5 Fungal Selenium Nanoparticles

Rasouli [41] explored *Nematospora coryli* yeast's ability to produce selenium nanoparticles (SeNPs). The SPR band of SeNPs was identified by UV–Vis spectroscopy examination as having an intensity peak at 561 nm. TEM images displayed spherical NPs of the size 50–250 nm, while X-ray diffraction (XRD) patterns indicated a non-crystalline structure of the SeNPs. Hussein et al. [42] utilized endophytic fungi—*Aspergillus quadrilineatus*, *A. terreus*, and *Fusarium equiseti*—to synthesize SeNPs. TEM analysis revealed that SeNPs synthesized by *A. quadrilineatus* and *Aspergillus ochraceus* exhibited a spherical shape, whereas those from *A. terreus* displayed variations in size. Additionally, SeNPs synthesized by *F. equiseti* exhibited a combination of spherical and rod-shaped structures, indicating morphological diversity among the fungal strains. Zhang et al. [43] used *Mariannaea* sp. HJ mycelia to create SeNPs under specified conditions of PH and SeO_2 concentrations. Se (IV) ions were eliminated in the late exponential phase. The spherical nanoparticles, which had an average size of 45.19 and 212.65 nm, were evenly distributed in the cytosol and outer wall of the fungal cells.

14.2.3 Algal Biosynthesis of Nanoparticles

Silver nanoparticles are synthesized using green algae and cyanobacteria. *Botryococcus braunii* was used to produce copper and silver nanoparticles [44], while *Cyanobacteria* extract of *Nostoc sp.Bahar* M for AgNPs synthesis [45], involving metal salt reduction with color changes indicating nanoparticle formation due to SPR. Fourier Transform Infrared Spectroscopy (FTIR) analysis suggests the involvement of aromatic compounds and proteins as reducing and capping agents, respectively. SEM and TEM imaging reveals spherical AgNPs ranging from 8.5 to 150 nm, while XRD patterns confirm their crystalline nature. Silver nanoparticles (AgNPs) from *Sargassum* seaweed were biosynthesized by Kavishri et al. [46], and XRD verified the AgNPs' outstanding crystalline structure. Images from TEM and field emission scanning electron microscopy (FESEM) revealed a largely homogeneous, spherical structure with slight variation in form and size owing to weaker connections. With *Sargassum natans* and *Sargassum fluitans*, López-Miranda synthesized AgNPs and discovered polyhedral forms with an average size of 26.39 nm [47]. The ZnO nanoparticles were biofabricated by utilizing an extract from *Spirogyra hyalina*. XRD analysis revealed that the nanoparticles had a crystalline structure of approximately 45 nm on average. SEM analysis revealed a size range from 50 to 80 nm. FTIR analysis indicated the presence of carbonyl and NH_2 groups, highlighting the importance of proteins in the stability of nanoparticles [48].

14.3 APPLICATION OF MICROBIAL NANOPARTICLES IN MEDICINE AND HEALTHCARE INDUSTRIES

This section discusses the application of microbial nanoparticles in medicine and healthcare as shown in Figure 14.2.

14.3.1 Antimicrobial Activities of Microbial-Synthesized Nanoparticles

This section delineates the antimicrobial activities, encompassing antibacterial, antifungal, and antiviral properties. The emergence of drug-resistant pathogens poses significant challenges to healthcare facilities and medical sectors, particularly in their ability to withstand conventional

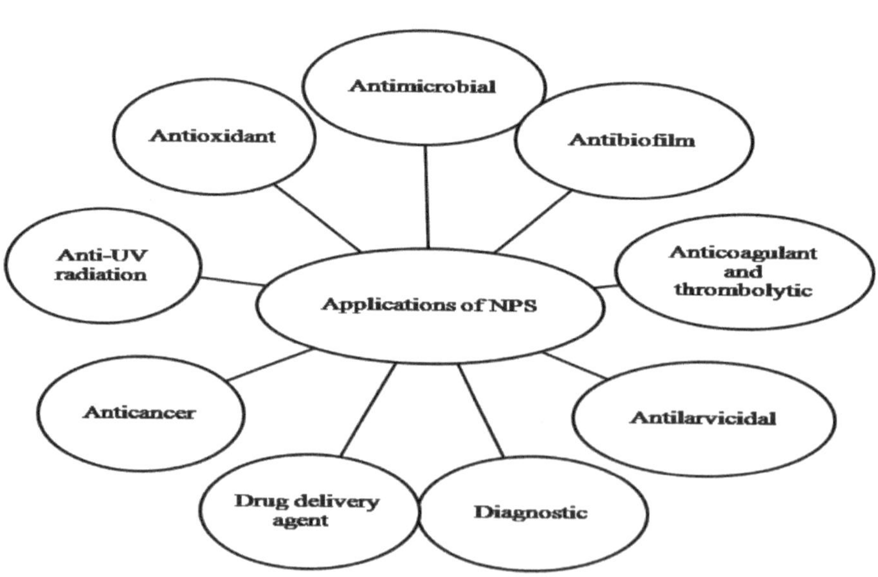

FIGURE 14.2 Application of microbial nanoparticles.

drugs. However, the utilization of eco-friendly synthesized nanoparticles offers a potential solution to combat drug resistance effectively.

Ag, Se, and Ag-Se nanoparticles derived from *B. paramycoides* were assessed for their antibacterial effects on *Escherichia coli*, *K. pneumoniae*, *Bacillus subtilis*, and *Pseudomonas aeruginosa*, along with their antifungal activity against *Aspergillus brasiliensis* and *Candida albicans*. The results revealed promising antifungal properties across all nanoparticles, particularly against *C. albicans*, which exhibited the highest vulnerability [15]. Sharma et al. [21] investigated the antibacterial properties of *T. purpureogenus*–derived AgNPs, revealing an minimum inhibitory concentration (MIC) of 1.5625 µg/mL against various bacteria, including *Listeria* sp., *E. coli*, *Shigella* sp., and *Salmonella typhi* showing inhibition zones of 13.00 ± 0.29 mm, 17.00 ± 0.14 mm, 18.00 ± 0.21 mm, and 14.00 ± 0.13 mm, respectively. Conversely, *T. purpureogenus*–derived AgNPs demonstrated the remarkable antibacterial activity on *E. coli* and *Staphylococcus epidermidis* with MIC values of 32 and 4 µg/mL and MBC values of 32 and 64 µg/mL, respectively [22]. Additionally, another study by Hu et al. [23] further highlighted the efficacy of these silver nanoparticles in suppressing the growth of bacterial pathogens at low concentrations, with MICs recorded at 16.12 µg/mL for gram-positive bacteria and 13.98 µg/mL for Gram-negative bacteria. The organism's zone of inhibition was 9, 11, and 13 mm for *Staphylococcus aureus*, *E. coli*, and *P. aeruginosa*, respectively, similar to antibiotic control (vancomycin). However, it had a weaker inhibition to *Bacillus cereus* (11 mm) and *Salmonella enterica* (11 mm) as compared to the antibiotic. Furthermore, the AgNPs demonstrated a strong cell wounding healing activity [23].

P. verrucosum derived AgNPs showed antifungal activity, inhibiting *F. chlamydosporum* and *A. flavus at* 41.94% *and* 44.58% respectively at concentration of 150 ppm whereas the percentage inhibition at 200 ppm was 56.67% and 59.13% respectively [25]. AgNPs derived from *A. terreus* was susceptible *to P. aeruginosa*, *Streptococcus faecalis* and *A. niger* minimally at 0.312 µg/mL, whereas *E. coli* and *C. albicans* was minimally inhibited by 1.25 µg/mL AgNPs [27]. Nanda et al. [29] discovered that *S. aureus*, *Vibrio parahaemolyticus*, *Shewanella putrefactions*, and *C. albicans* were susceptible to AgNPs from *A. tamari*. *S. aureus* (ZOI: 8–12 mm), *Shewanella putrefaciens* (ZOI: 8–10 mm), *C. albicans* (ZOI: 12–16 mm) and *Vibrio parahaemolyticus* (ZOI: 11–12 mm) was notably sensitive, while *E. coli* showed resistance. Silver nanoparticles (AgNPs) produced from *B. bassiana* effectively inhibited *S. aureus*, *P. aeruginosa*, and *E. coli* when combined with ciprofloxacin. The MICs achieved were 0.5, 0.4, and 0.4 ppm, respectively. This represents a reduction of about 0.1–0.2 ppm compared to using the antibiotic alone [28].

Majeed et al.'s [30] study on *P. italicum*–derived silver nanoparticles (AgNPs), when combined with moxifloxacin, demonstrated a potent inhibition against *S. enterica*, *E. coli*, *S. aureus*, and *B. cereus* with zone of inhibition ranging from 13 to 17 mm. Furthermore, when AgNPs were combined with streptomycin and moxifloxacin antibiotics, they led to a substantial increase of inhibition zones by 2–3 mm and 1–4 mm, respectively. This demonstrated the synergistic effects on the AgNPs and the antibiotics. Additionally, AgNPs synthesized by both archaeal and bacterial strains displayed antibacterial activity against *E. coli*, *S. aureus*, *P. aeruginosa*, and *B. subtilis*, with MIC_{50} values of 20, 10, 40, and 40 ppm for archaeal AgNPs and 20, 20, 40, and 40 ppm for bacterial AgNPs [5]. Similarly, selenium nanoparticles (SeNPs) demonstrated antibacterial effects, with MIC_{50} values of SeNP-A and SeNP-B against the same bacterial strains ranging from 100 to 200 ppm [5].

Gram-positive bacteria (*Bacillus mycoides*) and Gram-negative bacteria (*E. coli*) and *C. albicans* were inhibited by AgNPs from *A. fumigatus* with inhibition zones ranging from 15 to 19 mm, under different PH conditions of 6.0–12.0. Interestingly, *E. coli* displayed a zone of inhibition of 16 mm for pH 6.0, 8.0, and 10 and 15 mm at pH 12. *B. mycoides* had a ZOI of 18 mm at pH 6 and 8, whereas at pH 10 and 12, it displayed a ZOI of 19 and 16 mm, respectively. Moreover, *C. albicans* had a varying zone of inhibition of 15–18 mm depending of pH value [32]. The AgNPs exhibited comparable antimicrobial effectiveness when treated to fabrics [32].

T. oryzae AgNPs displayed potent antibacterial action against Beta-lactamase-producing clinical bacteria *S. epidermidis* (ZOI: 20.33 mm), *Enterobacter aerogenes* (ZOI: 20 mm), and *E. coli* (ZOI:

18 mm), whereas extended Beta-lactamase-producing *P. aeruginosa* (ZOI: 12mm) displayed intermediate susceptibility. The mechanism involved cell wall disruption, suggesting potential higher toxicity [33]. The mycosynthesized silver nanoparticles (AgNPs) derived from *P. atropurpurascens*, *P. expansum*, and *P. chrysosporium* demonstrated effective inhibition against *E. coli* with ZOI ranging from 12 to 30 mm but not *S. aureus*. The thicker peptidoglycan layer present in *S. aureus* may have led to its resistance [34].

Algae-AgNPs exhibited most pronounced antibacterial properties against *S. aureus* than in *E. coli*. *S. aureus* was inhibited at a low concentration of 10% v/v with algae-AgNPs exhibiting a zone of inhibition of 13 mm, whereas *E. coli* was minimally inhibited by 50% v/v with algae-AgNPs exhibiting a zone of inhibition of 12 mm. However, both of them were significantly inhibited at 100% and 50% v/v, respectively [49].

Balaraman et al. [50] reported that AgNPs synthesized from *Sargassum myriocystum* strongly inhibited both gram-positive and gram-negative bacteria. Notably, they achieved maximum inhibition zones of 18.6 mm for *S. epidermidis* and 17.6 mm for *P. aeruginosa*. Furthermore, significant inhibition of biofilm formation was observed at a concentration of 50 μg/mL. At low concentrations, bimetallic nanoparticles of Ag-SeNPs strongly posed antimicrobial activity against pathogenic fungi, bacteria, and yeast [51].

Ganesan et al. [4] tested antibacterial and antifungal activities of biosynthesized ZnONPs against *E. coli*, *S. aureus*, and *C. albicans*. High concentrations (150 μL) of ZnONPs showed significant inhibition zones: 16 mm for *S. aureus*, 14 mm for *E. coli*, and 24 mm for *C. albicans*. Hameed et al. [48] found that *Spirogyra hyaline*–biofabricated ZnONPs showed superior antibacterial activity against *P. aeruginosa*, *Bacillus pumilus*, *S. aureus*, and *E. coli* with zone of inhibition ranging from 15.33 to 24.33 mm in a concentration-dependent manner. Notably, *S. aureus* was exhibited the highest zone of inhibition (24.33 mm). Overall, algae ZnONPs were more powerful against all the bacteria than the algae extract alone.

Fouda et al.'s [52] study found that ZnONPs derived from *A. terreus* has potent antimicrobial properties, particularly against gram-positive bacteria like *B. subtilis* and *S. aureus* exhibiting a zone of inhibition of 11.0 and 10.5 mm, respectively, at an MIC value of 250 μg/mL. *E. coli* and *P. aeruginosa* were inhibited at 500 μg/mL at a low zone of inhibition of 9.2 and 9.3 mm, respectively. These nanoparticles also explicitly had a moderate antibacterial activity on cotton fabrics treated with a nano-sized zinc colloidal solution, indicating their potential in textile-based antimicrobial strategies.

AuNPs produced from *P. ostreatus* showed increased antibacterial activity against *B. subtilis* (ZOI: 20–30 mm), *S. aureus* (ZOI: 27–30 mm), *E. coli* (ZOI: 18–39 mm), and *C. albicans* (ZOI: 11–13 mm). There was a remarkable increase in the zone inhibition of biosynthesized nanoparticles when compared to standard antibiotics, indicating that nanoparticles have high potency than antibiotics [37].

Similarly, Ag-AuNPs significantly suppressed *A. niger*, *A. fumigatus*, and *A. flavus* growth at concentrations of 150 and 200 μg/mL, achieving inhibition rates of 70%–89%. These findings imply that bimetallic nanoparticles may possess greater potency as antifungal agents compared to their monometallic counterparts. Moreover, bimetallic nanoparticles (Ag-AuNPs) achieved the maximum inhibition at 100 μg/mL to drug-resistant strains of *E. coli* (90.7%), *Klebsiella granulomatis* (86.1%), *S. aureus* (78.5%), and *P. aeruginosa* (76.3%) [38].

El-Sayed and El-Sayed explored the susceptibility of multi-drug-resistant bacteria isolates to metal nanoparticles (AgNPs, CuNPs, and ZnONPs). Gram-negative bacteria, especially *P. aeruginosa*, were more susceptible to AgNPs than gram-positive bacteria. The inhibition level correlated with dosage, with the highest ZOI (22.4 mm) against *P. aeruginosa* observed with 15 μg of AgNPs. *Aspergillus awamori* was more susceptible to CuNPs at an MIC value of 31.67 μg/mL, whereas *S. aureus* was highly susceptible to 21.7 μg/mL ZnONPs [3].

L. paracasei–synthesized selenium nanoparticles (SeNPs) exhibited superior anti-Candidal activities across various *Candida* species compared to Che-SeNPs. The efficacy of *L. paracasei*

SeNPs increased with higher concentrations, with minimum fungicidal concentrations (MFCs) ranging from 80 to 130 mg/mL [17]. Similarly, Rasouli [41] demonstrated that SeNPs derived from *N. coryli* displayed anti-*Candida* activity, with an MIC of 80 ppm against *C. albicans*. Moreover, Hussein et al. [42] observed that SeNPs synthesized from *A. quadrilineatus* and *Aspergillus ochraceus* exhibited antifungal activities against *C. albicans*, *A. brasiliensis*, *A. alternate*, and *F. oxysporum* in a concentration-dependent manner with a ZOI of 11–57 mm. Their MIC values ranged from 62.50 to 500 μg/mL, with *C. albicans* exhibiting the lowest MIC values (62.50–250 μg/mL) for all synthesized nanoparticles, whereas other fungi exhibited MIC values from 250 for *A. brasiliensis*, *A. alternate*, and *F. oxysporum* to 125 μg/mL for *A. brasiliensis*. Additionally, these mycosynthesized SeNPs were susceptible to *E. coli*, *B. cereus*, *B. subtilis*, *S. aureus*, *P. aeruginosa*, and *K. pneumoniae* exhibiting ZOI ranging from 14 to 30 mm, 12 to 32 mm, 11 to 43 mm, 15 to 54 mm, 11 to 26 mm, and 11 to 38 mm, respectively, spanning around all the synthesized nanoparticles with MIC values ranging from 62.5 to 500 μg/mL. The ZOI of all the NPs superseded that of the common antibiotic amoxicillin/clavulanic acid (ZOI: 10–18.33 mm) at a concentration of 100 μg/mL.

Arya et al. [1] showed that green algae (*B. braunii*) copper and silver nanoparticles had a broad-spectrum antimicrobial activity on *E. coli*, *P. aeruginosa*, *K. pneumoniae* (gram-negative bacteria), and *S. aureus* (gram-positive bacteria). *S. aureus* was the most susceptible bacteria of all the bacteria at ZOI 22 mm with MIC values of 250 and 200 μg/mL for CuNPs and AgNPs, respectively.

Abdelhakim et al. [2] revealed that *A. tenuissima*–derived ZnONPs had broad-spectrum antibacterial efficacy against pathogenic bacteria *P. aeruginosa*, *K. pneumoniae*, and *S. aureus* with MIC values of 200 μg/mL, 100 μg/mL for *E. coli*, and 50 μg/mL for *C. albicans*. Furthermore, ZnONPs demonstrated superior antibacterial and antifungal efficacy compared to the standard antimicrobial drugs, amoxicillin/clavulanic acid and nystatin, respectively *R. endophytica* MAE 11 AgNPs prevented *Candida* growth by damaging cellular membranes, leading to the leakage of cellular contents and reducing cell viability [18]. Amanyamadan et al. [39] found that FeNPs synthesized from *A. flavus* D05 MCFF had antibacterial and antifungal effects against *S. aureus* (ZOI: 10 mm at 100 mg/mL), *A. fumigatus* (ZOI: 10 mm at 10 mg/mL), and *C. albicans* (ZOI: 11 mm at 10 mg/mL) (Table 14.3).

TABLE 14.3
Different Types of Microorganisms Inhibited by Microbial-Synthesized Nanoparticles

Microbial Source	Nanoparticle	Organisms Inhibited	ZOI (mm)	Reference
		Bacteria		
Haloferax and *Halogemetricum* strains	Ag	*Escherichia coli, Staphylococcus aureus, Pseudomonas aeruginosa, Bacillus subtilis*		[5]
Bacillus paramycoides	Ag,	*E. coli, Klebsiella pneumoniae, B. subtilis, P. aeruginosa, Candida albicans*	12–34	[15]
	Se	*E. coli, K. pneumoniae, B. subtilis, P. aeruginosa, Aspergillus brasiliensis, C. albicans*	14–35	
	Ag-Se	*E. coli, K. pneumoniae, B. subtilis, P. aeruginosa, A. brasiliensis, C. albicans*	11–42	
Rothia endophytica	Ag	*Candida*		[18]
		Fungi		
Alternaria tenuissima	ZnO	*P. aeruginosa, K. pneumoniae, E. coli, S. aureus, C. albicans*	1.8–16	[2]
Fusarium solani	Ag	*P. aeruginosa, E. coli, K. pneumoniae, S. aureus, Enterococcus* sp.*, Aspergillus awamori, Aspergillus fumigatus, Fusarium oxysporum*	8–24	[3]

(*Continued*)

TABLE 14.3 (*Continued*)
Different Types of Microorganisms Inhibited by Microbial-Synthesized Nanoparticles

Microbial Source	Nanoparticle	Organisms Inhibited	ZOI (mm)	Reference
Fusarium solani	ZnO	*P. aeruginosa*, *E. coli*, *K. pneumoniae*, *S. aureus*, *Enterococcus* sp., *A. awamori*, *A. fumigatus*, *F. oxysporum*	7–18	[3]
Fusarium solani	Cu	*P. aeruginosa*, *E. coli*, *K. pneumoniae*, *S. aureus*, *Enterococcus* sp., *A. awamori*, *A. fumigatus*, *F. oxysporum*	6–15	[3]
Periconium sp.	ZnO	*E. coli*, *S. aureus*, *C. albicans*	14–24	[4]
Talaromyces	Ag	*Listeria*, *Shigella*, *E. coli*, *S. typhi*.	13–18	[21]
Purpureogenus		*E. coli*, *S. epidermidis*		[22]
		S. aureus, *E. coli*, *P. aeruginosa*, *B. cereus*, *S. enterica*	*11–13*	[23]
Penicillium verrucosum	Ag	*F. chlamydosporum*, *A. flavus*	–	[25]
Aspergillus terreus BA6	Ag	*P. aeruginosa*, *Streptococcus faecalis*, *A. niger*	–	[27]
Beauveria bassiana	Ag	*S. aureus*, *P. aeruginosa*, *E. coli*	–	[28]
Aspergillus tamari	Ag	*S. aureus*, *Vibrio parahaemolyticus*, *Shewanella putrefactions*, *C. albicans*	8–16	[29]
Penicillium italicum	Ag	*Salmonella enterica*, *E. coli*, *S. aureus*, *Bacillus cereus*	13–17	[30]
Aspergillus fumigatus	Ag	*B. mycoides*, *E. coli*, *C. albicans*	15–19	[32]
Tritirachium oryzae W5H	Ag	*S. epidermidis*, *E. aerogenes*, *E. coli*, *S. aureus*	16–20	[33]
P. atropurpurascens, *P. expansum*, *and P. chrysosporium*	Ag	*E. coli*	12–30	[34]
Fusarium keratoplasticum	ZnO	*B. subtilis*, *S. aureus*, *E. coli*, *P. aeruginosa*	16–23	[35]
Aspergillus niger	ZnO	*B. subtilis*, *S. aureus*, *E. coli*, *P. aeruginosa*	15–22	[35]
Pleurotus ostreatus	Au	*B. subtilis*, *S. aureus*, *E. coli*, *C. albicans*	11–39	[37]
Aspergillus niger	Ag-Au	*A. niger*, *A. fumigatus*, *A. flavus*		[38]
Trichoderma longibrachiatum	Ag-Au	*E. coli*, *K. granulomatis*, *S. aureus*, *P. aeruginosa*		[38]
Aspergillus flavus	Fe	*S. aureus*, *A. fumigatus*	10–11	[39]
Nematospora coryli	Se	*C. albicans*	–	[41]
Aspergillus quadrilineatus,	Se	*C. albicans*, *A. brasiliensis*, *A. alternate*, *F. oxysporum*	11–57	[42]
Aspergillus terreus, *Fusarium equiseti*		*E. coli*, *B. cereus*, *B. subtilis*, *S. aureus*, *P. aeruginosa*, *K. pneumoniae*	11–54	
Aspergillus terreus	ZnO	*B. subtilis*, *S. aureus*, *E. coli*, *P. aeruginosa*	9.2–11	[52]
		Algae		
Botryococcus brauna	Cu, Ag	*E. coli*, *P. aeruginosa*, *K. pneumoniae*, *S. aureus*, *F. oxysporum*	12.0–21	[1]
Spirogyra hyaline	ZnO	*P. aeruginosa*, *Bacillus pumilus*, *S. aureus*, *E. coli*	15–17	[48]
Noctiluca scintillans	Ag	*S. aureus*, *E. coli*	12–18	[49]
Sargassum myriocystum	Ag	*S. epidermidis*, *P. aeruginosa*	17.6–18.6	[50]

14.3.2 Antibiofilm Potential of Microbial Nanoparticles

El-Behery et al. [15] found that 5.0 μg/mL of silver nanoparticles (AgNPs), selenium nanoparticles (SeNPs), and bimetallic Ag-SeNPs synthesized from *B. paramycoides* inhibited growth and biofilm production poorly. AgNPs displayed notable antibiofilm potential, with inhibition percentages of 87.57%, 86.71%, and 86.62% against *S. aureus*, *E. coli*, and *P. aeruginosa*, respectively. SeNPs exhibited lesser biofilm inhibition, ranging from 86.39% against *C. albicans* to 76.04% against *S. aureus*. Importantly, 5 μg/mL of bimetallic Ag-SeNPs showed strong antibiofilm potential, inhibiting biofilm formation by 90.88% in *C. albicans*, 90.70% in *E. coli*, and 90.62% in *S. aureus*.

14.3.3 Antioxidant Activities of Microbial Nanoparticles

Antioxidants are crucial for protecting cells against oxidative stress induced by free radicals, from harming cellular components. These nanoparticles, such as Ag-AuNPs, ZnO, and AgNPs, display notable antioxidant capabilities by effectively inhibiting free radicals like 2, 2-diphenyl-1-picrylhydrazyl (DPPH) and hydrogen peroxide (H_2O_2).

Elegbede et al. [38] demonstrated the antioxidant capacity of Ag-AuNPs, exhibiting appreciable DPPH radical inhibition activities ranging from 51.46% to 52.82% for NE Ag-AuNPs and from 48.51% to 53.79% for TE Ag-AuNPs at a dosage of 10–100 μg/mL. Moreover, they efficiently scavenged H_2O_2 within the range of 85%–95% for NE Ag-AuNPs and 80.5%–95.5% for TE Ag-AuNPs at a concentration of 1–40 μg/mL. Likewise, Ganesan et al. showcased the antioxidant activity of ZnO nanoparticles from the endophytic fungus *Periconium* sp., revealing dose-dependent radical scavenging with a peak of 85.52% at a concentration of 100 μg/mL comparable to ascorbic acid [4]. Sharma et al. [21] reported the antioxidant properties of mycosynthesized AgNPs, displaying DPPH percent inhibition ranging from 15.215% to 32.894% at varying concentrations. Additionally, Othman et al. [32] demonstrated dose-dependent radical scavenging of *A. tenuissima*-ZnONPs, wherein increased concentrations correlated with heightened scavenging activity. The antioxidant efficacy of AgNPs, SeNPs, and bimetallic Ag-SeNPs showed a dose-dependent reduction of DPPH free radicals with varying IC_{50} values [15].

Rasouli [41] demonstrated that SeNPs recovered from *N. coryli* effectively scavenged free radicals (DPPH) in a concentration-dependent manner. Hussein et al. [42] found that selenium SeNPs from four fungal strains increased DPPH inhibition activity, with IC_{50} values below 200 μg/mL. SeNPs from *A. quadrilineatus* recorded a low IC_{50} value of 91.50 μg/mL. Ag, Se, and Ag-Se nanoparticles from *B. paramycoides* demonstrated a dose-dependent decrease of DPPH free radicals and enhanced scavenging ability with increasing NP concentration, particularly when contrasted with ascorbic acid [15].

14.3.4 Larvicidal Activities of Microbial Nanoparticles

Aedes aegypti mosquito-borne viruses, including Zika, dengue, and chikungunya, have increased worldwide, causing a high burden of human diseases [53,54]. These mosquitoes, found in tropical, sub-tropical, and temperate areas, have shown resistance to conventional insecticidal drugs [54]. Nanoparticles, such as silver nanoparticles, have emerged as a potential solution to combat these resistance issues. *S. myriocystum*–synthesized AgNPs and extracts showed dose-dependent (5–25 mg/L) larvicidal activity of *A. aegypti* and *Culex quinquefasciatus* fourth instar larvae. The killing rate was 100% at concentration of 25 mg/L for the AgNPs, whereas the extract showed reduced mortality, achieving 81% and 88% rate for *A. aegypti* and *C. quinquefasciatus* respectively. The lethal concentration of AgNPs to kill 50% of *A. aegypti* and *C. quinquefasciatus* larvae (LC50) was 6.90 and 5.59 mg/L, respectively [50].

14.3.5 Antiviral Activity of Microbial Nanoparticles

Behbudi [55] showed that AgNPs interact with proteins from viruses such as SARS-COV-2, disrupting viral proteins on the angiotensin-converting enzyme and spike proteins, thereby lowering viral infectivity. Morris et al. [56] showed that AgNPs inhibit the replication of respiratory syncytial virus (RSV) in epithelial cell lines and BALB/c mice. Moreover, AgNPs treatment decreased pro-inflammatory cytokines significantly and raised the levels of CXCL1, granulocyte colony-stimulating factor (G-CSF), and granulocyte macrophage colony-stimulating factor (GM-CSF) in infected mice, indicative of heightened neutrophil recruitment and activation in lung tissues.

14.3.6 Anticoagulant and Thrombolytic Properties

Given that blood clots are associated with various health risks including cardiovascular disorders, injuries, autoimmune reactions, allergies, and cancer, controlling blood coagulation disorders becomes imperative. Elegbede et al. [38] explored the potential biomedical applications of biosynthesized Ag-AuNPs from *A. niger* L3 and *T. longibrachiatum*. They investigated the nanoparticles' ability to prevent blood clot formation and dissolve existing blood clots, crucial aspects in managing blood coagulation disorders using nanotechnology. The Ag-AuNPs acted as effective anticoagulant agents by inhibiting clot formation and rapidly dissolving preformed blood clots within 5 minutes as revealed by effective dispersion of disk-like red blood cells observed in optical microscopic due to reaction of blood clot with the biosynthesized Ag-AuNPs.

14.3.7 Anticancer Activities of Microbial-Synthesized Nanoparticles

Cancer is a major global health challenge, leading to numerous deaths. Nanomedicine has made significant advancements in detecting and treating cancer, with gold nanoparticles (AuNPs) being a versatile tool for diagnosis and therapy. These nanoparticles can be used in biosensors, targeted drug delivery, and assays and are crucial in established cancer therapies like chemotherapy-radiotherapy, thermo-chemotherapy, and gene therapy [57].

AgNPs synthesized from microbial sources like cyanobacteria and fungi have shown promising anticancer properties. These AgNPs exhibit cytotoxic effects on numerous cancer cell lines, like breast, colon, lung, skin, and prostate cancer cells, with minimal toxicity to normal cell lines. The mechanisms underlying their anticancer activity involve induction of apoptosis, necrosis, and protection against reactive oxygen species [22,31,32,45,57,58].

Silver nanoparticles derived from algae (AgNPs) show promise in cancer treatment by impeding tumor cell growth [49]. They have demonstrated effectiveness against A-549 cells [26] and HeLa cell lines [50], inhibiting cell growth and inducing apoptosis [26]. Additionally, they exhibit dose-dependent cytotoxicity against HeLa cells with an IC_{50} value of 73.66 μg/mL [50].

Fungal-made zinc oxide nanoparticles (ZnONPs) showed potent anticancer properties, outperforming Taxol in inhibiting various cancer cell types [32]. ZnONPs derived from *A. tenuissima*, an endophytic fungus, exhibited potent anticancer effects compared to Taxol. ZnONPs inhibited MCF-7 and HepG-2 cells at a concentration of 0.78 μg/mL and inhibited Hfb-4 at a concentration of 3.125 μg/mL, slightly higher than Taxol. ZnONPs also induce concentration-dependent cell death [2].

Fouda et al. [52] showed that ZnONPs derived from *A. terreus* AF-1 exerted cytotoxicity on cancer cell lines, specifically Caco-2, with an IC_{50} of 79.57 μg/mL. The treated cell lines exhibited loss of monolayer integrity, shrinkage, and granulation. Gold nanoparticles (AuNPs) derived from *F. solani* can induce apoptosis in HeLa and MCF-7 cells by halting the cell cycle while sparing HEK cells [36]. AuNPs from *P. ostreatus* displayed stronger anticancer effects against HepG2 and HCT-116 cells compared to commercial AuNPs, hindering cell proliferation and altering cell morphology [37].

14.3.8 Anti-UV Radiation Properties of Microbial Nanoparticles

UV radiation, including UVA and UVB, poses risks such as collagen fiber damage and skin aging, with UVB being particularly concerning due to its direct DNA damage and potential for causing skin cancer. Fouda et al. [52] investigated the impact of *A. terreus*–synthesized ZnO nanoparticles (ZnONPs) on UV protection in cotton fabrics. The treated fabrics demonstrated notably enhanced UV protection compared to untreated ones. ZnONPs-loaded cotton fabrics effectively blocked UVA and UVB radiation by 76.3% and 85.4%, respectively, surpassing the blocking capabilities of untreated fabrics, which blocked UVA and UVB by 66.9% and 79%, respectively. Mohamed et al. [35] revealed that ZnONPs' UV protection varies depending on their shape, with nanorods exhibiting better UPF protection than spherical and hexagonal nanoparticles. Higher UV blocking was demonstrated by cotton textiles coated with rod ZnONPs than by untreated textiles.

14.3.9 Diagnostic Application of Microbial Nanoparticles

Noble metal nanoparticles, such as gold and silver, possess unique SPR properties, making them highly sensitive and valuable for early disease diagnosis [59]. These nanoparticles can be functionalized with specific capping agents and exhibit catalytic properties akin to enzymes. In medical applications, nanotechnology has facilitated the development of intravascular probes, real-time cancer detection methods, and sensitive diagnostic techniques like enzyme-linked immunosorbent assay (ELISA) and microscopic cantilevers, utilizing ultra-small paramagnetic iron oxide particles [59].

Lateral flow assays (LFA) have witnessed profound usage for point-of-care diagnosis of infectious diseases based on the efficacy to detect biomolecules on disposable strips and provide rapid results without requiring expensive equipment or expert supervision [60]. For example, Xiang et al. [61] developed an LFA using gold nanoparticles on a plasmonic chip to detect SARS-CoV-2 IgG and IgM antibodies. The method relied on color changes: blue indicated antibody-gold binding due to SPR, while red signaled unbound gold nanoparticles. The assay's performance varied between antibodies, with IgM showing an accuracy of 0.69, a sensitivity of 0.57, and a specificity of 1.00, while IgG demonstrated higher values with 0.86 accuracy, 0.81 sensitivity, and 1.00 specificity. When considering both antibodies together, the overall sensitivity of the test reached 0.82, suggesting a reasonably effective detection method for SARS-CoV-2 antibodies.

Gold nanoparticles (AgNPs) have also shown promise in cancer diagnosis due to their distinct electrical, optical, thermal, chemical, and biological properties [62].

Vital for host defense, metabolism, and signaling, the content of hydrogen peroxide in bodily fluids can be an indicator of metabolic disorders [63]. Elgamouz et al. developed a temperature-sensitive, colorimetric detector for hydrogen peroxide (H_2O_2) using algae-silver nanoparticles (Algae-AgNPs). As temperature increased to 45°C, H_2O_2 detection efficiency decreased due to weakened attractive forces between Algae-AgNPs and H_2O_2. Calibration curves in three ranges (4.7–32 nM, 4.7–32 μM, and 4.7–32 mM) showed linear responses for absorbance (Abs) and change in absorbance (ΔAbs) with detection limits of 1.34–1.78 nM [49].

Alhogail et al. [64] used magnetic nanoparticles for developing biosensors capable of detecting *P. aeruginosa*. By incubating 10^7 cfu/mL on the functionalized gold surface, the manufactured sensor detected *P. aeruginosa* protease. The golden hue of the sensor is exposed upon the release and collection of the peptide segment-microbial nanoparticle (MNP) moiety. *P. aeruginosa* was quantitatively detected at various concentrations using the biosensing approach. As the concentration of bacteria increased, the sensor indicated a progressive increase in the visible bare gold area. In about 1 minute, the colorimetric biosensor showed a limit of detection of 10^2 cfu/mL.

In another innovative approach, the bacteria-induced click chemistry and gold nanoparticles were utilized to develop a point-of-care diagnostic device paired with a smartphone application for the rapid detection of pathogenic bacteria in complex artificial sepsis blood samples.

TABLE 14.4
Nanomaterial Diagnostic Approach

Approach	Nanomaterial	Diagnostic	Reference
LFA	Gold	SARS-CoV-2 IgG and IgM antibodies	[61]
Calorimetric sensor	Silver	H_2O_2	[49]
Biosensors	Magnetic	*Pseudomonas aeruginosa*	[64]
Smartphone app	Gold	Septicemia	[65]
Biosensor, imaging	Fe_3O_4 magnetosomes	Toxic assessment, molecular imaging	[4]

Varying concentrations (10^2–10^7 cfu/mL) of *E. coli*, *P. aeruginosa*, *B. subtilis*, and *S. aureus* were incubated with 100 nM Cu^{2+}, 5.4 nM azide, and alkyne-AuNPs for 60 minutes. This incubation caused a color change from red to blue, indicating aggregation with the AuNPs-Cu^{2+} complex. At lower bacteria concentrations (below 10^3 CFU/mL), no color change was observed, marking the limit of detection by the naked eye. To enhance detection, a smartphone app was developed using SWIFT and compiled to run on the IOS 9.3 operating system. The app employed RGB pixels with linear calibrations for all bacteria under study using a B/R pixel ratio. This allowed the app to scan and determine the color and concentration of bacteria, providing a portable platform for point-of-care sensing of pathogenic bacteria.

Bacterial Fe_3O_4 magnetosomes, characterized by high chemical purity, a narrow size range, and consistent crystal morphology, have found diverse applications in molecular imaging, cancer therapy, and the development of chip-based whole-cell biosensors for toxicity assessment [4] (Table 14.4).

14.3.10 Microbial Nanoparticles as Drug Delivery Agents

Accurate positioning of drug in a desired dosage to their target sites (in cells or tissues of choice) at right time is the main concern in designing and developing novel drug delivery systems. Therefore, total drug consumption and its side effects can be reduced significantly to make it precise, safe drug delivery to achieve the maximum therapeutic impact [66]. In pharmacology, nanoparticles are perfect drug delivery materials because they reduce pain, promote wound healing, and shield animals from infections. They have an effect on assimilation, absorption, digestion, and discharge of drugs, which makes it possible to observe drug dynamics, therapeutic action, bioavailability, stability, and decreased toxicity. Dendrimers, nanospheres, niosomes, and polymeric micelles are examples of polymer-based, lipid-based, and metal-based nanoparticle systems [67].

MNPs serve as effective nanodrug carriers as they possess special characteristics and biocompatibility. They have demonstrated success in targeting cancer cells, particularly lung and ovarian cancers. The blood-brain barrier and other conventional therapeutic barriers can be bypassed by MNPs, enabling the continuous supply of therapeutic medicines for disorders like brain tumors that are challenging to treat [68] (Figure 14.3).

14.4 CHALLENGES AND FUTURE PROSPECTS

Microbial nanotechnology, integrating nanotech and microbiology, faces key challenges. Stability and optimization are critical, requiring meticulous attention to pH, salt content, and temperature for consistent nanoparticle synthesis. Characterization of biologically manufactured nanoparticles is essential, demanding thorough examination of organic material filtrate. Additionally, certain nanoparticle features require customized optimization procedures, which are especially important for medical applications. Scaling up production sustainably poses a significant challenge, necessitating innovative strategies and investments in green infrastructure. Despite advancements,

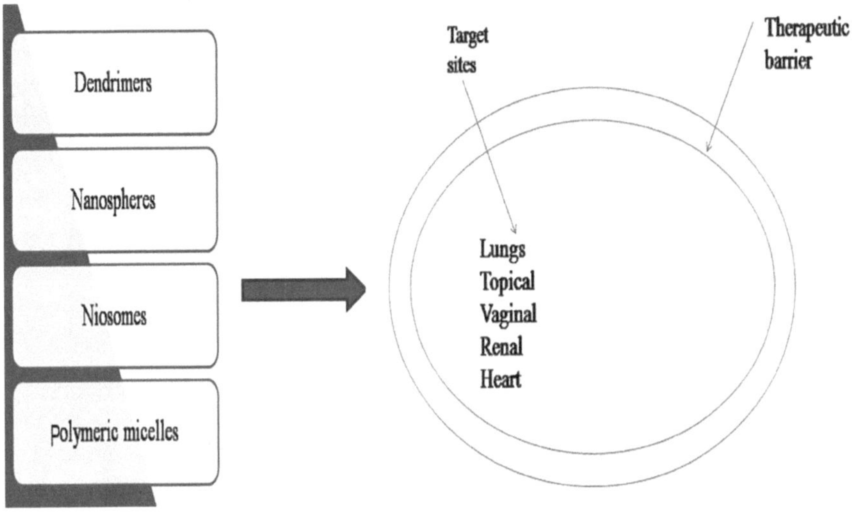

FIGURE 14.3 Nanocarriers target drug delivery.

understanding the mechanistic elements of green synthesis methods remains incomplete, urging further research for process optimization and novel techniques.

Interdisciplinary collaboration will drive innovation, leveraging expertise from diverse fields. Advanced characterization techniques will enhance nanoparticle design, while ongoing research into optimization strategies will improve efficiency and reproducibility. Sustainable production methods are paramount, requiring investment in green technology. Commercialization and widespread adoption hinge on overcoming technical challenges and demonstrating safety. To fully realize the potential of microbial nanotechnology and transform research into useful applications, cooperation between regulators, industry, and academia is essential [4,69].

14.5 CONCLUSION

Microbial nanotechnology represents a ground-breaking approach in medicine and healthcare, offering innovative solutions to overcome the limitations of conventional methods in nanoparticle production. By leveraging the unique capabilities of microorganisms such as bacteria, fungi, and algae, this field provides an eco-friendly method for synthesizing metallic nanoparticles with diverse biological activities. These nanoparticles exhibit properties essential for a wide range of applications in medicine, including antimicrobial, antibiofilm, antioxidant, anticancer, larvicidal, anticoagulant, and thrombolytic activities (crucial aspects in managing blood coagulation disorders, cardiovascular disorders, injuries, autoimmune reactions, and allergies). Moreover, the antiviral properties of microbial nanoparticles have been demonstrated, showing promise in reducing viral replication and modulating inflammatory responses.

The utilization of microbial synthesis techniques not only offers greater stability and reproducibility but also reduces reliance on harsh chemicals and pure starting materials, making the process more environmentally friendly. Various biological entities have been identified for their ability to serve as both reducing and capping agents in nanoparticle generation, resulting in a relatively simple, cost-effective, and stable synthesis process.

In the realm of healthcare, microbial nanoparticles hold immense promise. They play a crucial role in the development of point-of-care diagnostic kits and biosensors, offering affordable and user-friendly solutions for disease detection. Additionally, microbial nanoparticles serve as effective nanocarriers for drug delivery, overcoming traditional therapeutic limitations such as the blood-brain barrier. This enables sustained delivery of therapeutic agents, particularly for challenging diseases like brain tumors. These attributes make microbial nanotechnology an invaluable tool in managing various diseases, from early detection to providing therapeutic efficacy.

In conclusion, microbial nanotechnology represents a transformative approach in medicine and healthcare. Its potential spans from the development of advanced diagnostic tools to targeted drug delivery systems, offering novel solutions for disease management and improving patient outcomes. As research in this field continues to advance, microbial nanoparticles are poised to revolutionize healthcare practices, paving the way for more effective and sustainable treatments.

LIST OF ABBREVIATION

AFM:	atomic force microscopy
AgNP:	silver nanoparticles
AuNPs:	gold nanoparticles
CuNPs:	copper nanoparticles
DPPH:	2, 2-diphenyl-1-picrylhydrazyl
H_2O_2:	hydrogen peroxide
HRTEM:	high-resolution transmission electron microscopy
MIC:	minimum inhibitory concentration
NP(s):	nanoparticle(s)
SAED:	selected area electron diffraction
SEM:	scanning electron microscopy
SeNP:	selenium nanoparticles
SPR:	surface plasmon resonance
TEM:	transmission electron microscopy
ZnONPs:	zinc nanoparticles
RSV:	respiratory syncytial virus
SARS-COV-2:	Severe Acute Respiratory Syndrome Corona virus 2

REFERENCES

1. Arya, A., V. Mishra, and T.S. Chundawat. 2019. Green synthesis of silver nanoparticles from green algae (Botryococcus braunii) and its catalytic behavior for the synthesis of benzimidazoles. *Chemical Data Collections* 20: 100190. https://doi.org/10.1016/j.cdc.2019.100190.
2. Abdelhakim, H.K., E.-S.R. El-Sayed, and F.B. Rashidi. 2020. Biosynthesis of zinc oxide nanoparticles with antimicrobial, anticancer, antioxidant and photocatalytic activities by the endophytic alternaria tenuissima. *Journal of Applied Microbiology* 128, no. 6: 1634–1646. https://doi.org/10.1111/jam.14581.
3. El Sayed, M.T., and A.S.A. El-Sayed. 2020. Biocidal activity of metal nanoparticles synthesized by *Fusarium solani* against multidrug-resistant bacteria and mycotoxigenic fungi. *Journal of Microbiology and Biotechnology* 30, no. 2: 226–236. https://doi.org/10.4014/jmb.1906.06070.
4. Ganesan, V., H. Muruganandham, S. Vivekanandhan, and M. Sankaralingam 2020. Periconium sp. (endophytic fungi) extract mediated sol-gel synthesis of ZnO nanoparticles for antimicrobial and antioxidant applications. *Materials Science in Semiconductor Processing* 105: 104739. https://doi.org/10.1016/j.mssp.2019.104739.
5. Abdollahnia, M., A. Makhdoumi, M. Mashreghi et al. 2020. Exploring the potentials of halophilic prokaryotes from a solar saltern for synthesizing nanoparticles: The case of silver and selenium. *PLoS One* 15, no. 3: e0229886.
6. Della Flora, I.K., and C.J. de Andrade. 2023. Biosynthesis of metallic nanoparticles by bacterial cell-free extract. *Nanoscale* 15, 13886–13908.
7. Sachin, K., and S.K. Karn. 2021. Microbial fabricated nanosystems: Applications in drug delivery and targeting. *Frontiers in Chemistry* 9. https://doi.org/10.3389/fchem.2021.617353.
8. Deshmukh, S.P., S.M. Patil, S.B. Mullani, and S.D. Delekar. 2019. Silver nanoparticles as an effective disinfectant: A review. *Materials Science & Engineering. C, Materials for Biological Applications*, 97, 954–965. https://doi.org/10.1016/j.msec.2018.12.102.

9. Ball, A.S., S. Patil, and S. Soni. Introduction into nanotechnology and microbiology. In *Methods in Microbiology*, vol. 46, pp. 1–18. Academic Press, 2019. https://doi.org/10.1016/bs.mim.2019.04.003.
10. Lateef, A., O.M. Darwesh, and I.A. Matter. Microbial nanobiotechnology: the melting pot of microbiology, microbial technology and nanotechnology. In *Microbial Nanobiotechnology: Principles and Applications*, pp. 1–19, 2021. https://doi.org/10.1007/978-981-33-4777-9_1.
11. Yetisgin, A.A., S. Cetinel, M. Zuvin, et al. 2020. Therapeutic nanoparticles and their targeted delivery applications. *Molecules* 25, no. 9: 2193. https://doi.org/10.3390/molecules25092193.
12. Prasad, M., U.P. Lambe, B. Brar, et al. 2018. Nanotherapeutics: An insight into healthcare and multi-dimensional applications in medical sector of the modern world. *Biomedicine & Pharmacotherapy* 97: 1521–1537. https://doi.org/10.1016/j.biopha.2017.11.026.
13. Mubeen, B., A.N. Ansar, R. Rasool, et al. 2021. Nanotechnology as a novel approach in combating microbes providing an alternative to antibiotics. *Antibiotics* 10, no. 12: 1473. https://doi.org/10.3390/antibiotics10121473.
14. Lahiri, D., M. Nag, H.I. Sheikh, et al. 2021. Microbiologically-synthesized nanoparticles and their role in silencing the biofilm signaling cascade. *Frontiers in Microbiology* 12. https://doi.org/10.3389/fmicb.2021.636588.
15. El-Behery R.R., E.-S.R. El-Sayed, and G.S. El-Sayyad 2023.Gamma rays-assisted bacterial synthesis of bimetallic silver-selenium nanoparticles: powerful antimicrobial, antibiofilm, antioxidant, and photocatalytic activities. *BMC Microbiology* 23. no.1: 224. https://doi.org/10.1186/s12866-023-02971-1.
16. Mona, S., and K. Syoof. 2020. Synthesis of silver nanoparticles using Klebsiella pneumoniae and their activity. *Journal of Physics: Conference Series*, 1664: 012115. https://doi.org/10.1088/1742-6596/1664/1/012115.
17. El-Saadony, M.T., A.M. Saad, T.F. Taha, et al. 2021. Selenium nanoparticles from Lactobacillus paracasei HM1 capable of antagonizing animal pathogenic fungi as a new source from human breast milk. *Saudi Journal of Biological Sciences* 28, no. 12: 6782–6794. https://doi.org/10.1016/j.sjbs.2021.07.059.
18. Elbahnasawy, M.A., A.M. Shehabeldine, A.M. Khattab, et al. 2021. Green biosynthesis of silver nanoparticles using novel endophytic Rothia endophytica: Characterization and anticandidal activity. *Journal of Drug Delivery Science and Technology* 62: 102401. https://doi.org/10.1016/j.jddst.2021.102401.
19. Wadhwani, S.A., U.U. Shedbalkar, R. Singh, and B.A. Chopade. 2018. Biosynthesis of gold and selenium nanoparticles by purified protein from *Acinetobacter* Sp. SW 30. *Enzyme and Microbial Technology* 111: 81–86. https://doi.org/10.1016/j.enzmictec.2017.10.007.
20. Nassar, A.-R.A., H.M. Atta, M.A. Abdel-Rahman, et al. 2023. Myco-synthesized copper oxide nanoparticles using harnessing metabolites of endophytic fungal strain aspergillus terreus: An insight into antibacterial, anti-candida, biocompatibility, anticancer, and antioxidant activities. *BMC Complementary Medicine and Therapies* 23, no. 1. https://doi.org/10.1186/s12906-023-04056-y.
21. Sharma, A., A. Sagar, J. Rana, et al. 2022. Green synthesis of silver nanoparticles and its antibacterial activity using fungus talaromyces purpureogenus isolated from taxus baccata linn. *Micro and Nano Systems Letters* 10, no. 1. https://doi.org/10.1186/s40486-022-00144-9.
22. Bhatnagar, S., T. Kobori, D. Ganesh, et al. 2019. Biosynthesis of silver nanoparticles mediated by extracellular pigment from talaromyces purpurogenus and their biomedical applications. *Nanomaterials* 9, no. 7: 1042. https://doi.org/10.3390/nano9071042.
23. Hu, X., K. Saravanakumar, T. Jin, and M.H. Wang. 2019. Mycosynthesis, characterization, anticancer and antibacterial activity of silver nanoparticles from endophytic fungus *Talaromyces purpureogenus*. *International Journal of Nanomedicine* 14: 3427–3438. https://doi.org/10.2147/IJN.S200817.
24. Gemishev, O.T., M.I. Panayotova, N.N. Mintcheva, et al. 2019. A green approach for silver nanoparticles preparation by cell-free extract from Trichoderma reesei fungi and their characterization. *Materials Research Express* 6, no. 9: 095040. https://doi.org/10.1088/2053-1591/ab2e6a.
25. Yassin, M.A., A.M. Elgorban, A.E.-R.M.A. El-Samawaty, et al. 2021. Biosynthesis of silver nanoparticles using *Penicillium verrucosum* and analysis of their antifungal activity. *Saudi Journal of Biological Sciences* 28, no.4: 2123–2127. https://doi.org/10.1016/j.sjbs.2021.01.063.
26. Akther, T., V. Mathipi, N.S. Kumar, et al. 2019. Fungal-mediated synthesis of pharmaceutically active silver nanoparticles and anticancer property against A549 cells through apoptosis. *Environmental Science and Pollution Research* 26, no. 13: 13649–13657. https://doi.org/10.1007/s11356-019-04718-w.
27. Lotfy, W.A., B.M. Alkersh, S.A. Sabry, et al. 2021. Biosynthesis of silver nanoparticles by aspergillus terreus: Characterization, optimization, and biological activities. *Frontiers in Bioengineering and Biotechnology* 9. https://doi.org/10.3389/fbioe.2021.633468.

28. Tyagi, S., P.K. Tyagi, D. Gola, et al. 2019. Extracellular synthesis of silver nanoparticles using entomopathogenic fungus: characterization and antibacterial potential. *SN Applied Sciences* 1, no. 12. https://doi.org/10.1007/s42452-019-1593-y.
29. Nanda, A., B.K. Nayak, and M. Krishnamoorthy. 2018. Antimicrobial properties of biogenic silver nanoparticles synthesized from phylloplane fungus, aspergillus tamarii. *Biocatalysis and Agricultural Biotechnology* 16: 225–228. https://doi.org/10.1016/j.bcab.2018.08.002.
30. Majeed, S., M. Danish, A.H.B. Zahrudin, et al. 2018. Biosynthesis and characterization of silver nanoparticles from fungal species and its antibacterial and anticancer effect. *Karbala International Journal of Modern Science* 4, no. 1: 86–92. https://doi.org/10.1016/j.kijoms.2017.11.002.
31. Janakiraman, V., K. Govindarajan, and C.R. Magesh. 2019. Biosynthesis of silver nanoparticles from endophytic fungi, and its cytotoxic activity. *BioNanoScience* 9, no. 3: 573–579. https://doi.org/10.1007/s12668-019-00631-1.
32. Othman, A.M., M.A. Elsayed, N.G. Al-Balakocy, et al. 2019. Biosynthesis and characterization of silver nanoparticles induced by fungal proteins and its application in different biological activities. *Journal of Genetic Engineering and Biotechnology* 17, no. 1: 8. https://doi.org/10.1186/s43141-019-0008-1.
33. Al-limoun, M., H.N. Qaralleh, K.M. Khleifat, et al. 2020. Culture media composition and reduction potential optimization of mycelia-free filtrate for the biosynthesis of silver nanoparticles using the fungus tritirachium oryzae W5H. *Current Nanoscience* 16, no. 5: 757–769. https://doi.org/10.2174/1573413715666190725111956.
34. Sanguiñedo, P., R.M. Fratila, M.B. Estevez, et al. 2018. Extracellular biosynthesis of silver nanoparticles using fungi and their antibacterial activity. *Nano Biomedicine and Engineering* 10, no. 2. https://doi.org/10.5101/nbe.v10i2.p165-173.
35. Mohamed, A.A., A. Fouda, M.A. Abdel-Rahman, et al 2019. Fungal strain impacts the shape, bioactivity and multifunctional properties of green synthesized zinc oxide nanoparticles. *Biocatalysis and Agricultural Biotechnology* 19: 101103. https://doi.org/10.1016/j.bcab.2019.101103.
36. Clarance, P., B. Luvankar, J. Sales, et al. 2020. Green synthesis and characterization of gold nanoparticles using endophytic fungi Fusarium solani and its in-vitro anticancer and biomedical applications. *Saudi Journal of Biological Sciences* 27, no. 2: 706–712. https://doi.org/10.1016/j.sjbs.2019.12.026.
37. El Domany, E.B., T.M. Essam, A.E. Ahmed, et al. 2018. Biosynthesis physico-chemical optimization of gold nanoparticles as anti-cancer and synergetic antimicrobial activity using Pleurotus ostreatus fungus. *Journal of Applied Pharmaceutical Science* 8, no. 5: 119–128. https://doi.org/10.7324/japs.2018.8516.
38. Elegbede, J.A., A. Lateef, M.A. Azeez, et al. 2019. Silver-gold alloy nanoparticles biofabricated by fungal xylanases exhibited potent biomedical and catalytic activities. *Biotechnology Progress* 35, no. 5. https://doi.org/10.1002/btpr.2829.
39. Gouda, A., N.M. Sidkey, H.A. Shawky, et al. 2020. Biosynthesis, characterization and antimicrobial activity of iron oxide nanoparticles synthesized by fungi. *Al-Azhar Journal of Pharmaceutical Sciences* 62, no. 2: 164–179. https://doi.org/10.21608/ajps.2020.118382.
40. Mahanty, S., M. Bakshi, S. Ghosh, et al. 2019. Green synthesis of iron oxide nanoparticles mediated by filamentous fungi isolated from sundarban mangrove ecosystem, India. *BioNanoScience* 9, no. 3: 637–651. https://doi.org/10.1007/s12668-019-00644-w.
41. Rasouli, M. 2019. Biosynthesis of Selenium Nanoparticles using yeast *Nematospora coryli* and examination of their anti-candida and anti-oxidant activities. *IET Nanobiotechnology*, 13, no. 2: 214–218. https://doi.org/10.1049/iet-nbt.2018.5187.
42. Hussein, H.G., E.-S.R. El-Sayed, N.A. Younis, et al. 2022. Harnessing endophytic fungi for biosynthesis of selenium nanoparticles and exploring their bioactivities. *AMB Express* 12, no. 1 https://doi.org/10.1186/s13568-022-01408-8.
43. Zhang, H., H. Zhou, J. Bai, et al. 2019. Biosynthesis of selenium nanoparticles mediated by fungus mariannaea Sp. HJ and their characterization. *Colloids and Surfaces A: Physicochemical and Engineering Aspects* 571: 9–16. https://doi.org/10.1016/j.colsurfa.2019.02.070.
44. Anji, R., K. Chaitanya, D. Giri, et al. 2018. Green synthesis of copper and silver nanoparticles using Botryococcus braunii and their antimicrobial activity. *Materials Today: Proceedings*, 5, no. 1: 2322–2330. https://doi.org/10.1016/j.matpr.2018.02.114.
45. Bin-Meferij, M.M., and R.S. Hamida. 2019. Biofabrication and antitumor activity of silver nanoparticles utilizing novel nostoc Sp. Bahar M. *International Journal of Nanomedicine* 14: 9019–9029. https://doi.org/10.2147/ijn.s230457.
46. Kavishri, S., A. Geetha, I.G.K. Ilangovar, S. Vasugi, P. Sivaperumal, and S. Balachandran, 2024. Facile synthesis of silver nanoparticles from sustainable sargassum sp. seaweed material and its anti-inflammatory application. *Cureus* 16, no. 4: e57754.

47. López-Miranda, J.L., R. Esparza, M.A. González-Reyna, B.L. España-Sánchez, A.R. Hernandez-Martinez, R. Silva, and M. Estévez 2021. Sargassum influx on the mexican coast: A source for synthesizing silver nanoparticles with catalytic and antibacterial properties. *Applied Sciences* 11, no. 10: 4638.
48. Hameed, H., A. Waheed, M.S. Sharif, et al. 2023. Green synthesis of zinc oxide (ZnO) nanoparticles from green algae and their assessment in various biological applications. *Micromachines* 14, no. 5: 928.
49. Elgamouz, A., H. Idriss, C. Nassab, et al. 2020. Green synthesis, characterization, antimicrobial, anti-cancer, and optimization of colorimetric sensing of hydrogen peroxide of algae extract capped silver nanoparticles. *Nanomaterials* 10, no. 9: 1861. https://doi.org/10.3390/nano10091861.
50. Balaraman, P., B. Balasubramanian, D. Kaliannan, et al. 2020. Phyco-synthesis of silver nanoparticles mediated from marine algae Sargassum myriocystum and its potential biological and environmental applications. *Waste and Biomass Valorization* 11, no. 10: 5255–5271. https://doi.org/10.1007/s12649-020-01083-5.
51. Bankier, C., R.K. Matharu, Y.K. Cheong, et al. 2019. Synergistic antibacterial effects of metallic nanoparticle combinations. *Scientific Reports* 9, no. 1: 16074.
52. Fouda, A., S.E.-D. Hassan, S.S. Salem et al. 2018. In-vitro cytotoxicity, antibacterial, and UV protection properties of the biosynthesized zinc oxide nanoparticles for medical textile applications. *Microbial Pathogenesis* 125: 252–261. https://doi.org/10.1016/j.micpath.2018.09.030.
53. Elizondo-Quiroga, D., M. Ramírez-Medina, A. Gutiérrez-Ortega, et al. 2019. Vector competence of Aedes aegypti and Culex quinquefasciatus from the metropolitan area of Guadalajara, Jalisco, Mexico for Zika Virus." *Scientific Reports* 9, no. 1. https://doi.org/10.1038/s41598-019-53117-1.
54. Estrada-Franco, J.G., N.A. Fernández-Santos, et al. 2020. Vertebrate-Aedes aegypti and Culex quinquefasciatus (diptera)-arbovirus transmission networks: Non-human feeding revealed by meta-barcoding and next-generation sequencing. *PLOS Neglected Tropical Diseases* 14, no. 12: e0008867. https://doi.org/10.1371/journal.pntd.0008867.
55. Behbudi, G. 2021. Effect of silver nanoparticles disinfectant on covid-19. *Advances in Applied NanoBio-Technologies* 2, no. 2: 63–67.
56. Morris, D., M. Ansar, J. Speshock, T. Ivanciuc, et al. 2019. Antiviral and immunomodulatory activity of silver nanoparticles in experimental RSV infection. *Viruses* 11, no. 8: 732. https://doi.org/10.3390/v11080732.
57. Hammami, I., N.M. Alabdallah, A. Al jomaa, et al. 2021. Gold nanoparticles: Synthesis properties and applications. *Journal of King Saud University - Science* 33, no. 7: 101560. https://doi.org/10.1016/j.jksus.2021.101560.
58. Anandaradje, A., V. Meyappan, I. Kumar, et al. 2020. Microbial synthesis of silver nanoparticles and their biological potential. In *Nanoparticles in Medicine*, pp. 99–133. https://doi.org/10.1007/978-981-13-8954-2_4. Bhuvaneshwari, V., D. Vaidehi, and S. Logpriya. 2018. Green synthesis of copper oxide nanoparticles for biological applications. *Microbiology: Current Research* 02, no. 01. https://doi.org/10.4066/2591-8036.e105.
59. Mohanta, Y.K., A. Hashem, E.F.A. Allah, et al. 2020. Bacterial synthesized metal and metal salt nanoparticles in biomedical applications: An up and coming approach. *Applied Organometallic Chemistry* 34, no. 9: e5810.
60. Geng, H., S.V. Pedersen, Y. Ma, T. Haghighi, et al. 2022. Noble metal nanoparticle biosensors: From fundamental studies toward point-of-care diagnostics. *Accounts of Chemical Research* 55, no. 5: 593–604.
61. Xiang, J., M. Yan, H. Li, T. Liu, C. Lin, S. Huang, and C. Shen, 2020. Evaluation of enzyme-linked immunoassay and colloidal goldimmunochromatographic assay kit for detection of novel coronavirus (SARS-Cov-2) causing an outbreak of pneumonia (COVID-19). MedRxiv. https://doi.org/10.1101/2020.02.27.20028787.
62. Kalimuthu, K., B.S. Cha, S. Kim, et al. 2020. Eco-friendly synthesis and biomedical applications of gold nanoparticles: A review. *Microchemical Journal* 152: 104296. https://doi.org/10.1016/j.microc.2019.104296.
63. Giaretta, J.E., H. Duan, F. Oveissi, S. Farajikhah, F. Dehghani, and S. Naficy. 2022. Flexible sensors for hydrogen peroxide detection: A critical review. *ACS Applied Materials & Interfaces* 14, no. 18: 20491–20505. https://doi.org/10.1021/acsami.1c24727.
64. Alhogail, S., G.A.R.Y. Suaifan, F.J. Bikker, et al. 2019. Rapid colorimetric detection of Pseudomonas aeruginosa in clinical isolates using a magnetic nanoparticle biosensor. *ACS Omega* 4, no. 26: 21684–21688.
65. Mou, X.Z., X.Y. Chen, J. Wang, et al. 2019. Bacteria-instructed click chemistry between functionalized gold nanoparticles for point-of-care microbial detection. *ACS Appl Mater Interfaces* 11: 23093–23101.

66. Purohit, J., A. Chattopadhyay, and N.K. Singh. 2019. Green synthesis of microbial nanoparticle: Approaches to application. In *Microbial Nanobionics*, vol. 2, pp. 35–60. https://doi.org/10.1007/978-3-030-16534-5_3.
67. Youssef, F.S., H.A. El-Banna, H.Y. Elzorba, et al. 2019. Application of some nanoparticles in the field of veterinary medicine. *International Journal of Veterinary Science and Medicine* 7, no. 1: 78–93.
68. Omar, R.A., S. Afreen, N. Talreja, et al. 2019. Impact of nanomaterials on the microbial system. In *Microbial Nanobionics*, pp. 141–158. https://doi.org/10.1007/978-3-030-16383-9_6.
69. Pandit, C., A. Roy, S. Ghotekar, A. Khusro, et al. 2022. Biological agents for synthesis of nanoparticles and their applications. *Journal of King Saud University-Science* 34, no. 3: 101869.

15 Ethical and Environmental Considerations of Microbial Nanotechnology

Applications in Industrial Settings

Rohit Shendkar, Dr. Rashmi Ranjan Guru, and Swayamprava Dalai

15.1 INTRODUCTION

15.1.1 The Emergence of Microbial Nanotechnology in Industry

The emergence of microbial nanotechnology represents a revolutionary convergence of biology and engineering, offering unprecedented opportunities for innovation across diverse industries [1]. At its core, microbial nanotechnology harnesses the remarkable capabilities of microorganisms to manipulate material properties at the nanoscale, enabling the production of novel materials, energy sources, and therapeutic agents with unparalleled precision and efficiency. This rapidly evolving discipline presents significant potential for tackling urgent societal issues and fostering sustainable economic advancement [2,3].

In industry, microbial nanotechnology is catalyzing transformative advancements in numerous sectors. One of the most notable application areas is the production of nanomaterials. Microorganisms such as bacteria, fungi, and viruses are adept at synthesizing a wide array of nanomaterials, including nanoparticles (NPs), nanowires, and nanostructures, through processes such as biomineralization and bioreduction [4,5]. These microbial-derived nanomaterials exhibit unique properties and functionalities that can be tailored for diverse industrial applications, ranging from electronics and catalysis to environmental remediation and healthcare [6].

Furthermore, microbial nanotechnology is revolutionizing the production of renewable energy sources [7]. Microbial fuel cells (MFCs), for instance, leverage the electrochemical activities of microorganisms to generate electricity from organic matter, offering a sustainable alternative to conventional energy sources. Similarly, engineered microbial systems are being explored for the production of biofuels, bioplastics, and biochemicals, contributing to the transition toward a bio-based economy with reduced reliance on fossil fuels and mitigated environmental impact [8].

In the realm of healthcare, microbial nanotechnology is driving innovations in drug delivery, diagnostics, and therapeutics. Engineered microbial NPs, or "bionanoparticles," hold promise as targeted drug delivery vehicles capable of transporting therapeutic payloads to specific tissues or cells with enhanced precision and efficacy. Moreover, microbial nanobiosensors offer rapid and sensitive detection of pathogens, toxins, and biomarkers, enabling early diagnosis and personalized treatment strategies for various diseases [9].

Overall, the emergence of microbial nanotechnology is revolutionizing the industry by unlocking new frontiers in materials science, energy production, and healthcare. Research in this field continues to advance, but the potential for disruptive innovation and sustainable development remains limitless, paving the way for a future powered by microbial ingenuity and nanotechnological prowess.

DOI: 10.1201/9781003487142-15

15.1.2 The Need for Ethical and Environmental Assessments

The escalating field of nanotechnology offers immense promise for transformative innovation across many industries, from healthcare to the environment. However, with these opportunities come significant responsibilities. The imperative for ethical and environmental assessments in nanotechnology research cannot be overstated [10]. Ethical considerations are paramount due to the potential implications on human health, privacy, and equity. NPs, at the heart of many nanotechnology applications, pose unknown risks to human health, necessitating thorough evaluation to ensure safety. Moreover, the development of nanoscale surveillance technologies raises concerns about privacy infringement, necessitating ethical frameworks to safeguard individual rights and autonomy [11]. Equally important is the need to address societal disparities, ensuring that the benefits of nanotechnology are equitably distributed across diverse populations. On the environmental front, the impact of nanomaterials on ecosystems, resource depletion, and energy consumption cannot be ignored. NPs released into the environment may disrupt ecosystems and harm wildlife, highlighting the need for robust environmental assessments to guide sustainable practices in production, usage, and disposal [12]. Furthermore, the resource-intensive nature of nanotechnology demands careful consideration of sustainable sourcing and energy-efficient processes to mitigate environmental degradation. Integrated assessments that combine ethical and environmental perspectives are essential for responsible nanotechnology research. Such assessments should engage multidisciplinary expertise and foster public dialogue to address complex ethical dilemmas and environmental challenges. Ultimately, by prioritizing ethical principles and environmental stewardship in nanotechnology research, we can maximize the benefits of this transformative technology while minimizing potential risks and ensuring a more sustainable future for humanity and the planet.

In this chapter, ethical and environmental considerations of microbial nanotechnology applications will be discussed in industrial settings.

15.2 MICROBIAL NANOTECHNOLOGY IN INDUSTRIAL APPLICATIONS

Microbial nanotechnology offers multifaceted opportunities for innovation and sustainability in industrial applications, ranging from NP synthesis and pollution control to sustainable manufacturing and biofuel production. By harnessing the inherent capabilities of microorganisms at the nanoscale, this interdisciplinary field promises to revolutionize industrial processes and minimize environmental impact, and is a new way toward a more sustainable and resource-efficient future [13–15].

Microbial-assisted nanotechnology revolutionizes industrial settings, enabling precise NP synthesis and production while promoting sustainability. It aids in pollution control and remediation, offering efficient methods for environmental cleanup. Moreover, it fosters sustainable manufacturing, biocatalysis, and biofuel production. Ethical frameworks guide its responsible application, ensuring societal benefit and environmental stewardship (Figure 15.1).

15.2.1 Nanoparticle Synthesis and Production

Microbial nanotechnology has revolutionized NP synthesis and production, offering efficient, sustainable, and cost-effective methods for generating a wide range of nanomaterials with tailored properties and functionalities. Microorganisms, including bacteria, fungi, and viruses, serve as natural nano-factories, capable of producing NPs through various biological mechanisms [16–18].

Biomineralization is one such mechanism, where microorganisms control the nucleation and growth of inorganic materials, leading to the formation of NPs with precise sizes, shapes, and compositions [4,19]. Bioreduction is another approach, where microorganisms use enzymes and metabolites to convert metal ions into nanoscale particles, offering environmentally friendly routes for NP synthesis.

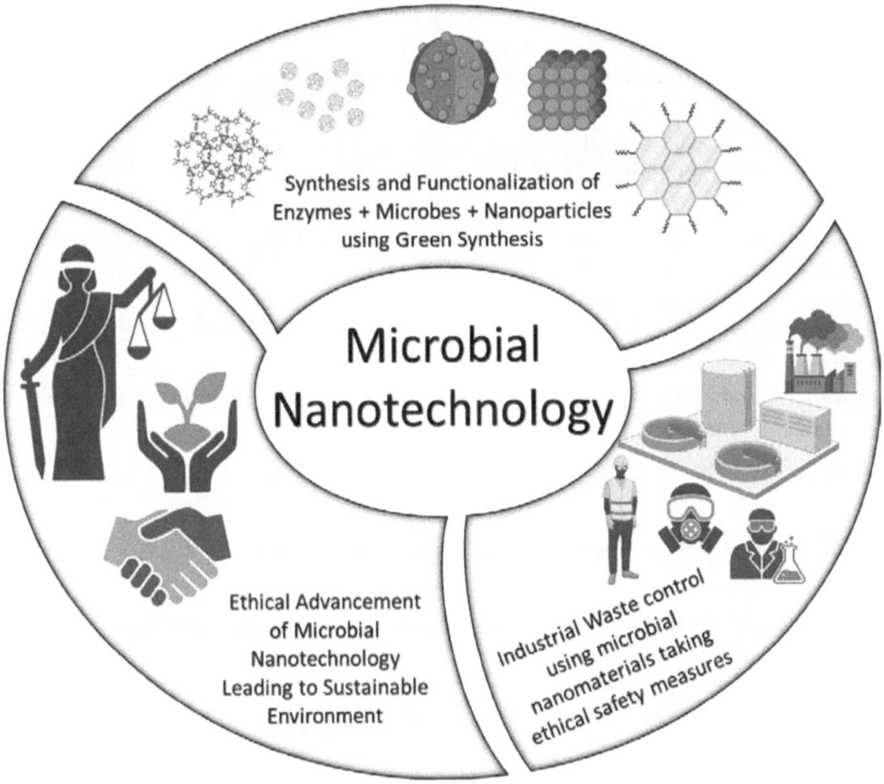

FIGURE 15.1 Schematic showing the various aspects of microbial nanotechnology.

Microbial nanotechnology also enables the production of complex nanostructures and hybrid materials by genetically engineering microorganisms or manipulating their growth conditions [20,21]. This allows for the synthesis of multifunctional NPs with applications in catalysis, sensing, drug delivery, and more.

Moreover, microbial-mediated NP synthesis processes are scalable, sustainable, and compatible with green chemistry principles, reducing energy consumption, minimizing waste generation, and utilizing renewable resources. Overall, microbial nanotechnology has significantly advanced NP synthesis and production, unlocking new possibilities for innovation across various industrial sectors while promoting environmental sustainability [22].

15.2.2 Pollution Control and Remediation

Microbial nanotechnology is a promising approach for pollution control and environmental remediation, offering state-of-the-art solutions to address the challenges of environmental contamination. Leveraging the unique properties of microorganisms and nanomaterials, this interdisciplinary field provides effective and sustainable strategies for mitigating pollutants in soil, water, and air [23,24].

The main application of microbial nanotechnology in pollution control is the development of engineered microorganisms, or bioremediation, capable of degrading or sequestering various contaminants. These bioremediators are designed to utilize specific metabolic pathways or enzymes to detoxify pollutants, including heavy metals, organic compounds, and hazardous chemicals [25,26]. Microbial nanotechnology enhances the effectiveness of bioremediation processes by incorporating nanomaterials, such as NPs and nanocomposites, which serve as carriers for microbial cells, provide surface area for pollutant adsorption, and facilitate the transport of contaminants to microbial enzymes for degradation [27].

Furthermore, microbial NPs themselves exhibit characteristic properties that make them effective agents for pollutant removal and remediation. For example, metal NPs, such as zero-valent iron (ZVI) NPs, have been widely used for the remediation of contaminated groundwater through processes such as chemical reduction and adsorption [28]. Similarly, engineered NPs, including carbon-based nanomaterials like graphene and carbon nanotubes (CNTs), demonstrate high surface area and reactivity, enabling efficient adsorption and degradation of organic pollutants [29]. Additionally, microbial nanotechnology offers novel approaches for addressing emerging pollutants and recalcitrant contaminants that are challenging to remediate using conventional methods. Engineered microorganisms equipped with specific enzymes or pathways can target these pollutants, while nanomaterials enhance their performance and stability in harsh environmental conditions.

Microbial nanotechnology presents substantial potential for enhancing pollution management and environmental restoration initiatives by offering novel, sustainable, and economically viable methods for the remediation of contaminated ecosystems. As investigations in this domain persist in their advancement, microbial nanotechnology is anticipated to assume an essential function in tackling ecological issues and fostering global sustainability.

15.2.3 Sustainable Manufacturing

Microbial nanotechnology offers promising avenues for sustainable manufacturing practices by leveraging the unique capabilities of microorganisms and nanomaterials [30]. One key aspect is the development of eco-friendly synthesis routes for nanomaterials using microbial-mediated processes. Fungi are unique in mediating the eco-friendly and sustainable NP synthesis. In addition to exhibiting a high surface area to mass ratio and the capacity to facilitate nutrient translocation, fungal hyphae engage in growth, branching, and fusion to create filamentous exploratory mycelia. The hydrated mucilaginous sheath that often surrounds hyphae acts as a matrix that is conducive to geochemical interactions [31]. The prevalence of metal-binding functional groups within the cell wall, along with associated extracellular polymeric substances (EPSs), which function as nucleation sites for the formation of NPs and nanominerals, renders the branched network an efficient template for these processes.

Moreover, due to their diverse active and incidental mechanisms to mitigate metal toxicity, fungi represent exceptional candidates for metal immobilization, mineral dissolution, and formation, with many species exhibiting resilience in environments characterized by elevated metal concentrations [31,32]. By means of both direct and indirect mechanical and metabolic interactions, microorganisms are capable of affecting metal speciation, toxicity, mobility, and the processes of mineral formation, dissolution, or degradation [31]. The use of fungi for the synthesis of nanomaterials appears to have benefits, especially in light of the increased interest in biological methods for environmentally benign and sustainable means of producing NPs [31].

Metals and metal oxides are examples of inorganic NPs known as nanozymes, which imitate the characteristics of enzymes in redox processes. Oxidase (OXD), peroxidase (POD), catalase (CAT), and superoxide dismutase (SOD) activities are involved in the majority of catalytic processes mediated by nanozymes [33]. Fungi are very special as they can mediate the creation of reactive NPs, such as metal oxides.

Because of its widespread distribution on Earth's surface and its well-known catalytical activities mimicking natural peroxidase, ferrihydrite ($Fe10\ O14.(OH)_2$) is one of the most interesting nanozymic minerals. To counteract oxidative stress and preserve redox balance, this antioxidant enzyme eliminates excess reactive oxygen species (ROS), such as superoxide ($O_2^{\cdot -}$) and hydrogen peroxide (H_2O_2). The fungal transformation of hematite results in ferrihydrite NPs, which can scavenge hydrogen peroxide to reduce cytotoxicity and guarantee the acquisition of necessary iron. Overall, microbial nanotechnology offers promising opportunities for advancing sustainable manufacturing practices, contributing to environmental stewardship and resource conservation in industrial sectors.

15.2.4 Biocatalysis and Biofuel Production

Nanotechnology and nanomaterials have emerged as a promising niche of biofuel research, offering cost-effective methods to enhance production quality. Several advantages are associated with the use of NPs in biofuel synthesis compared to other sources, including their size, unique properties (e.g., high surface area-to-volume ratio), special characteristics (e.g., high crystallinity), and catalytic activity (catalytic, adsorptive, and stability) [34–36]. For the synthesis of biofuels, nanotechnology has shown promise when combined with other procedures such as gasification, pyrolysis, hydrogenation, and anaerobic digestion [37,38]. CNT and metal oxide (MOOP) NPs are commonly used as nanocatalysts in biofuel production due to their additional properties that support high potential recovery [39]. CNTs are carbon allotropes created by rolling graphene sheets into a cylindrical form [40]. These nanotubes are mostly utilized in the creation of biosensors and MFCs because of their capacity to carry redox reactions and electron transfer kinetics [41,42].

According to Liu et al., adding 100 mg/L CNTs to anaerobic sludge blanket (UASB) reactors improved the synthesis of biohydrogen, resulting in a hydrogen yield of 2.45 mol/mol glucose and a production rate of 5.55 L/L/d [41]. Magnetic micro-ferrites doped with calcium have been found to have a notable impact on biodiesel synthesis, increasing yield from soybean cooking oils by over 85% [43]. Utilizing waste-activated sludge, researchers have also looked into the use of nano zero-valent iron (nZVI) and Fe_2O_3 to improve the production of biogas [44]. Protein synthesis, nitrogen metabolism, physiological functions, and microbial growth are all enhanced by the addition of NPs to the growth medium. In *Chlorella vulgaris*, for example, optimal concentrations of Ag and Au NPs have been shown to boost photosynthetic activity [45]. It has been proven that the addition of zero-valent iron NPs enhances biogas generation from waste matter [46,47]. Nickel NPs have also been widely used in hydrogenation to convert glucose into sorbitol [48].

15.3 ETHICAL FRAMEWORKS AND CONSIDERATIONS

15.3.1 Ethical Principles and Their Applicability

The ethical principles of beneficence and non-maleficence are crucial in guiding decision-making across various industries, including the microbial nanotechnology sector. Within the realm of nanotechnology, these principles hold particular significance due to the potentially far-reaching impact that the development and utilization of nanomaterials can have on human health, the environment, and society. The core focus of these principles lies in promoting positive outcomes while simultaneously preventing any potential harm.

15.3.1.1 Beneficence

The advancement of nanomedicines and their applications in therapeutics constitutes the focal point of investigation. Scientists aspire to enhance healthcare outcomes by developing pioneering treatments that exhibit augmented effectiveness, precise administration, and diminished adverse effects. The application of microbial nanotechnology within environmental frameworks, specifically in the realm of bioremediation, is noteworthy. In this context, researchers can employ genetically modified microorganisms to effectively eliminate harmful substances, thereby significantly improving environmental well-being and promoting sustainable practices. The agriculture field has the potential to improve the productivity of crops, the absorption of nutrients, and the ability to resist pests. The principle of beneficence serves as a guiding force in developing technologies that positively impact food production and the overall security of global food resources. The contributions to public health initiatives through the advancement of microbial nanotechnologies can be utilized for the purpose of water purification, disease detection, and various other preventive measures. It is evident in the development of advanced imaging techniques for more acute and early diagnosis of diseases [9].

15.3.1.2 Non-maleficence

The principle of non-maleficence holds great importance in the realm of nanomaterial design, as it guarantees the safety of these materials when exposed to humans and the environment. The principle of non-maleficence also encompasses the ethical use of data derived from microbial nanotechnology research. To prevent any possible harm to individuals and maintain trust, it is essential to ensure the privacy and confidentiality of sensitive information. Responsible communication with the public, transparently addressing microbial nanotechnologies' potential risks, uncertainties, and limitations, prevents unwarranted fear and enables individuals to make informed decisions.

15.3.1.3 Autonomy and Informed Consent

Autonomy and informed consent are integral to the ethical practice of microbial nanotechnology. Whether in research, medical applications, environmental deployments, or workplace safety, respecting individuals' autonomy and obtaining informed consent are essential for fostering trust, transparency, and ethical conduct in the industry.

15.3.1.4 Justice and Fair Distribution

Justice and fair distribution are essential ethical considerations in the microbial nanotechnology industry, particularly regarding nanomaterials and technologies' development, application, and potential impacts. Justice requires establishing and adhering to rigorous health and safety standards in developing and using microbial nanotechnologies. Ensuring the well-being of workers, researchers, and the public is paramount. Companies should consider access to healthcare innovations, environmental benefits, and other positive outcomes to avoid disproportionately favoring specific populations. Environmental risks and benefits should be distributed fairly among different people.

Fair distribution in the context of nanomedicine involves ensuring that innovative treatments and therapies are accessible to a broad population. Efforts should be made to avoid creating healthcare disparities. Companies should adopt practices that promote environmental sustainability, avoiding overexploitation and ensuring that benefits are distributed over the long term. Industries should communicate ethically, avoiding sensationalism or downplaying risks. Open and honest communication helps manage fears and fosters a more informed public discourse.

15.3.2 Ethical Dilemmas in Industrial Applications

15.3.2.1 Worker Safety and Well-Being

In the industry of microbial nanotechnology, workers' safety and well-being have paramount importance. It requires careful attention to potential biological and nanomaterial hazards and the implementation of robust safety measures. Complete risk assessments should be conducted to identify potential hazards associated with microbial nanotechnology. The toxicity, pathogenicity, and other relevant characteristics of the microorganisms involved should be assessed. Good personal protective equipment (PPE) kit, including lab coats, gloves, masks, and goggles, should be provided to protect workers from exposure to microorganisms and nanomaterials. PPE is regularly inspected, maintained, and replaced as needed. Training on safe handling practices, emergency response procedures, and the proper use of protective equipment should be provided. Engineering controls, such as containment systems and ventilation, should be implemented to minimize the release of microorganisms and nanomaterials into the workplace environment. Regular medical surveillance is conducted to monitor workers' health when exposed to hazardous microbes or nanomaterials. Emergency response plans that outline procedures for spills, accidents, or unexpected releases of microorganisms should be developed and regularly reviewed. It should be ensured that workers are trained on these procedures. Workspace design is important to minimize the risk of accidental exposure. Workers' safety and well-being also contribute to the responsible and sustainable development of industry.

15.3.2.2 Intellectual Property and Open Access

Intellectual property (IP) and open access in the industry involve navigating the balance between fostering innovation through IP protection and promoting the sharing of knowledge and information for the greater good. In IP, securing patents can encourage innovation by providing inventors with exclusive rights to their inventions and fostering investment in research and development. IP protection allows companies to commercialize their innovations, attract investors, and have market exclusivity, incentivizing private investment in research and development. Licensing IP can facilitate collaboration between industry and academia, enabling the transfer of technology and knowledge.

Open access promotes the free exchange of information and knowledge, accelerating scientific progress by allowing researchers to build upon each other's work. Collaboration and interdisciplinary research break down barriers and allow researchers from diverse backgrounds to contribute to a field. It enhances public trust by providing transparent access to research outcomes and engagement. Consider dual-licensing models, where certain aspects of technology are protected by IP, while others are made available under open-access licenses. Investigating the establishment of patent pools or collaborative frameworks that enable multiple stakeholders to access and utilize critical technologies collectively, while ensuring the protection and respect of IP rights. Finding the right balance between IP protection and open access in microbial nanotechnology requires thoughtful consideration of the specific context, goals, and values of the involved stakeholders. A nuanced and collaborative approach can help drive innovation while addressing concerns related to accessibility, fairness, and responsible knowledge sharing.

15.4 ENVIRONMENTAL IMPACTS AND SUSTAINABILITY

15.4.1 Environmental Assessment of Microbial Nanotechnology

Conducting comprehensive environmental assessments of microbial nanotechnology is essential for understanding and mitigating its environmental impacts. By considering factors such as carbon footprint, life cycle assessment, resource utilization, waste generation, and sustainability in industrial processes, stakeholders can promote responsible innovation and ensure that microbial nanotechnology contributes to sustainable development goals.

The carbon footprint of microbial nanotechnology refers to the amount of greenhouse gas emissions, particularly carbon dioxide (CO_2), generated throughout the lifecycle of nanotechnology products or processes. This includes raw material extraction, manufacturing, transportation, use, and disposal emissions. Assessing the carbon footprint helps quantify the environmental impact of microbial nanotechnology and identify opportunities to reduce emissions through measures such as energy efficiency improvements, renewable energy adoption, and process optimization [49]. Life cycle assessment (LCA) is a systematic approach for evaluating the environmental impacts of a product, process, or service throughout its entire life cycle, from raw material extraction to end-of-life disposal. LCA considers various environmental indicators such as energy consumption, water use, greenhouse gas emissions, air and water pollution, and resource depletion. Conducting an LCA of microbial nanotechnology enables stakeholders to identify environmental hotspots, prioritize areas for improvement, and make informed decisions to minimize environmental impacts across the entire life cycle. Assessing resource utilization involves evaluating the consumption of raw materials, energy, water, and other resources in microbial nanotechnology processes. Efficient resource utilization is essential for minimizing environmental impacts and promoting sustainability. Strategies to improve resource efficiency may include recycling and reuse of materials, optimizing production processes, and adopting cleaner production technologies. Waste generation in microbial nanotechnology encompasses the generation of solid waste, wastewater, and emissions during production and use. Minimizing waste generation is crucial for reducing environmental pollution and conserving resources. Waste management strategies such as waste reduction, recycling, treatment, and safe disposal can help mitigate the environmental impacts of microbial nanotechnology processes. Promoting sustainability in industrial processes includes integrating environmental, social, and economic considerations into decision-making. Sustainable practices in microbial nanotechnology include reducing environmental impacts, endorsing social responsibility, and ensuring economic viability. This may involve adopting green chemistry principles, using renewable resources, minimizing pollution and waste, adhering to ethical standards, and promoting equitable access to technological benefits.

15.5 CASE STUDIES: ETHICAL CHALLENGES IN REAL-WORLD APPLICATIONS

Improper handling and disposal of engineered nanoparticles (ENPs) can lead to several health challenges and environmental risks. The type, characteristics, and concentration of ENPs exposed to an environment play a crucial role in determining its impact [50]. According to Tang et al., the age and health state of exposed people—such as small children and those with pre-existing medical conditions—determine the distribution and clearance of ENPs in the body [51].

Smaller NPs are more likely to be absorbed by cells and cause cytotoxicity [52–54] such as hepatobiliary toxicity [55]. Furthermore, the NPs' surface charge and aggregation state determine how they distribute, accumulate, and interact with other molecules at the target site. The smaller size effect, unique magnetic properties, quantum size effect, and macroscopic quantum tunneling effect are among the unique characteristics of NPs [56,57]. Because of all these qualities, NPs have extensive applications [58].

The unexpected nano-hazards in domestic and industrial settings are unclear and hard to measure. However, in vitro and in vivo nano-toxicity of ENPs has been vastly reported in different parts of the human body, such as the brain, immunological and circulatory systems, reproductive systems, and developmental systems [51]. Humans may be exposed to ENPs by direct contact with NPs, ingestion, inhalation, intravenous injection, or skin contact with contaminated food, drink, or air [41,51,59,60]. Wu et al. showed that Fe_3O_4-NPs can cause mitochondrial damage and reduce macrophage viability after 48 hours treatment, inducing a shift of macrophage polarization to the M1 phenotype [61].

Several nations have also reported on real-world incidents and fatalities caused by occupational exposure to ENPs. Human toxicity can be caused by inhaling particulate matter from construction sites (e.g., SiO_2, ZnO, and TiO_2) [62], using sprays containing nano aerosols indoors (e.g., Ag emission from throat spray) [60], and releasing carbon black particulate matter and CNTs from recycled vehicle padding. It has been reported that SiO_2-NPs can induce mitochondrial membrane potential (MMP) depolarization in cardiomyocytes and reduce ATP production, resulting in Ca^{2+} damage [63]. Li et al. showed that nanosized carbon black, zinc dioxide, and silica could all cause ROS accumulation and mitochondrial damage in human corneal epithelial cells and human conjunctival epithelial cells to varying degrees, but TiO_2-NPs have no toxic effect [64].

According to Weir et al., ENPs can be consumed through nanofoods (such as TiO_2 from candies, sweets, gums, and cookies) [65], through the food chain (such as through the bioaccumulation of ZnO in edible plants), through the deposition of airborne NPs on food and beverages [60], through maternal-fetal transmission through the placenta, and through exposure during breastfeeding [51]. Dermal contact with ENPs can happen when skin and hair are treated with sunscreen lotions and cosmetics (e.g., TiO_2 and ZnO release), when clothing is worn (e.g., application of nano-Ag in textile industries as an antimicrobial agent), and when surfaces and products containing ENPs—e.g., computer processors containing Si NPs—are inadvertently touched [60]. To determine the most likely pathways of ENP exposure from consumer products containing ENPs included in the CPI, Vance et al. [60] conducted research. For 58% of the goods, skin is the predominant route of exposure; the remaining 16% are likely inhaled and 25% are swallowed. In juvenile zebrafish co-exposed to PCBPA, Zhu et al. demonstrated that SiO_2-NPs, even at safe concentrations, enhanced the thyroid glands by enhancing the bioaccumulation and bioavailability of PCBPA hormone disruption [66].

When people misuse aerosolized sprays containing ENPs, it can immediately negatively impact the respiratory system. There are no strict laws restricting the usage of nanomaterials indoors. For this reason, using nano-aerosol sprays requires ethics. The nanomaterials' size and form determine the respiratory system's deposition ability. The smaller the nanomaterial, the deeper the penetration into the lungs. Metal or metal oxide NPs are potentially aerosolized nanomaterials. More than 190 products (such as hair dryers and sprays) are listed in the CPI as potentially aerosolizing ENPs during usage.

Aerosol emissions from Ag NP–containing products and TiO_2-containing nanofilm products have been documented. Case studies of individuals who were exposed to ENPs over an extended period showed significant negative effects on their lungs and other respiratory system components. Two of the seven young female Chinese workers who were exposed to polyacrylate spray died, while the other three all developed strange pleuropulmonary symptoms [67]. ENPs were found in pleural and pulmonary cells.

Later, in 2011, Song et al. revealed that silica ENPs were present in the spray paints. Thirteen days after inhaling Ni NPs, a 38-year-old man passed away, and his lung macrophages contained Ni NPs smaller than 25 nm [68,69]. The Ni NPs brought on the adult respiratory distress syndrome. In a different study, 3 months of exposure to a polyester powder paint containing TiO_2 NPs resulted in the development of bronchiolitis obliterans, an inflammatory disease affecting the bronchioles in a 58-year-old male [70]. When handling nickel NP powder, a 26-year-old female chemist experienced facial flushing, throat irritation, and a symptom known as "post nasal drip" [71].

According to another case [72], a 33-year-old female was exposed to toner dust and suffered from abdominal pain, weight loss, and diarrhea. Tissue specimens included submesothelial aggregates of carbon nanoparticles (CNPs), ranging in size from 31 to 67 nm. According to Theegarten et al., inhaled ENPs can travel through the blood and lymphatic vessels and impact other organs including the kidneys. These case reports show that ENP exposure in the workplace needs to be acknowledged, even if they do not prove or disprove any specific risk posed by ENPs [72].

15.6 REGULATORY AND GOVERNANCE FRAMEWORKS

15.6.1 Current Regulatory Landscape for Microbial Nanotechnology

The regulatory landscape for microbial nanotechnology involves multiple facets including regulations related to nanotechnology, microbial products, and their intersection [73].

Many countries have specific regulations or guidelines governing nanotechnology. These regulations focus on ensuring the safety of nanomaterials, including those produced using microbial processes. They may cover aspects such as manufacturing practices, labeling requirements, risk assessment, and environmental impact. Microbial products, including those produced using nanotechnology, are subject to regulations related to their safety, efficacy, and environmental impact. These regulations vary by jurisdiction but often involve assessment by regulatory agencies such as the U.S. Environmental Protection Agency (EPA), the European Food Safety Authority (EFSA), or the Food and Drug Administration (FDA) in the United States [74].

Microbial nanotechnology often falls under the umbrella of biotechnology regulations, which govern the use of living organisms or their products in various applications. These regulations may include requirements for risk assessment, containment measures, and approval processes for commercialization. Regulatory agencies also assess microbial nanotechnology products' potential health and safety risks. This includes evaluating their potential toxicity, allergenicity, and other health effects [75]. Microbial nanotechnology products may have environmental implications, such as their impact on ecosystems or their potential to persist in the environment. Regulatory agencies may require environmental risk assessments and mitigation measures to address these concerns.

There are ongoing efforts to harmonize regulations related to nanotechnology and biotechnology at the international level. Organizations such as the International Organization for Standardization

(ISO) and the Organisation for Economic Co-operation and Development (OECD) work to develop common guidelines and standards to facilitate global trade while ensuring safety and efficacy [76].

As microbial nanotechnology continues to advance, new regulatory challenges may arise. These may include concerns related to the use of emerging technologies such as CRISPR for microbial engineering, the development of novel nanomaterials with unique properties, and the potential for unintended consequences in complex biological systems [77]. Given the rapidly evolving nature of both nanotechnology and biotechnology, the regulatory landscape for microbial nanotechnology is likely to continue evolving as well, with regulatory agencies adapting to address emerging challenges and technologies. Researchers and industry stakeholders need to stay informed about relevant regulations and engage with regulatory agencies to ensure compliance and facilitate responsible innovation.

15.6.2 International Agreements and Standards

International agreements and standards help harmonize regulations across different countries and regions. This harmonization is essential for facilitating global trade and ensuring a level playing field for businesses operating in multiple jurisdictions. For microbial nanotechnology, harmonized regulations can streamline the approval process for products and promote consistency in safety and efficacy assessments [78].

International standards provide guidance on risk assessment and management for microbial nanotechnology products. These standards help regulatory agencies and industry stakeholders evaluate the potential risks associated with new technologies and develop appropriate mitigation measures. By adhering to internationally recognized standards, companies can demonstrate the safety and reliability of their products to regulators and consumers worldwide.

International standards establish requirements for product quality and safety in microbial nanotechnology. These standards cover aspects such as manufacturing practices, product testing, labeling, and documentation. By following established standards, companies can ensure that their products meet minimum quality and safety requirements, which is essential for gaining regulatory approval and building consumer trust.

International agreements and standards can facilitate innovation in microbial nanotechnology by providing a common framework for research and development. By establishing common terminology, methodologies, and best practices, standards help researchers and industry stakeholders communicate and collaborate more effectively. This collaboration can accelerate the pace of innovation and enable the development of new products and technologies that meet global regulatory requirements.

International standards contribute to consumer protection by setting minimum requirements for product safety and information disclosure. These standards help ensure that microbial nanotechnology products are safe for consumers to use and that relevant information, such as potential risks and proper usage instructions, is communicated. By adhering to international standards, companies should demonstrate their commitment to consumer safety and build consumer trust in their products.

15.6.3 The Need for Proactive Governance

Proactive governance involves identifying potential risks and challenges associated with microbial nanotechnology before they become widespread. By anticipating these risks, policymakers and regulators can develop appropriate regulations and mitigation strategies to minimize harm to public health, the environment, and society. Emerging technologies like microbial nanotechnology often involve inherent uncertainties regarding their potential impacts. Proactive governance acknowledges this uncertainty and seeks to manage it through adaptive and flexible regulatory approaches.

This may include mechanisms for ongoing monitoring, risk assessment, and stakeholder engagement to update regulations as new information becomes available [79].

Microbial nanotechnology raises ethical considerations related to issues such as environmental sustainability, social equity, and human health. Proactive governance involves integrating ethical principles into decision-making processes to ensure that the benefits of technology are distributed equitably and that potential harms are mitigated. This may require engaging with diverse stakeholders, including ethicists, community groups, and affected populations. Proactive governance seeks to balance the need for regulation with the imperative to foster innovation. By providing clear regulatory pathways and incentives for responsible innovation, policymakers can encourage the development of beneficial applications of microbial nanotechnology while minimizing risks. This may involve supporting research and development, providing funding for technology assessment, and creating mechanisms for technology transfer and commercialization [80].

Proactive governance of microbial nanotechnology requires international cooperation and coordination. Given the global nature of emerging technologies, harmonizing regulations and standards across different countries and regions is essential to ensure consistency and prevent regulatory arbitrage. International collaboration can enable information sharing, capacity building, and the development of common principles and best practices. It involves actively engaging with the public to build trust and confidence in the regulatory process. This may include providing opportunities for public input, transparency in decision-making, and education and outreach efforts to increase awareness of microbial nanotechnology and its potential impacts. By involving the public in the governance process, policymakers can ensure that regulatory decisions reflect societal values and preferences.

15.6.4 Balancing Innovation with Safety and Ethics

Balancing innovation with safety and ethics is paramount when considering emerging technologies like microbial nanotechnology. Innovation should be accompanied by rigorous risk assessment to identify potential hazards and uncertainties associated with microbial nanotechnology. By understanding the risks upfront, measures can be put in place to mitigate them effectively, ensuring that innovation proceeds safely and responsibly [81].

Ethical considerations should be integrated into the innovation process from the outset. This involves identifying and addressing ethical implications such as environmental impact, social equity, and human health. Ethical frameworks can guide decision-making by outlining principles and values that should be upheld throughout the development and deployment of microbial nanotechnology. Engaging with diverse stakeholders, including scientists, policymakers, industry representatives, ethicists, community groups, and the public, is essential for balancing innovation with safety and ethics. By soliciting input from different perspectives, policymakers can better understand the potential implications of microbial nanotechnology and develop regulations and guidelines that reflect a broad range of interests and concerns. Transparency in decision-making processes and accountability for outcomes are essential for ensuring that innovation is conducted responsibly. Policymakers and industry stakeholders should be transparent about the risks and benefits of microbial nanotechnology and accountable for the impacts of their actions on society and the environment [82].

Increasing awareness and understanding of microbial nanotechnology among policymakers, scientists, industry stakeholders, and the public can foster a culture of responsible innovation. Education initiatives can help people make informed decisions about the development and use of microbial nanotechnology, empowering them to advocate for ethical and sustainable practices. Effective regulatory oversight is crucial for ensuring that innovation in microbial nanotechnology adheres to safety and ethical standards. Regulatory agencies should have the necessary authority, resources, and expertise to evaluate the safety and efficacy of microbial nanotechnology products and enforce compliance with applicable regulations. Given the global nature of emerging technologies, international collaboration is important for addressing safety and ethical concerns associated with microbial nanotechnology. Collaborative efforts can facilitate information sharing, harmonize regulatory approaches, and promote best practices across borders. By integrating risk assessment, ethical frameworks, stakeholder engagement, transparency, education, regulatory oversight, and international collaboration, policymakers and industry stakeholders can strike a balance between fostering innovation and ensuring safety and ethics in developing and deploying microbial nanotechnology.

15.7 FUTURE CHALLENGES AND OPPORTUNITIES

Emerging ethical issues in microbial nanotechnology encompass concerns over environmental impact, including unintended consequences of engineered microorganisms, potential ecosystem disruption, and genetic modification risks. Additionally, ethical dilemmas arise regarding equitable access to nanotechnology-based solutions, socio-economic disparities, and dual-use applications for military or surveillance purposes. Ensuring transparency, public engagement, and responsible governance frameworks are crucial to address these ethical challenges. Moreover, considerations of privacy, informed consent, and societal implications of nanotechnology research and development must be integrated into regulatory policies and ethical guidelines to promote the responsible advancement of microbial nanotechnology.

Future opportunities in microbial nanotechnology involve leveraging engineered microorganisms for targeted pollutant remediation, sustainable agriculture, and renewable energy production. Evolving environmental considerations emphasize the need for stringent risk assessment, ecosystem monitoring, and public engagement to ensure the responsible deployment of microbial nanotechnology while minimizing ecological impacts. Integrating ethics and sustainability into research and development (R&D) processes is becoming increasingly crucial in microbial nanotechnology. Future challenges and opportunities in this domain lie at the intersection of ethical considerations, environmental sustainability, and technological innovation. Promoting public understanding of microbial nanotechnology and its potential benefits and risks to foster informed decision-making and responsible innovation.

15.8 CONCLUSION

Ethical considerations are paramount in microbial nanotechnology, given its potential impact on the environment and human health. Industry and researchers bear a dual responsibility to ensure that the development and application of microbial nanotechnology adhere to ethical principles. Firstly, the industry must prioritize the responsible use of microbial nanotechnology by integrating ethical considerations into product development, manufacturing processes, and waste management practices. This involves transparent communication of potential risks and benefits, adherence to regulatory frameworks, and proactive measures to minimize adverse effects on ecosystems and public health. Additionally, the industry should invest in research to understand microbial nanotechnology's long-term environmental and societal implications. Secondly, researchers play a crucial role in advancing microbial nanotechnology ethically by conducting rigorous risk assessments, fostering interdisciplinary collaborations, and promoting open dialogue on ethical dilemmas. They should strive for transparency in their methodologies, data sharing, and reporting of findings to facilitate informed decision-making by stakeholders. Moreover, researchers must advocate for policies that promote responsible innovation and equitable access to microbial nanotechnology solutions.

The ethical advancement of microbial nanotechnology requires a collaborative effort between industry and researchers, guided by sustainability, safety, and social responsibility principles. By upholding ethical standards, stakeholders can harness the potential of microbial nanotechnology to address pressing global challenges while minimizing unintended consequences and safeguarding the well-being of current and future generations.

REFERENCES

1. Ng CA, Chew SN, Bashir MJ, et al. Enhancing microbial fuel cell performance for sustainable treatment of palm oil mill wastewater using carbon cloth anode coated with activated carbon. *International Journal of Hydrogen Energy*. 2024;52:1092–1104.
2. Lateef A, Darwesh O, Matter I. Microbial nanobiotechnology: The melting pot of microbiology, microbial technology and nanotechnology. In: *Microbial Nanobiotechnology: Principles and Applications*. 2021: pp.1–19. https://doi.org/10.1007/978-981-33-4777-9_1.
3. François M, Lin K-S, Rachmadona N, et al. Utilization of carbon-based nanomaterials for wastewater treatment and biogas enhancement: A state-of-the-art review. *Chemosphere*. 2023;350:141008.
4. Li Q, Liu F, Li M, et al. Nanoparticle and nanomineral production by fungi. *Fungal Biology Reviews*. 2022;41:31–44.
5. Zhang D, Liu P. Biosynthesis of metal nanoparticles: Bioreduction and biomineralization. *Nanotechnology Reviews*. 2023;12(1):20230170.
6. Mazari SA, Ali E, Abro R, et al. Nanomaterials: Applications, waste-handling, environmental toxicities, and future challenges - A review. *Journal of Environmental Chemical Engineering*. 2021;9(2):105028.
7. Vasantha V, Sharvari S, Alfia N, et al. Application of nanotechnology toward improved production of sustainable bioenergy. In: *Nanomaterials*: Elsevier; 2021: pp. 445–479. https://doi.org/10.1016/B978-0-12-822401-4.00010-6.
8. Ingle AP, Antunes F, Paula A, et al. Nanotechnological interventions for sustainable production of microbial biofuel and bioenergy. In: *Microbial Biotechnology for Renewable and Sustainable Energy*: Springer; 2022: pp. 191–226.
9. Kaur H, Shorie M. Nanomaterial based aptasensors for clinical and environmental diagnostic applications. *Nanoscale Advances*. 2019;1(6):2123–2138.
10. Molahalli V, Shetty A, Sharma A, et al. Risks and ethics of nanotechnology: An overview. In: *Nanoparticles and Plant-Microbe Interactions*. 2023: pp. 35–68. https://doi.org/10.1016/B978-0-323-90619-7.00014-X.
11. Atalla K, Chaudhary A, Eshaghian-Wilner MM, et al. Ethical, privacy, and intellectual property issues in nanomedicine. In: *Wireless Computing in Medicine: From Nano to Cloud with Ethical and Legal Implications*. 2016: pp. 567–600. https://doi.org/10.1002/9781118993620.ch20.
12. Babatunde DE, Denwigwe IH, Babatunde OM, et al. Environmental and societal impact of nanotechnology. *IEEE Access*. 2019;8:4640–4667.
13. Gohain A. Microbial nanotechnology: Current development and potential applications in the field of biotechnology. In: *Modern Nanotechnology*: Volume 2: Green Synthesis, Sustainable Energy and Impacts: Springer; 2023: pp. 27–43.
14. Zango ZU, Khoo KS, Garba A, et al. A review on titanium oxide nanoparticles modified metal-organic frameworks for effective CO_2 conversion and efficient wastewater remediation. *Environmental Research*. 2024;252:119024.
15. Zango ZU, Khoo KS, Garba A, et al. A review on carbon-based biowaste and organic polymer materials for sustainable treatment of sulfonamides from pharmaceutical wastewater. *Environmental Geochemistry and Health*. 2024;46(4):145.
16. Qureshi A, Blaisi NI, Abbas AA, et al. Prospectus and development of microbes mediated synthesis of nanoparticles. In: *Microbial Nanotechnology: Green Synthesis and Applications*. 2021: pp. 1–15. https://doi.org/10.1007/978-981-16-1923-6_1.
17. Ray MK, Mishra AK, Mohanta YK, et al. Nanotechnology as a promising tool against phytopathogens: A futuristic approach to agriculture. *Agriculture*. 2023. https://doi.org/10.3390/agriculture13091856.
18. Mohanta YK, Mishra AK, Panda J, et al. Promising applications of phyto-fabricated silver nanoparticles: Recent trends in biomedicine. *Biochemical and Biophysical Research Communications*. 2023;688:149126.
19. Ray MK, Panda J, Panda BP, et al. Mycoremediation of heavy metals and/or metalloids in soil. In: Sarma H, Joshi S, editors. *Land Remediation and Management: Bioengineering Strategies*. Singapore: Springer Nature; 2023: pp. 161–190.
20. Iravani S, Varma RS. Biofactories: Engineered nanoparticles via genetically engineered organisms. *Green Chemistry*. 2019;21(17):4583–4603.
21. Iravani S, Varma RS. Genetically engineered organisms: Possibilities and challenges of heavy metal removal and nanoparticle synthesis. *Clean Technologies*. 2022;4(2):502–511.
22. Ardo FM, Khoo KS, Sobri MZA, et al. Modelling photoperiod in enhancing hydrogen production from *Chlorella vulgaris* sp. while bioremediating ammonium and organic pollutants in municipal wastewater. *Environmental Pollution*. 2024;346:123648.

23. Aliyari Rad S, Nobaharan K, Pashapoor N, et al. Nano-microbial remediation of polluted soil: A brief insight. *Sustainability*. 2023;15(1):876.
24. Mandeep, Shukla P. Microbial nanotechnology for bioremediation of industrial wastewater. *Frontiers in Microbiology*. 2020;11:590631.
25. Pham MN, Nishimura F, Lan JCW, et al. Recent advancement of eliminating antibiotic resistance bacteria and antibiotic resistance genes in livestock waste: A review. *Environmental Technology & Innovation*. 2024;36:103751.
26. Rawindran H, Khoo KS, Ethiraj B, et al. Environmental impact assessment via life cycle analysis on ultrafiltration membrane fabricated from polyethylene terephthalate waste to treat microalgal cultivation wastewater for reusability. *Environmental Research*. 2024;251:118687.
27. Janssen DB, Stucki G. Perspectives of genetically engineered microbes for groundwater bioremediation. *Environmental Science: Processes & Impacts*. 2020;22(3):487–499.
28. Pandey K, Sharma S, Saha S. Advances in design and synthesis of stabilized zero-valent iron nanoparticles for groundwater remediation. *Journal of Environmental Chemical Engineering*. 2022;10(3):107993.
29. Madima N, Mishra S, Inamuddin I, et al. Carbon-based nanomaterials for remediation of organic and inorganic pollutants from wastewater. A review. *Environmental Chemistry Letters*. 2020;18:1169–1191.
30. Yadav S, Chander S, Gupta A, et al. Biogenic engineered zinc oxide nanoparticle for sulfur black dye removal from contaminated wastewater: Comparative optimization, simulation modeling, and isotherms. *Bioengineered*. 2024;15(1):2325721.
31. Gadd GM, Raven JA. Geomicrobiology of eukaryotic microorganisms. *Geomicrobiology Journal*. 2010;27(6–7):491–519.
32. Dhillon GS, Brar SK, Kaur S, et al. Green approach for nanoparticle biosynthesis by fungi: Current trends and applications. *Critical Reviews in Biotechnology*. 2012;32(1):49–73.
33. Meng Y, Li W, Pan X, et al. Applications of nanozymes in the environment. *Environmental Science: Nano*. 2020;7(5):1305–1318.
34. Dikshit PK, Kumar J, Das AK, et al. Green synthesis of metallic nanoparticles: Applications and limitations. *Catalysts*. 2021;11(8):902.
35. do Nascimento RO, Rebelo LM, Sacher E. Physicochemical characterizations of nanoparticles used for bioenergy and biofuel production. In: *Nanotechnology for Bioenergy and Biofuel Production*. 2017: pp.173–191. https://doi.org/10.1007/978-3-319-45459-7_8.
36. Vijayaraghavan P, Veeramanikandan V, Pradeep BV, et al. Enhanced aerobic naphthalene degradation utilizing indigenous microbial flora as a biocatalyst in oil-contaminated wastewater. *Topics in Catalysis*. 2024;67:1–11.
37. Ali S, Shafique O, Mahmood S, et al. Biofuels production from weed biomass using nanocatalyst technology. *Biomass and Bioenergy*. 2020;139:105595.
38. Hussain ST, Ali SA, Bano A, et al. Use of nanotechnology for the production of biofuels from butchery waste. *International Journal of Physical Sciences*. 2011;6(31):7271–7279.
39. Saoud K. Nanocatalyst for biofuel production: A review. In: *Green Nanotechnology for Biofuel Production*. 2018: pp. 39–62. https://doi.org/10.1007/978-3-319-75052-1_4.
40. Singh N, Dhanya B, Verma ML. Nano-immobilized biocatalysts and their potential biotechnological applications in bioenergy production. *Materials Science for Energy Technologies*. 2020;3:808–824.
41. Liu Z, Lv F, Zheng H, et al. Enhanced hydrogen production in a UASB reactor by retaining microbial consortium onto carbon nanotubes (CNTs). *International Journal of Hydrogen Energy*. 2012;37(14):10619–10626.
42. Usman A, Khoo KS, Ariffin MM, et al. Adsorption of terbutaline B-agonists from wastewater by mechano-synthesized iron oxide nanoparticles modified copper (Ii) isonicotinate metal-organic framework. Available at SSRN 4850341. 2024.
43. Dantas J, Leal E, Mapossa A, et al. Magnetic nanocatalysts of $Ni_{0.5}Zn_{0.5}Fe_2O_4$ doped with Cu and performance evaluation in transesterification reaction for biodiesel production. *Fuel*. 2017;191:463–471.
44. Wang T, Zhang D, Dai L, et al. Effects of metal nanoparticles on methane production from waste-activated sludge and microorganism community shift in anaerobic granular sludge. *Scientific Reports*. 2016;6(1):25857.
45. Eroglu E, Eggers PK, Winslade M, et al. Enhanced accumulation of microalgal pigments using metal nanoparticle solutions as light filtering devices. *Green Chemistry*. 2013;15(11):3155–3159.
46. Karri S, Sierra-Alvarez R, Field JA. Zero valent iron as an electron-donor for methanogenesis and sulfate reduction in anaerobic sludge. *Biotechnology and Bioengineering*. 2005;92(7):810–819.

47. Su L, Shi X, Guo G, et al. Stabilization of sewage sludge in the presence of nanoscale zero-valent iron (nZVI): Abatement of odor and improvement of biogas production. *Journal of Material Cycles and Waste Management*. 2013;15:461–468.
48. Kobayashi H, Hosaka Y, Hara K, et al. Control of selectivity, activity and durability of simple supported nickel catalysts for hydrolytic hydrogenation of cellulose. *Green Chemistry*. 2014;16(2):637–644.
49. Kumar A, Sadhu SD, Singh R. Nanotechnology: Key for sustainable future. In: *Nanotechnology in Environmental Science*. 2018: pp.759–804.
50. Nel AE, Parak WJ, Chan WC, et al. Where are we heading in nanotechnology environmental health and safety and materials characterization? *ACS Nano*. 2015;9:5627–5630.
51. Tang S, Wang M, Germ KE, et al. Health implications of engineered nanoparticles in infants and children. *World Journal of Pediatrics*. 2015;11:197–206.
52. Manzanares D, Ceña V. Endocytosis: The nanoparticle and submicron nanocompounds gateway into the cell. *Pharmaceutics*. 2020;12(4):371.
53. Means N, Elechalawar CK, Chen WR, et al. Revealing macropinocytosis using nanoparticles. *Molecular Aspects of Medicine*. 2022;83:100993.
54. Varma S, Dey S, Palanisamy D. Cellular uptake pathways of nanoparticles: Process of endocytosis and factors affecting their fate. *Current Pharmaceutical Biotechnology*. 2022;23(5):679–706.
55. Recordati C, De Maglie M, Bianchessi S, et al. Tissue distribution and acute toxicity of silver after single intravenous administration in mice: Nano-specific and size-dependent effects. *Particle and Fibre Toxicology*. 2015;13:1–17.
56. Albanese A, Tang PS, Chan WC. The effect of nanoparticle size, shape, and surface chemistry on biological systems. *Annual Review of Biomedical Engineering*. 2012;14(1):1–16.
57. Ma Y, Zhang Z, Xu Y, et al. A bright carbon-dot-based fluorescent probe for selective and sensitive detection of mercury ions. *Talanta*. 2016;161:476–481.
58. Missaoui WN, Arnold RD, Cummings BS. Toxicological status of nanoparticles: What we know and what we don't know. *Chemico-Biological Interactions*. 2018;295:1–12.
59. Smolkova B, El Yamani N, Collins AR, et al. Nanoparticles in food. Epigenetic changes induced by nanomaterials and possible impact on health. *Food and Chemical Toxicology*. 2015;77:64–73.
60. Vance ME, Kuiken T, Vejerano EP, et al. Nanotechnology in the real world: Redeveloping the nanomaterial consumer products inventory. *Beilstein Journal of Nanotechnology*. 2015;6(1):1769–1780.
61. Wu C, Shen Z, Lu Y, et al. p53 promotes ferroptosis in macrophages treated with Fe_3O_4 nanoparticles. *ACS Applied Materials & Interfaces*. 2022;14(38):42791–42803.
62. Van Broekhuizen P, van Broekhuizen F, Cornelissen R, et al. Use of nanomaterials in the European construction industry and some occupational health aspects thereof. *Journal of Nanoparticle Research*. 2011;13:447–462.
63. Guerrero-Beltrán CE, Bernal-RamÃrez J, Lozano O, et al. Silica nanoparticles induce cardiotoxicity interfering with energetic status and Ca^{2+} handling in adult rat cardiomyocytes. *American Journal of Physiology-Heart and Circulatory Physiology*. 2017;312(4):H645–H661.
64. Li S, Ma Y, Ye S, et al. Polystyrene microplastics trigger hepatocyte apoptosis and abnormal glycolytic flux via ROS-driven calcium overload. *Journal of Hazardous Materials*. 2021;417:126025.
65. Weir A, Westerhoff P, Fabricius L, et al. Titanium dioxide nanoparticles in food and personal care products. *Environmental Science & Technology*. 2012;46(4):2242–2250.
66. Zhu B, Han J, Lei L, et al. Effects of SiO_2 nanoparticles on the uptake of tetrabromobisphenol A and its impact on the thyroid endocrine system in zebrafish larvae. *Ecotoxicology and Environmental Safety*. 2021;209:111845.
67. Song Y, Li X, Du X. Exposure to nanoparticles is related to pleural effusion, pulmonary fibrosis and granuloma. *European Respiratory Journal*. 2009;34(3):559–567.
68. Phillips JI, Green FY, Davies JC, et al. Pulmonary and systemic toxicity following exposure to nickel nanoparticles. *American Journal of Industrial Medicine*. 2010;53(8):763–767.
69. Song Y, Li X, Wang L, et al. Nanomaterials in humans: Identification, characteristics, and potential damage. *Toxicologic Pathology*. 2011;39(5):841–849.
70. Cheng T-H, Ko F-C, Chang J-L, et al. Bronchiolitis obliterans organizing pneumonia due to titanium nanoparticles in paint. *The Annals of Thoracic Surgery*. 2012;93(2):666–669.
71. Journeay WS, Goldman RH. Occupational handling of nickel nanoparticles: A case report. *American Journal of Industrial Medicine*. 2014;57(9):1073–1076.
72. Theegarten D, Boukercha S, Philippou S, et al. Submesothelial deposition of carbon nanoparticles after toner exposition: Case report. *Diagnostic Pathology*. 2010;5:1–4.

73. Ali F, Neha K, Parveen S. Current regulatory landscape of nanomaterials and nanomedicines: A global perspective. *Journal of Drug Delivery Science and Technology*. 2023;80:104118.
74. Ridge S. A regulatory framework for nanotechnology. Homeland Security Affairs. 2018. https://hdl.handle.net/10945/58356.
75. Gupta N. Nanotechnology in agri-food sector: Ethical, regulatory, and governance landscape in EU. In: *Ethics in Nanotechnology*. 2021: pp. 25–58. https://doi.org/10.1515/9783110719932-002.
76. Allan J, Belz S, Hoeveler A, et al. Regulatory landscape of nanotechnology and nanoplastics from a global perspective. *Regulatory Toxicology and Pharmacology*. 2021;122:104885.
77. Demirer GS, Silva TN, Jackson CT, et al. Nanotechnology to advance CRISPR-Cas genetic engineering of plants. *Nature Nanotechnology*. 2021;16(3):243–250.
78. Marchant GE, Sylvester DJ, Abbott KW, et al. International harmonization of regulation of nanomedicine. *Studies in Ethics, Law, and Technology*. 2010;3(3). https://doi.org/10.2202/1941-6008.1120.
79. Kuzma J. Moving forward responsibly: Oversight for the nanotechnology-biology interface. *Journal of Nanoparticle Research*. 2007;9:165–182.
80. Pelley J, Saner MA. International approaches to the regulatory governance of nanotechnology. Regulatory Governance Initiative, Carleton University; 2009. https://ssrn.com/abstract=1532390.
81. Gottardo S, Mech A, DrbohlavovÃ¡ J, et al. Towards safe and sustainable innovation in nanotechnology: State-of-play for smart nanomaterials. *NanoImpact*. 2021;21:100297.
82. Trump B, Cummings C, Klasa K, et al. Governing biotechnology to provide safety and security and address ethical, legal, and social implications. *Frontiers in Genetics*. 2023;13:1052371.

Index

Note: *Italic* page numbers refer to figures.

FSC
www.fsc.org
MIX
Papier | Fördert gute Waldnutzung
FSC® C083411

Kolibri 360 GmbH
Ferdinand-Jühlke-Straße 7
99095 Erfurt, Deutschland
produktsicherheit@kolibri360.de